Condensed MATTER THEORIES

VOLUME 3

Condensed MATTER THEORIES

VOLUME 3

Edited by

Jouko S. Arponen
University of Helsinki
Helsinki, Finland

R. F. Bishop
University of Manchester Institute of Science and Technology
Manchester, United Kingdom

and

Matti Manninen
Helsinki University of Technology
Espoo, Finland

Plenum Press • New York and London

ISBN-13:978-1-4612-8271-6 e-ISBN-13: 978-1-4613-0971-0
DOI: 10.1007/978-1-4613-0971-0

LC 87-656591

Proceedings of the 11th International Workshop on Condensed Matter Theories held
July 27-August 1, 1987, in Oulu, Finland

Softcover reprint of the hardcover 1st edition 1988

A Division of Plenum Publishing Corporation
233 Spring Street, New York, N.Y. 10013

PREFACE

This book is the third volume in an approximately annual series which comprises the proceedings of the ***International Workshops on Condensed Matter Theories.*** The first of these meetings took place in 1977 in São Paulo, Brazil, and successive workshops have been held in Trieste, Italy (1978), Buenos Aires, Argentina (1979), Caracas, Venezuela (1980), Mexico City, Mexico (1981), St. Louis, USA (1982), Altenberg, Federal Republic of Germany (1983), Granada, Spain (1984), San Francisco, USA (1985), and Argonne, USA (1986). The present volume contains the proceedings of the Eleventh Workshop which took place in Oulu, Finland during the period 27 July - 1 August, 1987.

The original motivation and the historical evolution of the series of Workshops have been amply described in the preface to the first volume in the present series. An important objective throughout has been to work against the ever-present trend for physics to fragment into increasingly narrow fields of specialisation, between which communication is difficult. The Workshops have traditionally sought to emphasise the unity of physics. By bringing together scientists working in many different areas of condensed matter theory, for the dual purpose of fostering collaborations between them and promoting the exchange of ideas between various disciplines, a common language has been exposed and developed. The Editor of the first volume in the series, F.B. Malik, expressed it thus: "Given a proper forum, scientists working in such diverse areas as band structure calculations and neutron stars can still come together, understand each others' research and borrow ideas from one area to another. In these days of specialization, this is an uncommon thing but important for the cross-fertilization of different fields and the overall understanding of the physical world around us".

From their Pan-American origins, the Workshops in this series rapidly developed into the significant international meetings that they have now become. This last Workshop in Finland, for example, had participants from thirteen different countries. A particularly pleasing development was the presence for the first time of scientists from the Soviet Union. The Workshops have thus successfully fostered truly international collaborations between scientists divided not only by interdisciplinary barriers but also by geographical and political boundaries.

The 31 invited papers given at the *Eleventh International Workshop on Condensed Matter Theories* comprise the bulk of the present Volume. Two additional contributions are included from invited speakers who were intending to participate but who had to cancel at the last moment. These two papers were approved for inclusion by the Series Editorial Board, following the recommendations of independent referees.

In keeping with the aims and spirit of the series, the individual articles that comprise this book belong to condensed matter theory interpreted very broadly. Most of the articles deal either with such methods as the various cluster techniques, Jastrow variational approaches, density functional methods and Green function techniques, or with such physical systems as quantum fluids, nuclear matter, correlated electronic systems, and superfluid or superconductive systems. However, some of the papers also deal with subjects and techniques that, while nonstandard in this subject grouping, open new perspectives on related phenomena in standard condensed matter physics. Examples are provided by the articles dealing with quantum chromodynamics as a many-body problem, quark degrees of freedom in nucleon - antinucleon scattering, nonlinear phenomena in quantum optics, and neural networks in chaos or in equilibrium. Particularly topical examples are the contributions dealing with such urgent and exciting new discoveries as the supernova 1987A and high-temperature superconductivity.

For ease of use the articles do not follow the same ordering as the talks given at the Workshop. The Editors have attempted to group them instead. The largest such group consists of those articles where the formal methodology itself plays an important role. The techniques discussed include the parquet technique in the Green function formulation, various paired-fermion and bosonization methods, the coupled-cluster method, and other approaches of methodological interest. The other groups consist of articles in which the main emphasis lies more with applications of the available techniques to such systems as quantum fluids, and electronic systems in solids, metals, semiconductors and the new ceramic materials displaying superconductivity at high temperatures; and to many-body systems or phenomena of interest in nuclear physics and high-energy physics, or in which highly collective nonlinear dynamical behaviour plays a key role. The reader, however, should be warned against placing too strict or too narrow an interpretation on our classification scheme.

It is not possible to record in this book all of the other events and activities which contributed so much to the success and ethos of the Workshop. Two events are particularly worthy of mention in this regard. The first is the talk entitled "Interacting Boson Model: A Short Survey" by Igal Talmi, on the development and background of the collective model of the nucleus. The second was the lively panel discussion on high-temperature superconductivity, chaired by H. Glyde, with R. Klemm, R.N. Silver, R. Kalia, J. Keller, E. Bashkin and S. Kusmartsev as the main participants.

The Editors wish to express their gratitude to many people who contributed to the choice of scientific programme and in other ways to the success of the Workshop. In particular we acknowledge the support and advice of John W. Clark, Manuel de Llano,

Alpo Kallio, and F. Bary Malik. We also thank Manfred L. Ristig and various members of the International Advisory Committee for their helpful ideas and recommendations. The meeting was made possible by financial support from the Research Institute for Theoretical Physics, Helsinki, the Finnish Ministry of Education, the Finnish Cultural Foundation and NORDITA, Copenhagen. The contribution of each of these sponsors is gratefully acknowledged. It is also a pleasure to record the particular assistance of Alpo Kallio and Erkki Pajanne, who contributed so much to the smooth organisation of the Workshop, and of Mrs. Maila Volanen for her efficient work as the Workshop Secretary.

Helsinki, Finland
Manchester, England

Jouko S. Arponen
Raymond F. Bishop
Matti Manninen

CONTENTS†

FORMAL METHODS

†Asterisk (*) next to name identifies the speaker

QUANTUM FLUIDS

ELECTRONIC SYSTEMS AND SOLIDS

NUCLEAR AND HIGH-ENERGY MANY-BODY PROBLEMS

DYNAMICS OF NONLINEAR MANY-BODY SYSTEMS

PARQUET THEORY: THE DIAGRAMS

Roger Alan Smith

Center for Theoretical Physics
Physics Department
Texas A&M University
College Station, TX 77843

Alexander Lande

Physics Department
University of Groningen
Groningen, Netherlands

INTRODUCTION

The summation of parquet diagrams provides an interesting and powerful approach to many-body theory. In this paper, we present several results on the diagrammatic structure of parquet theory. In comparison with previous diagrammatic discussions at this series of workshops[1] or elsewhere [2,3], this paper will derive the final form of the parquet equations (a new result) using simpler methods than the earlier work. The presentation will also be self-contained. As an illustration of the power of the present approach, we also derive the equations which would form the starting point for three-body parquet. Ultimately, we feel that the three-body parquet will be useful in obtaining more accurate results in physical systems, as well as being intimately related to the generation of better vertex approximations following Baym and Kadanoff[4] and Baym[5]. While the discussion here is self-contained, it is presented solely in terms of the diagrammatic structure. There is certainly much more to doing parquet theory than knowing the diagrams, and these other considerations will be discussed in the final section.

ONE-BODY PARQUET

Let's start by introducing one-body parquet. Imagine that we wish to construct a one-body Green's function from a non-interacting one-body Green's function and a proper self-energy (the origin of which is external to one-body parquet). We can think of the proper self-energy as a box with two points to which may be attached one one incoming line and one outgoing line. A series of boxes may be strung together by a sequence of Green's functions, each of which connects the outgoing point of one box to the incoming point of another box. Each such diagram is one-body reducible, in the sense that any string will be cut into two pieces if one intermediate line is severed. If we denote the proper self-energy by Σ^* and the full self-energy by Σ, the equation representing the sum of all diagrams for Σ which can be obtained with a given Σ^* is

$$\Sigma = \Sigma^* + \Sigma^* G \Sigma^* + \Sigma^* G \Sigma^* G \Sigma^* + ... \tag{1}$$

This may be formally summed as

$$\Sigma = \Sigma^* + \Sigma^* G \Sigma, \tag{2}$$

and the reducible part of Σ is

$$\Sigma_r = \Sigma^* G(\Sigma^* + \Sigma_r). \tag{3}$$

Since any diagram has one incoming point and one outgoing point, breaking the diagram divides these points in only one way. In addition, because there is only one operation which can be used to construct reducible diagrams, it suffices in eqs. 1-3 to represent this operation by the letter G.

In the more general case, we will be dealing with boxes which have $2n$ points at which may be connected n ingoing lines and n outgoing lines. We label the incoming points 1 to n and understand that if we were to trace through the lines internal to a box that the incoming point i would ultimately lead to the outgoing point i. We now consider what will happen if we connect two boxes together with n lines in such a way that each line goes from one box to the other box. The resulting object also has n incoming points and n outgoing points, but in general some incoming points will be associated with one of the two boxes and some with the other. The number of different ways that this may be done is $N_n = \frac{(2n)!}{2n!n!}$. In general, N_n is the number of different reducibility channels. N_1 is 1, in accordance with the (obvious) fact that the overall incoming point must be associated with one Σ^* and the overall outgoing point with the other. The next cases are $N_2 = 3$ and $N_3 = 10$.

Because there are multiple channels for $n > 1$, we look for a generalization of the Dyson equation 3. We introduce our approach for the case $n = 1$; in subsequent sections we generalize this to other n. We first write

$$\Sigma_x = (\Sigma^* + \Sigma_x) G (\Sigma^* + \Sigma_x). \tag{4}$$

Σ_x includes all diagrams in Σ_r, but it clearly generates most diagrams too many times. The reason that it does is that the operation G is associative. For any diagrams A, B, and C, we have

$$AG(BGC) = (AGB)GC. \tag{5}$$

Any diagram generated using the Σ_x to the left of the G in eq. 4 is of the form of $(AGB)GC$. According to eq. 5, the same diagram would come from $AG(BGC)$, and hence we have double-counting. We avoid double-counting by keeping everything to the right of G in eq. 4 and removing from the left of G all terms which could be generated some other way using the associative law. This leaves only the Σ^* on the left-hand side of G; the result of this is eq. 3.

The virtue of this method is that it is easily applied when there are many different reducibility channels. We will apply method in the next two sections for the two-body and three-body parquet equations.

TWO-BODY PARQUET

The two-body parquet theory deals with methods for constructing Feynman diagrams for the effective interaction Γ. The lowest order contribution is the bare interaction V, and we take the lines connecting the boxes to be the full one-body G's dressed with the self-energy.

Since $N_2 = 3$, there are three different ways of dividing the external points of a reducible diagram into two classes by cutting two lines. Let us label the incoming points as (1,2) and the outgoing points as (1',2') in such a way that the lines internal to a box enter at an unprimed point and leave at a primed point with the same number. Then diagrams may be divided into classes I, S, T and U according to whether they are irreducible or the separation leaves the overall points (1,2), (1,1') or (1,2') connected. In the case of the T diagrams, there are in fact several different ways of performing the internal connections. The various possibilities are illustrated in fig. 1.

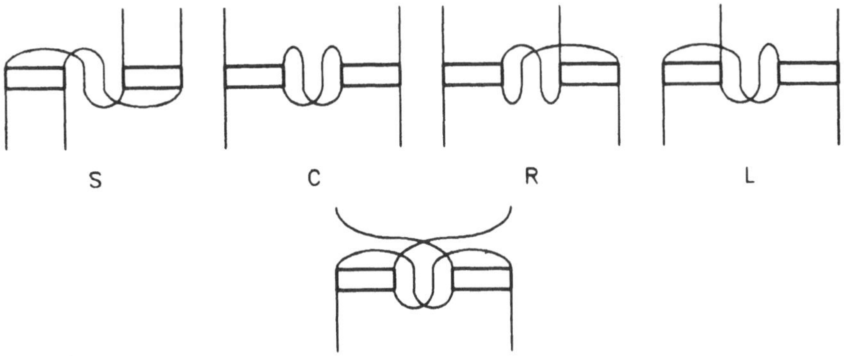

Fig. 1. The reducible two-body diagrams.

While it is easy to refer to the diagrams by their letters, it is useful to keep in mind that S corresponds to ladders, C to chains (or rings), R and L to right-hand and left-hand vertex corrections, and U to particle-hole ladders. The corresponding combining operations are referred to as s, c, r, l and u. Γ denotes $I + S + C + R + L + U$.

Let us now derive the correct Dyson equations for this problem. As before, we start out by writing incorrect equations and rejecting terms which cause double-counting. Since there are several operations, the question of associativity is more complicated. The correct generalization of eq. 5 is to look for equalities of the form

$$(AO_1B)O_2C = A'O_3(B'O_4C') \tag{6}$$

for some sets of operators O_i. The bad Dyson equations are

$$\begin{aligned}
S &= (I+S+C+R+L+U)s(I+S+C+R+L+U)\\
C &= (I+S+C+R+L+U)c(I+S+C+R+L+U)\\
R &= (I+S+C+R+L+U)r(I+S+C+R+L+U)\\
L &= (I+S+C+R+L+U)l(I+S+C+R+L+U)\\
U &= (I+S+C+R+L+U)u(I+S+C+R+L+U).
\end{aligned} \tag{7}$$

One may determine the class of associativity relationships of the type of eq. 6 by directly constructing diagrams for all pairs of operations and checking equalities. A (not-so-obvious) equivalent alternative is to look at a diagram constructed by two operations according to either side of eq. 6. If the resulting diagram has the overall external points on only two diagrams, then an associativity relation of the type of eq. 6 holds, since the overall diagram can be taken apart in two ways. In that case, diagrams of the type O_1 should not deleted from the left-hand side of the equation for diagrams of type O_2. Otherwise the overall diagram can only be taken apart in one way and no further restrictions are imposed. After carrying out these examinations, the right Dyson equations are found to be

$$\begin{aligned}
S &= (I+C+R+L+U)s(I+S+C+R+L+U)\\
C &= (I+S+U)c(I+S+C+R+L+U)\\
R &= (I+S+U)r(I+S+C+R+L+U)\\
L &= (I+S+C+R+L)l(I+S+C+R+L+U)\\
U &= (I+S+C+R+L)u(I+S+C+R+L+U).
\end{aligned} \tag{8}$$

The apparent asymmetry in the equations is just that. We have chosen in all cases to carry out the operations in such a way that the full Γ appears on the right side of the operation. This is the final result of ref.2, but it has been derived in a much simpler manner.

The next step is to rearrange these equations to restore the symmetry and delineate the structure more clearly. The S, R, and U equations are fine as they stand. The C diagrams include diagrams which have vertex corrections on the right-hand side, and the L diagrams include diagrams which may have intermediate chains and/or right-hand vertex corrections. We write

$$L = L' + L_C + L_R \tag{9}$$

and

$$C = C' + C_R, \tag{10}$$

where C_R and L_R end in a right-vertex correction, C' does not. L_C ends in a C', and L' does not end in either a C' or an R.

Some rearrangement of eqs. 8 then give

$$C' = (I + S + U)b(I + S + U + C'), \tag{11}$$

which is the form of a chain equation but with a new operation b and with the L and R removed from the right-hand side. The operation b is like the c operation, except that instead of the two boxes being connected directly to each other as by a c, the connection is a line which leaves one box, passes through the *full* vertex on one side, passes through the other box, and passes through the *full* vertex on the other side to return to the first box. In physical terms, the Lindhard bubble is competely dressed by having everything possible go across it. The mnemonic b is for "bubble".

Similarly, the L' and L_R equations can be rearranged to give diagrams which correspond to $(I + S + U + C')$ on the left or both sides. The penultimate form for the equations is obtained by defining

$$\begin{aligned} S'' &= S \\ U'' &= U \\ C'' &= C' \\ R'' &= R + C_R \\ L'' &= L' + L_C \\ A'' &= L_R. \end{aligned} \tag{12}$$

With this notation, the penultimate form of the two-body parquet equations is

$$\begin{aligned} S'' &= (I + C'' + R'' + L'' + A'' + U'')s(I + S'' + C'' + R'' + L'' + A'' + U'') \\ C'' &= (I + S'' + U'')b(I + S'' + U'' + C'') \\ R'' &= (I + S'' + U'' + C'')r(I + S'' + C'' + R'' + L'' + A'' + U'') \\ L'' &= (I + S'' + C'' + R'' + L'' + A'' + U'')l(I + S'' + C'' + U'') \\ A'' &= (I + S'' + C'' + R'' + L'' + A'' + U'')l(I + S'' + C'' + U'') \\ &\quad r(I + S'' + C'' + R'' + L'' + A'' + U'') \\ U'' &= (I + S'' + C'' + R'' + L'' + A'')u(I + S'' + C'' + R'' + L'' + A'' + U''). \end{aligned} \tag{13}$$

The ultimate form is obtained by dropping all of the double primes (and hence giving new meaning to some symbols). The symbol A is chosen to stand for "ambidextrous", since it has vertex corrections on both sides.

Note that this form of the equations is now completely symmetric in the left and right terms. Further, the quantity $(I + S + U)$ is both the thing to be chained and the thing to which vertex corrections are made on either (R and L) or both (A) sides. In addition, the bubble operation b automatically includes all vertex corrections which can be constructed in the chaining process.

The applications of these equations will be discussed in the last section.

THREE-BODY PARQUET

In three-body parquet, one is looking at boxes with three pairs of legs. There are $N_3 = 10$ different channels, and careful enumeration indicates that there are 28 different types of diagrams. These are shown in fig. 2.

The first four diagrams are generalizations of the S and U types in the two-body problem. The remaining fall into 4 general classes: chains (C), right and left vertex corrections (R) and (L), and "triple whammy" (W) diagrams, in which one line runs back and forth between the subdiagrams three times.

A computer program was written on the Commodore AMIGA to study the associativity properties of these diagrams. By either of the two methods indicated, the correct Dyson equations were derived. Rather than expressing the terms present on the right-hand side, we write the equations in the form

$$X = (\Gamma_3 - D_X)X(\Gamma_3), \tag{14}$$

where Γ_3 is the sum of all irreducible and reducible diagrams.

The following table shows which diagrams are in D_X for all X:

X	D_X	X	D_X
1	1	2	2
3	3	4	4
5c	5c, 5r, 5l, 5w	5r	5c, 5r, 5l, 5w
5l	3	5w	5c, 5r, 5l, 5w
6c	6c, 6r, 6l, 6w	6r	6c, 6r, 6l, 6w
6l	4	6w	8c, 8r, 8l, 8w
7c	5c, 5r, 5l, 5w	7r	7c, 7r, 7l, 7w
7l	3	7w	5c, 5r, 5l, 5w
8c	6c, 6r, 6l, 6w	8r	8c, 8r, 8l, 8w
8l	4	8w	8c, 8r, 8l, 8w
9c	9c, 9r, 9l, 9w	9r	2
9l	9c, 9r, 9l, 9w	9w	10c, 10r, 10l, 10w
10c	9c, 9r, 9l, 9w	10r	10c, 10r, 10l, 10w
10l	2	10w	10c, 10r, 10l, 10w

While some symmetries are apparent in this table, we have not yet achieved a form comparable to that of eq. 13 for the three-body case.

DISCUSSION

This contribution has concentrated on the topological properties rather than the physical ones. The final form of the two-body parquet equations is of immediate interest

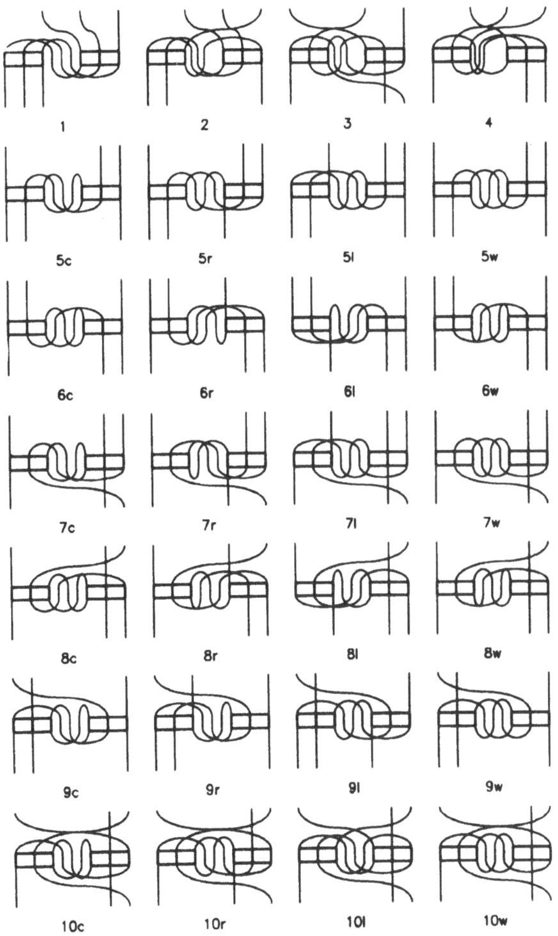

Fig. 2. The three-body diagrams.

for carrying out many-body calculations. We have previously defined a set of local *approximations* for boson systems and shown that they lead to the same equations as the boson hypernetted-chain optimized variational calculations[6-7]. The same approach has been extended to deal with spin-dependent forces in boson systems[8]. There remain deep questions about how best to incorporate spin dependence[9].

For fermion systems, one must necessarily make tradeoffs with consistency in the Baym and Kadanoff sense[10]. The basic problem is that there are several features that one would like to get correct in an approximate calculation. Unfortunately, getting them exactly correct requires an exact calculation. We are still in the process of analyzing how best to sacrific some principles and use the violation as a gauge of unreliability of the calculation.

We would like to emphasize that while the details of implementing a tractable fermion parquet theory have not been explicitly laid out, the parquet diagrams are an essential element in any sensible perturbation-theoretic approach to realistic many-body problems. One must include the ladder diagrams to handle hard-core interactions and one must include the chains to handle screening properly. While there is apparent freedom in dealing with other diagrams, including self-energy insertions, most of that will be eliminated by determining the most important consistency conditions.

One feature that we have previously shown to be important in the boson system is including sets of diagrams related by Ward identities. Some preliminary work indicates that this may be even more crucial in fermion systems, where we are being led to include more statistical correlations in defining the local approximations. This looks like it will be relevant for both spin-independent and spin-dependent quantities.

The three-body parquet theory is in its relative infancy. We believe that it may be useful in constructing a new two-body vertex "beyond parquet" for the two-body Green's function, in much the way that the use of the two-body Γ can be used to construct the self-energy needed to make a better one-body Green's function. Ultimately, solving some kind of three-body diagrams is probably needed to improve quantitative reliability for most problems of interest.

ACKNOWLEDGEMENTS

Discussions with A. D. Jackson and E. Krotscheck have been particularly fruitful in many areas of parquet theory. The work done here was supported in part by the NSF under grants 8507157 and 8605979. R. A. S. is grateful to the U. S. Army Research Office for helping to make participation in this conference possible.

REFERENCES

1. R. A. Smith, *Planar Theory Made Plainer*, Proceedings of the IX International Workshop on Condensed Matter Theories, San Francisco, Aug. 1985, ed. F. B. Malik, (Plenum Press, New York, 1986), 9-18.

2. A. D. Jackson, A. Lande and R. A. Smith, *Variational and perturbation theory made planar*, Phys. Report **86**, 55-111 (1982).

3. A. Lande and R. A. Smith, *Crossing-symmetric rings, ladders, and exchanges*, Phys. Lett. **131B**, 253-256 (1983).

4. G. Baym and L. P. Kadanoff, Phys. Rev. **124**, 287(1961).

5. G. Baym, Phys. Rev. **127**, 1391 (1962).

6. A. D. Jackson, A. Lande and R. A. Smith, *Planar theory made variational*, Phys. Rev. Lett. **54**, 1469-1471 (1985).

7. A. D. Jackson, A. Lande, R. W. Guitink and R. A. Smith, *Application of parquet perturbation theory to ground states of boson systems*, Phys. Rev. **B31**, 403(1985).

8. R. A. Smith and A. D. Jackson, *Planar Theory With Spin and Tensor Forces*, Nucl. Phys., in press.

9. E. Krotscheck, private communication.

10. A. D. Jackson and R. A. Smith, *The High Cost of Consistency in Green's-Function Expansions*, Phys. Rev. **A**, in press.

GENERALIZED SELF-CONSISTENT FIELD EQUATIONS FOR MULTIPLE PAIRS OF FERMIONS *

J. Y. Shapiro
Physics Department, Fordham University
Bronx, NY 10458, U.S.A.

F. B. Malik
Physics Department, Southern Illinois University
Carbondale, IL 62901, U.S.A.
and
Institutionen för Fysik, Åbo Akademi
(SF-20500) Turku, Finland

INTRODUCTION

The Hartree-Fock [1,2] self-consistent field (SCF) method was developed to deal with fermion systems that could be described reasonably well by single particle orbitals. The electronic structure of atoms is such a system. Here, because of the strong Coulomb field of the nucleus, in the first approximation electrons move in a central field, making the single particle description a reasonable one. In atoms, the contribution of the central Coulomb potential to the total energy is dominant. The Hartree-Fock SCF approach, which provides the best set of single particle orbitals, is a natural choice and has been very successful. It also serves as a starting point for evaluating higher order corrections.

On the other hand, there are systems such as nuclei which do not have a central potential to start with. For such systems, one would like to start with a wave function that includes correlations between pairs of nucleons. This is particularly of interest if the mean free path of the constituents is of the order of the dimensions of the system under consideration, which is the case for nuclei. For fermion systems like these, one would like to expand the total wave function as a linear combination of products of pair functions and then derive the SCF equations using a variational method.

The key stumbling block to this approach has been to find a proper linear combination that makes the total wave function antisymmetric under the exchange of any two of its coordinates. In the cluster expansion scheme[3,4] and Jastrow[5][1] approach, antisymmetrization is achieved by writing the total wave function as a determinant of single particle orbitals multiplied by a product of symmetric state-independent pair functions. This does not represent the most general expansion of the total wave function as a linear combination of products of pair functions in a fashion that ensures antisymmetrization.

*Supported in part by a grant from the U.S. Army Research Office.

[1]The original form was suggested by N. F. Mott for a hard sphere Bose gas and applied by R. B. Dingle[6].

Recently, however, we have obtained an expansion not only in multi-pair functions but for any multi-cluster functions [7] by generalizing the earlier work of Malik [8,9]. In this paper we restrict ourselves to the multi-pair formulation.

THEORY

To expand the wave function of a system of n fermions in terms of a single pair function, one can write [8]

$$\Psi(1,2,3,\ldots,n-1,n) = \frac{1}{\sqrt{n!}} \sum_{i=1}^{n-1} \sum_{j=i+1}^{n} (-1)^{i+j} \Phi(i,j) D(\bar{i},\bar{j}) \tag{1}$$

where the arguments $(\bar{i},\bar{j})$ of D mean that the ith and jth coordinates in D are missing. The wave function Ψ is antisymmetric if both Φ and D are antisymmetric functions – for the antisymmetrization it is not necessary for D to be a determinant of single particle orbitals.

One can introduce a second pair function into our wave function by reusing the expansion of Eq.(1) for the factor $D(\bar{i},\bar{j})$. Repeated use of this process for a system with an even number $2n$ of particles leads to a completely antisymmetric expansion of Ψ as a linear combination of products of n pair functions.

Formally, this expansion can be written as

$$\Psi = \frac{1}{\sqrt{(2n)!}} \sum_P \epsilon^P P \Phi_{\alpha_1}(1,2)\Phi_{\alpha_2}(3,4)\cdots\Phi_{\alpha_n}(2n-1,2n) \tag{2}$$

with the understanding that the expression $\sum \epsilon^P P$ appearing in Eq.(2) is a sum over permutations of the particle coordinates defined as follows: the permutations entering into the product of antisymmetrized pair functions in Eq.(2) form a subgroup $g = Sn(2) \otimes Sn(2) \cdots \otimes Sn(2)$ (with n factors) of the complete permutation group $Sn(2n)$ of the coordinates of the 2n particles. The sum over permutations is done in such a way as to contain one element from each coset of g. The number ϵ^P is $+1$ if P is even and -1 if P is odd. The total number of terms in the summation is $(2n)!/2^n$. The pair functions are assumed to be normalized according to Eq.(5) below.

Given a Hamiltonian H and appropriate orthonormality conditions, the SCF equation for a pair function can be derived by minimizing the total energy ϵ with respect to variations of the pair function, i.e.,

$$\delta\epsilon = \delta\frac{(\Psi, H\Psi)}{(\Psi,\Psi)} = 0. \tag{3}$$

For our purpose, the auxiliary conditions under which the variation is to be performed are the condition of antisymmetrization

$$\Phi_\alpha(2,1) = -\Phi_\alpha(1,2), \tag{4}$$

and the conditions of orthonormality

$$\int\int \Phi_\alpha^*(1,2)\Phi_\beta(1,2)\, d1\, d2 = 2\delta_{\alpha\beta}, \tag{5}$$

and

$$\int \Phi_\alpha^*(1,2)\Phi_\beta(1,3)\,d1 = 0 \tag{6}$$

for $\beta \neq \alpha$ and for both states α and β occupied.

Although H might contain three-body or higher interactions [9], here we consider only Hamiltonians with one- and two-body operators, i.e.,

$$\mathcal{H} = \sum_{i=1}^{2n} T(i) + \sum_{i=1}^{2n-1}\sum_{j=i+1}^{2n} V(i,j) \tag{7}$$

In Eq.(7) $T(i)$ is a one-body operator which is usually the kinetic energy term $(-\hbar^2/2m)\nabla^2$ but could contain an additional single particle potential $U(i)$.

The SCF equations are obtained (for details, see ref.[7]) by the variational condition (3) using the method of Lagrange multipliers to incorporate the auxiliary conditions (5) and (6). We get

$$[T(1) + T(2) + V(1,2) - \epsilon_\alpha]\Phi_\alpha(1,2)$$

$$+\frac{1}{2}\sum_{\beta=1}^{n}\int\int \Phi_\beta^*(3,4)V_{eff}\Phi_\alpha(1,2)\Phi_\beta(3,4)\,d3\,d4 = 0 \tag{8}$$

where $\beta \neq \alpha$ in the sum. The Lagrange multiplier ϵ_α is the eigenvalue associated with a pair in the state α and V_{eff} is given by

$$\begin{aligned} V_{eff} &= V(1,3)[I - X(1,3)] + V(1,4)[I - X(1,4)] \\ &+ V(2,3)[I - X(2,3)] + V(2,4)[I - X(2,4)] \end{aligned} \tag{9}$$

In Eq.(9) the exchange operator $X(i,j)$ transposes all coordinates of i and j, including spins. Clearly Eq.(8) represents n coupled integrodifferential *eigenvalue* equations that are to be solved self-consistently for the n pair functions Φ_α with $\alpha = 1, 2, \cdots, n$.

It is also possible to write down the set of SCF field equations (8) in integral form. However, this form depends upon the choice of unperturbed basis. Defining K and W by

$$K(1,2)\Phi_\alpha(1,2) \equiv \frac{1}{2}\sum_{\beta=1}^{n}\int\int \Phi_\beta^*(3,4)V_{eff}\Phi_\alpha(1,2)\Phi_\beta(3,4)\,d3\,d4 \tag{10}$$

and

$$[V(1,2) + K(1,2)]\,\Phi_\alpha(1,2) \equiv W(1,2)\Phi_\alpha(1,2) \tag{11}$$

we can rewrite Eq.(8) as

$$[T(1) + T(2) + W(1,2)]\,\Phi_\alpha(1,2) = \epsilon_\alpha\Phi_\alpha(1,2) \tag{12}$$

If one chooses as the unperturbed Hamiltonian $T^{(0)} = T(1) + T(2)$ and projects Φ_α onto the eigenfunctions ϕ_α of $T^{(0)}$ so that

$$(I - Q_\alpha)\,\Phi_\alpha(1,2) = \phi_\alpha(1,2), \tag{13}$$

then the integral equation for Φ_α is

$$\Phi_\alpha(1,2) = \phi_\alpha(1,2) + \frac{1}{\epsilon_\alpha - T^{(0)}} Q_\alpha W(1,2)\Phi_\alpha(1,2) \tag{14}$$

Multiplying Eq.(14) by $W(1,2)$, one can define an equivalent matrix equation for an operator $B_\alpha(1,2)$ that yields the energy shift

$$\Delta E = \epsilon_\alpha - \varepsilon_\alpha^{(0)} = (\phi_\alpha, B_\alpha \phi_\alpha) \tag{15}$$

with

$$B_\alpha(1,2) = W(1,2) + W(1,2)\frac{1}{\epsilon_\alpha - T^{(0)}} Q_\alpha B_\alpha(1,2) \tag{16}$$

and where the $\varepsilon_\alpha^{(0)}$ are the eigenvalues of the equation

$$T^{(0)}\phi_\alpha = \varepsilon_\alpha^{(0)}\phi_\alpha. \tag{17}$$

The alternative choice of unperturbed Hamiltonian

$$H^{(0)} = T^{(0)} + K(i,j)$$

leads to the integral equation

$$\Phi_\alpha(1,2) = \psi_\alpha(1,2) + \frac{1}{\epsilon_\alpha - H^{(0)}} q_\alpha V(1,2)\Phi_\alpha(1,2) \tag{18}$$

with $H^{(0)}\psi_\alpha = E_\alpha^{(0)}\psi_\alpha$ and $(I - q_\alpha)\Phi_\alpha = \psi_\alpha$. The corresponding energy shift and the equation for the energy shift operator J_α are, respectively, given by

$$\Delta E = \epsilon_\alpha - E_\alpha^{(0)} = (\psi_\alpha, J_\alpha \psi_\alpha) \tag{19}$$

and

$$J_\alpha(1,2) = V(1,2) + V(1,2)\frac{1}{\epsilon_\alpha - H^{(0)}} q_\alpha J_\alpha(1,2) \tag{20}$$

It is interesting to relate the total energy ϵ of the system to the eigenvalues ϵ_α of the pairing equation (8). For this purpose, we define the unperturbed eigenvalues ϵ_α^0 as

$$\epsilon_\alpha^o = \frac{1}{2}\int\int \Phi_\alpha^*(1,2)[T(1) + T(2) + V(1,2)]\Phi_\alpha(1,2)\, d1\, d2 \tag{21}$$

(Note that ϵ_α^0 is not the same as $\varepsilon_\alpha^{(0)}$ defined by Eq.(17).) We also define the interaction energy $V_{\alpha\beta}^0$ between the pairs α and β by

$$V_{\alpha\beta}^o = \frac{1}{4}\int\int\int\int \Phi_\alpha^*(1,2)\Phi_\beta^*(3,4)V_{eff}\Phi_\alpha(1,2)\Phi_\beta(3,4)\, d1\, d2\, d3\, d4 \tag{22}$$

A short calculation then gives for the total energy ϵ

$$\epsilon = \frac{1}{2}\sum_{\alpha=1}^{n}(\epsilon_\alpha + \epsilon_\alpha^o) \tag{23}$$

$$= \sum_{\alpha=1}^{n}\epsilon_\alpha - \frac{1}{2}\sum_{\alpha=1}^{n}\sum_{\beta=1}^{n} V_{\alpha\beta}^o \tag{24}$$

where $\beta \neq \alpha$ in the last sum. The eigenvalue ϵ_α is given by

$$\epsilon_\alpha = \epsilon_\alpha^0 + \sum_{\beta\neq\alpha} V_{\alpha\beta}^0. \tag{25}$$

DISCUSSION AND PROPOSED APPLICATION

A. Discussion

The derivation of the SCF equation (8) for the pair function is valid for all two-body interactions provided the integrals (22) exist, which is the case for most of the realistic nucleon-nucleon potentials. Hence these equations are an ideal starting point for calculating the total energy of nuclei.

Since the equations are derived from a variational principle, the calculated total energy is always an upper bound to the exact energy of the system. The square integrability condition insures the validity of these equations for a finite system.

The SCF equations are also valid for the case of nuclear matter because one can introduce a box normalization for the wave function and thereby fulfil the orthonormality conditions of Eqs.(5) and (6).

As noted in the single pair case [8], the structure of these SCF equations in their integral forms (14) or (18) is different from the usual Bethe-Goldstone[10] (BG) equations or various versions of G-matrix equations that are used in calculations of the binding energy per nucleon in nuclear matter and of the total energy or the excitation energy of a pair in finite nuclei. The key differences are:

(a) The pair equation derived here involves hermitian operators so that the eigenvalues are real, which is not the case for the BG or the usual G-matrix equation. The non-hermiticity of the BG and G-matrix equations quite often present difficulties in actual evaluation, as discussed by Cooper[11], Lüders[12], Emery[13], and Van Hove[14].

(b) The q_α operator in Eq.(18) is different from the projection operator in the BG equation. This $(I - q_\alpha)$ merely projects the complete pair function Φ_α onto the unperturbed states ψ_α and is not a projection operator that prevents scattering to occupied states. The entire Pauli exclusion principle in our scheme is incorporated into the non-local operator X in V_{eff} and hence in K.

It is to be noted that the set of equations (8) goes well beyond the Hartree-Fock (HF) or Brueckner-Hartree-Fock[14] (BHF) approximation. Eq.(8) includes pair-pair interactions, which is not the case in the HF or BHF method.

In addition, clearly one should not attempt to calculate the total energy by summing eigenvalues. The correct approach is to use Eq.(23) or (24).

Whereas the SCF equation (8) differs from the BG or BHF approach, it is in complete accord with the type of equations used in shell model calculations with configuration admixture for nuclei and the configuration interaction approach in atomic and molecular physics. The latter approaches have indeed achieved considerable success in explaining experimental facts in their respective areas.

The appropriate application of Eq.(8) is to calculate the total binding energies of finite nuclei. Although the set of equations (8) is valid for pairs in infinite nuclear matter, the binding energy per nucleon is not a well defined observable. Usually, the coefficient of the mass number A in the mass formula is taken to be the binding energy per nucleon in nuclear matter. However, for finite nuclei, the separation of volume and surface terms cannot be done in an unambiguous way. This can be illustrated by adopting a trapezoidal energy-density distribution for nuclei (this is a fairly good approximation to the observed density distribution for medium-mass and heavy nuclei.[15]). The distribution

$$\rho = \begin{cases} \rho_0 & \text{for } r < c \\ \rho_0 \left[\frac{d-r}{d-c}\right] & \text{for } c \leq r < d \\ 0 & \text{for } r \geq d \end{cases} \quad . \tag{26}$$

gives for the total energy

$$E = \frac{\pi \rho_0}{3}(c + d)(c^2 + d^2) \tag{27}$$

The coefficient of A in this equation depends critically on what is assumed for the A-dependence of the parameters c and d. Since there is considerable freedom in this choice, one can get different values for the coefficient of A.

In addition to this problem, one can describe nuclear masses by other appropriate mass distribution functions instead of Eq.(26), as well as by the mass formula based on a liquid drop. Thus one must be extremely cautious in taking the coefficient of A in a mass formula based on a liquid drop to be "the experimental binding energy per nucleon in nuclear matter".

With this view in mind, we now present an outline of the kind of application that might be done in the near future.

B. Proposed Application

As a first application of the pairing equation (8), we plan to calculate ground state properties of some light even-A nuclei, for example, ^4He, ^6He, ^6Li and ^6Be. These calculations will include the Coulomb potential, so that the isotopic spin formulation will *not* be used. Some simple v_6 and v_8 potentials will be used, e.g., those of Massey[16] and of Eikemeier and Hackenbroich[17].

One needs to find self-consistent solutions of the set of n pair equations

$$[T(1) + T(2) + V(1,2) - \epsilon_\alpha]\Phi_\alpha(1,2)$$

$$+\frac{1}{2}\sum_{\beta=1}^{n}\int\int \Phi^*_\beta(3,4)V_{eff}\Phi_\alpha(1,2)\Phi_\beta(3,4)\, d3\, d4 = 0$$

We plan to cycle through these equations for different Φ_α's until these pair functions stabilize. In solving for a particular pair function Φ_α, it will be assumed that the functions Φ_β, $\beta \neq \alpha$, are known from the latest iteration of the pair equations for Φ_β. Thus at each step one will be solving an equation that is homogeneous and linear in Φ_α. At each step the solution Φ_α will be orthonormalized to the set of pair functions Φ_β, $\beta \neq \alpha$. To begin this procedure one requires initial guesses for $n - 1$ of the pair functions.

As an example, consider the six-particle nucleus ^6He. We have three pair functions Φ_1, Φ_2 and Φ_3, where Φ_1 could be taken as a pair of neutrons and Φ_2 as a pair of protons in a core, and Φ_3 as a pair of valence neutrons. As initial guesses, we could assume that Φ_1 and Φ_2 are in $(1s)^2$ shell model configurations. We would then solve pair equations in turn for $\Phi_3, \Phi_1, \Phi_2, \Phi_3, \Phi_1, \Phi_2, \ldots$, until stability is attained. We first solve for Φ_3, assuming that Φ_1 and Φ_2 are in $(1s)^2$ configurations. The solution of Eq.(8) for Φ_3 would then be orthonormalized to the assumed Φ_1 and Φ_2. Next we solve the pair equation for Φ_1 using the just-obtained solution for Φ_3 and assuming that Φ_2 is in the $(1s)^2$ configuration. The solution for Φ_1 would be orthonormalized to Φ_3 and Φ_2. Next we solve the equation for Φ_2 using the solutions already obtained for Φ_3 and Φ_1, again orthonormalizing. This procedure is continued until the solutions stabilize. The equations (2)-(9) require slight modifications in this application because the total wave function Ψ is not antisymmetric in all particles, but is separately antisymmetrized for neutrons and protons. Thus, for example, the interpair potential V_{eff} between a nn-pair and a pp-pair would not contain exchange terms, i.e., it would be given by

$$V_{eff} = V(1,3) + V(1,4) + V(2,3) + V(2,4)$$

rather than Eq.(9).

For an np-pair, the pair function Φ_α would *not* be antisymmetric, and exchange terms would appear in V_{eff} only between the identical particles. For example, suppose that $\Phi_\alpha(1,2)$ were an nn-pair function and $\Phi_\beta(3,4)$ an np-pair function with particle 3 being the neutron. Then one would have

$$V_{eff} = V(1,3)[I - X(1,3)] + V(1,4) + V(2,3)[I - X(2,3)] + V(2,4)$$

in this case.

In our initial calculations we will expand the pair functions in terms of a truncated, orthonormal set of functions $\{\chi_i\}$. For this set we will use either spherical or axially symmetric oscillator functions. The expansion will take the form

$$\Phi_\gamma(1,2) = \sum_{ij} c_{ij}^\gamma \left[\chi_i(1)\chi_j(2) - \chi_i(2)\chi_j(1)\right] \tag{28}$$

where

$$c_{ij}^\gamma = 0 \text{ for } i \geq j \tag{29}$$

for a nn or pp pair, and

$$\Phi_\gamma(1,2) = \sum_{ij} c_{ij}^\gamma \chi_i(1)\chi_j(2) \tag{30}$$

for an np pair. In both cases we normalize by

$$\sum_{i,j} \left|c_{ij}^\gamma\right|^2 = 1.$$

Substituting these expressions into the pair equation (8) and using the orthonormality of the single particle functions gives a set of equations for the coefficients c_{ij}^γ. For example, consider the case of an even-even nucleus for which we assume only nn and pp pairs. Substituting the expansion (30) into the pair equation and taking the scalar product with $\chi_r(1)\chi_s(2) - \chi_r(2)\chi_s(1)$ leads to the set of equations

$$\sum_k \left[T_{rk}\left(c_{ks}^\alpha - c_{sk}^\alpha\right) - T_{sk}\left(c_{kr}^\alpha - c_{rk}^\alpha\right)\right] + \sum_{k,l} V_{rs,kl}\left(c_{kl}^\alpha - c_{lk}^\alpha\right)$$

$$+ \sum_{\beta\neq\alpha}\sum_k \left\{\left[\sum_{i,j} b_{ij}^\beta \left(V_{ri,kj} - V_{ri,jk}\right)\right]\left(c_{ks}^\alpha - c_{sk}^\alpha\right) - \left[\sum_{i,j} b_{ij}^\beta \left(V_{si,kj} - V_{si,jk}\right)\right]\left(c_{kr}^\alpha - c_{rk}^\alpha\right)\right\}$$

$$= \epsilon_\alpha \left(c_{kr}^\alpha - c_{rk}^\alpha\right) \tag{31}$$

where

$$b_{ij}^\beta = \sum_t \left(c_{it}^{\beta*} - c_{ti}^{\beta*}\right)\left(c_{jt}^\beta - c_{tj}^\beta\right) \tag{32}$$

Only one of the coefficients c_{ij} or c_{ji} in each parenthesis $(c_{ij} - c_{ji})$ in the equations (33) and (34) is different from zero because of Eq.(31). Thus we have a set of homogeneous algebraic eigenvalue equations to solve for the coefficients c_{ij} and the pairing energy ϵ_α.

To get the total energy of the nucleus, we must also calculate the unperturbed energies ϵ_α^0. However, these are just the expectation values of $T(1) + T(2) + V(1,2)$ for our assumed pair function wave function. Using the expansion (30), one obtains

$$\epsilon_\alpha^0 = 2\sum_{i,k} b_{ik}^\alpha T_{ik} + \sum_{i,j,k,l} \left(c_{ij}^{\alpha *} - c_{ji}^{\alpha *}\right)\left(c_{kl}^\alpha - c_{lk}^\alpha\right) V_{ij,kl} \tag{33}$$

where b_{ik}^α is given by Eq.(34).

Similar expressions can be obtained for the case where one also includes np pairs.

One purpose of performing computations on light nuclei with simple potentials is to gain familiarity and to do some experimenting with methods of solving the pair equation (8) in a self-consistent fashion. Results will be different with different choices of pairs. For example, one would expect different results for ^4He with two np pairs as opposed to one nn and one pp pair. It is, of course, of interest to determine which of these two choices gives the lowest energy.

After having gained some experience on these nuclei with only a few pairs, we plan to proceed to more realistic forces such as the Argonne v_{14} potential [18] and to heavier nuclei of interest, such as ^{18}O. Here, of course, we will be interested in reproducing the low-lying spectra and transition probabilities. As a reasonable first approximation, we will assume that low-lying excited states consist of the excitation of a single pair.

References

[1] D. R. Hartree. *Proceedings of the Cambridge Philosophical Society*, 24:89, 1927.

[2] V. Fock. *Zeitschrift fur Physik*, 61:126, 1930.

[3] J. E. Mayer and M. G. Mayer. *Statistical Mechanics*. John Wiley, New York, 1940.

[4] J. De Boer. *Reports on Progress in Physics*, 12:305, 1948.

[5] Robert Jastrow. *Physical Review*, 98:1479, 1955.

[6] R. B. Dingle. *Philosophical Magazine*, 40:573, 1949.

[7] F. Bary Malik, Roger H. Richardson, and Joseph Y. Shapiro. 1987. Submitted to Journal of Physics G: Nuclear Physics.

[8] F. Bary Malik. *Annals of Physics (New York)*, 36:86, 1966.

[9] F. Bary Malik. 1987. Submitted to Journal of Physics G: Nuclear Physics.

[10] Hans A. Bethe and G. Goldstone. *Proceedings of the Royal Society (London)*, A228:551, 1956.

[11] Leon N. Cooper. *Physical Review*, 104:1189, 1956.

[12] L. Luders. *Zeitschrift fur Naturforschung*, 14a:1014, 1959.

[13] V. J. Emery. *Nuclear Physics*, 12:69, 1959.

[14] L. Van Hove. *Physica*, 25:849, 1959.

[15] F. Bary Malik and I. Reichstein. In F. B. Malik, editor, *Condensed Matter Theories, Vol. 1*, Plenum Publishing Corp., New York, 1986.

[16] S. Hochberg, H. S. W. Massey, H. Robertson, and L. H. Underhill. *Proceedings of the Physical Society*, A68:746, 1955.

[17] H. Eikemeier and H. H. Hachenbroich. *Nuclear Physics*, A169:407–416, 1971.

[18] R. B. Wiringa, R. A. Smith, and T. L. Ainsworth. *Physical Review*, C29:1207, 1984.

THE ANHARMONIC OSCILLATOR REVISITED

H.G. Kümmel

Institut für Theoretische Physik II
Ruhr-Universität Bochum
463o Bochum 1, West-Germany

The quartic anharmonic oscillator with the Hamiltonian

$$\underline{H} = \frac{1}{2}\underline{p}^2 + \frac{1}{2}\underline{q}^2 + \frac{\lambda}{4}\underline{q}^4$$

is a well established testing ground for various approximation methods. On the one hand it has non trivial features: ordinary perturbation theory (expansion into powers of λ) breaks down; it may be used as a starting point for ϕ^4 field theories and for testing Monte Carlo methods. On the other hand, it can be solved exactly either by matrix diagonalization in the harmonic oscillator basis or by solving the corresponding differential equation.

Several techniques for rearranging the perturbation series have been applied successfully[1]. The coupled cluster method (CCM) is one of them. It has been applied to the anharmonic oscillator by Hsue et al [2] and by Kaulfuß and Altenbokum [3] (quoted as KA). This method is well established in many body theory, both in nuclear physics and especially in quantum chemistry [4,5].

The present paper - some part's of which are due to R. Mentz - is intended to present new results which shed some light on the former ones. Indeed there were two major aspects in KA, which are worth considering again: The 'non-existence' of the truncated wave function and the fact that the 'correlation amplitudes' S_n are not vanishing for large n.

Before discussing it let me remind you that the ground state CCM uses the wave function (WF) in the form

$$|\Psi_0\rangle = \exp \underline{S}\, |\phi_a\rangle\,, \tag{1}$$

where

$$\underline{S} = \Sigma\, \underline{S}_{2n}\,, \quad \underline{S}_n = \frac{1}{\sqrt{n!}} S_n (\underline{a}^+)^n\,, \tag{2}$$

(for the anharmonic oscillator only even labels occur). $\underline{a}^+,\underline{a}$ are the usual ladder operators and $|\phi_a\rangle$ is the ground state of the harmonic oscillator,

$$\underline{a}\, |\phi_a\rangle = 0\,. \tag{3}$$

The CCM equations follow from the Schrödinger equation by projecting from the left with $\langle\phi_a|e^{-\underline{S}}$, $\langle\phi_a|\underline{a}^2 e^{-\underline{S}}$, etc.:

$$\begin{aligned} &\langle\phi_a| e^{-\underline{S}} \underline{H}\, e^{\underline{S}} |\phi_a\rangle = E_0\,, && \text{(a)} \\ &\langle\phi_a| \underline{a}^2 e^{-\underline{S}} \underline{H}\, e^{\underline{S}} |\phi_a\rangle = 0\,, && \text{(b)} \\ &\langle\phi_a| \underline{a}^4 e^{-\underline{S}} \underline{H}\, e^{\underline{S}} |\phi_a\rangle = 0\,. && \text{etc. (c)} \end{aligned} \tag{4}$$

It is convenient to optimize first the ladder operators $\underline{a},\underline{a}^+$ and the ground state $|\phi_a\rangle$ going with them. The most general transformation leading from $\underline{a},\underline{a}^+,|\phi_a\rangle$ to $\underline{b},\underline{b}^+,|\phi_b\rangle$ with

$$[\underline{b},\underline{b}^+] = 1\,, \quad \underline{b}\,|\phi_b\rangle = 0 \tag{5}$$

is a Bogolybov transformation

$$\underline{b} = \frac{\underline{a} + t\, \underline{a}^+}{\sqrt{1-t^2}}\,. \tag{6}$$

t can be determined from several principles:

i) optimize the energy by $\langle\phi_b|\underline{H}|\phi_b\rangle = \text{min.}$ (7)
(Hartree condition)

ii) optimize the 'starting wave function' by

$$|\langle\phi_b|\Psi_0\rangle| = \text{max.} \tag{8}$$

Details are given in KA. The Hamiltonian in terms of $\underline{b},\underline{b}^+$ is of the form

$$H = \mathcal{E}_0(t) + \varepsilon(t)\, \underline{b}^+ \underline{b} + \alpha(t)\left(\underline{b}^2 + \underline{b}^{+2}\right) + \gamma(t) :(\underline{b}+\underline{b}^+)^4: \qquad (9)$$

(: : means normal ordering).

The Hartree condition fixes t (via $\alpha(t) = 0$), whereas the maximum overlap condition leads to $S_2=0$ and t is determined in conjunction with S_4, S_6, ... via the CCM equations (4).

These equations can be written down explicitly [3]. They have been solved in KA for SUB(n)-truncations with n up to 3o (S_{32}, S_{34}, ... thrown away). Their results can be summarized as follows:

i) the errors for the energy are smaller than 0.1 % for $0 < \lambda \leq 1000$ if one sticks to low order truncations

ii) the truncated WF's are not normalizable

iii) the S_n for large S_n behave rather irregularly and do not go down to zero for large n. Also the quality of the energy starts to deteriorate.

Let me first dwell upon i) and extend it a bit. One remarkable feature of these low order ($n \leq 6$) CCM equations is their extreme stability, see fig. 1. We have plotted the

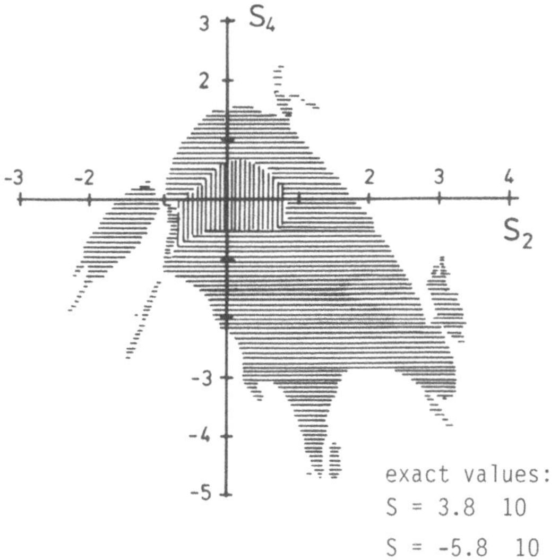

Fig. 1 Convergence region for S_2 and S_4. Outer/inner region corresponds to convergence after 4o/10 iterations. (λ =1.0).

starting values for S_2^o and S_4^o for a certain (Brueckner-like) iteration map for the nonlinear CCM equations. S_2^o or S_4^o may be wrong by a factor of 1000 and one still arrives at the exact values of the order 10^{-3} to 10^{-4} for S_4 and S_2. Outside this region typically there is no convergence except for very few far away starting values. In other words: normally one will arrive either at the correct values or have no convergence at all. Furthermore, we have applied SUB(4) CCM to resonant states for $0 > \lambda \geq -0.3$: it worked pretty well again - see fig. 2.

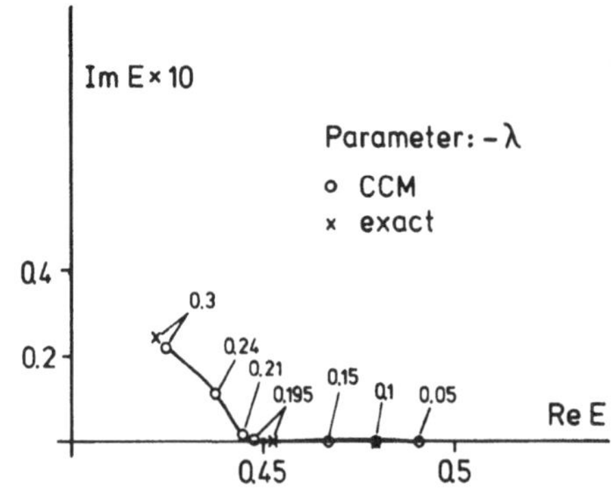

Fig. 2 Resonance energies for $0 > \lambda \geq -0.3$.

We now discuss ii), the non normalizability of the truncated wave function

$$|\Psi_0\rangle_N \equiv \exp\left(\sum^N \underline{S}_n\right)|\phi_b\rangle. \tag{10}$$

It is easily seen that this is not serious. Clearly, if we 'truncate the exponential' by defining

$$\exp_M\left(\sum^N \underline{S}\right) = \left(1 + \underline{S}_2 + \frac{1}{2}\underline{S}_2^2 + \dots + \frac{1}{(\frac{M}{2})!}\underline{S}_2^{\frac{M}{2}} + \underline{S}_4 + \dots + \underline{S}_4^{\frac{M}{4}}\right),$$

<u>where all terms with powers of $\underline{b}^+$ larger than M are omitted,</u> the resulting WF

$$\exp_M\left(\sum^N \underline{S}\right)|\phi_b\rangle$$

is normalizable. Yet, if we chose M sufficiently large the CCM equations (4) do not change at all in a given truncation. A trivial example is the energy

$$E_0 = \langle \phi_b | \underline{H}(1+\underline{S}_2 + \tfrac{1}{2}\underline{S}_2^2 + \underline{S}_4|\phi_b\rangle$$

which is exact (for any truncation). The two-body equation (4b) remains intact in the SUB(4) approximation even if we use only

$$\exp_8 \sum_n^4 \underline{S}_n |\phi_b\rangle = (1+\underline{S}_2 + \cdots \tfrac{1}{4}\underline{S}_2^4 + \underline{S}_4 + \cdots + \underline{S}_4^2)|\phi_b\rangle$$

as WF. This 'double truncation' not only restores the normalizability of the WF, it also does not change the CCM equations if we don't keep the 'truncation of the exponential' too low. Moreover, this wave function gives correct matrix elements for operators which do not test too highly excited states. We have compared the energy expectation values with the exact energy for several exponential function truncations, see fig. 3.

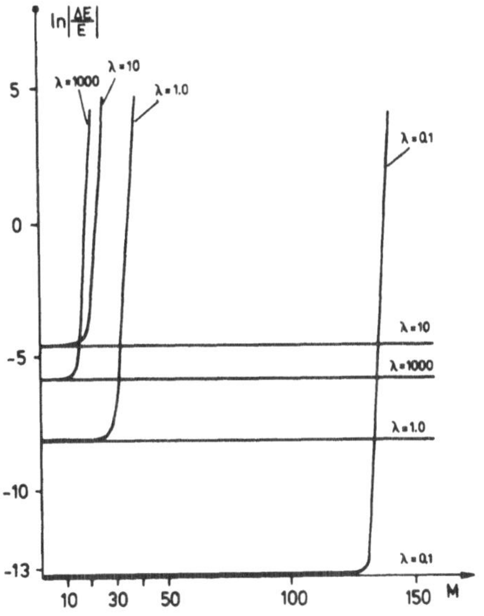

Fig. 3 Energy expectation values for different double truncation schemes. Horizontal lines are the exact energies.

For simplicity only the SUB(4) approximation in the maximum overlap scheme (S_2=0) is used in this figure. Take λ =1.0 as an example: The wave function $\exp_{100}\underline{S}_4|\phi_b\rangle = (1 + \underline{S}_4 + .. \underline{S}_4^{25})|\phi_b\rangle$

reproduces the energy with very small errors, whereas $\exp_{120}S_4 |\phi_b\rangle$ ceases to be a good wave function.

I now come to the points iii), the undesirable features of S_n and truncation scheme at large n. In Fig. 4 I have

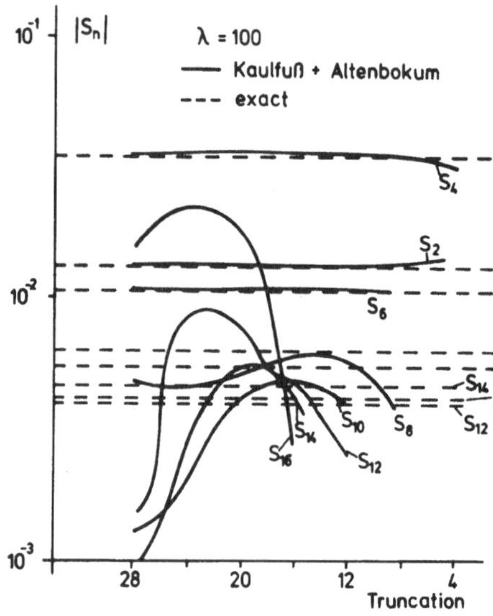

Fig. 4 $|S_n|$ as function of truncation schemes.

plotted the S_n's from KA for different truncations. Up to n=6 things look pretty good: the S_n's are almost stable and very near to the exact values. S_8 oscillates around the exact value. Afterwards things become more irregular. One gets the suspicion that the truncation scheme breaks down. One of the reasons for this may be the fact that theexact S_n become large - a fact that would prohibit the standard SUB(n) truncation as used here. Luckily, one can solve the anharmonic oscillator problem exactly and reconstruct the exact S_n's from the wave function. In a preliminary study we have done this and indeed found that the S_n first go down and then up again. This is shown for λ=1oo (the worst case) in Fig. 5. Thus there is little hope that a SUB(n) truncation will work well in this case beyond n=1o. Of course, this is no surprise as we anyway expect only asymptotic convergence. One may turn around this argument and say that the fact that the S_n go down first makes the CCM so powerful. But it is not a method to improve things systematically and forever.

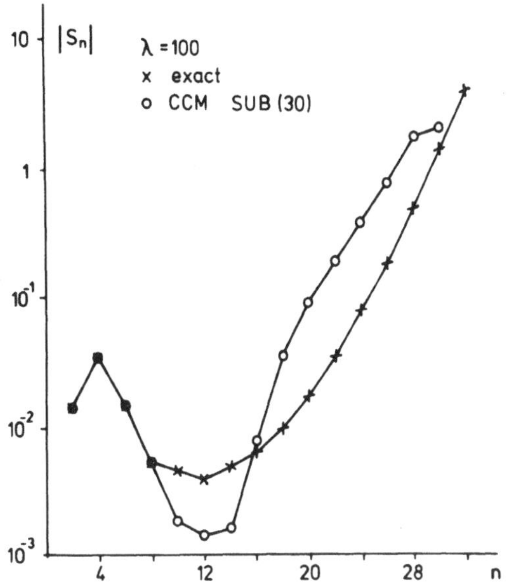

Fig. 5 Exact $|S_n|$ and $|S_n|$ obtained via CCM in SUB(30)

Two questions arise: How is this result related to the various partial summation methods explored in the literature? After all, what I have presented here may be called 'computer experiments', the mathematics behind it still has to be understood. We are planning to investigate this. The second question is: are these results relevant for realistic many body systems? I personally doubt that they are: Firstly the arguments about cancellations and/or 'healing' as used in many body theory certainly have no meaning in the anharmonic oscillator. Secondly, in many body systems calculations beyond SUB(4) anyway are out of question because of the unsurmountable numerical problems involved. So, as a result of this investigation there remains just a warning sign telling us that we should be careful - and avoid extreme optimism concerning CCM.

Literature

1. J.Killingbeck,J.Phys.A14 1oo5(1981) (further refs. therein)
2. C.S. Hsue and J.L. Chern, Phys.Rev.D29 643(1984)
3. U.B. Kaulfuß and M. Altenbokum, Phys.Rev.D33 3658(1986)
4. H. Kümmel, K.H. Lührmann and J.G. Zabolitzky, Phys.Rep. 36C 1(1978)
5. V. Krasnicka, V. Laurinc and S. Biscupic, Phys.Rep.9o 159(1982)

BOSON, SYMPLECTIC AND OTHER REPLICAS FOR SIMPLE HAMILTONIANS

A.P. Zuker, M. Dufour, and C. Pomar[1]

Laboratoire de Physique Nucléare Théorique, C.R.N.
BP 20, 67037 Strasbourg Cedex, France

ABSTRACT. We present a linearization method for some simple (naturally tridiagonal) Hamiltonians. For the ground states it is equivalent to the lowest approximation in the coupled cluster formalism and its extension to excited states is straightforward. Then we construct sets of equivalent Hamiltonians (boson or symplectic replicas) that produce the same secular problem. In general they are not manifestly Hermitian. We show how to deal with this problem and we extract mean fields that describe both the normal and symmetry breaking regimes and at the same time incorporate variationally terms usually thought of as correlations.

I. INTRODUCTION

When applied to a simple class of Hamiltonians, the coupled cluster formalism (CCF) of Coester and Kümmel (KLZ77, LÜ77) yields remarkably good results for the ground state (GS) energies, even in the lowest approximation. In general the formalism does not extend in any obvious or natural way to excited states but in the cases we are going to treat it does, and the method that emerges, suggests in turn, ways to deal with more general situations.

We shall start by describing very schematically the class of naturally tridiagonal Hamiltonians and single out the LMG model (LMG65) which contains all the features common to the members of the class (Section I). Then we introduce the geometrical characterization of the GS solution in CCF and show how to deal with excited states through a linearization method (Section II). As the method has been presented in great detail in a

1)Permanent address : TANDAR, CNEA Av. Libertador 8250,
1429 Buenos Aires, Argentina

recent reference (DPZ87) , the first two sections only contain a sketch of ideas and results that suggest, and will be useful in understanding, the symplecticreplicas.

In section III we deal with the replicas. They consist in ways to write different Hamiltonians that define the same eigenvalue problem. The Hamiltonians are given in terms of creation and annihilation operators that are not necessarily Hermitian conjugates of each other but obey boson of fermion-like commutation rules. Using transformations that conserve these rules (i.e. sympletic), each replica will yield a mean field that will be particularly well adapted to different regions of the spectrum. In section IV we give some examples of how the method works for the LMG model.

I. LMG AND OTHER SIMPLE TRIDIAGONAL MODELS

Consider the Hamiltonian

$$H_{LMG}= \varepsilon S_o + V\left(S_+^2+S_-^2\right) \quad , \tag{I.1}$$

constructed with generators of SU2 :

$$\left[S_+,S_-\right]= S_o \quad , \quad \left[S_o,S_+\right]= 2S_+ \quad , \quad \left[S_o,S_-\right]= -2S_- \quad . \tag{I.2}$$

The general wavefunction has the form

$$|\tilde{k}>= \sum_m A_m^{(k)} \frac{S_+^m}{m!}|0> \quad , \quad S_-|0>=0 \quad , \quad S_o|0>=-M \quad , \quad <0|\frac{S_-^m}{m!}\frac{S_+^m}{m!}|0>= \binom{M}{m} , \tag{I.3}$$

where the second and thirds equalities define the vacuum and the fourth one follows from (I.2) and uses the usual notation for the combinational coefficient. From (I.3) it is elementary to calculate

$$0= <0|S_-^m\left(H-E_k\right)|\tilde{k}> =$$

$$Vm(m-1)A_{m-2}^{(k)}+\left[(2m-M)\varepsilon-E_k\right]A_m^{(k)}+V(M-m)(M-m-1)A_{m+2}^{(k)}= 0 \quad , \tag{I.4}$$

where E_k is the k-th eigenvalue. The problem is defined for m=0,2...M+y or m=1,3,...M-1+y, where y=0/1 if M is even/odd.

The difference eigenvalue problem (I.4) belongs to a class defined by the generic equation

$$W\, x(x+A)A_{x-1}+\left[V_0+2V_1x+V_2x^2-E\right]A_x+W(N-x)(N+B-x)A_{x+1}=0 \quad , \qquad (I.5)$$

where we can assume with no loss of generality that we are interested in solutions obeying boundary conditions $A_{-1}=A_{N+1}=0$; i.e. $x=0,1,\ldots N$. By setting $m=2x+y$, $W=4V$, $A=B=1/2-\delta_{y_0}$, $N=M/2-\delta_{y_1}$ (if M even), $V_1=2\varepsilon$, we can bring eq. (I.4) to the general form. For other examples of Hamiltonians belonging to the class we refer to DPZ87.

The wavefunction $|\tilde{k}>$ in (I.3) has been defined on an orthogonal basis with an arbitrary choice of normalization. A crucial point in our argument is that we should take advantage of this arbitrariness to recast eq. (I.4) (or more generally (I.5)) and choose the special form that will suit best our purpose. The useful alternatives are the following :

If $A_x=(x+A)!x!B_x$, then

$$WB_{x-1}+[\ ]B_x+W(x+1)(x+A+1)(N-x)(N+B-x)B_{x+1}=0 \quad . \qquad (I.6)$$

If $A_x=(x+A)!C_x$, then

$$WxC_{x-1}+[\ ]C_x+W(x+A+1)(N-x)(N+B-x)C_{x+1}=0 \quad . \qquad (I.7)$$

If $A_x=(x+A)!(N+B-x)!D_x$, then

$$Wx(N+B-x+1)D_{x-1}+[\ ]D_x+W(N-x)(x+A+1)D_{x+1}=0 \quad . \qquad (I.8)$$

And finally, if $A_x=\left[x!(x+A)!(N-x)!(N+B-x)!\right]^{1/2}F_x$, we have a manifestly symmetric problem and rather than write the difference equation we give the secular matrix elements :

$$H_{x\ x+1}=\left[(x+1)(x+A+1)(N-x)(N+B-x)\right]^{1/2} \quad , \qquad (H-E)_{xx}=[\ ] \quad . \qquad (I.9)$$

The square brackets $[\ ]$ in eqs. (I.6) to (I.9) correspond always to the coefficient of A_x in eq. (I.5).

Form (I.9) is clearly indicated for numerical diagonalization, while form (I.6) is the best to apply the linearization method we now discuss. More on the other forms later.

II. THE LINEARIZATION METHOD

In fig. 1 we show the following things :

i) The coefficient of WB_{x+1} in eq. (I.6) (a quartic in x) for H_{LMG} with M=30 (N in the figure) and y=0.

ii) The parabola that intersects the quartic at points $x=0,\pm 1$.

iii) The density of the GS wavefunction for $R=\frac{MV}{\varepsilon}=-0.8$ and y=0 (even matrix).

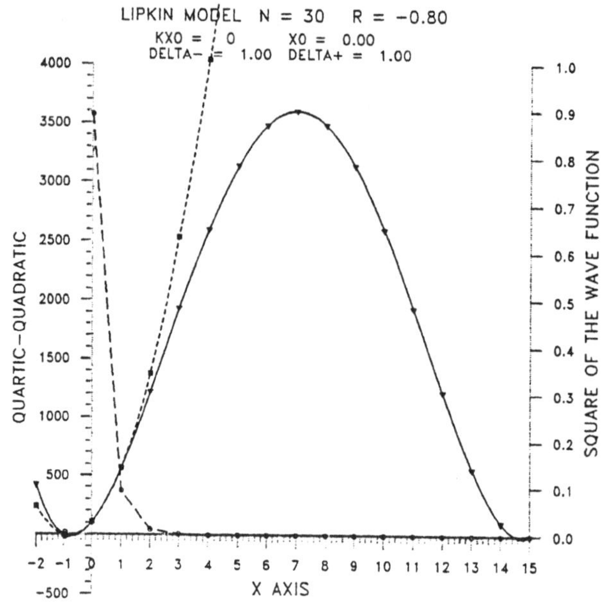

LIPKIN MODEL N = 30 R = -0.80
KXO = 0 XO = 0.00
DELTA- = 1.00 DELTA+ = 1.00
QUARTIC-QUADRATIC
SQUARE OF THE WAVE FUNCTION
X AXIS

Fig. 1

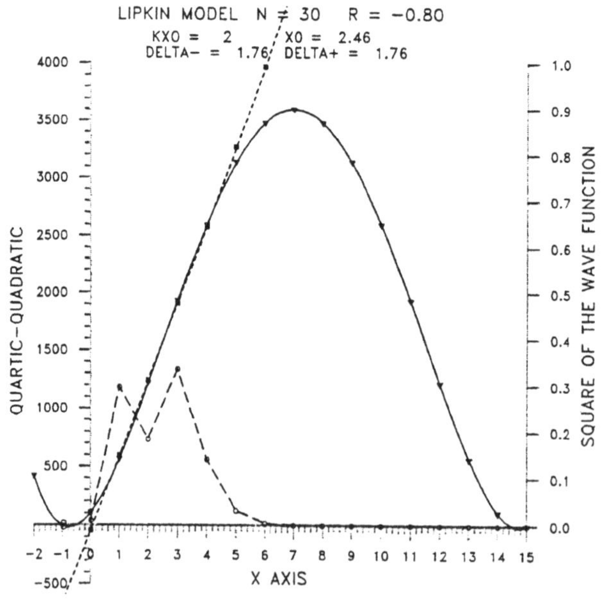

LIPKIN MODEL N = 30 R = -0.80
KXO = 2 XO = 2.46
DELTA- = 1.76 DELTA+ = 1.76
QUARTIC-QUADRATIC
SQUARE OF THE WAVE FUNCTION
X AXIS

Fig. 2

Notice that the parabola approximates almost perfectly the quartic at the points where the wavefunction is non negligible.

It means that, if we had solved approximately eq. (I.6) by replacing the quartic by the parabola, we would have obtained an extremely good approximation for the ground state. In DPZ87 it is shown that this procedure is strictly identical to the lowest approximation of the coupled cluster formalism.

If now we examine fig. 2 we see that for the second excited state in the same problem, the parabola of fig. 1 would be a poor approximation, but another one is drawn that is likely to be much better.

The idea of the linearization method becomes quite evident : with a quadratic instead of a quartic we know how to solve analytically eq. (I.6). Furthermore we know that the coupled cluster solution is excellent and we have an obvious hint to generalize it to excited states. Before we sketch the method, let us examine two more examples.

In figure 3 we have the GS density for R=-3 and we see again that a good quadratic approximant exists, while fig. 4 shows a situation in which the wavefunction has spread in such a way that no satisfactory parabola can be found.

To solve the approximant equation we assume it has been brought to the form (in particular $\underline{\varepsilon}=2\varepsilon$ in (I.4))

$$WB_{x-1}+(2\underline{\varepsilon}x+\underline{\varepsilon}D-E)B_x+\bar{W}(x-x_1+1)(x-x_2)B_{x+1}=0 \qquad \text{(II.1)}$$

(the diagonal term has to be linearized if necessary).

By changing variables, $n=x-x_1$, and setting $C_n=(x-x_1)!B_x$ we obtain

$$WnC_{n-1}+(2\underline{\varepsilon}n+\bar{D}-E)C_n+\bar{W}(n+\bar{N})C_{n+1}=0 \quad , \qquad \text{(II.2)}$$

with $\bar{N}=x_2-x_1$, $\bar{D}=\underline{\varepsilon}D+2\underline{\varepsilon}x_1$. If we return to the figures we identify x_1 with the zero of the parabola closest to the zero of the quartic at x=-1. Three things may happen

o) $\bar{N} \approx 1/2$ and $W\bar{W} > 0$, as in fig. 1.

u) $N \gg 0$ and $W\bar{W} > 0$, as in fig. 2.

c) $\bar{N} \approx -N$ and $W\bar{W} < 0$, as in figs. 3 and 4.

(Note that $\bar{W}$ is W times the coefficient of x^2 in the quadratic).

Now we introduce three Hamiltonians and their associated bases

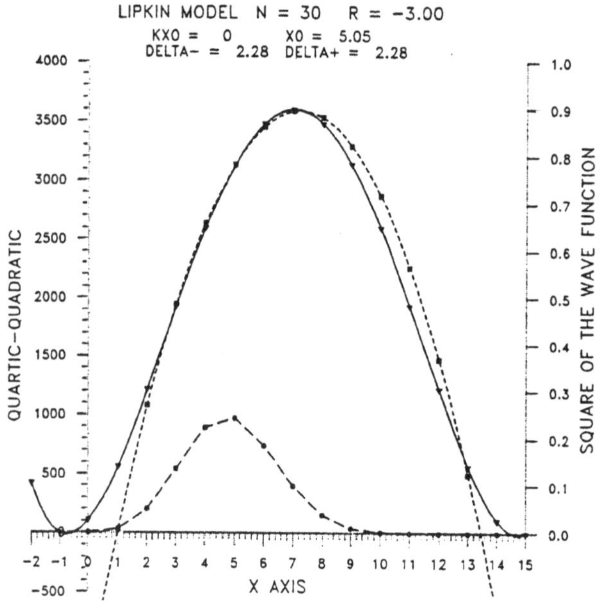

LIPKIN MODEL N = 30 R = -3.00
KXO = 0 XO = 5.05
DELTA- = 2.28 DELTA+ = 2.28
QUARTIC-QUADRATIC
SQUARE OF THE WAVE FUNCTION
X AXIS
4000
3500
3000
2500
2000
1500
1000
500
-500
1.0
0.9
0.8
0.7
0.6
0.5
0.4
0.3
0.2
0.1
0.0
-2 -1 0 1 2 3 4 5 6 7 8 9 10 11 12 13 14 15

Fig. 3

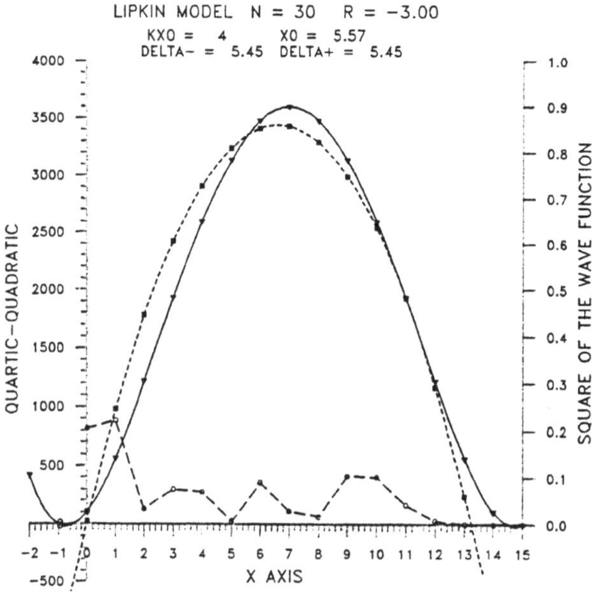

LIPKIN MODEL N = 30 R = -3.00
KXO = 4 XO = 5.57
DELTA- = 5.45 DELTA+ = 5.45
QUARTIC-QUADRATIC
SQUARE OF THE WAVE FUNCTION
X AXIS
4000
3500
3000
2500
2000
1500
1000
500
-500
1.0
0.9
0.8
0.7
0.6
0.5
0.4
0.3
0.2
0.1
0.0
-2 -1 0 1 2 3 4 5 6 7 8 9 10 11 12 13 14 15

Fig. 4

$$H_o=\underline{\varepsilon}a^+a+\bar{D}+\frac{\bar{W}}{2}aa+\frac{W}{2}a^+a^+ \qquad |n\rangle=\frac{(a^+)^{2n}}{n!2^n}|0\rangle \tag{II.3}$$

$$H_u=2\underline{\varepsilon}a^+a+\bar{D}+\bar{W}ab+Wa^+b^+ \qquad |n\rangle=\frac{(a^+)^n(b^+)^{\bar{N}-1}}{n!}|0\rangle \tag{II.4}$$

$$H_b=2\underline{\varepsilon}a^+a+\bar{D}-\bar{W}b^+a+Wa^+b \qquad |n\rangle=\frac{(a^+)^n(b^+)^{|\bar{N}|-n}}{n!(|\bar{N}|-n)!}|0\rangle \tag{II.5}$$

The operators obey $[a,a^+]=[b,b^+]=1$, but a^+/b^+ is NOT the Hermitian conjugate of a/b (we shall come back to this point in section III). There is no problem in checking that for wavefunctions of the form $\sum C_n|n\rangle$, the coefficients C_n obey exactly the difference equation (II.2) for the corresponding cases o (oscillator), u (unbound spectrum) and b (bound spectrum). Notice the profound difference between the b spectrum that contains a finite number of states (because of the boundary conditions at n=0 and $n=|\bar{N}|$) and the other two cases for which there is only one boundary at n=0 (we are interested in boundaries for integer $n \geqslant 0$, other values are unphysical). Let us now borrow from DPZ87 the expression for the energies and centroïds and widths of the wavefunction distributions.

$$E_k=\bar{D}+\bar{N}(\omega-\underline{\varepsilon})+2k'\omega \qquad \omega=(\underline{\varepsilon}^2-W\bar{W})^{1/2} \tag{II.6}$$

$$n_o=\frac{\varepsilon}{2\omega}(2k'+\bar{N})-\frac{\bar{N}}{2}=x_o-x_1 \tag{II.7}$$

$$\sigma^2=W\bar{W}\left[2k'(N+k')+\bar{N}\right]/(2\omega)^2 \tag{II.8}$$

The parameter k' is related to the number of nodes in the wavefunction. The subject is rater delicate and we only state the two possibilities of choice.

P approximation. Then $k'=k-x_1$, where k is the number of nodes (changes in sign). It applies to states in the normal phase.

L approximation. Then k'=k. It applies to states that have undergone a phase transition (critical behaviour will be discussed in section IV).

Equations (II.6) to (II.8) are sufficient to set up an iterative scheme. Assume we are interested in the k-th state. Then

1°) Guess the centroïd and width of the wavefunction

2°) Find a quadratic approximant that does well in the region where the wavefunction is large.

3°) Solve the equations and compare the input values of n_o and σ^2 with the output (II.7) and (II.8). If they are not close enough reinitialize. Iterate until convergence is achieved.

In practice convergence is very fast since the equations provide educated guesses. Let us give a few hints of their workings : (the reader is advised to keep sight of the equations and the figures).

For cases o and u, $W\bar{W}>0$, and when W is large enough ω can become imaginary. It means we are in the wrong region of x for the given k. To fix ideas assume we start with k=0 and W small enough. Then E_o differs very little (O(1)) from the unperturbed position $\bar{D}(O(N))$ since $\bar{N}=O(1)$. The wavefunction is peaked at the origin (II.7) and has a width O(1) (II.8). We are in the o regime. Increase W. The situation remains unchanged until ω goes imaginary. The wavefunction will move rapidly to larger n_o to make $W\bar{W}$ smaller. Very soon we enter the b regime for which ω is always real. But now $\bar{N}$ has become O(N) : there is a large gain in energy and σ^2 becomes O(N). This is what is seen in figs. 1 and 2.

In figs. 5 we present results of calculations of the scaled correlation energy $[E_k-(M-2k)\varepsilon]/M$ for LMG and for different values of M (N in the figure). $R=MV/\varepsilon$. (ε as in (I.4)).

In fig. 6 we have the total scaled energy for $|R|=2$, a region where the lowest states have undergone a phase transition. In LMG the sign of R is irrelevant and we only show states for E<0.

This very schematic presentation of the linearization method has served to introduce the ideas of open (unbound) and closed (bound) spectra and study some of their properties. But the main purpose is to raise the following question : we have seen that single particle Hamiltonians (II.3) to (II.5) can do a very good job. Is it not possible to transform the original Hamiltonian so as to make the approximate ones appear as mean fields, while keeping the residual terms to estimate their importance ?

The next sections will show how to implement transformations that achieve this aim.

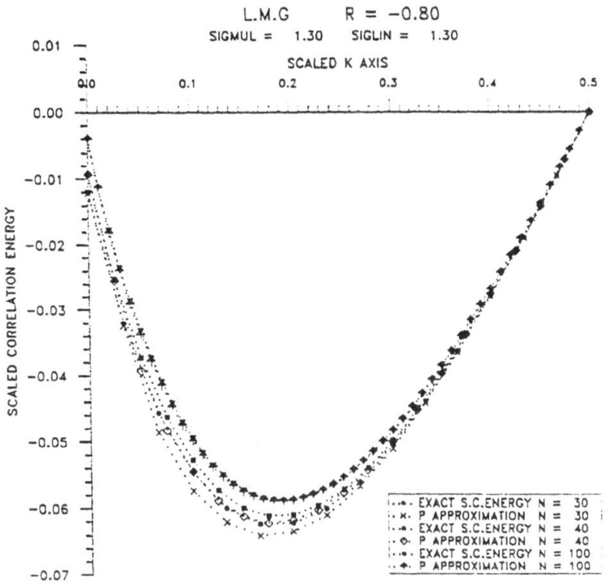
L.M.G R = -0.80
SIGMUL = 1.30 SIGLIN = 1.30
SCALED K AXIS
0.0
0.1
0.2
0.3
0.4
0.5
0.01
0.00
-0.01
-0.02
-0.03
-0.04
-0.05
-0.06
-0.07
SCALED CORRELATION ENERGY
EXACT S.C.ENERGY N = 30
P APPROXIMATION N = 30
EXACT S.C.ENERGY N = 40
P APPROXIMATION N = 40
EXACT S.C.ENERGY N = 100
P APPROXIMATION N = 100

Fig. 5

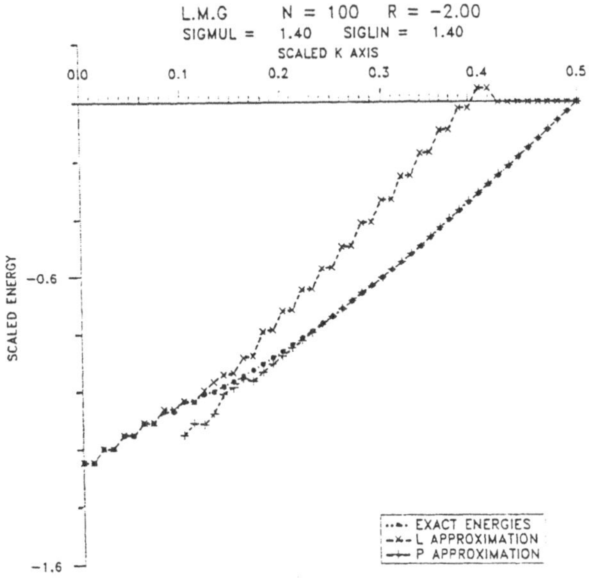
L.M.G N = 100 R = -2.00
SIGMUL = 1.40 SIGLIN = 1.40
SCALED K AXIS
0.0
0.1
0.2
0.3
0.4
0.5
-0.6
-1.6
SCALED ENERGY
EXACT ENERGIES
L APPROXIMATION
P APPROXIMATION

Fig. 6

When we wrote H_{LMG} in eq. (I.1) we introduced the generators of SU2 through their commutation rules (I.2). This information proved sufficient to derive the secular equation (I.4), but we may still ask what kind of objects S_o, S_+ and S_- actually are. Let us consider the following possibilities (j is a half integer)

$$S_o = \sum_{0<m\leq j} a^+_{jm} a_{jm} - b^+_{jm} b_{jm} \qquad S_+ = \sum_m a^+_{jm} b_{jm} \qquad S_- = \sum_m b^+_{jm} a_{jm} \tag{III.1}$$

$$S_o = j+1/2 - \sum_{|m|\leq j} a^+_{jm} a_{jm} \qquad S_+ = \sum_{m>0} a_{j-m} a_{jm} (-)^{j-m} \qquad S_- = \sum_{m>0} a^+_{jm} a^+_{j-m} (-)^{j-m} \tag{III.2}$$

$$S_o = a^+a - b^+b \qquad S_+ = a^+b \qquad S_- = b^+a \tag{III.3}$$

$$S_o = N - 2a^+a \qquad S_+ = (N=a^+a)a \qquad S_- = a^+ \tag{III.4}$$

$$S_o = N - 2a^+a \qquad S_+ = \sqrt{(N-a^+a)}\,a \qquad S_- = a^+\sqrt{(N-a^+a)} \tag{III.5}$$

In (III.1) and (III.2) the operators are fermions (i.e. $\{a^+_{jm}, a_{jm}\} = \{b^+_{jm}, b_{jm}\} = 1$, all other anticommutators vanish). In (III.3) to (III.5) (called respectively Schwinger, Dyson and Holstein-Primakoff *realizations*) the operators obey boson commutation rules. In the Dyson case it is obvious that a and a^+ cannot be Hermitian conjugates if S_+ and S_- are. We have kept the usual boson notation a, a^+ whenever $[a, a^+] = 1$, even if the operators are not bosons (e.g. the pair x, ∂_x obey $[\partial_x, x] = 1$, but they are not boson operators). We shall see in a moment that no confusion will arise from this practice.

Inserting (III.1) (III.5) in (I.1) will produce five apparently different Hamiltonians that are in fact identical and lead to the same eq. (I.4). We call them *replicas*.

The SU(1,1) algebra can be realized as

$$S_o = (a^+a + aa^+) \qquad S_+ = a^+a^+ \qquad S_- = aa \tag{III.6}$$

or

$$S_o = (a^+a + b^+b) \qquad S_+ = a^+b^+ \qquad S_- = ab \tag{III.7}$$

The commutation rules are as those of SU2 except that now $[S_+, S_-] = -S_o$. It is easy to see that our approximant Hamiltonians (II.3) and (II.4) are written in terms of generators of SU(1,1) realized as in (III.6) and

(III.7), while for (II.5) we have SU2 under its Schwinger guise.

The realizations (III.1) and (III.2) are standard, but as physicists prefer to deal with bosons rather than fermions, attention has focused on the possibility of using the boson realizations (references on the subject can be found in RS80).

Our own view of replicas will bypass realizations of the algebra(s) and focus directly on the difference equation in its general form (I.5) and the procedure will be exactly the same that the one leading from eq. (II.2) to eqs. (II.3-5). In implementing it we shall take advantage of the freedom offered by the norm variants of (I.5) in (I.6) to (I.8).

The difference replicas of a tridiagonal secular problem can take 4 different forms : Symmetric, Asymmetric, Open and Closed, denoted respectively by S, A, O, C.

S)) Consider the basis

$$|n\rangle=\frac{(a^+)^{n_a}(b^+)^{n_b}(c^+)^{n_c}(d^+)^{n_d}}{n_a!n_b!n_c!n_d!}|0\rangle\equiv\frac{(a^+c^+)^n(b^+d^+)^{N-n}(c^+)^A(d^+)^B}{n!(n+A)!(N-n)!(N+B-n)!}|0\rangle \qquad \text{(III.8)}$$

and the Hamiltonian

$$H=V_o+V_1a^+a+V_2(a^+a)^2+W(a^+c^+bd+b^+d^+ac) \qquad \text{(III.9)}$$

The difference equation will be exactly (I.5). In the special cases when A and/or B are half integers we have to take $a^+=c^+$ and/or $b^+=d^+$. In particular for LMG the interaction term is $V(a^+a^+bb+b^+b^+aa)$ and we recover the Schwinger picture.

A)) Consider the basis

$$|n\rangle=(a^+)^n|0\rangle \quad , \quad a^+|n-1\rangle=|n\rangle \quad , \quad a|n+1\rangle=(n+1)|n\rangle \qquad \text{(III.10)}$$

with a Hamiltonian of the form

$$H=2\underline{\varepsilon}a^+a+\underline{D}+W(a^+a+B)(a^+a+A)a^++\bar{W}(a^+a+C)a \qquad \text{(III.11)}$$

which leads to

$$W(n+B)(n+A)G_{n-1}+\left[2\underline{\varepsilon}n+\underline{D}-E\right]G_{n+1}+\bar{W}(n+1)(n+C)G_{n+1}=0 \qquad \text{(III.12)}$$

By replacing $B_n=(n+B)!(n+A)!G_n$ we recover an equation of type (I.6).

If we have chosen this way of presenting the problem it is achieve minimal asymmetry in the replica : we have up to terms of type $a^+a\, a^+a\, a^+$, but only terms of the type $a^+a\, a$. An even more asymmetric replica would have obtained if we had worked directly with eq. (I.6) : 1 term of the type a^+ against a, $a^+a\, a$, $a^+a\, a^+aa$, etc.!

Notice that for each normalization we obtain a different replica.

0)) Consider the basis

$$|n\rangle = \frac{(a^+b^+)^n (b^+)^P}{n!}|0\rangle \quad , \quad a^+b^+|n-1\rangle = n|n\rangle \quad , \quad ab|n+1\rangle = (n+P+1)|n\rangle \qquad (III.13)$$

and

$$H = (2\underline{\varepsilon}a^+a + \underline{D}) + W\left(\frac{a^+a + b^+b}{2} + A\right)a^+b^+ + \bar{W}\left(\frac{a^+a + b^+b}{2} + B\right)ab \qquad (III.14)$$

leading to

$$Wn\left(n + \frac{P}{2} + A\right)A_{n-1} + \left[2\underline{\varepsilon}n + \underline{D} - E\right]A_n + \bar{W}(n+P+1)\left(n + \frac{P}{2} + B\right)A_{n+1} = 0 \qquad (III.15)$$

which, to within notational changes, is again (I.5) <u>OR</u> (I.8).

In the particular case

$$|n\rangle = \frac{(a^+)^{2n}}{2^n n!}|0\rangle \quad , \quad a^+a^+|n-1\rangle = 2n|n\rangle \quad , \quad aa|n+1\rangle = 2\left(n + \frac{1}{2}\right)|n\rangle \qquad (III.16)$$

we have

$$H = \underline{\varepsilon}a^+a + \underline{D} + \frac{W}{2}\left(\frac{a^+a}{2} + A\right)a^+a^+ + \frac{\bar{W}}{2}\left(\frac{a^+a}{2} + B\right)aa \qquad (III.17)$$

and

$$Wn(n+A)D_{n-1} + (2\underline{\varepsilon}n + \underline{D} - E)D_n + \bar{W}\left(n + \frac{1}{2}\right)(n+B)D_{n+1} = 0 \qquad (III.18)$$

which is definitely (I.8) for LMG ($A = -N - \frac{1}{2}, B = -N, 4V = -W = -\bar{W}$, $N = \frac{M}{2}$, $\underline{\varepsilon} = 2\varepsilon$).

C)) The closed spectrum comes with basis

$$|n\rangle = \frac{(a^+)^n (b^+)^{N-n}}{n!(N-n)!} \quad , \quad a^+b|n-1\rangle = n|n\rangle \quad , \quad b^+a|n+1\rangle = (N-n)|n\rangle, \qquad (III.19)$$

$$H = \underline{\varepsilon}(a^+a - b^+b) + W\left(\frac{a^+a - b^+b}{2} + A\right)a^+b + \bar{W}\left(\frac{a^+a - b^+b}{2} + B\right)b^+a \quad , \qquad (III.20)$$

leading to difference equation

$$Wn\left(n-\frac{N}{2}+A\right)A_{n-1}+\left[(2n-N)\varepsilon-E\right]A_n+\bar{W}(N-n)\left(n-\frac{N}{2}+B\right)A_{n+1}=0 \qquad \text{(III.21)}$$

which is of the form (I.5) or (I.8) depending on the choice of parameters.

Except for the S variety, all our Hamiltonians are not manifestly Hermitian. Since the original problem can be assumed to be always Hermitian, a paradox arises that has hindered the development of the field for a long time. Our proposal to solve it has been already hinted at : do not pretend a^+ and a are boson operators if they are not. In the same way that in introducing the SU2 algebra everything we need are the commutation rules, in introducing " boson " operators the only thing we need are the commutation rules. The bras and kets in our vector (or Hilbert) spaces must be normalizable but there is no reason to assume that they are Hermitian conjugates.

If we start from boson fields and we subject them to unitary or orthogonal transformations we keep both the commutation rules and the Hermitian conjugation of the operators. The most general transformations that conserve only the commutation rules are called symplectic. We shall study two examples that will be useful later on.

Assume we have boson operators a^+,a. The most general homogeneous linear transformation has the form (*denotes complex and/or Hermitian conjugation)

$$\begin{aligned} A^+&=\alpha a^+ +\beta a \quad , \qquad \left[A,A^+\right]=1 \implies \alpha\bar{\alpha}-\beta\bar{\beta}=1 \\ A&=\bar{\alpha}a+\bar{\beta}a^+ \quad , \qquad A=(A^+)^\star \implies \bar{\alpha}=\alpha^\star, \quad \bar{\beta}=\beta^\star \end{aligned} \qquad \text{(III.22)}$$

If we start with 4 independent real parameters we see that the commutation rule fixes one of them and the Hermiticity condition fixes two. Of particular interest is the choice $(A^+)^\star=A^+=x \quad A^\star=-A=-\partial_x$ achieved for $\alpha=\beta=\bar{\alpha}=-\bar{\beta}=1/\sqrt{2}$.
Notice that $a^+=\lambda a^+$, $A=\frac{1}{\lambda}a$ is a symplectic transformation and so is $A^+=a^++\gamma$, $A=a+\bar{\gamma}$ (unitary if $\bar{\gamma}=\gamma^\star$), although now it is inhomogeneous.

It is always important to be able to vary the parameters of the transformation independently and we study now a case that will be of direct use. Assume we are after the general symplectic transformation that mixes separately the creation (or annihilation) operators of two fields a^+,b^+(a,b). Let us introduce the sequence of transformations

$$a^+=\alpha C^++\beta D^+ \qquad a=\alpha C+\beta D$$
$$b^+=\alpha D^+-\beta C^+ \qquad b=\alpha D-\beta C \Bigg\} \quad \alpha^2+\beta^2=1 \tag{III.23,1}$$

and

$$C^+=\bar{\alpha}A^++\bar{\beta}B^+ \qquad C=\bar{\alpha}A-\bar{\beta}B$$
$$D^+=\bar{\alpha}B^++\bar{\beta}A^+ \qquad D=\bar{\alpha}B-\bar{\beta}A \Bigg\} \quad \bar{\alpha}^2-\bar{\beta}^2=1 \tag{III.23,2}$$

Then $a^+=(\alpha\bar{\alpha}+\beta\bar{\beta})A^++(\alpha\bar{\beta}+\bar{\alpha}\beta)B^+$ and similar expressions for the other operators. Defining $S_1=S_++S_-$ and $S_{-1}=S_+-S_-$ we obtain the following transformation rules for the SU2 generators

$$S_o=a(\bar{a}S'_o-\bar{b}S'_{-1})+bS'_1 \qquad a=\alpha^2-\beta^2 \qquad b=2\alpha\beta \qquad a^2+b^2=1$$
$$S_1=aS'_1-b(\bar{a}S'_o-\bar{b}S'_{-1}) \qquad \bar{a}=\bar{\alpha}^2+\bar{\beta}^2 \qquad \bar{b}=2\bar{\alpha}\bar{\beta} \qquad \bar{a}^2-\bar{b}^2=1 \tag{III.24}$$
$$S_{-1}=\bar{a}S'_{-1}-\bar{b}S'_o$$

' denotes transformed.

Please do not confuse the parameters a,b with operators. As in (III.22) we have 3 parameters at our disposal. We have defined two (α and $\bar{\alpha}$ or a and $\bar{a}$, say). The missing one is an overall dilation ($A^+\to\lambda A^+$, $A\to A/\lambda$; similar for B,B^+) that may be easily incorporated. Note that

$$Wa^+b+\bar{W}b^+a=\sqrt{(W\bar{W})}\left[\left(\frac{W}{\bar{W}}\right)^{1/2}a^+b+\left(\frac{\bar{W}}{W}\right)^{1/2}b^+a\right] \tag{III.25}$$

which means that $a^+\to\sqrt{W}a^+$, $a=a/\sqrt{W}$, etc., makes the form manifestly Hermitian (as in eqs. (II.3) to (II.5) or non Hermitian (if $W\bar{W}<0$).

The same sequence of transformations can be introduced to parametrize the general symplectic in (III.22).

IV. EXAMPLES

NOTE : equations are numbered as if we were in section III.

Now we come to an interesting question : why is it that physicists prefer to work with bosons ? The answer that it is easier, applies to systems of real fermions and bosons. When we deal with replicas we pay a price if the original problem is a fermion one : either we end up with too many bosons (S case above) or with " non Hermitian " Hamiltonians (all the other cases). Symplectic transformations will be useful in dealing with the latter, to the extend that their added complexity will also bring in some extra benefit. In what follows we see that this will be the case.

Let us examine how the S, A, O and C replicas fare in the LMG model. NOTE : IN ALL THAT FOLLOWS NOTATION AS IN(I.4) BUT N=M.

S)) We have

$$H=\varepsilon S_o+V(a^+a^+bb+b^+b^+aa) \qquad \text{(III.26)}$$

Introducing bosons expands enormously the algebra. From a modest SU(2) we move to Sp(2,2) whose generators are all the quadratic forms in a,a^+, b,b^+. The associated transformation $a^+=\alpha A^++\beta B^++\gamma A+\delta B+\sigma$ and similar expressions for the other operators, is quite impressive and we may wonder whether something is gained by using it. We shall not explore the question in detail but concentrate on two special cases.

SU2) We simply keep to a transformation of type (III.23,1). It applies identically to the fermion replica (eq. (III.1)) or the bosonic one in (III.3). Using (III.24) with $\bar{a}=1$, and dropping the primes we arrive at a transformed H of the form

$$H=\varepsilon a S_o-|V|\left[\frac{3}{4}S_o^2-\frac{1}{4}N(N+2)\right]+\frac{1}{2}V(a^2+1)(S_+^2+S_-^2)$$
$$+\varepsilon b S_1-\frac{V}{2}ab\left(S_oS_1+S_1S_o\right) \qquad \text{(III.27)}$$

The original even and odd matrices now mix and the tridiagonal form has become pentadiagonal. The k-th matrix element associated with the second line in (III.27) can be cancelled through the choice

$$b\left[\varepsilon-|V|a\left(N-2k-\delta_{ko}\right)\right]=0 \qquad \text{(III.28)}$$

There are obviously two solutions. If b=0, we are back in the untransformed case and we have gained nothing. Otherwise we have a non trivial solution beyond the critical value $R=r=(N-1)V/\varepsilon=1$. Making the k dependent choice in (III.28) and neglecting the terms in S_1 in eq. (III.27) we can solve for each k in the P approximation. The results for the case already illustrated in fig. 6 show that this procedure (CTN) extends the domain of validity of the L approximation. As the corresponding figure (and much more on this subject) can be found in DPZ87 we prefer to exhibit CTN results for a very large R in fig. 7

As the SU2 transformation represents the standard treatment we shall use the GS energy results as reference for future work :

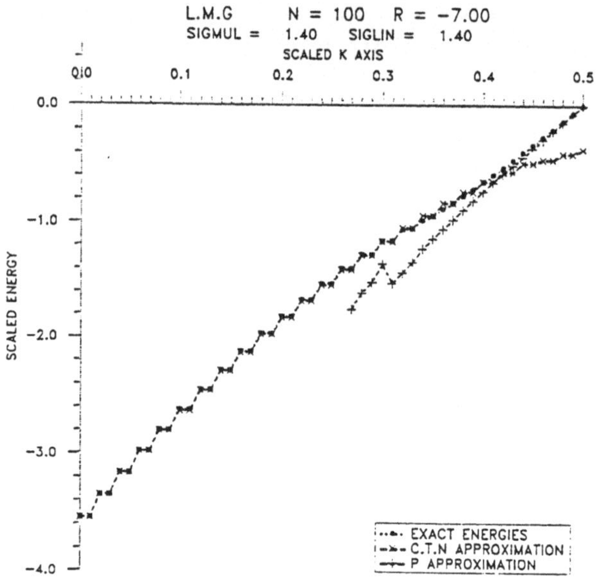

Fig. 7

$$E=-N\varepsilon-\varepsilon\left[r-\sqrt{1-r^2}\right], \quad r\leqslant 1 \qquad E=-\frac{N\varepsilon}{2}\left(|r|+\frac{1}{|r|}\right) \text{ if } |r|\geqslant 1 \qquad \text{(III.29)}$$

For $r\leqslant 1$ we have added in brackets the correction energy ((II.6), assuming $\bar{N}\cong 1/2$) although it is not included in the SU2 mean field. As we know the estimates (III.29) to be good, our alternative treatments should (at least !) reproduce them.

SU(1,1) x SU(1,1)). Here we examine what happens if we mix separately the a^+,a and b^+,b fields through an inhomogeneous transformation of type (III.22). First we note that since the problem is Hermitian we can restrict ourselves to an orthogonal transformation. Then we argue that

$$b^+=B^++\gamma \quad , \quad a^+=A^+ \qquad \text{(III.30)}$$

is quite sufficient to have an idea of what will happen.

The fisrt thing to notice is that $\hat{n}=a^+a+b^+b=N$ is a conserved quantity and that A^+A+B^+B will not be conserved. So we choose to transform $H-N\varepsilon$ instead of H to avoid some ugly linear terms in B^+. We obtain

$$H-N\varepsilon=2\varepsilon A^+A+VN(A^+A^++AA)+V\sqrt{N}(A^+A^+B+B^+AA)+V(A^+A^+BB+B^+B^+AA) \qquad \text{(III.31)}$$

The new vacuum is now such that

$$0=B|0>=(b-\gamma)|0> \quad , \quad |0>=\sum_n C_n (b^+)^n|0> \quad ; \quad A|0>=a|0>=0 \quad . \qquad \text{(III.32)}$$

The equation for $|0>$ is easily solved to give

$$C_{n+1}=-\frac{\gamma C_n}{n} \, , \quad C_n=\frac{(-\gamma)^n}{n!} \, , \quad <0|\hat{n}|0>=\gamma^2 \, , \quad <0|\hat{n}^2-\gamma^4|0>=\gamma^2 \qquad \text{(III.33)}$$

The obvious choice is $\gamma^2=N$ (we would obtain it by transforming H instead of $H-N\varepsilon$, and imposing conservation of $\hat{n}$ in the average through a Lagrange multiplier ; left as exercise).

It is quite obvious that the one body terms in eq. (III.31) are dominant and we can diagonalize them to obtain an oscillator of frequency $2\sqrt{\varepsilon^2-(NV)^2}$.

By the extraordinarily simple device (III.30) we have extracted the RPA solution as a mean field

We have not done a full variational calculation (which would request a more general SU(1,1) x SU(1,1) transformation) but we can read quite easily from (III.31) what would be the looks and the order of magnitude of the successive terms. The $a^+\to\alpha A^+ +\beta A$ transformation that diagonalizes the one body term will produce residual contributions in B^+ and B^+B^+ (+hc) demanding a new transformation of the B,B^+ field etc. By then we would be providing contributions of the next order in 1/N to the RPA etc. We do not pursue the subject further since we are going to find an even better replica to deal with the situation.

A)) The Hamiltonians of type (III.11) or the even more asymmetric ones can be dealt with an inhomogeneous symplectic transformation. The procedure is similar to what is done in general : eliminate dangerous terms (i.e. in a^+,a, a^+a^+,aa). We have done it already in (III.28) although the dangerous term was associated with a^+b+b^+a. In C)) we shall see how to deal with a strongly non Hermitian case. In the case at hand we'll need 4 paramters, available as $\alpha,\bar{\alpha}$ in (III.22) plus the two shifts γ and $\bar{\gamma}$.

We shall say no more about this case except that its simplicity and flexibility makes it very appealling.

0)) The open case is very nice in that it contains all the features of the SU(1,1) x SU(1,1) formulation but needs only one boson. Furthermore, the Hamiltonian is almost Hermitian. Recasting (III.18) in terms of the generators of SU(1,1) in (III.6) we obtain ($S_1=a^+a^++aa$, $S_{-1}=a^+a^+-aa$)

$$H=-(N+1)\varepsilon+\varepsilon S_o+(N+1)VS_1-\frac{V}{4}(S_oS_1+S_1S_o)-\frac{V}{2}S_{-1} \qquad \text{(III.34)}$$

The S_{-1} non Hermiticity is manifestly very small compared to the S_{-1} term. <u>If we drop it we obtain exactly the version of ref. KL81 which relied on approximating a Holstein-Primakoff replica.</u>
The proof is an elementary exercise that demands going over to an x,∂_x (or x,p_x) representation. Before calculating anything we note that the S_1 term is much larger than the S_oS_1 one for small number of quanta (near GS) and that again the RPA is staring at us as part of the central field.

Now we introduce the inhomogeneous boson transformation

$$a^+=\alpha A^++\beta A+\sigma \quad , \quad a=\alpha A+\beta A^++\sigma \quad , \quad \alpha^2-\beta^2=1 \qquad \text{(III.35)}$$

which induces

$$\left.\begin{array}{ll} S_1\to aS_1+bS_o+2\sigma(\alpha+\beta)(A^++A)+2\sigma^2 & a=\alpha^2+\beta^2 \\ S_o\to aS_o+bS_1+2\sigma(\alpha+\beta)(A^++A)+2\sigma^2 & b=2\alpha\beta \\ S_{-1}\to S_{-1} & a^2=b^2=1 \quad . \end{array}\right\} \qquad \text{(III.36)}$$

The transformed of H in (III.34) is now

$$H=H_o+H_{res}+H_1+H_2, \quad \text{with}$$

$$H_o=\left[a\varepsilon+Vb(N+1)-3V\sigma^2(a+b)\right]S_o-\frac{V}{4}ab(3S_o^2+1)+2\sigma^2\left[\varepsilon+(N+1)V\right]-2V\sigma^4-(N+1)\varepsilon$$

$$H_{res}=-\frac{V}{2}ab(S_+^2+S_-^2)-V(\alpha+\beta)(a+b)\sigma(A^+A^+A^++AAA)-\frac{V}{2}S_{-1}$$

$$H_1=(A^++A)\left\{2\sigma(\alpha+\beta)\left[\varepsilon+(N+1)V\right]-4V\sigma^3(\alpha+\beta)\right\}-\frac{V}{2}(\alpha+\beta)(a+b)\sigma\left[(A^++A)S_o+hc\right]$$

$$H_2=S_1\left[b\varepsilon+aV(N+1)-3V\sigma^2(a+b)\right]-(S_oS_1+S_1S_o)\,\frac{V}{4}(a^2+b^2) \qquad \text{(III.37)}$$

The dangerous terms H_1 and H_2 are eliminated through the prescription

$$<k|H_1|k+1>=<k|H_2|k+2>=0, \qquad |k>=(A^+)^k|0> \quad , \tag{III.38,1}$$

known to be variational for the GS (k=0), in which case we obtain

$$\sigma\left[\varepsilon+\left(N+1-\frac{a+b}{2}\right)V-2V\sigma^2\right]=0=\varepsilon b+aV\left(N+1-\frac{a^2+b^2}{a}\right)-3V\sigma^2(a+b) \tag{III.38,2}$$

leading to the approximate solutions $\left(\text{assume } V<0,\ \varepsilon>0,\ |r|=\frac{N|V|}{\varepsilon}\right)$

$$\begin{aligned}&\sigma=0 \quad , \qquad a=(1-r^2)^{-1/2} \ , \quad b=|r|(1-r^2)^{-1/2}, \quad |r|\leqslant 1 \ ,\\ &\sigma^2=\frac{N}{2}\left(1-\frac{1}{|r|}\right), \quad a=\frac{3|r|-1}{\sqrt{8(r^2-1)}} \ , \quad b=\frac{|r|-3}{\sqrt{8(r^2-1)}} \ , \quad |r|\geqslant 1 .\end{aligned} \tag{III.38,3}$$

Note : for V>0 we must introduce imaginary values of α and β, otherwise $\sigma^2=\frac{N}{2}\left(1+\frac{1}{|r|}\right)$ does not ensure continuity at $|r|=1$, which is nonsense. It is easy to check that the GS energies,

$$\begin{aligned}&E\simeq a\varepsilon+Vb(N+1)-(N+1)\varepsilon=-N\varepsilon-(\omega-\varepsilon), \qquad \omega=\varepsilon\sqrt{1-r^2}, \qquad |r|\leqslant 0 \ ,\\ &E\simeq 2\sigma^2\left[\varepsilon+NV\right]-2V\sigma^4-N\varepsilon=-\frac{\varepsilon N}{2}\left(|r|+\frac{1}{|r|}\right) \qquad |r|\geqslant 0 \ ,\end{aligned} \tag{III.39}$$

agree with the reference values (III.29).

We shall comment on these results after we study the C case.

C)) The closed spectrum Hamiltonian (III.20) when specialized to LMG, assuming the (I.8) version of the difference equation, yields

$$H=2\varepsilon S_o+VNS_{+1}-V(S_oS_{-1}+S_{-1}S_o) \quad , \tag{III.40}$$

where now S_i are the SU2 generators. The similarity with eq. (III.34) is formal since now we have a badly non Hermitian problem. It is here that transformation (III.23) is needed in full.

It leads to

$$\begin{aligned}H=&2\varepsilon\left[a\left(\bar{a}S_o-\bar{b}S_{-1}\right)+bS_1\right]+NV\left[aS_1-b\left(\bar{a}S_o-\bar{b}S_1\right)\right]\\ &-V\left[-2a\bar{a}\bar{b}(S_o^2+S_{-1}^2)+a(\bar{a}^2+\bar{b}^2)(S_oS_{-1}-hc)-b\bar{b}(S_oS_1+hc)+\bar{b}(b^2-a^2)(S_1S_{-1}-hc)\right]\end{aligned} \tag{III.41}$$

and cancellation of dangerous terms amounts to request

$$\begin{aligned}&<k|\ (2\varepsilon b+NVa)S_1+Vb\bar{b}(S_oS_1+S_1S_o)\ |k+1>=0\\ &<k+1|(-2\varepsilon a\bar{b}+NVb\bar{b})S_{-1}-Va(\bar{a}^2+\bar{b}^2)(S_oS_{-1}+S_{-1}S_o)|k>=0 \quad .\end{aligned} \tag{III.42}$$

Setting $<S_o S_i + S_i S_o> = \bar{N}<S_i>$, we obtain (III.43)

$$b = -\frac{NVa}{2\varepsilon + \bar{N}V\bar{b}} \quad \text{from the first equation, which} \qquad \text{(III.44)}$$

when replaced in the second yields

$$2\varepsilon\bar{N}V + \bar{b}\left[(2\varepsilon)^2 + (NV)^2 + (\bar{N}V)^2\right] + \bar{b}^2\left[6\varepsilon\bar{N}V\right] + 2(\bar{N}V)^2\bar{b}^3 = 0 \qquad \text{(III.45)}$$

Specializing to the GS ($\bar{N} \approx -N$) we obtain the factorisation

$$\left(\bar{b} - \frac{1}{r}\right)\left[\left(\bar{b} - \frac{1}{r}\right)^2 - \left(1 - \frac{1}{r^2}\right)\right] = 0 \ , \qquad \text{(III.46)}$$

a remarkably elegant outcome. For $|r| \leqslant 1$ we have $\bar{b} = +\frac{1}{r}(1 - \sqrt{1-r^2})$ (vanishing at small r dictates sign unambiguously), furthermore

$$a^2 = \frac{1}{2}\left(1 + \sqrt{1-r^2}\right) \qquad b^2 = \frac{1}{2}\left(1 - \sqrt{1-r^2}\right) \qquad a^2 b^2 = \frac{1}{4} r^2$$

$$\bar{a}^2 = \frac{2}{r^2}\left[1 - \sqrt{1-r^2}\right] \qquad \bar{b}^2 = \frac{1}{r^2}\left[1 - \sqrt{1-r^2}\right]^2 \ , \quad |\bar{a}a| = 1, \quad |\bar{a}b| = |\bar{b}| \text{ etc.} \qquad \text{(III.47)}$$

For the dominant diagonal terms in (III.41) we find

$$H_o \approx 2\varepsilon|a\bar{a}|S_o - |NV\bar{a}b|S_o - |2Va\bar{a}\bar{b}|S_o^2 = -N\varepsilon \qquad \text{(III.48)}$$

(the signs of a and $\bar{a}$ are definite positive by definition, the sign of $\bar{b}$ has been fixed and then that of b follows from (III.44)). Note that S_o in the GS is now $-N/2$ <u>not</u> $-N$.

The next order contribution must come from $V\left[2a\bar{a}\bar{b}(S_{-1}^2) - \bar{a}b(S_1 S_{-1} - hc)\right]$, whose GS expectation value turns out to be $VN\bar{b} = \varepsilon(1 - \sqrt{1-r^2})$, which agrees with the standard result.

For $|r| \geqslant 0$ we have $\bar{b} = 1/r$, $b = -ar$, $a\bar{a} = 1/|r|$ and the leading terms in H_o are now

$$H_o = -2\varepsilon|a\bar{a}S_o| - |NV\bar{a}bS_o| + |Va\bar{a}\bar{b}|2S_o^2 = -\frac{N\varepsilon}{2}\left(\frac{1}{|r|} + |r|\right) \ , \qquad \text{(III.49)}$$

naturally in agreement with the standard result.

From these results we can draw some general conclusions. As a rule, all the replicas we have exhibited, *except* the standard one, yield a mean field in the precritical regime. In lowest order it coincides with the RPA result and there is little doubt that if we work out more carefully the equations, extremely accurate values will be available. The very idea that the RPA could be extended *variationally* is in itself a valuable corollary.

The Schwinger (S) replica can be given either an SU(1,1) x SU(1,1) or and SU(2) interpretation. The former is quite close to the almost Hermitian 0 formulation. It is important to notice that the SU(1,1) x SU(1,1) transformation will violate the conservation of N and that the GS is now (in lowest approximation) a coherent state of centroïd N and width $\sqrt{N}$ (III.33). There is no reason to have this extra problem when a simpler replica that does very much the same job is available.

It is quite clear that each of the replicas corresponds quite closely to one of the linear approximants that we have introduced heuristically through geometrical considerations. The advantage of the replicas is that the transformation methods generalize in a straightforward way, while the geometry of more general matrices will be far more obscure than in the naturally tridiagonal cases.

Our work on linearization also has the advantage to illustrate the specificity of each replica. It is clear that the 0 cases are ideally adapted to the neighbourhood of the GS in precritical conditions. On the other hand, the loss of the boundary condition at n=N may raise problems as we move up in the spectrum. For the C case the situation is reversed as we can see by inspection. In (III.31) or (III.34) the one body term is O(1) and the two body term is O(1/N) for S_0 small (near GS). For excited states the latter grows. In eq. (III.41) both one and two body are of O(1) at first but as we move up in the spectrum the former becomes dominant *and* H *increasingly Hermitian.* It is easily understood if we refer to figs. 1 to 4. What is far less clear from the figures is how (III.41) can be of any use near the GS. A closed spectrum will apparently yield always a correlation energy of O(N) (from (II.6)). It is only through its strong non-hermiticity that (III.41) can make it : the matrix element $\langle 0|S\_H|0\rangle$ is cancelled to O(1/N) while $\langle 0|HS_+|0\rangle$ is reinforced and remains O(1). This explains why the correlation energy remains O(1) and leads to some puzzling consequences : if we neglect altogether $\langle 0|S\_H|0\rangle$ the transformation will produce an RPA correction to the energy but the wavefunction will apparently be the unperturbed one ($|0\rangle$). This is indeed the case for *the wavefunction to the left* $\langle 0|$, but not for the wavefunction to the right. ($|0\rangle$+ other terms).

CONCLUSIONS

We have come to understand that boson formulations are useful in that they can exhibit as mean field, features that are usually catalogued as correlations. It is interesting to be able to exhibit several competing schemes that lead to the same approximations in low order, even when they do not seem a priori well adapted to the case at hand. It leads us to expect a fairly decent behaviour over large portions of the spectrum.

The technique of building (*difference*) replicas seems to us simpler to implement directly on the difference equations instead of using realizations of the underlying algebras as an intermediate step. Of course the possibility of doing so hinges on our capacity to write difference equations. But then we can argue that algebraic eigenvalue problems deal with matrices and it is unlikely that operator techniques can be of much help if we can't write the matrices.

Although it is certainly useful to have as many replicas as possible of the same problem it is even more important to know how to deal with them. This is why we attach some importance to the introduction of symplectic transformations which put in proper perspective the " non-Hermiticity " problem. To our knowledge the idea has been explored only in a contribution to this volume (CD87). We hope the present work will be of some help, to those who read it, in making them feel more friendly to non Hermitian matrices.

REFERENCES

CD87 M.C. Cambiaggio, J. Dukelsky, this volume

DPZ87 M. Dufour, C. Pomar and A.P. Zuker, CRN preprint available. To be submitted to Annals of Physics.

KL81 A. Klein and C.T. Li, Phys. Rev. Lett. 46 (1981) 895

KLZ77 H. Kümmel, K.H. Lührmann and J.G.Zabolitzky, Phys. Rep. 36C (1978) 188

LMG65 H.J. Lipkin, M. Meshkov and J.A. Glick, Nucl. Phys. 62 (1965) 188

LÜ77 K.H. Lührmann, Ann. of Phys. 103 (1977) 253

RS80 P. Ring, P. Shuck, The Nuclear Many Body Problem (Springer-Verlag) 1980)

TOWARDS A COUPLED CLUSTER GAUGE-FIELD APPROACH TO QUANTUM HYDRODYNAMICS

J. Arponen*, R.F. Bishop[†], E. Pajanne[‡] and N.I. Robinson[†]

*Department of Theoretical Physics, University of Helsinki
Siltavuorenpenger 20C, SF-00170 Helsinki, Finland

[†]Department of Mathematics
University of Manchester Institute of Science and Technology
P.O. Box 88, Manchester M60 1QD, England

[‡]Research Institute for Theoretical Physics
Siltavuorenpenger 20C, SF-00170 Helsinki, Finland

1. INTRODUCTION

The extended coupled cluster method (ECCM)[1,2] has been discussed in an earlier volume of this series,[3] where it was also applied to provide an effective gauge-field description of a charged impurity in a polarisable medium (e.g. positron in a metal).[4] The purpose of the present paper is to apply the ECCM and comparable gauge field techniques to the zero-temperature condensed Bose fluid, in the most general case of arbitrary spatial inhomogeneity and arbitrary time dependence. Our principal aim is to derive the appropriate hydrodynamical balance equations for such local observables as the number density, current density and energy density. This is achieved by coupling the system to scalar and vector gauge fields, so that the theory may then be formulated in a completely gauge-invariant fashion to take fully into account the underlying U(1) symmetry imposed by number conservation.

The ECCM formalism[1-3,5,6] is based on the equations of motion for a set of linked-cluster amplitudes which completely characterise the many-body system. At the lowest level of truncation, we shall show explicitly that the formalism degenerates to the mean-field description of Gross[7] and Pitaevskii[8] for the condensate wavefunction or one-body order parameter. We particularly examine the decompositions of the general (off-diagonal) matrix elements of the one- and two-body density operators, in terms of these linked-cluster ECCM amplitudes. The various hydrodynamical balance equations are shown to follow straightforwardly. Furthermore, we show how these equations are also exactly satisfied by most practical approximation schemes which might be applied to the otherwise complete ECCM description. It is also of considerable practical importance that each of the ECCM amplitudes obeys the exact cluster property of vanishing when some subset of the cluster of particles described by the given amplitude becomes infinitely far removed from the remainder. The resulting "quasi-locality" of the basic ECCM amplitudes then implies that our gauge-invariant formalism is in principle capable of a complete hydrodynamical description of the system at zero temperature, including possible states of topological excitation or deformation, and of broken symmetry.

2. ELEMENTS OF THE ECCM

The exp(S) similarity transformation of the normal coupled cluster method (CCM)[9,10] has proven to be extraordinarily powerful and versatile in the development of simple and finite algorithms for the ground-state energy of a wide variety of quantum many-body systems.[11] The desire to obtain approximations for the expectation values of other operators, which are properly compatible with the energy values, led to the introduction[1] of another operator S" composed of a set of linked-cluster amplitudes. This correspondingly led to the consideration of a double similarity transformation and to the development of the so-called ECCM. We have shown[6] how the method may be viewed as a biorthogonal formulation for a rather general quantal many-body problem. It is therefore best not to restrict ourselves from the start to Hamiltonians H or their relevant subsequent transforms that are necessarily manifestly hermitian.

Arbitrary time-dependent states $|\Psi(t)\rangle$ and $\langle\Psi'(t)|$ are now parametrised in the ECCM as,

$$
\begin{aligned}
|\Psi(t)\rangle &= e^{\phi(t)}e^{S(t)}|\Phi\rangle = e^{\phi(t)}e^{S(t)}e^{-S''(t)}|\Phi\rangle \quad , \\
\langle\Psi'(t)| &= e^{-\phi(t)}\langle\Phi|e^{S''(t)}e^{-S(t)} \quad ,
\end{aligned}
\tag{1}
$$

where $|\Phi\rangle$ is some suitable model state such that the algebra of all operators in the many-body Hilbert space is spanned by the two Abelian subalgebras of creation and destruction operators defined with respect to $|\Phi\rangle$. Thus, the subalgebra of creation (destruction) operators and the state $|\Phi\rangle$ ($\langle\Phi|$) are assumed to be cyclic in the sense that essentially all of the ket (bra) states of the space can be constructed by acting on $|\Phi\rangle$ ($\langle\Phi|$) with appropriate elements of the creation (destruction) operator subalgebra. The operators $S(t)$ and $S''(t)$ contain respectively only creation and only destruction operators (defined again with respect to the cyclic vector $|\Phi\rangle$),

$$
S''(t)|\Phi\rangle = 0 = \langle\Phi|S(t) \quad , \tag{2}
$$

and with corresponding amplitudes that are linked-cluster quantities. The factor $\phi(t)$ in Eq.(1) is a, largely inconsequential, c-number scale factor. The normalisation condition,

$$
\langle\Psi'(t)|\Psi(t)\rangle = 1 \quad , \tag{3}
$$

is manifestly preserved for all times, and the average-value functional $\langle A\rangle$ of an arbitrary operator A,

$$
\langle A\rangle \equiv \bar{A}(t) \equiv \langle\Psi'(t)|A|\Psi(t)\rangle \tag{4}
$$

may evidently be written as the model-state expectation value of an operator $\hat{A}(t)$ which is the double similarity transform already mentioned of the operator A,

$$
\langle A\rangle = \langle\Phi|\hat{A}(t)|\Phi\rangle \quad ; \quad \hat{A}(t) \equiv e^{S''(t)}e^{-S(t)}Ae^{S(t)}e^{-S''(t)} \quad . \tag{5}
$$

The linked-cluster amplitudes pertaining to the operators S and S" are now regarded as independent variables which may be determined variationally by demanding stationarity of the action functional,

$$
\mathcal{A} \equiv \int dt \, \langle\Phi|e^{S''(t)}e^{-S(t)}(i\partial/\partial t - H)e^{S(t)}|\Phi\rangle \quad . \tag{6}
$$

The stationarity conditions then yield the eigenvalue equations for the ground-state energy E_o in the time-independent case,

$$\hat{H}|\Phi\rangle = E_0|\Phi\rangle \quad ; \quad \langle\Phi|\hat{H} = E_0\langle\Phi| \quad , \tag{7}$$

and the dynamical evolution equations for the operators S and S" in the general case. The time-derivatives in Eq.(6) are easily taken, using the fact that the creation and destruction subalgebras are Abelian,

$$\mathcal{A} = \int dt[i\langle\Phi|e^{S''}\dot{S}|\Phi\rangle - \bar{H}] = \int dt[-i\langle\Phi|\dot{S}''e^{S''}S|\Phi\rangle - \bar{H}] \quad , \tag{8}$$

and where in the second form we have performed an integration by parts. The structure of Eq.(8) strongly suggests a change of independent operators from S(t) and S"(t) to a new pair $\Sigma(t)$ and $\tilde{\Sigma}(t)$, defined as

$$\Sigma(t)|\Phi\rangle \equiv (1 - |\Phi\rangle\langle\Phi|)e^{S''(t)}S(t)|\Phi\rangle \quad ; \quad \tilde{\Sigma}(t) \equiv S''(t) \quad , \tag{9}$$

in terms of which most of the resulting formalism has a more symmetric appearance. Thus, for example, the action takes the form,

$$\mathcal{A} = \int dt[i\langle\Phi|\tilde{\Sigma}(t)\dot{\Sigma}(t)|\Phi\rangle - \bar{H}] = \int dt[-i\langle\Phi|\dot{\tilde{\Sigma}}(t)\Sigma(t)|\Phi\rangle - \bar{H}] \quad . \tag{10}$$

Henceforth we specialise to the case where the cyclic vector $|\Phi\rangle$ is the bare boson vacuum, and where the many-boson configuration space is parametrised by real-space coordinates $x \equiv (x^1,x^2,x^3)$. The basic single-boson creation and destruction field operators, $a_x^{\dagger}$ and a_x respectively, obey the usual canonical commutation relations,

$$[a_x,a_y^{\dagger}] = \delta^{(3)}(x-y) \equiv \delta(x-y) \quad . \tag{11}$$

We shall consistently denote coordinate-space three-vectors as r,r',x_1, x_2 ... , and their corresponding infinitesimal volume elements as dr,dr', dx_1,dx_2, ... The basic ECCM operators $\Sigma(t)$ and $\tilde{\Sigma}(t)$ are now parametrised as,

$$\begin{aligned}\Sigma(t) &= \sum_{n=1}^{\infty} \frac{1}{n!} \int dx_1 \ldots \int dx_n \sigma_n(x_1 \ldots x_n;t)a_{x_1}^{\dagger} \ldots a_{x_n}^{\dagger} \\ \tilde{\Sigma}(t) &= \sum_{n=1}^{\infty} \frac{1}{n!} \int dx_1 \ldots \int dx_n \tilde{\sigma}_n(x_1 \ldots x_n;t)a_{x_n} \ldots a_{x_1} \quad ,\end{aligned} \tag{12}$$

in which the fundamental amplitudes σ_n and $\tilde{\sigma}_n$ are all linked-cluster quantities, and which may, without loss of generality be assumed to be symmetric under arbitrary permutations of their coordinate-space arguments. We note that with these parametrisations, the states of Eq.(1) are not particle-number eigenstates, and we thus work from the outset in a number non-conserving formulation of the type introduced by Bogoliubov.[12] If we constrain ourselves to a Hilbert space of definite particle number, N, the identity operator has the following useful resolution

$$I = \sum_{n=0}^{\infty} \frac{1}{n!} \int dx_1 \ldots \int dx_n\, a_{x_1}^{\dagger} \ldots a_{x_n}^{\dagger} |\Phi\rangle\langle\Phi| a_{x_n} \ldots a_{x_1} \quad , \tag{13}$$

wherein the n = 0 term is simply the model state projector $|\Phi\rangle\langle\Phi|$.

The ECCM formulation is now completely specified by the set of c-number amplitudes $\{\sigma_n(x_1 \ldots x_n;t),\tilde{\sigma}_n(x_1 \ldots x_n;t)\}$, which may be regarded as a complete set of dynamic variables. It is not difficult to see from Eqs.(5),(9) and (12) that the lowest (one-body) amplitudes σ_1 and $\tilde{\sigma}_1$ are precisely the usual *"condensate wavefunctions"*, $\langle a_x\rangle = \sigma_1(x;t)$ and $\langle a_x^{\dagger}\rangle = \tilde{\sigma}_1(x;t)$, appropriate to a general non-uniform and temporally-varying macroscopically occupied condensate. All of the average-value functionals of Eqs.(4) and (5) may, for example, be fully specified in terms of the basic ECCM amplitudes, $\bar{A} = \bar{A}[\sigma_n,\tilde{\sigma}_n]$. Stationarity of the action of Eqs.(10) and (12) then also easily leads to the fundamental evolution equations,

$$
\begin{aligned}
i\dot{\sigma}_n(x_1 \ldots x_n;t) &= \delta\bar{H}/\delta\tilde{\sigma}_n(x_1 \ldots x_n;t) \quad , \\
-i\dot{\tilde{\sigma}}_n(x_1 \ldots x_n;t) &= \delta\bar{H}/\delta\sigma_n(x_1 \ldots x_n;t) \quad ;
\end{aligned}
\tag{14}
$$

where the functional derivatives of an arbitrary expectation value $\bar{A} = \bar{A}[\sigma_n, \tilde{\sigma}_n]$ are defined with respect to infinitesimal changes in accord with the earlier notation and the symmetry under permutations of the arguments of the basic ECCM amplitudes, as

$$
\delta\bar{A} = \sum_{n=1}^{\infty} \frac{1}{n!} \int dx_1 \ldots \int dx_n \left[\frac{\delta\bar{A}}{\delta\sigma_n(x_1 \ldots x_n;t)} \delta\sigma_n(x_1 \ldots x_n;t) + \frac{\delta\bar{A}}{\delta\tilde{\sigma}_n(x_1 \ldots x_n;t)} \delta\tilde{\sigma}_n(x_1 \ldots x_n;t) \right] . \tag{15}
$$

We discussed in an earlier volume in this series[3] how the ECCM exactly maps an arbitrary quantal many-body problem into an equivalent classical Hamiltonian field theory in some well-defined complex symplectic phase space. The underlying classical fields in this formulation are just the c-number configuration-space amplitudes $\sigma_n(x_1 \ldots x_n;t)$ and $\tilde{\sigma}_n(x_1 \ldots x_n;t)$. It has been explicitly demonstrated that what underpins this equivalence is the exact ECCM result (which may be derived directly from the definitions already given[3]) that the average-value functional for the commutator of two arbitrary operators A and B, may be expressed in terms of a *generalised Poisson bracket* $\{\bar{A},\bar{B}\}$,

$$
\langle \Psi'(t) | [A,B] | \Psi(t) \rangle = i\{\bar{A},\bar{B}\} \quad , \tag{16}
$$

where

$$
i\{\bar{A},\bar{B}\} \equiv \sum_{n=1}^{\infty} \frac{1}{n!} \int dx_1 \ldots \int dx_n \left[\frac{\delta\bar{A}}{\delta\sigma_n(x_1 \ldots x_n;t)} \frac{\delta\bar{B}}{\delta\tilde{\sigma}_n(x_1 \ldots x_n;t)} - \frac{\delta\bar{A}}{\delta\tilde{\sigma}_n(x_1 \ldots x_n;t)} \frac{\delta\bar{B}}{\delta\sigma_n(x_1 \ldots x_n;t)} \right] \tag{17}
$$

We stress again that all of the basic amplitudes $\{\sigma_n,\tilde{\sigma}_n\}$, which now completely specify the system, are linked-cluster quantities, and are hence quasi-local in the sense of obeying the cluster property.

Finally, in order to simplify the notation, we shall henceforth often drop the time label from, and thus leave implicit the time-dependence of, our basic amplitudes $\{\sigma_n,\tilde{\sigma}_n\}$ and other dynamical variables constructed from them.

3. THE GAUGE-FIELD APPROACH

We now consider the application of the ECCM formalism to a Bose fluid composed of N identical uncharged bosons of mass m each, and interacting (in units with $\hbar = 1$) via two-body forces described by a Hamiltonian,

$$
h_{(o)} = -\frac{1}{2m} \sum_{j=1}^{N} \nabla_j^2 + \sum_{i=1}^{N} \sum_{j=1}^{i-1} v(x_i - x_j) \quad . \tag{18}
$$

Since the vacuum boson field theory already described does not conserve particle number, it is convenient to introduce a chemical potential, μ, and work with the grand-canonical Hamiltonian $H_{(o)} \equiv h_{(o)} - \mu N$, with N the number operator. Furthermore, since the basic Hamiltonian of Eq.(18) does conserve particle number, we wish to formulate the theory in a *gauge-invariant* fashion so that the underlying U(1) symmetry is preserved.

This is most naturally achieved from the outset by introducing *external gauge fields* and coupling the system to them.[13] These fields are the usual scalar and vector potentials, $\phi(x,t) \equiv \phi(x)$ and $\vec{A}(x,t) \equiv \vec{A}(x)$ respectively, familiar from electrodynamics where they arise similarly due to charge conservation, and which couple respectively to the operators for the particle number density $\rho(x)$ and the canonical current density $\vec{j}(x)$,

$$\begin{aligned} &\rho(r) \equiv \rho(r,r) \quad ; \quad \rho(r,r') \equiv a^{\dagger}_{r} a_{r'} , \\ &\vec{j}(r) \equiv \lim_{\vec{r}' \to \vec{r}} \frac{i}{2m} (\vec{\nabla}_r - \vec{\nabla}_{r'}) \rho(r,r') \quad . \end{aligned} \tag{19}$$

In this way we are led to study the Hamiltonian $H_{(o)} \to H$,

$$H = \sum_{j=1}^{N} \frac{1}{2m} [-i\vec{\nabla}_j - \vec{A}(x_j)]^2 + \sum_{j=1}^{N} [\phi(x_j) - \mu] + \sum_{i=1}^{N} \sum_{j=1}^{i-1} v(x_i - x_j) \quad , \tag{20}$$

which can be equivalently written in second-quantised form as

$$\begin{aligned} H &= H_{(o)} + \int dr \left[\left\{ \phi(r) + \frac{1}{2m} \vec{A}^2(r) \right\} \rho(r) - \vec{A}(r) \cdot \vec{j}(r) \right] \quad , \\ H_{(o)} &= \frac{1}{2m} \int dr (\vec{\nabla}_r a^{\dagger}_r) \cdot (\vec{\nabla}_r a_r) + \tfrac{1}{2} \int dr \int dr' v(r-r') a^{\dagger}_r a^{\dagger}_{r'} a_{r'} a_r - \mu \int dr a^{\dagger}_r a_r \quad . \end{aligned} \tag{21}$$

The essence of the gauge-field approach is that by suitable choices of the gauge fields we can put the system and the observer into arbitary relative motion. The coupling to the gauge fields may be thought of as inducing *local or differential Galilei transformations* in the system. Particularly simple examples are: (i) a global Galilei transformation to put the system into uniform motion with constant relative velocity $\vec{V}_o$ can be induced by the choice $\vec{A}(r) = m\vec{V}_o$, $\phi(r) = -\frac{1}{2}mV_o^2$; and this leads in the case of translationally-invariant systems to the wavefunction achieving an extra overall phase factor $\exp(it\vec{V}_o \cdot \vec{P})$ where $\vec{P}$ is the total momentum operator which generates spatial translations; and (ii) the system can be put into uniform rotation with angular velocity $\vec{\Omega}$ by the choice $\vec{A}(r) = m\vec{\Omega} \times \vec{r}$, $\phi(r) = -\frac{1}{2}m|\vec{\Omega} \times \vec{r}|^2$; and this leads in the case of a rotationally-invariant system to the wavefunction acquiring an extra overall phase factor $\exp(it\vec{\Omega} \cdot \vec{L})$ where $\vec{L}$ is the total angular momentum operator which generates spatial rotations.

For the case of the more general local U(1) gauge transformations, the Schrödinger wavefunction transforms in the usual way by acquiring an extra overall phase factor as,

$$\Psi(x_1 \ldots x_N) \longrightarrow \Psi'(x_1 \ldots x_N) = \exp\left[-i \sum_{j=1}^{N} \Lambda(x_j,t)\right] \Psi(x_1 \ldots x_N) \quad , \tag{22}$$

where $\Lambda(r,t) \equiv \Lambda(r)$ is some arbitrary gauge parameter or phase field. As usual, one easily sees from Eq.(20) that the transformed wavefunction Ψ' satisfies the usual time-dependent Schrödinger equation, with a Hamiltonian of the same form as Eq.(20) except with gauge fields which are correspondingly transformed as,

$$\vec{A} \longrightarrow \vec{A}' = \vec{A} - \vec{\nabla}\Lambda \quad ; \quad \phi \longrightarrow \phi' = \phi + \partial\Lambda/\partial t \quad . \tag{23}$$

It is clear that under these local Galilei transformations induced by the gauge fields, the (gauge-invariant) physical forces try to create local translational motions which vary both temporally and spatially throughout the system. A proper hydrodynamical description must then evidently be able to distinguish in a gauge-invariant fashion between, for example, the local hydrodynamical kinetic energy density of translation and the intrinsic kinetic energy density in the local rest frame. In the remainder of this paper we show how the ECCM can be used in this gauge-field approach to provide a complete (and, in principle, exact) hydrodynamical account of the system. Within the ECCM we have seen above that any aspect of the system can be described in terms of the linked-cluster amplitudes $\{\sigma_n, \tilde{\sigma}_n\}$. The gauge transformation of Eq.(22) may be equivalently specified in terms of them as,

$$\sigma_n(x_1 \ldots x_n) \rightarrow \sigma'_n(x_1 \ldots x_n) = \exp\left[-i \sum_{j=1}^{n} \Lambda(x_j)\right] \sigma_n(x_1 \ldots x_n) \quad ,$$
$$\tilde{\sigma}_n(x_1 \ldots x_n) \rightarrow \tilde{\sigma}'_n(x_1 \ldots x_n) = \exp\left[+i \sum_{j=1}^{n} \Lambda(x_j)\right] \tilde{\sigma}_n(x_1 \ldots x_n) \quad . \qquad (24)$$

4. EVOLUTION OF THE ECCM AMPLITUDES: GENERALISED GROSS-PITAEVSKII THEORY

We now wish to examine more closely the dynamic equations (14). The average-value functional $\bar{H}$ for the Hamiltonian of Eqs.(20)-(21) is readily given in terms of the one- and two-body density matrices. The one-body density matrix is the expectation value of the one-body density operator of Eq.(19), $\bar{\rho}(r,r') \equiv \langle\Psi'(t)|\rho(r,r')|\Psi(t)\rangle$. Use of Eqs.(1) and (9) leads to the result,

$$\bar{\rho}(r,r') = \langle\Phi|\tilde{\Sigma} a_r^\dagger I a_{r'} \Sigma|\Phi\rangle$$
$$= \sum_{n=0}^{\infty} \frac{1}{n!} \int dx_1 \ldots \int dx_n \tilde{\sigma}_{n+1}(r x_1 \ldots x_n) \sigma_{n+1}(r' x_1 \ldots x_n) \quad , \qquad (25)$$

where the second equality follows from the first by insertion of a resolution of the identity I of the form of Eq.(13) in the place so indicated.

The two-body density matrix is also needed to evaluate the expectation value $\bar{V}$ of the potential energy,

$$\bar{V}[\sigma_n, \tilde{\sigma}_n] = \tfrac{1}{2}\int dr \int dr' v(r-r') D(r,r') \quad ; \quad D(r,r') \equiv \langle a_r^\dagger a_{r'}^\dagger a_{r'} a_r \rangle \quad . \qquad (26)$$

For present purposes we shall not need the explicit functional form of $\bar{V} = \bar{V}[\sigma_n, \tilde{\sigma}_n]$, although for specific further applications the general two-body density matrix is probably most conveniently presented in terms of the *reduced subsystem amplitudes*[1,14] χ_n and ϕ_n,

$$\langle a_{r_1}^\dagger a_{r_2}^\dagger a_{r_3} a_{r_4} \rangle = \sum_{n=0}^{\infty} \frac{1}{n!} \int dx_1 \ldots \int dx_n \phi_{n+2}(r_1 r_2; x_1 \ldots x_n) \chi_{n+2}(r_3 r_4; x_1 \ldots x_n) , \qquad (27)$$

where we have used a similar insertion of the unit operator as in Eq.(25), and where

$$\chi_{n+2}(rr'; x_1 \ldots x_n) \equiv \langle\Phi|a_{x_1} \ldots a_{x_n} \hat{a}_r \hat{a}_{r'}|\Phi\rangle \quad ,$$
$$\phi_{n+2}(rr'; x_1 \ldots x_n) \equiv \langle\Phi|\hat{a}_r^\dagger \hat{a}_{r'}^\dagger a_{x_1} \ldots a_{x_n}|\Phi\rangle \quad . \qquad (28)$$

Explicit expressions for these reduced subsystem amplitudes in terms of the basic ECCM amplitudes $\{\sigma_n, \tilde{\sigma}_n\}$ are not needed here, but may be found using the functional derivative techniques described elsewhere[3,5] to evaluate various matrix elements involving double similarity-transformed operators of the form of Eq.(5).

After some straightforward manipulations, the equations of motion (14), with the Hamiltonian of Eqs.(20)-(21), can be more explicitly written as,

$$
\begin{aligned}
i \frac{d}{dt} \sigma_n(x_1 \ldots x_n) &= \sum_{j=1}^{n} h(x_j)\sigma_n(x_1 \ldots x_n) + \frac{\delta \bar{V}}{\delta \tilde{\sigma}_n(x_1 \ldots x_n)} , \\
-i \frac{d}{dt} \tilde{\sigma}_n(x_1 \ldots x_n) &= \sum_{j=1}^{n} \tilde{h}(x_j)\tilde{\sigma}_n(x_1 \ldots x_n) + \frac{\delta \bar{V}}{\delta \sigma_n(x_1 \ldots x_n)} ,
\end{aligned}
\tag{29}
$$

where the reduced one-body Hamiltonians are defined as,

$$
\begin{aligned}
h(r) &\equiv -\frac{1}{2m}\nabla_r^2 + \frac{i}{m}\vec{A}(r)\cdot\vec{\nabla}_r + \frac{i}{2m}\left[\vec{\nabla}_r\cdot\vec{A}(r)\right] + \frac{1}{2m}\vec{A}^2(r) + \phi(r) - \mu , \\
\tilde{h}(r) &\equiv -\frac{1}{2m}\nabla_r^2 - \frac{i}{m}\vec{A}(r)\cdot\vec{\nabla}_r - \frac{i}{2m}\left[\vec{\nabla}_r\cdot\vec{A}(r)\right] + \frac{1}{2m}\vec{A}^2(r) + \phi(r) - \mu .
\end{aligned}
\tag{30}
$$

We note that Eqs.(29)-(30) comprise a coupled set of nonlinear and non-local equations for the basic ECCM amplitudes. They may be regarded as a formally exact generalisation of the approximate nonlinear Gross-Pitaevskii equations[7,8] for a weakly interacting condensed Bose fluid. The Gross-Pitaevskii equations are actually obtained as the lowest-order (SUB1) truncation of our equations (29), in which all of the amplitudes $\{\sigma_n, \tilde{\sigma}_n\}$ with $n > 1$ are set to zero. The resulting equations are precisely the self-consistent time-dependent Hartree equations for the amplitudes $\sigma_1(r)$ and $\tilde{\sigma}_1(r)$, which, as we have seen, are just the condensate wavefunctions. By approximating the two-body potential by a repulsive delta function form, the equilibrium (time-independent) version of the resulting local equation is the usual Gross-Pitaevskii equation with the well-known kink-soliton solutions in a one-dimensional geometry and comparable vortex solutions in a cylindrical geometry.[15]

5. THE ONE-BODY DENSITY MATRIX

The fundamental equations of motion (29) and (30) suffice to construct the time-evolution of the expectation value of an arbitrary operator. Special interest attaches to the one-body density matrix $\bar{\rho}(r,r')$. We note, for example, from Eq.(25) and the fact that each of the basic ECCM amplitudes $\{\sigma_n, \tilde{\sigma}_n\}$ obeys the cluster property, that $\bar{\rho}(r,r')$ behaves at large separations as,

$$
\bar{\rho}(r,r') \xrightarrow[|\vec{r}-\vec{r}'|\to\infty]{} \tilde{\sigma}_1(r)\sigma_1(r') , \tag{31}
$$

and hence exhibits the off-diagonal long-range order typical of superfluid systems.[16] Its equation of motion is readily constructed, using the representation of Eq.(25), to give

$$
\begin{aligned}
\frac{\partial}{\partial t}\bar{\rho}(r,r') &= \frac{i}{2m}(\nabla_{r'}^2 - \nabla_r^2)\bar{\rho}(r,r') + \frac{1}{m}\left[\vec{A}(r)\cdot\vec{\nabla}_r + \vec{A}(r')\cdot\vec{\nabla}_{r'}\right]\bar{\rho}(r,r') \\
&+ i\left[\phi(r) - \phi(r') + \frac{1}{2m}\{\vec{A}^2(r) - \vec{A}^2(r')\}\right]\bar{\rho}(r,r') \\
&+ \frac{1}{2m}\left[\{\vec{\nabla}_r\cdot\vec{A}(r)\} + \{\vec{\nabla}_{r'}\cdot\vec{A}(r')\}\right]\bar{\rho}(r,r') + iC(r,r'),
\end{aligned}
\tag{32}
$$

where the function $C(r,r')$ may be expressed in terms of the generalised

Poisson bracket defined in Eqs.(16) and (17) as,

$$C(r,r') = i\{\bar{V},\bar{\rho}(r,r')\} \quad . \tag{33}$$

We also note that the result expressed in Eqs.(32) and (33) can be more directly derived by starting from the Heisenberg equation of motion for the operator $\rho(r,r')$, and taking its expectation value in the state $|\Psi(t)\rangle$. We can thus show either in this way or, more laboriously, by using the explicit form of the functional $\bar{V}[\sigma_n,\sigma_n]$, that the function $C(r,r')$ can be expressed in terms of the two-body density matrix as,

$$\begin{aligned} C(r,r') &= \tfrac{1}{2}\int dx_1 \int dx_2\, v(x_1 - x_2)\langle\left[a^\dagger_{x_1} a^\dagger_{x_2} a_{x_2} a_{x_1} , a^\dagger_r a_{r'}\right]\rangle \quad , \\ &= \int dx[v(r-x) - v(r'-x)]\ \langle a^\dagger_x a^\dagger_r a_{r'} a_x\rangle \quad . \end{aligned} \tag{34}$$

We readily see from Eq.(25) that under the gauge transformation (24) previously discussed, the one-body density matrix transforms as,

$$\bar{\rho}(r,r') \quad \rightarrow \bar{\rho}'(r,r') = e^{i[\Lambda(r)-\Lambda(r')]}\bar{\rho}(r,r') \quad , \tag{35}$$

and that both the diagonal term $\bar{\rho}(r,r)$ and, from Eqs.(26)-(28), the functional $\bar{V}[\sigma_n,\sigma_n]$ are gauge-invariant. It is now convenient to transform to relative and centre-of-mass coordinates,

$$\vec{\xi} \equiv \vec{r}-\vec{r}' \quad ; \quad \vec{R} \equiv \tfrac{1}{2}(\vec{r}+\vec{r}') \quad , \tag{36}$$

since our goal of deriving the hydrodynamical balance equations for the most important local physical field densities may most readily be attained by expanding the equation of motion (32) for $\rho(r,r') \equiv \bar{\rho}(R|\xi)$ in powers of ξ up to a given order, and with coefficients whose specific combinations in each order can be identified as being gauge-invariant.

The straightforward Taylor expansion,

$$\bar{\rho}(R|\xi) = \bar{\rho}(R) + \beta^a(R)\xi^a + \tfrac{1}{2}\gamma^{ab}(R)\xi^a\xi^b + \tfrac{1}{6}\delta^{abc}(R)\xi^a\xi^b\xi^c + O(\xi^4) \quad , \tag{37}$$

where $\bar{\rho}(R) \equiv \bar{\rho}(R,R)$ is the number density at space-point R, contains coefficients which are certainly not gauge-invariant. The roman superscripts a, b,c, ... in Eq.(37) are Cartesian three-vector indices; and the various coefficients $\gamma^{ab}, \delta^{abc}$, ... are assumed to be completly symmetric tensors, with no loss of generality. Summation over repeated indices is also assumed in Eq.(37) and henceforth. We note, for example, from Eq.(19) that the first order coefficient $\beta^a(R)$ may be written as,

$$\beta^a(R) = -imj^a(R) \quad , \tag{38}$$

in terms of the local canonical current density $j^a(R)$, which is definitely not gauge-invariant. By considering the gauge transformation properties expressed in Eqs.(35) and (23), it is not difficult to show that the following combinations of tensors, of first, second and third order respectively, *are* gauge-invariant,

$$J^a(R) = J^a \equiv j^a - (\bar{\rho}/m)A^a \quad , \tag{39a}$$

$$\Gamma^{ab}(R) = \Gamma^{ab} \equiv -(1/m^2)\gamma^{ab} - (1/\bar{\rho})j^a j^b \quad , \tag{39b}$$

$$\begin{aligned} \Delta^{abc}(R) = \Delta^{abc} &\equiv (1/2im^2)\delta^{abc} + (1/2m\bar{\rho})(j^a\gamma^{bc}+j^b\gamma^{ca}+j^c\gamma^{ab}) \\ &+ (m/\bar{\rho}^2)j^a j^b j^c + (\bar{\rho}/24m^2)(\partial^a\partial^b A^c+\partial^b\partial^c A^a+\partial^c\partial^a A^b) \quad , \end{aligned} \tag{39c}$$

where $\partial^a Z^b \equiv \partial Z^b/\partial R^a$ for an arbitrary vector $Z^a = Z^a(R)$. We note in particular that Eq.(39a) expresses the true (invariant) current density $\vec{J}(R)$ in terms of the (non-invariant) canonical current density $\vec{j}(R)$. In terms of these gauge-invariant combinations, the one-body density matrix has the convenient expansion,

$$\ln[\bar{\rho}(R|\xi)/\bar{\rho}(R)] = -i\xi^a[A^a + m\bar{\rho}^{-1}J^a] - \tfrac{1}{2}m^2\bar{\rho}^{-1}\xi^a\xi^b\Gamma^{ab}$$

$$+ \frac{i}{3}\xi^a\xi^b\xi^c\left[-\frac{1}{24}(\partial^a\partial^b A^c + \partial^b\partial^c A^a + \partial^c\partial^a A^b) + m^2\bar{\rho}^{-1}\Delta^{abc}\right] + O(\xi^4) \quad . \qquad (40)$$

Apart from the last term in Eq.(32) we are now ready to expand this equation of motion in powers of ξ^a. For this last term we may use the representation in Eq.(34). Expanding in gauge-invariant quantities, we may readily derive the expansion,

$$C(r,r') = -\xi^a\bar{\rho}\, F^a_{int} + \tfrac{1}{2}i\xi^a\xi^b\bar{\rho}\left[A^a F^b_{int} + A^b F^a_{int} + mW^{ab}\right] + O(\xi^3) \quad , \qquad (41)$$

with the following definitions of the new gauge-invariant entities $\vec{F}_{int} = \vec{F}_{int}(R)$ and $W^{ab} = W^{ab}(R)$,

$$F_{int}(R) \equiv -\frac{1}{\bar{\rho}(R)}\int dx\left[\vec{\nabla}_R v(R-x)\right]\langle a^\dagger_R a^\dagger_x a_x a_R\rangle \quad ,$$

$$W^{ab}(R) \equiv -\frac{1}{\bar{\rho}(R)}\int dx\left[\frac{\partial v(R-x)}{\partial R^a}\langle a^\dagger_x J^b(R) a_x\rangle + \frac{\partial v(R-x)}{\partial R^b}\langle a^\dagger_x J^a(R) a_x\rangle\right]. \qquad (42)$$

The operator $\vec{J}(R)$ in Eq.(42) is the gauge-invariant total current density operator, given by analogy with Eqs.(19) and (39a) as,

$$\vec{J}(R) = -\frac{i}{2m}\left[a^\dagger_R(\vec{\nabla}_R a_R) - (\vec{\nabla}_R a^\dagger_R)a_R\right] - \frac{1}{m}\vec{A}(R)a^\dagger_R a_R \quad . \qquad (43)$$

It is useful for later purposes to consider the physical meaning of the above coefficients. From their definitions in Eq.(42) we see that: (i) $\vec{F}_{int}(R)$ represents the average force per particle due to the internal (pairwise) interactions in the presence of correlations; and (ii) since the *gauge-invariant velocity field* $\vec{u}(R)$ may be defined as,

$$\vec{u}(R) \equiv J(R)/\bar{\rho}(R) \quad , \qquad (44)$$

then $W^{ab}(R)$ is the *symmetrised velocity-force correlation function*. Its trace, $W^{aa}(R)$, thus represents twice the average rate of increase of internal kinetic energy per particle at space-point R, due to the interparticle forces.

Finally, by inserting Eqs.(37) and (41) into the equation of motion (32) for the one-body density matrix, a comparison of the coefficients of the terms of zeroth, first and second order in ξ^a respectively, yields the following *local* evolution equations,

$$\frac{\partial\bar{\rho}}{\partial t} = -\partial^a j^a + \frac{1}{m}\bar{\rho}(\partial^a A^a) + \frac{1}{m}A^a(\partial^a\bar{\rho}) \quad , \qquad (45a)$$

$$\frac{\partial j^a}{\partial t} = \frac{1}{m^2}\partial^b\gamma^{ab} + \frac{1}{m}\left[(\partial^a A^b)j^b + (\partial^b j^a)A^b\right] - \frac{\bar{\rho}}{m}\left[\partial^a\phi + \frac{1}{m}A^b(\partial^a A^b)\right]$$

$$+ \frac{1}{m}j^a(\partial^b A^b) + \frac{\bar{\rho}}{m}F^a_{int} \quad , \qquad (45b)$$

and

$$
\begin{aligned}
\frac{\partial \gamma^{ab}}{\partial t} = & - \frac{i}{m} \partial^c \delta^{abc} + \frac{1}{m}[\gamma^{ac}(\partial^b A^c) + \gamma^{bc}(\partial^a A^c) + \gamma^{ab}(\partial^c A^c) + (\partial^c \gamma^{ab}) A^c] \\
& + j^a [m \partial^b \phi + A^c(\partial^b A^c)] + j^b [m \partial^a \phi + A^c (\partial^a A^c)] + \frac{1}{4m} \partial^c (\bar{\rho} \partial^a \partial^b A^c) \\
& - \bar{\rho}[A^a F^b_{int} + A^b F^a_{int} + mW^{ab}] \quad . \qquad (45c)
\end{aligned}
$$

We shall now analyse Eqs.(45a-c) in some detail, and show that they lead to the *hydrodynamical balance equations* for the most important local physical observables.

6. HYDRODYNAMICS AND THE CONTINUITY EQUATIONS

In a hydrodynamic description, the most fundamental quantities that parametrise the system are the number, momentum and energy densities and the corresponding fluxes or currents that describe their flows. We show now how Eqs.(45a-c) relate respectively to the continuity or balance equations for these quantities.

6.1. Equation of Motion for the Number Density

Taking into account the definition (39a) of the total (gauge-invariant) current density $\vec{J}(R)$, we see easily that Eq.(45a) takes the usual form,

$$\partial\bar{\rho}/\partial t + \vec{\nabla}\cdot\vec{J} = 0 \quad , \qquad (46)$$

of the *current continuity equation*. As is well known, this equation simply expresses the local form of the global number conservation law, $d\bar{N}/dt = 0$. It is clearly gratifying to obtain this equation so readily since its correct imposition has been our primary objective in introducing the external gauge fields. We also stress that not only is Eq.(46) exactly true in the complete, untruncated formalism but, equally importantly, it also holds in most practical truncations that need to be made in real implementations of the ECCM. For example, a well-used approximation hierarchy is the so-called SUBn scheme in which the ECCM amplitudes $\{\sigma_i, \tilde{\sigma}_i\}$ with $i > n$ are set to zero, and where the remaining amplitudes with $1 \le i \le n$ are obtained as the solutions to the 2n coupled equations of motion (14) or (29)-(30). Since our derivation of Eq.(45a) proceeded via Eqs.(32) and (25), it should be immediately clear that Eq.(46) is also true in an arbitrary SUBn truncation.

6.2. Equation of Motion for the Momentum Density

It is not difficult to re-express Eq.(45b) in terms of gauge-invariant tensor quantities as,

$$\frac{\partial J^a}{\partial t} + \partial^b (J^a J^b/\bar{\rho}) = \frac{\bar{\rho}}{m} F^a \quad , \qquad (47)$$

where we have also made use of the lower-order continuity equation (46). The vector field $\vec{F}$ in Eq.(47) is seen to comprise three separate pieces,

$$\vec{F}(R) = \vec{F} \equiv \vec{F}_{int} + \vec{F}_{ext} + \vec{F}_{kin} \quad , \qquad (48)$$

where: (i) $\vec{F}_{int}$, defined in Eq.(42), has already been explained to be the average force per particle due to the internal interparticle forces of interaction; (ii) $\vec{F}_{ext}$ is the external force per particle due to the gauge fields, and which has the familiar (gauge-invariant) Lorentz form from the analogous and better-known application to electrodynamics,

$$\vec{F}_{ext} \equiv \vec{E} + \vec{u} \times \vec{B} \quad ; \quad \vec{B} \equiv \vec{\nabla} \times \vec{A} \ , \ \vec{E} \equiv -\vec{\nabla}\phi - \partial\vec{A}/\partial t \quad , \qquad (49)$$

in terms of the velocity field $\vec{u} = \vec{u}(R)$ defined in Eq.(44); and (iii) $\vec{F}_{kin}$ is defined to be,

$$\vec{F}_{kin} \equiv -m\bar{\rho}^{-1}\partial^b\Gamma^{ab} \quad , \tag{50}$$

and is hence proportional to the divergence of the (gauge-invariant) *kinetic stress tensor,* $\Gamma^{ab} = \Gamma^{ab}(R)$, which can be readily seen from Eq.(40) to be proportional to the second-order cumulant coefficient in the expansion for the one-body density matrix. The vector field $\vec{F} = \vec{F}(R)$ thus has the interpretation of being the total average force per particle at some space-point R in the fluid.

Equation (47) is clearly the equation of motion for the current or momentum density. Its physical content is perhaps more readily appreciated by rewriting it, using the usual definition of the convective derivative, d/dt,

$$d/dt \equiv \partial/\partial t + u^b\partial^b \quad , \tag{51}$$

as

$$m\,[\bar{\rho}^{-1}d\vec{J}/dt + \vec{u}(\vec{\nabla}\cdot\vec{u})] = \vec{F} \quad . \tag{52}$$

A further combination of Eq.(52) with the lower-order continuity equation (46) then yields the very simple result,

$$md\vec{u}/dt = \vec{F} \quad , \tag{53}$$

which is precisely the correct Newtonian equation of motion for the system.

Before proceeding we point out again here that the equation of motion (47) or (53) is also true in SUBn approximation for the same basic reason as was discussed in the case of the lower-order continuity equation (46). Indeed both continuity equations remain true in even more drastic sub-approximations where some specific class of terms is further neglected from those which rightfully contribute to the potential energy expectation value $\bar{V} = \bar{V}[\sigma_n,\tilde{\sigma}_n]$. Further approximations of this type are often made in such realistic applications of the ECCM to physical systems with, for example, strongly repulsive (hard-core) interparticle forces. In these cases each such term contributing to $\bar{V}$ transforms under the gauge transformations in precisely the same way. The symmetry is thereby obeyed by each term separately, and the conservation laws hence continue to remain valid.

It is also worth pointing out that the equation of motion (47) can profitably be thought of as a conservation law which is the natural *generalisation of the well-known f-sum rule* away from the linear response regime. Thus, we can show by acting on the system at equilibrium with an infinitesimal sudden impulsive change in the external scalar potential, $\delta\phi(r,t) = \psi(r)\delta(t)$, that the corresponding infinitesimal change in the number density $\delta\bar{\rho}(r,t)$ is continuous, but its time-derivative changes discontinuously at $t = 0$ by an amount that can be calculated from Eqs.(46)-(50). By further relating the change in density, $\delta\bar{\rho}$, as the system responds (linearly) from its equilibrium to an arbitrary small change $\delta\phi$ in the external scalar potential via the usual density-density response function, we can readily show that the linear response of our system obeys the f-sum rule. We point out again that the validity of the f-sum rule remains true even in the practical truncation or approximation schemes already discussed.

6.3. Equation of Motion for the Energy Density

We finally turn our attention to Eq.(45c), which we again wish first to express in terms of gauge-invariant tensors. It turns out to be especially convenient to express the second-order equation of motion in terms of the particular gauge-invariant kinetic stress tensor K^{ab}, which is related to the kinetic energy density as we see below, and which is defined as

$$K^{ab}(R) = K^{ab} \equiv \tfrac{1}{2}m(\Gamma^{ab}+\bar{\rho}u^a u^b) \quad . \tag{54}$$

A straightforward but rather lengthy calculation employing the lower-order equations of motion (46) - (47), yields the conservation law,

$$\frac{\partial K^{ab}}{\partial t} + \partial^c\Big[\Delta^{abc} + u^a K^{bc} + u^b K^{ac} + u^c K^{ab} - m\bar{\rho}u^a u^b u^c + \frac{\bar{\rho}}{24m^2}(\varepsilon^{bcd}\partial^a B^d + \varepsilon^{acd}\partial^b B^d)\Big]$$
$$= \tfrac{1}{2}\bar{\rho}(W^{ab} + u^a E^b + u^b E^a) + m^{-1}(\varepsilon^{bcd}K^{ac} + \varepsilon^{acd}K^{bc})B^d \quad , \tag{55}$$

where ε^{abc} is the usual antisymmetric unit tensor of third rank, and $\vec{E}$ and $\vec{B}$ are the *external force fields* of Eq.(49).

We finally show how Eq.(55) can be used to study the conservation law of total energy in differential form. The global energy conservation law is easily obtained by calculating the time derivative of $\bar{H} = \bar{H}[\sigma_n,\tilde{\sigma}_n;t]$ from Eq.(15) as,

$$\frac{d\bar{H}}{dt} = \frac{\partial\bar{H}}{\partial t} + \sum_{n=1}^{\infty}\frac{1}{n!}\int dx_1\cdots\int dx_n\left[\frac{\delta\bar{H}}{\delta\sigma_n(x_1\ldots x_n)}\dot{\sigma}_n(x_1\ldots x_n) + \frac{\delta\bar{H}}{\delta\tilde{\sigma}_n(x_1\ldots x_n)}\dot{\tilde{\sigma}}_n(x_1\ldots x_n)\right] \quad . \tag{56}$$

From the basic equations of motion (14), it immediately follows that $d\bar{H}/dt = \partial\bar{H}/\partial t$, and hence that the total energy is conserved, $d\bar{H}/dt = 0$, in the case when the gauge fields are time-independent. From the corresponding local form which we now investigate, we shall identify the energy current density and the local energy source density in the general case when the external gauge fields are time-dependent.

The total Hamiltonian (20) or (21) is the sum, $H = H_1+V$, of a one-body part H_1 and a two-body part V due to pairwise interactions. The one-body contribution to the expectation value $\bar{H}$ is expressed in terms of a local energy density $\varepsilon_1(R)$,

$$\bar{H}_1 = \int dR\,\varepsilon_1(R) \quad , \tag{57}$$

which may readily be evaluated to give the explicit form,

$$\varepsilon_1(R) = \frac{1}{8m}\nabla_R^2\,\bar{\rho}(R) + \tfrac{1}{2}m\Gamma^{aa}(R) + \tfrac{1}{2}m\bar{\rho}(R)\vec{u}^2(R) + \bar{\rho}(R)[\phi(R)-\mu] \quad . \tag{58}$$

The third term, $\tfrac{1}{2}m\rho\vec{u}^2$, represents the hydrodynamical kinetic energy density of average translational motion; while the second term, $\tfrac{1}{2}m\mathrm{Tr}(\Gamma)$, together with the first term, represents the kinetic energy density in the local rest frame. (We note that since the first term in Eq.(58) is a perfect divergence, it is strictly speaking redundant, since its contribution to $\bar{H}_1$ is zero. Its inclusion does however ensure the positivity of the kinetic energy density.) We note also that each of the terms in Eq.(58) is separately gauge-invariant, except for the potential energy term, $\rho\phi$, due to the external scalar potential.

Although the two-body energy $\bar{V}$ cannot simply be similarly expressed in terms of a local energy density, $\varepsilon_2(R)$, we now show that its time-derivative may be comparably decomposed. From Eq.(26) we first calculate,

$$\partial D(r,r')/\partial t = -i\langle [a_r^\dagger a_{r'}^\dagger a_{r'} a_r, H]\rangle$$
$$= -\vec{\nabla}_r \cdot \langle a_{r'}^\dagger \vec{J}(r) a_{r'}\rangle - \vec{\nabla}_{r'} \cdot \langle a_r^\dagger \vec{J}(r') a_r\rangle \quad , \tag{59}$$

using the Heisenberg equation of motion. We thus find the result,

$$\partial\bar{V}/\partial t = \tfrac{1}{2}\int dr \int dr' v(r-r')\partial D(r,r')/\partial t = (\partial/\partial t)\int dr\, \varepsilon_2(r) \quad , \tag{60}$$

where we have defined $\varepsilon_2(r)$ in terms of its time-derivative as,

$$\frac{\partial\varepsilon_2(r)}{\partial t} \equiv - \int dr' v(r-r')\vec{\nabla}_r \cdot \langle a_{r'}^\dagger \vec{J}(r) a_{r'}\rangle$$
$$= -\tfrac{1}{2}\,\bar{\rho}(r) W^{aa}(r) - \vec{\nabla}_r \cdot \vec{P}(r) \quad , \tag{61}$$

and where we have used the definition (42) and the new definition,

$$\vec{P}(r) \equiv \int dr' v(r-r')\langle a_{r'}^\dagger \vec{J}(r) a_{r'}\rangle \quad . \tag{62}$$

The first term in Eq.(61) represents the average rate of decrease of the kinetic energy density due to the interparticle interactions, as already discussed above; while the local field $\vec{P}$ in the second term is clearly an energy flux due to the internal forces. It is defined so that the interaction energy of any pair of particles is asymmetrically associated wholly with one of them.

We finally combine the above results as,

$$\partial\bar{H}/\partial t \equiv \int dr\, \partial\varepsilon(r)/\partial t = \int dr[\partial\varepsilon_1(r)/\partial t + \partial\varepsilon_2(r)/\partial t] \quad . \tag{63}$$

The total energy density $\varepsilon(r)$ now satisfies a continuity equation which can readily be found from Eqs.(58) and (61), and by employing our previous gauge-invariant equations of motion (55) and (46), in the form,

$$\frac{\partial\varepsilon}{\partial t} + \vec{\nabla}\cdot\vec{J}_\varepsilon = S_\varepsilon \quad , \tag{64}$$

where the total energy flux vector, $\vec{J}_\varepsilon$, is given by,

$$\vec{J}_\varepsilon(r) = \vec{J}_\varepsilon \equiv \frac{1}{8m}\nabla^2\vec{J} + \left(\frac{m}{2\bar{\rho}}\,\mathrm{tr}\,\Gamma + \tfrac{1}{2}m\vec{u}^2 + \phi - \mu\right)\vec{J}$$
$$+ (\Delta^{aab} + mu^a\Gamma^{ab})\hat{r}^b - \frac{\bar{\rho}}{12m^2}\vec{\nabla}\times\vec{B} + \vec{P} \quad , \tag{65}$$

with $\hat{r}^b$ a unit vector in the b-direction; and where the energy source density is given by,

$$S_\varepsilon(r) = S_\varepsilon \equiv \bar{\rho}\,\frac{\partial\phi}{\partial t} - \vec{J}\cdot\frac{\partial\vec{A}}{\partial t}$$
$$= \bar{\rho}\,\frac{d\phi}{dt} + \vec{J}\cdot\vec{E} \quad . \tag{66}$$

In the second form for S_ε above we have used the convective derivative of Eq.(51). We note finally that due to the appearance of terms involving the scalar potential ϕ, none of the quantities $\varepsilon(r)$, $\vec{J}_\varepsilon(r)$ and $S_\varepsilon(r)$ is gauge-invariant. We may trivially rewrite the energy density balance equation in the equivalent form however,

$$\frac{\partial}{\partial t}(\varepsilon - \bar{\rho}\phi) + \vec{\nabla}\cdot(\vec{J}_\varepsilon - \phi\vec{J}) = \vec{J}\cdot\vec{E} \quad , \tag{67}$$

by combining it with the current continuity equation (46), where now the subtracted energy density $(\varepsilon-\bar{\rho}\phi)$ and flux $(\vec{J}_\varepsilon-\vec{J}\phi)$ are fully gauge-invariant, as is the source term $\vec{J}\cdot\vec{E}$.

7. SUMMARY AND DISCUSSION

The ECCM formally decomposes a quantum many-body problem into a non-local classical field theory for a set of interacting, non-local, n-body fields $\{\sigma_n, \tilde{\sigma}_n\}$, $n = 1,2, \ldots,$ which in the coordinate-space representation considered here are n-point functions of arguments $(x_1, x_2, \ldots x_n)$.[3,5] The lowest-order amplitudes, $\tilde{\sigma}_1(x) = \langle a_x^\dagger \rangle$ and $\sigma_1(x) = \langle \tilde{a}_x \rangle$, are just the condensate wavefunctions for the Bose fluid,[15] and are clearly local by definition, since they depend on a single space coordinate. However, the higher-order amplitudes are also of a conceptually similar character due to their linked-cluster or connected property, which ensures that they vanish as the relative separation of any two spatial arguments approaches infinity. For example, $\tilde{\sigma}_2(x,y) = \langle a_x^\dagger a_y^\dagger \rangle_{connected} \equiv \langle a_x^\dagger a_y^\dagger \rangle - \langle a_x^\dagger \rangle \langle a_y^\dagger \rangle$. Then, for sufficiently long-wavelength phenomena in the infra-red limit, the relevant values of the relative separations will be small in comparison with any natural scale lengths. In this way the basic ECCM amplitudes $\{\sigma_n, \tilde{\sigma}_n\}$ may be regarded as a set of *generalised quasi-local order parameters*, by analogy with the more conventional and strictly local (one-body) order parameters of condensed matter theory. The total energy expectation value, $\bar{H} = \bar{H}[\sigma_n, \tilde{\sigma}_n]$ is thus an almost local functional of these classical (c-number) quasi-local order parameters.

From the above viewpoint it is natural to compare the ECCM with both the *phenomenological Ginzburg-Landau theory*[15] and the *density-functional theory*.[17,18]. In Ginzburg-Laudau phenomenology, the total energy (or free energy at non-zero temperatures) is expressed as a strictly local fourth-order functional of the order parameters $\langle a_x^\dagger \rangle$ and $\langle a_x \rangle$. We stress however that whereas this approach is intended only to be valid close to the relevant phase transition where the order parameters are small, the present ECCM by contrast provides an exact microscopic description of a pure state at zero temperature. Similarly, in the density-functional theory of Hohenberg, Kohn and Sham,[17,18] the total energy is again constructed as a functional of another local parameter, namely the particle number density $\bar{\rho}(x)$. In its simplest manifestatations, such as the Thomas-Fermi approximation, the total energy is taken to be a completely local functional of $\bar{\rho}(x)$. We may thus view the ECCM as a rather different approach to incorporate in a very precise and exact manner the non-local corrections to these simple local approximations, and with which virtually all of the later work in density-functional theory has been concerned.

In the present work we have described an application of the ECCM to a condensed Bose fluid. We have concentrated on the properties of a single *trajectory in the ECCM phase space*, where a representative point in this *symplectic ECCM phase space* is fully characterised by the set of amplitudes $\{\sigma_n, \tilde{\sigma}_n\}$. The statistical mechanics of a physical system is nowadays often described in terms of the totality of all such possible trajectories -- namely the whole *phase portrait*. The present treatment may thus be viewed as the first step towards a more comprehensive statistical-mechanical description. In this first step however we have already shown how the individual trajectories fully comply with the pertinent gauge symmetries and the local hydrodynamical conservation laws for such physically relevant variables as the local densities for particle number, momentum and energy.

We have seen both how the ECCM in principle exactly generalises the earlier Gross-Pitaevskii formalism for the condensate wavefunctions, and how it provides a complete hydrodynamical description of the zero-temperature condensed Bose fluid. In this latter regard, the locality features inherent to the ECCM make it powerful enough to incorporate a proper description of *topological excitations* as the vortex lines observed in liquid ^{4}He. Such excitations are created by specific topological boundary conditions on the amplitudes $\{\sigma_n, \tilde{\sigma}_n\}$ which prevent their decay by any quasi-local processes. These boundary conditions may only be properly imposed in such a theory as the ECCM in which the cluster property is exactly obeyed by all relevant amplitudes. The boundary condition is thus typically imposed at some sufficiently large distance -- from the vortex core in the case of a vortex line, for example, and where the phases of σ_n and $\tilde{\sigma}_n$ increase or decrease respectively by $2\pi n$ when the centre-of-mass coordinate, $R \to \infty$, winds once about the core. In this way, all physical observables for the excitation are ultimately describable in terms of the lowest amplitudes σ_1 and $\tilde{\sigma}_1$ -- just as in Ginzburg-Landau theory -- since the higher amplitudes cannot obtain long-range contributions due to their observance of the cluster property.

In conclusion, we have demonstrated that the ECCM leads to a complete and correct hydrodynamical description of a zero-temperature condensed Bose liquid, which incorporates the possibility of describing topological excitations and non-equilibrium dynamics. The hydrodynamical continuity equations are valid not only in the complete description, but, most importantly, also in the usual practical truncation schemes necessary to implement it. Their gauge-invariant form has further enabled us to provide a description which, for example, correctly separates the average (hydrodynamical) kinetic energy of translation from the average kinetic energy in the local rest frame. By contrast, other microscopic treatments have often found real difficulty in obtaining this separation.

The present approach has clearly demonstrated the applicability of the ECCM into manifestly non-perturbative areas, far removed from its origins.[9,10] We hope that further work along these lines will focus on the geometrical aspects of the underlying ECCM phase space in order to develop, for example, new approximation schemes that go beyond such truncations described here which still hark back to the perturbation-theoretical roots.

REFERENCES

1. J. Arponen, Ann.Phys. (NY) 151: 311 (1983).
2. J. Arponen and E. Pajanne, in: "Recent Progress in Many-Body Theories," H. Kümmel and M. L. Ristig (eds.), Lecture Notes in Physics Vol. 198, Springer-Verlag, Berlin (1984), p.319.
3. J. Arponen, R. F. Bishop and E. Pajanne, in: "Condensed Matter Theories," Vol.2, P. Vashishta, R. Kalia and R. F. Bishop (eds.), Plenum, New York (1987), p. 357.
4. J. Arponen, R. F. Bishop and E. Pajanne, in: "Condensed Matter Theories," Vol. 2, P. Vashishta, R. Kalia and R. F. Bishop (eds), Plenum, New York (1987), p. 373.
5. J. Arponen, R. F. Bishop and E. Pajanne, "Extended coupled cluster method: I. Generalised coherent bosonisation as a mapping of quantum theory into classical Hamiltonian mechanics," Phys.Rev. A36: 2519 (1987).
6. J. Arponen, R. F. Bishop and E. Pajanne, "Extended coupled cluster method: II. Excited states and generalised random phase approximation," Phys.Rev. A36: 2539 (1987).
7. E.P. Gross, Ann.Phys. (NY) 4: 57 (1958); Nuovo Cim.20:454 (1961).
8. L.P. Pitaevskii, Sov. Phys.-JETP 13: 451 (1961).
9. F. Coester, Nucl.Phys. 7: 421 (1958).
10. F. Coester and H. Kümmel, Nucl.Phys. 17: 477 (1960).
11. R.F. Bishop and H. Kümmel, Physics Today 40 (No.3): 52 (1987).
12. N.N. Bogoliubov, J.Phys. USSR 11: 23 (1947).

13. Ta-Pei Cheng and Ling-Fong Li, "Gauge Theory of Elementary Particle Physics," Clarendon Press, Oxford (1984), p.229.
14. H. Kümmel, K. H. Lührmann and J. G. Zabolitzky, Phys.Rep. 36C: 1 (1978).
15. A. L. Fetter and J. D. Walecka, "Quantum Theory of Many-Particle Systems," Academic Press, New York (1965).
16. G. Rickayzen in: "The Helium Liquids," J. G. M. Armitage and I. E. Farquhar (eds.), Academic Press, London (1975), p.95.
17. P. Hohenberg and W. Kohn, Phys.Rev. 136: B864 (1964).
18. W. Kohn and L. J. Sham, Phys.Rev. 140: A1133 (1965).

ON THE EXISTENCE AND REALIZATION OF SIZE-EXTENSIVE EFFECTIVE HAMILTONIAN THEORIES FOR GENERAL MODEL SPACES

Debashis Mukherjee

Department of Physical Chemistry
Indian Association For The Cultivation of Science
Calcutta 700-032, India

INTRODUCTION

The concept of effective Hamiltonian* is central to many branches of physics and chemistry.[1] In the construction of the eigenfunctions of a many-fermion hamiltonian H, one customarily chooses a set of determinants to approximately span the desired eigenfunctions. In the effective hamiltonian formalisms, one constructs an operator H_{eff} that acts on only a small subset of these determinants but reproduces some selected eigenvalues of H obtainable within the space of all the determinants. The space spanned by the subset of determinants is called the model space (denoted as P and the eigenfunctions of H_{eff} are called model eigenfunctions. One may look upon the corresponding eigenfunctions of H as being obtained by the action of a wave-operator Ω acting on the model eigenfunctions. The effective hamiltonian approach is particularly suited for describing the set of energy levels which are quasi-degenerate, as is typical of the open-shell systems.

For a many-body theory to be of viable predictability over a wide range of particle number, it must be size-extensive, i.e. it must generate values for the extensive properties (such as energy) that scale properly with the number of particles. For the many-fermion open-shell states, one strives to ensure size-extensivity by formulating an H_{eff} that consists of connected terms only (vide sect. 2 for the details). Till very recently construction of a connected H_{eff} was considered possible only for complete model spaces (CMS).[2] An n-fermion CMS is spanned by a set of n-particle determinants having a set of common orbitals ("core"), completely occupied by n_c particles and a set of partially occupied orbitals ("valence") in which the n_v fermions are allocated in all possible ways, such that $n = n_c + n_v$. For each choice of n_v, one has an n_v-valence CMS.

The principal emphasis of our overview is to demonstrate that the existence and the realization of a connected H_{eff} is in no way predicated by the choice of a CMS. The essential requirement is the multiplicative separability of Ω in the asymptotic sense, when the many-fermion system behaves as a collection of non-interacting subsystems. It is a non-trivial

* We shall underline the frequently used terms and the mathematical entitles when they appear first in our exposition

point that the normalization chosen for Ω, which specifies its model space projection, must not conflict with the multiplicative separability, i.e. it must be size-extensive. As we shall show, the main reason behind the failure to get a connected H_{eff} for an incomplete model space (IMS) lay in the choice of the intermediate normalization (IN) convention for Ω, i.e. the model space projection of Ω, Ω_P is a unit operator 1_P, which is not a size-extensive normalization for an IMS. By foregoing the IN, a connected H_{eff} can be obtained for any general IMS. The other crucial requirement for getting the connected H_{eff} is to use all the operators consistently on the Fock-space, and demand that Ω is the wave-operator not only for a particular n_v-valence IMS of interest (which is a subset of an n_v-valence CMS) but also of all the lower m_v-valence IMS obtained by deleting (n_v-m_v) fermions form the valence orbitals of the n_v-valence IMS. We shall call such a Fock space Ω as valence-universal.

2. GENERAL CONDITIONS OF SEPARABILITY VS SIZE-EXTENSIVITY

The basic approach to generate H_{eff} may be succinctly summarized as follows: we look for a similarity transformation on H induced by,

$$H \rightarrow L = \Omega^{-1} H \Omega \tag{2.1}$$

such that

$$QLP = 0; \ (PLQ = 0) \tag{2.2}$$

where the equation in the paranthesis is also automatically valid if Ω is chosen to be unitary. Eqn. (2.2) sufficient to ensure that all the eigenvalues of $H_{eff} = PLP$ will generate a set of selected eigenvalues of H. To guarantee the size-extensivity of H_{eff}, it is essential that H_{eff} should be additively separable over the various possible subsystems 'a' in the limit of asymptotically vanishing interaction between the subsystems[3]:

$$H_{eff} \rightarrow \sum_a H^a_{eff} \tag{2.3}$$

Using the subsystem wave-operator Ω^a, we can construct H^a_{eff}'s as

$$H^a_{eff} = (\Omega^a)^{-1} H^a \Omega^a \tag{2.4}$$

where H^a is the subsystem hamiltonian for the component a. If Ω can be expressed under the asymptotic limit as

$$\Omega \rightarrow \prod_a \Omega^a \tag{2.5}$$

then it follows easily that eqn. (2.3) holds good. Thus the explicit maintenance of the multiplicative separability of Ω in the asymptotic limit of non-interacting subsystems (all possible decompositions) guarantees that H_{eff} is additively separable in the asymptotic sense. In such a situation, the matrix-elements of the various k-body operators of H_{eff} are all connected in the sense that their algebraic expressions are not factorizable into fragments:

$$\langle p_1 \cdots p_k | H^k_{eff} | q_1 \cdots q_k \rangle \not\rightarrow \langle p_1 p_2 | X | q_1 q_2 \rangle \langle p_3 p_4 p_5 | Y | q_3 q_4 q_5 \rangle \cdots \tag{2.6}$$

For, if it would have been so, then with subsystems defined as $X = (p_1 p_2; q_1 q_2)$, $Y = (p_3 p_4 p_5; q_3 q_4 q_5)$, etc., the k-body part of H_{eff} would have given a nonvanishing contribution in the limit of no interaction between these fragments, and as a result

$$H_{eff} \not\to \sum_{a=x,y,..}' H^a_{eff} \qquad (2.7)$$

In a diagrammatic representation, the matrix-elements of a connected H_{eff} can be depicted as connected diagrams. Thus any additively separable operator will be a connected operator, depictable as a connected diagram.

Eqn. (2.5) implies that an exponential type of representation of Ω, $\Omega \sim \exp(T)$, will automatically ensure

$$T \to \sum_a T^a \qquad (2.8)$$

Thus T is an additively separable operator and is connected. Conversely, and this is the most crucial observation, an exponential representation of Ω involving a connected T automatically enforce eqn. (2.5), and hence eqn. (2.3). Thus there is an intimate connection between the connectedness of T and the size-extensivity of H_{eff}.

We call T a cluster-operator, and the exponential representation of Ω, as a coupled-cluster ansatz (CC). Such an ansatz was first introduced in nuclear physics (mainly in the closed-shell context) by Coester and Kümmel[4] and since has been transcribed in quantum chemistry also[5], where it is very successful.[6] The foregoing analysis indicates that a CC ansatz is very natural for the open-shell situation as well. The additional aspects for the open-shells are: (a) choice of the operator T that has the requisite flexibility to generate the desired eigenstates of H by acting on the model space functions; (b) use of decoupling conditions, eqn. (2.2), such that the connectedness of T is ensured; and (c) choice of a normalization for Ω, Ω_p such that T remains connected. All the above three requirements can be satisfied in several different ways for CMS theories. Particularly, one may adopt either an n_v-valence Hilbert space approach and utilize a wave-operator correlating the n_v-valence model space only or a Fock space approach and choose a valence-universal Ω for all m_v-valence model spaces with $0 \leq m_v \leq n_v$. Both the intermediate normalization and a particular variant of unitary normalization for Ω are size-extensive and have been used extensively. For an IMS, all the three conditions above pose nontrivial problems, whose resolution requires mandatory use of a Fock-space valence universal Ω, and abandoning the customary intermediate normaization which is no longer size-extensive. To motivate towards these aspects, we shall first briefly review the general aspects of the CMS open-shell effective hamiltonian theories, using a Fock-space Ω.

3. RESUME OF THE FOCK-SPACE OPEN-SHELL THEORIES FOR CMS

We consider the valence (or active) orbitals to be of either particle-type or hole-type (diagrammatically depicted with double arrows). The rest of the particle or hole orbitals are inactive (depicted by single arrows). We call an operator as closed, A_{cl}, if it contains only active lines. An operator is open, A_{op}, if it generates exitations out of model space kets; they thus have active lines on the right and at least one inactive line on the left. Examples of open and closed operators are shown in Fig. 1. By definition, H_{eff} is closed, i.e. $H_{eff} = L_{cl}$. This is guaranteed by requiring that L_{op} vanishes, as in eqn. (2.2). In the Fock-space approach with a valence-universal Ω, the projectors P and Q are, however to be interpreted as $P^{(m)}$ and $Q^{(m)}$, for all $0 \leq m_v \leq n_v$. We also define the valence rank 'k' of an operator $A^{(k)}$ as the number of active lines (particle valence or hole valence) that they destroy. On Fock-space, all the operators $A^{(k)}$ for different k are linearly independent, and the decoupling conditions for L can be symbolically written as

$$L^{(k)}_{op} = 0, \forall \; 0 \leq k \leq n_v \tag{3.1}$$

Further, any operator $A^{(k)}$ consists of several p-body operators, with $p \geq k$, and these are also linearly independent. More elaborately, we thus have

$$\left[L^{(k)}_{op}\right]_{(p)} = 0, \forall \; 0 \leq k \leq n_v; \quad p \geq k \tag{3.2}$$

then the decoupling conditions, eqn. (3.2), are the determining equation for T for $\Omega \sim \exp(T)$, $T \equiv T_{op}$.

Mukherjee et al.[7] chose a valence-universal Ω of the exponential structure:

$$\Omega = \exp(T) \tag{3.3}$$

and proved the connectedness of T from the connected nature of the multi-commutator expansion of $L = \exp(-T)\, H \exp(T)$ by the Hausdorff formula. Then it follows that $L_{cl} = H_{eff}$ is also connected using the same argument. With the ansatz eqn. (3.3), all the equations for various $T^{(k)}$'s are coupled. This can be somewhat inconvenient in practice. Offermann et al. and Ey[8] chose to generate a wave-operator $\Omega^{(n_v)}$ for the n_v-valence problem recursively in the following manner:

$$\Omega^{(o)} P^{(o)} = \exp(T^{(o)}) P^{(o)} \tag{3.4a}$$

$$\Omega^{(1)} P^{(1)} = \exp(T^{(o)}) (1+T^{(1)}) P^{(1)} \tag{3.4b}$$

$$\Omega^{(2)} P^{(2)} = \exp(T^{(o)}) (1+T^{(1)} + \frac{1}{2} T^{(1)} \otimes T^{(1)} + T^{(2)}) P^{(2)} \tag{3.4c}$$

and so on. $\otimes$ stands for an outer product. At each valence level k, the only unknown cluster amplitudes are those for valence-rank k, $T^{(k)}$. They are solved from the $Q^{(k)}$ projection of Bloch equation[9] for the k-valence problem:

$$H \Omega^{(k)} P^{(k)} = \Omega^{(k)} P^{(k)} H^{[k]}_{eff} P^{(k)} \tag{3.5}$$

with

$$H^{[k]}_{eff} = \sum_{l=0}^{k} H^{(l)}_{eff} \tag{3.6}$$

where all $H^{(l)}_{eff}$ for $l < k$ contain only $T^{(i)}$ operators with $i < k$ and are thus constant quantities at the k-valence level. Is should be noted that

$$\Omega^{(nv)} P^{(k)} = \Omega^{(k)} P^{(k)}, \quad 0 \leq k \leq n_v \tag{3.7}$$

so that $\Omega \equiv \Omega^{(nv)}$ is also implicitly a Fock-space valence universal wave-operator. Hence the $Q^{(k)}$ projections of eqn. (3.5) are closely related to eqn. (3.2), and $T^{(k)}$ operators are necessarily connected. The $P^{(k)}$ projections eqn. (3.6) furnish the $H^{(k)}_{eff}$ (since all other $H^{(l)}_{eff}$, l k, are determined from the lower valence problems). Note that we now need to specify the normalization of $\Omega^{(k)}$'s. Choosing the operators $T^{(k)}$ as open, we automatically have the intermediate normalization for each $\Omega^{(k)}$:

$$P^{(k)} \Omega^{(k)} P^{(k)} \equiv P^{(k)} \Omega^{(n_v)} P^{(k)} = P^{(k)}, \; 0 \leq k \leq n_v \tag{3.8}$$

The Bloch equation in Fock space then can be written in the form

$$H\Omega = \Omega L_{cl} \tag{3.9}$$

with

$$\Omega_{cl} = 1_{cl} \tag{3.10}$$

Note that eqn. (3.9) leads automatically to eqn. (3.1) or (3.2), and although eqn. (3.10) is implied in the treatment by Mukherjee et al.[7], this has never been explicitly used to get $H_{eff} = L_{cl}$, since a pre-multiplication of eqn. (3.9) by Ω^{-1} followed by the projection of the closed part gets rid of the Ω operator from the right side of eqn. (3.9) irrespective of the chosen normalization for Ω_{cl}. This aspect will turn out to be of crucial importance for the IMS-developments.

Lindgren [10] chose a wave-operator $\Omega^{(n_v)}$ as a normal-ordered exponential with respect to the closed-shell core-problem:

$$\Omega^{(n_v)} P^{(n_v)} = \{\exp(T)\} P^{(n_v)} \tag{3.11}$$

and used sufficiency conditions like

$$\{\overline{H \Omega}\}^{(k)} = \{\overline{\Omega H_{eff}}\}^{(k)} \tag{3.12}$$

to ensure connectedness of T. Here $\{\overline{H\Omega}\}$ stands for $\sum_n \frac{1}{n!}\{\overline{H\, TT\ldots T}\}$, with nT's. Lindgren included in T all the operators $T_{(p)}^{(k)}$ of various valence ranks $0 \leq k \leq n_v$. They are linearly dependent on the n_v-valence model space, and one needs sufficiency conditions as eqns. (3.12) to determine all of them. Calling $\{\overline{H\Omega}\}$ as Z and $\{\overline{\Omega H_{eff}}\}$ was W, $T^{(k)}$'s can be determined from the open part of eqn. (3.12):

$$Z_{op}^{(k)} = W_{op}^{(k)}, \forall\, 0 \leq k \leq n_v \tag{3.14}$$

H_{eff} is determined from

$$Z_{cl}^{(k)} = W_{cl}^{(k)}$$

and, with T chosen as open, (i.e. $\Omega_{cl} = 1_{cl}$), we have

$$Z_{cl}^{(k)} = H_{eff}^{(k)} \tag{3.15}$$

Haque and Mukherjee[11] and later more elaborately, Lindgren and Mukherjee[12] showed that, if $\Omega^{(h_v)}$ of the form eqn. (3.11) be interpreted as valence universal, than the $T^{(k)}$ operators are all linearly independent, and eqn. (3.12) can be derived from the Fock space Bloch equation, eqn. (3.9). It should be noted that the normal ordered ansatz in eqn. (3.11) is entirely analogous structurally to the $\Omega^{(h_v)}$ of Offermann et al.[8], defined recusively.

We note here several interesting aspects of these Fock-space theories:

(a) Ω has the full flexibility to correlate all the model sapces $P^{(k)}$, which are all CMS (b) the use of the Fock space Bloch equation (or its equivalent generated by the pre-multiplication by Ω^{-1}) guarantees that $T^{(k)}$'s are all connected (c) the choice of $T^{(k)}$'s as open automatically enforces intermediate normalization for Ω at all the k-valence levels. Thus all the three aspects (a) to (c) of the open-shell situation discussed in Sec. 2 for a CMS are met. The above discussion shows that a Fock-space approach is both convenient and transparent to arrive at the connectedness of H_{eff} for a CMS. For other works in the nuclear physics literature, we refer to ref. (34,35).

4. THE OPEN-SHELL EFFECTIVE HAMILTONIAN THEORIES WITH IMS

If the orbitals in a CMS are rather spread out in energy then there may well be several virtual space functions which are quasi-dengerate in energy with some functions in the model space. In this case the cluster amplitudes connecting these model and virtual functions will be very large making the CC-equations unstable. Such offending virtual functions are called intruder states[13]. It seems plausible, that intruder states can be avoided by working with an IMS.

The first many-body formalism utilizing an IMS was put forward in an MBPT framework by Hose and Kaldor.[14] This was a Hilbert-space theory and it used the intermediate normalization for Ω. The expression for H_{eff} contained disconnected terms and was thus not size-extensive. Jeziorski and Monkhorst[15] generalized their Hilbert-space CC formalism to an IMS and using the intermediate normalization reached conclusions essentially similar to those of Hose and Kaldor. Brandow showed[16] that a model space with valence particles as well as valence holes, where the model space is complete with respect to active particles and active holes separately, entails disconnected terms in H_{eff} even at the second order. These model spaces are incomplete, and Lindgren showed that they are some special cases of a particular type of IMS, which he termed as quasicomplete model spaces (QMS); (see sec. 5). Lindgren attempted to prove the connectedness of H_{eff} for QMS,[17(a)] but this leads to disconnected terms.[17(b)]

Mukherjee[18] traced the origin of the appearance of the disconnected terms of H_{eff} in an IMS to the use of intermediate normalization (IN), which is generally not size-extensive for an IMS. By abandoning IN, and by choosing a different normalization for Ω which is size-extensive, Mukherjee proved[18,19] the connectedness of H_{eff} for both an ordinary exponential and the normal ordered exponential ansatz. Although IN is the normalization almost exclusively used for closed-shells and with CMS for open-shells it now appears in retrospect that this choice has delayed considerably a proper size-extentisve formulation of H_{eff} for IMS.

Taking the cue from the Fock space strategy for the CMS we have now to generalize the concept of open and closed operators for an IMS. Since an IMS is a subset of a CMS, we have to take care explicitly of three types of spaces: (i) the IMS, characterized by the projector P; (ii) the complementary active space of P, characterized by the projector R, such that the union of P and R is the CMS and (iii) the inactive virtual space, with the projector Q. Any operator will be called open, A_{op} if leads to a transition to Q space by acting on P. Diagrammatically these operators

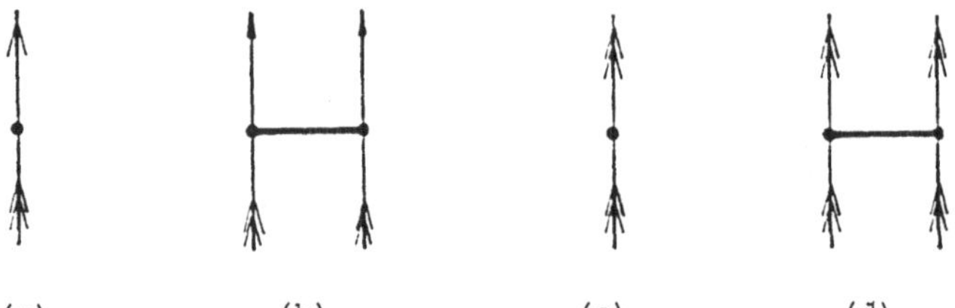

(a) (b) (c) (d)

Fig. 1. Examples of open (a,b) and closed (c,d) operators for CMS. For IMS, (c,d) generates both closed and quasi-open operators.

have active lines on the right and at least one inactive line on the left and the set of labels on the active lines on the right must be such that they belong to P. They have similar diagrammatical representations as the open operators of CMS Fig. 1 (a,b), with the additional restrictions on the labels of the active lines on the right, as discussed above. Any operator will be called quasi-open. A_{q-op}, if it leads to a transition from P space to R space. A_{q-op} operators have only active lines attached to them, as in Figs. 1(c,d), but those on the left are labelled in such a way that the set belong to R. The lines on the right are likewise so labelled that they belong to P. An operator is called closed, A_{cl}, if it leads to transitions within P by acting on P. The A_{cl} operators also have only active as in Figs. 1 (c,d); but their labels are such that they belong to P. It should be clear from the definitions, that products of open operators are open, and product of closed operators is closed, but the product of quasi-open operators can be closed as well as quasi-open depending on the actual valence labels attached to them. Moreover, a quasi-open operator may have a nonvanishing projection $P^{(n_v)} A_{q-op} P^{(n_v)}$. We illustrate this with an example. For a p-h model IMS with active labels p and α, as valence particle and valence hole repsectively, transition to the p-h vacuum Φ (core) by an operator T can be dipicted as in Fig. 2 (a). Since Φ is outside the model space, this operator is quasi-open. Similarly, an operator of the type shown in Fig. 2(b) induces excitation to the 2p-2h determinant, which is in the complementary active space R, and thus is also quasi-open. However their product is closed operator, since this leads to scattering from one p-h determinant to another, as is clear from Fig. 2(c). That the product of quasi-open operators can be closed is the main reason why the intermediate normalization is not size-extensive for an IMS.

Let us first take a particular n_v-valence IMS, (denoted as $P^{(n_v)}$) where the valence lines are either particle-type of hole-type. We classify all the operators as open, quasi-open and closed depending on its action on $P^{(n_v)}$. We then invoke the Bloch equation in Fock-space:

$$H\,\Omega = \Omega L_{cl} = \Omega H_{eff} \tag{4.1}$$

where

$$L = \Omega^{-1} H\,\Omega \tag{4.2}$$

with the property

$$L_{op}^{(k)} = 0; \; L_{q-op}^{(k)} = 0, \forall\; 0 \leq k \leq n_v \tag{4.3a,b}$$

The decoupling conditions (4.3) guarantee that $H_{eff} = L_{cl}$ is our desired effective hamiltonian for the n_v-valence IMS. If use a Fock space wave-operator Ω, with an exponential structure, to correlate $P^{(n_v)}$, then the cluster operator T can be written as

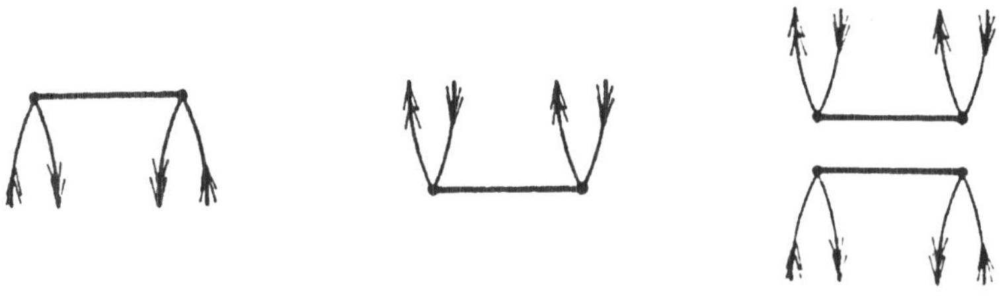

(a) (b) (c)

Fig. 2. Illustration that product of quasi-open operators can be closed for IMS.

$$T = \sum_{k=0}^{n_v} \sum_{p \geq k} \left(\left[T_{(p)}^{(k)}\right]_{op} + \left[T_{(p)}^{(k)}\right]_{q-op} \right) \tag{4.4}$$

If we now use the ansatz $\Omega = \exp(T)$, then eqns. (4.3a,b) show that T consists of only connected operators as a consequence of the Hausdorff-formula, just as in the CMS case. Alternatively, we may use the ansatz $\Omega = \{\exp(T)\}$, and derive

$$Z^{(k)} = \{ \overbrace{H\Omega} \}^{(k)} = W^{(k)} = \{ \overbrace{\Omega H_{eff}} \}^{(k)}, \forall\ 0 \leq k \leq n_v \tag{4.5}$$

The $T^{(k)}$'s are then obtained from the equations

$$Z_{op}^{(k)} = W_{op}^{(k)};\ Z_{q-op}^{(k)} = W_{q-op}^{(k)} \quad 0 \leq k \leq n_v \tag{4.6a,b}$$

Equations (4.6a,b) show that $T^{(k)}$'s are again connected. H_{eff} is obtained from

$$Z_{cl}^{(k)} = W_{cl}^{(k)} \tag{4.7}$$

and since products of $T_{q-op}^{(k)}$ will generally have closed operators and also $T_{q-op}^{(k)}$'s have in general non-vanishing projections on $P^{(n_v)}$, $P^{(n_v)} \Omega P^{(n_v)} \neq P^{(n_v)}$, and in general we have

$$W_{cl}^{(k)} = \{ \overbrace{\Omega H_{eff}} \}_{cl}^{(k)} \tag{4.8}$$

Eqn. (4.8) generalizes the expression (3.15) for an IMS, defining H_{eff}. It should be noted that for an IMS such as p-h model space, T_{q-op} operators like Figs. 2(b) are such that they have vanishing $P^{(n_v)}$ projections; but even then, their powers are closed as in Fig. 2 (c). In any case we thus do not have IN for Ω.

Is Ω a valence-universal wave-operator in the same sense as in a CMS? The answer is yes, and there are additional interesting aspects having no counterparts in CMS. Let us construct lower valence model spaces $P^{(m)}$, $0 \leq m \leq n_v$, by deleting $(n_v - m)$ valence orbitals from the determinants of $P^{(n_v)}$.

We may then make the following observations:

(a) An operator which is open with respect to $P^{(m)}$ is also, open with respect to $P^{(n_v)}$. This follows from the fact that an open-operator with respect to $P^{(m)}$ will have at least one inactive on its left (to excite to $Q^{(m)}$), and hence will act as an open operator on $P^{(n_v)}$.

(b) Similarly, an operator which is quasi-open on $P^{(m)}$ is also quasi-open on $P^{(n_v)}$.

(c) An operator which is closed with respect to $P^{(m)}$ is not necessarily closed with respect to $P^{(n_v)}$. In fact, an operator which is quasi-open with respect to $P^{(n_v)}$ may behave as a closed operator on $P^{(m)}$. But closed operator on $P^{(n_v)}$ is a closed operator on $P^{(m)}$.

(d) The open, quasi-open and closed classifications are done with respect to $P^{(n_v)}$, unless otherwise specified.

Thus, the conditions (4.3a,b) imply that L_{op} operators have still vanishing projections on $Q^{(m)}$ and $R^{(m)}$:

$$Q^{(m)} L_{op}^{[m]} P^{(m)} = Q^{(m)} \sum_{i=0}^{m} L_{op}^{(i)} P^{(m)} = 0 \tag{4.9a}$$

and also

$$R^{(m)} L^{[m]}_{q\text{-}op} P^{(m)} = R^{(m)} \sum_{i=0}^{m} L^{(i)}_{q\text{-}op} P^{(m)} = 0 \qquad (4.9b)$$

This hows that L is decoupled from the left with respect to $Q^{(m)}$ and $R^{(m)}$, and consequently Ω is indeed a valence-universal wave operator. However, within a particular m-valence model space, the matrix $P^{(m)} L P^{(m)}$ will generally have serveral vanishing matrix-elements. This happens when labels on L are such that it is closed on $P^{(m)}$, but is actually a quasi-open component of L, and vanishes by condition (4.3b). Similar conclusion holds good for the normal-ordered choice for Ω. Clearly, this sparser structure of $P^{(m)} L P^{(m)}$ is implied by the decoupling conditions (4.3a,b) which are more fundamental than (4.9a,b). For more detailed discussions, we refer to our forthcoming papers[12,20].

As an example, consider the IMS consisting of the functions $(1\sigma_g^2)$ and $(1\sigma_g, 2\sigma_g)$ for a two-valence H_2-problem. The quasi-open operators induce transition to $R^{(2)} \equiv (2\sigma_g^2)$, and are of the type $\langle 2\sigma_g^2 | O_2 | 1\sigma_g^2 \rangle \{E^{2\sigma_g^2}_{1\sigma_g^2}\}$, $\langle 2\sigma_g^2 | O_2 | 1\sigma_g 2\sigma_g \rangle \{E^{2\sigma_g^2}_{1\sigma_g 2\sigma_g}\}$ and $\langle 2\sigma_g | O_1 | 1\sigma_g \rangle \{E^{2\sigma_g}_{1\sigma_g}\}$ where E's are the respective operator parts. Thus the corresponding matrix-elements of L are zero. Consider now the one-valence IMS obtained from the above two-valence IMS. It has functions $1\sigma_g$ and $2\sigma_g$. (Incidentally, $P^{(1)}$ is always complete by construction). The matrix of $P^{(1)} L P^{(1)}$ has the following structure:

$$P^{(1)} L P^{(1)} \equiv \begin{bmatrix} \langle 1\sigma_g | L_1 | 1\sigma_g \rangle & \langle 1\sigma_g | L_1 | 2\sigma_g \rangle \\ 0 & \langle 2\sigma_g | L_1 | 2\sigma_g \rangle \end{bmatrix} \qquad (4.10)$$

which shows a partial factorization since, although $\langle 2\sigma_g | L_1 | 1\sigma_g \rangle$ is closed on $P^{(1)}$, it is quasi-open on $P^{(2)}$ and must thus vanish.

We note an importance difference between an IMS and a CMS. For an IMS, a choice of T as $[T_{op} + T_{q\text{-}op}]$ automatically leads to connected H_{eff} for an exponential choice of Ω[18]. For an normal ordered ansatz, care should be exercized to ensure that $P\Omega P = P$ is not used, as emphasized in ref. (19), Failure to do this leads to disconnected H_{eff}, as in refs. (14-17).

5. SOME SPECIAL IMS

The foregoing discussion leads to the realization that there is a dichotomy peculiar to IMS, viz the classification of the operators depends generally on the particular IMS chosen, although the concept of connectedness of T and H_{eff} follows from a Fock-space strategy which does not specify the number of valence fermions. There are, however, two special types of IMS where the classification of the operators does not depend on the number of valence fermions. We now discuss these two cases[20].

If we classify the active orbitals into various groups α, β, γ and collect all the determinants in which the occupancies n_α, n_β, n_γ ... within each group are fixed then we generate a quasi-complete model space (QMS). Each QMS is fully characterized by a vector $\vec{n}_I$ containing the occupancies. By taking a suitable (say, lexicographic) ordering convention, the various QMS' with different vectors $\vec{n}_I$ can be uniquely ordered. The closed operators in any QMS are those which preserve the occupancies within the groups i.e. which keep the vector n_I for any QMS unchanged. Those which transfer fermions between the groups are quasi-open.

Specifically, the quasi-open operators which move any QMS up in the lexicographic order may be called '+', and those which move it down may be called '-'. The quasi-open operators for QMS have vanishing model space projections. But products like +- or -+ will have generally closed operators. Thus, for QMS, L satisfies

$$L_{op}^{(k)} = 0; \; L_{+}^{(k)} = 0; \; L_{-}^{(k)} = 0, \forall_k \qquad (5.1a,b,c)$$

and a valence-universal Ω, having a T with $T_{op}^{(k)}$, $T_{+}^{(k)}$ and $T_{-}^{(k)}$, will yield a connected H_{eff} and IN is not satisfied since $T_{+}^{(k)}T_{-}^{(k)}$ etc. may be closed. We emphasize again that we need not specify in QMS the number of valence fermions.

A particularly useful QMS is a model space in which the active orbitals are both particles and holes; and we take the complete set of np-nh determinants for these active particles and holes.Fig. 3 depects our classification scheme for the orbitals. This consistutes a QMS with active particle and hole orbitals bunched in two groups. This type of choice for QMS is natural for calculating excitation energies (the most common situation is $\tilde{n}$=1). In this case, it is convenient to denote the particle-hole valence rank for the active orbitals as suffices; $A^{(k,l)}$, for example destroys k active particle and l active hole labels acting on a QMS. For n=1, the working equations for the normal ordered ansatz may be written down in terms of the following quantities

$$Z^{(k,l)} = \{\overline{H\Omega}\}^{(k,l)}; \; W^{(k,l)} = \{\overline{\Omega H_{eff}}\}^{(k,l)} \qquad (5.2a,b)$$

we have

$$Z_{op}^{(k,l)} = W_{op}^{(k,l)} \quad \forall \; k=0,1,; \quad l=0,1 \qquad (5.3a)$$

$$Z_{\pm}^{(k,l)} = W_{\pm}^{(k,l)} \quad \forall \; k=0,1,; \quad l = 0,1 \qquad (5.3b)$$

$$Z_{cl}^{(0,0)} = E_{cl}^{(0,0)}; \; Z_{cl}^{(0,1)} = H_{eff}^{(0,1)}; \; Z_{cl}^{(1,0)} = H_{eff}^{(1,0)};$$

$$Z_{cl}^{(1,1)} = H_{eff}^{(1,1)} \qquad (5.3c)$$

The last equality may appear surprizing in view of the fact that

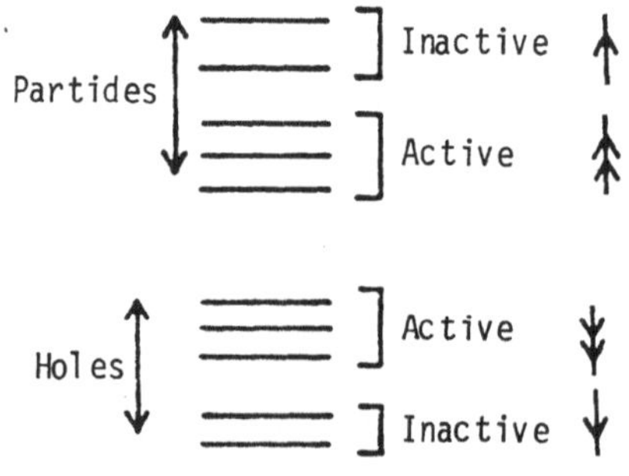

Fig. 3. Classification of the orbitals in a QMS calculation

$P^{(1,1)}\Omega P^{(1,1)} \neq P^{(1,1)}$, as depicted in Fig. 2 (c). But, let us note that $W_{cl}^{(1,1)} = \{\Omega_{cl} H_{eff}\}_{cl}$ and it is impossible to join the two T operators in Fig. 2 (c) with H_{eff}. As a result, the working equations are exactly the same as in a CMS. Haque and Kaldor[21] and Kaldor[22] calculated EE's of several atoms and molecules using Lindgren's version of the QMS[17(a)], which use IN for Ω. There will be disconnected diagrams in H_{eff} in that case, but they ignored them. Their working equations are precisely the same as ours, but our analysis shows them to be rigorously correct. For an extensive discussion, we refer to a paper by Sinha et al.[23] and for applications and further extension, we refer to two papers by Pal et al.[24].

An IMS which consists of the union of the first few or the last few of QMS' in lexocographic order is such that the quasi-open operators are either of '+' or of '-' variety. As a result, one has only T_+ or T_--operators in Ω, and thus one never encounters T_+T_- or T_-T_+ type products in the expansion for Ω. In this case IN is compatible with connectivity of T and H_{eff}. We call this type of model space as an <u>isolated incomplete</u> model space (IIMS). The classification of operators for an IIMS does not again need the specification of the number of valence occupancies, since the lower valence model spaces obtained from a particular IIMS will again be IIMS. (This may be verified as an exercise. Consider as example, the IIMS with vectors [4,2,1], [4,1,2], [4,0,3]. Assume these successive vectors in reverse lexicographic order. Show the lower valence model spaces are also successive vectors in reverse lexicographic order). As an example, we may consider and IIMS consisting of the ground state Hartree-Fock function $\Phi(P^{(0,0)})$ and a QMS of particle-hole determinants $(P^{(1,1)})$. Ω for this model space supports IN as a size-extensive normalization. This again provides IP, EA, EE and the energy of the ground state.

6. GENERATION OF MULTIPLE SOLUTIONS

We now point out an interesting structural aspect of the non-linear eqns. (5.3a,b,c) which allows us to generate multiple solutions from them. Owing to the normal ordering in Ω, only one $T^{(k,l)}$ at the most can contribute to $Z^{(k,l)}$. Similarly, only one $T^{(k,l)}$ can contribute to H_{eff} in $W^{(k,l)}$, and at the most two $T^{(k,l)}$'s can contribute to $W^{(k,l)}$ (one from Ω and one from H_{eff} in $\{\Omega H_{eff}\}^{(k,l)}$). By projecting eqns. (5.3a,b,c) onto spaces Q, R for an np-nh QMS, we get a set of simultaneous quadratic equations for the cluster amplitudes of $T^{(k,l)}$. Writing these amplitudes in a column t^{kl}, and decomposing $Z^{(k,l)}$ and $W^{(k,l)}$ into quantities where the dependence on t^{kl} is explicitly demonstrated, we have.

$$X_{21}^{kl} + X_{22}^{kl}\, t^{kl} = \Delta_{21}^{kl}\left[X_{11}^{kl} + X_{12}^{kl}\, t^{kl}\right] + t^{kl}\left[X_{11}^{kl} + X_{12}^{kl}\, t^{Xl}\right] \qquad (6.1)$$

The matrices X_{ij}^{kl} and Δ_{21}^{kl} are independent of t^{kl}, although they depend on vectors t^{pq} containing $T^{(p,q)}$ amplitudes of the lower valence ranks obtained earlier. The projection of eqn. (5.3c) onto P gives the matrix of the part of H_{eff} at the valence rank (k,l):

$$L_{11}^{kl} = X_{11}^{kl} + X_{12}^{kl}\, t^{kl} \qquad (6.2)$$

Let us now define the matrices d^{kl} and c^{kl} by

$$d^{kl} = t^{kl}\, c^{kl} \qquad (6.3)$$

where c^{kl} is chosen to bring L_{11}^{kl} in the diagonal form E_d^{kl}:

$$L_{11}^{kl}\, c^{kl} = c^{kl}\, E_d^{kl} \tag{6.4}$$

Defining the matrices $\bar{X}^{kl}$ by,

$$\bar{X}_{11}^{kl} = X_{11}^{kl}\ ;\ \bar{X}_{12}^{kl} = X_{12}^{kl};\ \bar{X}_{21}^{kl} = X_{21}^{kl} - \Delta_{21}^{kl}\, X_{11}$$

$$\bar{X}_{22}^{kl} = X_{22}^{kl} - \Delta_{21}^{kl}\, X_{12}^{kl}\ , \tag{6.5}$$

we find that

$$\begin{bmatrix} \bar{X}_{11}^{kl} & \bar{X}_{12}^{kl} \\ \bar{X}_{21}^{kl} & \bar{X}_{22}^{kl} \end{bmatrix} \begin{bmatrix} c^{kl} \\ d^{kl} \end{bmatrix} = \begin{bmatrix} c^{kl} \\ d^{kl} \end{bmatrix} E_d^{kl} \tag{6.6}$$

which is a set of eigenvalue equations determining n_{11}^{kl} roots E_d^{kl}, where n_{11}^{kl} is the dimension of L_{11}^{kl}. c^{kl} has the same dimension, and d^{kl} is of dimension ($n_t^{kl} X n_{11}^{kl}$), where n_t^{kl} is the number of entries in t^{kl}. Clearly, depending on the starting iterates (c_0^{kl}, d_0^{kl}), the corresponding converged vectors (c^{kl}, d^{kl}) from eqn. (6.6) will generate different sets of roots E_d^{kl}. We can exploit this, with suitable root-seeking and root-homing procedures[25], to generate various different sets of t^{kl} from eqn. (6.3) corresponding to the different solution vectors (c^{kl}, d^{kl}). In sec. 7, we have shown how this can be exploited to generate main as well satellite peaks for ionization and Auger spectrum. This also throws light on the number of distinct roots obtainable, which cannot exceed($n_t^{kl}+n_{11}^{kl}$).

Let us also note that there are direct diagonalization strategies[25] where the large matrix $\bar{X}^{kl}$ in eqn. (6.6) need not be stored, so that expanding the dimension does not lead to any storage problem.

7. SAMPLE APPLICATIONS

We report here some sample numerical results on electronic structure studies, obtained by us and other collaborating groups. Excited states, IP, EE and Auger kinetic energies (related to double IP's) are presented. For excited state calculations, QMS and IIMS have been utilized. For EE calculations a QMS of p-h determinants has been used. IP results appear as by-products, since once $T^{(0,1)}$ amplitudes are determined, we can construct $H_{eff}^{(0,1)}$ which furnishes the IP's on diagonalization. Also, if we neglect $T^{(0,2)}$ amplitudes in a two-hole model space sitation for double ionization studies, then the same $T^{(0,1)}$ amplitudes may be used to construct $H_{eff}^{(0,2)}$ which furnishes double IP's on diagonalization. The kinetic energies of the Auger electrons are then obtainable from

$$K.E. = IP_{core} - DIP \tag{7.1}$$

Where IP_{core} is the core ionization potential. We have always used the normal ordered ansatz for Ω, and have approximated T as

$$T \simeq T_2^{(0,0)} + T_2^{(0,1)} + T_2^{(1,0)} \tag{7.2}$$

For IP calculations, we have used the eqns. (6.6) for k = 0 and l=1. The main IP results are obtained for those starting itera tes where the vector (c^{01}, d^{01}) are dominated by c^{01}. For the shake-up roots, d^{01} components are rather large, and a prior diagonalization of a part of the matrix $\bar{X}$ which contains the dominant configurations has been found necessary to get the good guesses for such a situation. All these different solution -sets generate different t^{01} vectors, and we get different L_{11}^{01} which gives main as well as satellite IP's. These t^{01} vec-

Table 1. The starred results indicate satellite peaks.

System	State	Energy	Other theor. results	Expt. results
	2'S(QMS)	-2.1458 au		-2.1458
He	1'S (IIMS)	-2.9019		-2.9037
	2'S (IIMS)	-2.1458		-2.1458
	IP		(ref. 27)	(ref. 28)
	$3\sigma_g^{-1}$	14.74 eV	15.89	15.6
	$2\sigma_g^{-1}$	31.73		29.4
		39.93*		38.0
		42.27*		
		43.20*		
	$1\pi_u^{-1}$	17.41	16.68	17.0
	$2\sigma_u^{-1}$	17.90	19.66	18.8
		27.68*		25.2
		38.18*		
N_2	EE			
	$B\,^3\Pi_g$	7.8 eV	9.6(ref.26)	8.1(ref.29)
	$a\,'\Pi_g$	8.7	11.5	9.3
	$A\,^3\Sigma_u^+$	7.7	8.4	7.8
	$a''\Sigma_u^-$	11.0	11.3	9.9
	$w\,'\Delta_u$	10.7	12.0	10.3
	$W\,^3\Delta_u$	9.3	10.1	8.9
	IP			
	$'\Pi^{-1}$	15.08 eV	14.28(ref.30)	16.1(ref. 31)
	$3\sigma^{-1}$	19.43	18.29	19.9
	$2\sigma^{-1}$	39.65	37.99	39.7
		42.78*	40.97*	hump around
		44.73*	43.00*	44-47
HF	Auger K.E.			
	$^3\Sigma^-$	647.69 eV	(ref.32)	(ref.33)
	$'\Delta$	644.28		644.29
	$^3\Pi$	644.53	643.81	
	$^1\Sigma^+$	643.76	642.64	642.36
	$^1\Pi$	641.14	640.90	
	$^1\Sigma^+$	636.49	635.89	636.92
	$^3\Pi$	624.86	624.60	625.10
	$^3\Pi$	622.23*	619.36	
	$^3\Sigma^+$	621.42	621.17	
	$^3\Sigma^+$	618.71	615.57	

tors also give different $H_{eff}^{(0,2)}$ for double IP's and hence both main and satellite Auger kinetic energies.

(a) He-atom 2'S excited state: We have used two IMS' for the calculations. One is the QMS; (1s2s) and the other is IIMS ($1s^2$,1s2s). A gaussian (13s, 6p,2d) basis has been used[26]. The orbitals are calculated using a scaled Silverstone-Yin potential.

(b) IP and EE of N_2: Calculations are done for the equilibrium distance of R_{eq} = 2.0693 a.u., using a Huzinaga-Dunning (9s5p→4s2p/3s) basis. Both the main and the satellite IP's are computed, using eqn. (6.6). All the holes and particles are taken as active.

(c) IP and Auger energies of HF: A Huzinaga-Dunning (9s5p→ 5s3p/3s) basis is used, with R_{eq} = 1.7325 a.u.. Approximation eqn. (7.2), and the formula eqn. (7.1) have been utilized to get Auger energies. IP_{core} for HF is taken to be 693.55 eV. t_{01} amplitudes for both the main and the satellite IP-peaks have been used to generate main and satellite Auger energies.

All the results are diplayed in Table 1. The agreement with experimental values, and other theoretical results - when available - is very good. This lends credence to the confidence that open-shell CC formalisms with general model spaces can provide a potentially powerful approch to electronic structure calculations. Kaldor and his group[21,22] have also made several applications using QMS - with the proviso of an approximation, using an IN for Ω, as explained in sec. 5.

8. ACKNOWLEDGEMENTS

The financial support of the DST (India) and of the Humboldt-Foundation (Germany) is gratefully acknowledged. Thanks are due to Professor W. Kutzelnigg for providing warm hospitality in the University of Bochum, where a part of the paper was written. Special thanks go to Ms. U. Krupinski for her help in preparing the typescript.

REFERENCES

1. B. Brandow, Int. J. Quantum Chem. 15, 207 (1979) for pertinent references
2. B. Brandow, Rev. Mod. Phys. 39, 771 (1967)
3. N.M. Hugenholtz, Physica 23, 481 (1957)
   H. Primas, in Modern Quantum Chemsitry, Vol. II (Ed.: O. Sinanoglu), Acad. Press, 1965
   R.J. Bartlett and G.D. Purvis, Int. J. Quantum Chem. 14, 561 (1978)
4. F. Coester, Nucl. Phys. 1, 421 (1958)
   F. Coester and H. Kümmel, Nucl. Phys. 17, 477 (1960). H. Kümmel, Nucl. Phys. 22, 177 (1969)
   H. Kümmel, K.H. Lührmann and J.G. Zabolitzky, Phys. Rep. 36c, 1 (1978)
5. J. Cizek. J. Chem. Phys. 45, 4256 (1966);
   Adv. Chem. Phys. 14, 35 (1969)
   J. Paldus, J. Cizek and I. Shavitt, Phys. Rev. A5, 50 (1972)
6. See, e.g. (a) R. J. Bartlett, Ann. Rev. Phys. Chem. 32, 359 (1981) and (b) J. Paldus in New Horizons Of Quantum Chemistry (Ed.: P. Löwdin and B. Pullman), Reidel, 1983 for extensive surveys.
7. D. Mukherjee, R.K. Moitra and A. Mukhopadhyay, Mol. Phys. 33, 955 (1977).
   D. Mukherjee, Pramana 12, 203 (1979)

8. R. Offermann, W. Ey and H. Kümmel, Nucl. Phys. A273, 349 (1976)
R. Offermann, Nucl. Phys. A273, 368 (1976). W. Ey, Nucl. Phys. A296, 189 (1976)
9. See, e.g. C. Bloch, Nucl. Phys. 6, 329 (1958) for an early exposition
10. I. Lindgren, Int. J. Quantum Chem. S12, 33 (1978)
11. M.A. Haque and D. Mukherjee, J. Chem. Phys. 80, 5058 (1984)
12. I. Lindgren and D. Mukherjee, Phys. Rep., 151 , 93 (1987)
13. T.H. Schucan and H.A. Weidenmüller, Ann. Phys. 73, 108 (1972); 76, 483 (1973)
14. G. Hose and U. Kaldor, J. Phys. B 12, 3827 (1979); Physica Scripta 21, 357 (1980)
15. B. Jeziorski and H.J. Monkhorst, Phys. Rev. A24, 1668 (1981)
16. B. Brandow in New Horizons Of Quantum Chemistry (Ed.: P. Löwdin and B. Pullman), Reidel, 1983
17. (a) I. Lindgren, Physica Scripta 32, 291 (1985)
(b) I. Lindgren, Physica Scripta 32, 611 (1985)
18. D. Mukherjee, Proc. Ind. Acad. Sci. 96, 145 (1986)
19. D. Mukherjee, Chem. Phys. Lett. 125, 207 (1986); Int. J. Quantum Chem. S20, 409 (1986).
20. W. Kutzelnigg, D. Mukherjee and S. Koch, submitted to J. Chem. Phys.
D. Mukherjee, W. Kutzelnigg and S. Koch, submitted to J. Chem. Phys.
21. M. Haque and U. Kaldor, Chem. Phys. Lett 128, 45 (1986)
U. Kaldor and M. Haque, Int. J. Quantum Chem. 29, 425 (1986)
22. U. Kaldor, Int. J. Quantum Chem. S20, 445 (1987)
U. Kaldor, submitted to J. Chem. Phys. (preprint)
23. D. Sinha, S. Mukhopadhyay and D. Mukherjee, Chem. Phys. Lett. 129, 369 (1986)
24. S. Pal, M. Rittby, R.J. Bartlett, D. Sinha and D. Mukherjee, Chem. Phys. Lett. 137, 273 (1987); J. Chem. Phys., submitted
25. K. Hirao and H. Nakatsuji, J. Comput. Phys. 45, 246 (1982)
26. S. Koch and D. Mukherjee, to be published
27. L.S. Cederbaum and W. Domcke, Adv. Chem. Phys. 36, 205 (1977)
28. S. Krummacher, V. Schmidt and F. Wuillenmier, J. Phys. B13, 2993 (1980)
29. C.W. McCurdy, T.H. Rscigno, D.L. Yager and V. McKoy, in Modern Theoretical Chemistry, Vol. 3 (Ed.: H.F. Schaefer), Plenum, 1977
30. W. von Niessen, L.S. Cederbaum, W. Domcke and G.H.F. Diercksen, Chem. Phys. 56, 43 (1981)
31. C.E. Brion, I.E. McCarthy, I.H. Suzuki and E. Wiegold, Chem. Phys. Lett. 67, 115 (1979)
32. O.M. Kvalheim and K. Faegri, Chem. Phys. Lett. 67, 127 (1979)
33. See e.g. C.M. Liegener, Chem. Phys. Lett. 90, 188 (1982)
34. K. Emrich, Nucl. Phys. A351, 379, 397 (1981)
35. J. Arponen, Ann. Phys. 151, 311(1983)
J. Arponen and J. Rantakivi, Nucl. Phys. A407, 141 (1983)

OPEN-SHELL COUPLED-CLUSTER STUDIES OF ATOMIC AND MOLECULAR SYSTEMS*

Uzi Kaldor

School of Chemistry
Tel Aviv University
69 978 Tel Aviv, Israel

INTRODUCTION

The exp(S) or coupled-cluster (CC) method[1-5] has been used widely in recent years for ab initio electronic structure calculations in closed-shell, non-degenerate systems, with highly satisfactory results.[6] The CCSD approximation,[7] in which single and double excitations are included to all orders, is usually employed; a few calculations including the effect of triple excitations (CCSDT) have appeared recently.[8] The theory becomes considerably more complicated when the system of interest cannot be described in terms of a closed-shell structure. A variety of multi-reference, open-shell (OSCC) formulations, designed to handle such situations, have been described.[9-23]

Recently we reported[24-31] the application of OSCC to the direct calculation of electron affinities (EA), ionization potentials (IP) and excitation energies (EE). These applications are summarized below, and new results are described. The method used largely follows Lindgren's normal-ordered formalism.[14] Single and double virtual excitations are included to all orders, while triple excitations with one or more electrons excited out of the valence shell are calculated to lowest order, using the converged CCSD amplitudes (the CCSD+T approximation[25]). Excitations of three closed-shell electrons do not contribute to electronic transition energies in this approximation, and are therefore ignored.

*Supported in part by the U.S.-Israel Binational Science Foundation.

METHOD

The Hamiltonian of the system is separated in the usual way into a zero-order operator H_0, with known eigenfunctions, and a perturbation V,

$$H = H_0 + V \tag{1}$$

$$H_0|\alpha\rangle = E_0^{\alpha}|\alpha\rangle \ . \tag{2}$$

A d-dimensional model space P and its complement Q are defined by projection operators,

$$P = \sum_{\alpha \in P} |\alpha\rangle\langle\alpha| \ , \qquad Q = 1 - P \ . \tag{3}$$

There will usually be d eigenfunctions of H with major components in the model space,

$$H\Psi^a = E^a\Psi^a \ , \tag{4}$$

$$P\Psi^a = \Psi_0^a \ , \quad a=1,2,...,d \tag{5}$$

where Ψ_0^a are linear combinations of $|\alpha\rangle$ $\alpha \in P$. The wave operator Ω transforms the model functions into exact ones,

$$\Omega\Psi_0^a = \Psi^a \ , \ a=1,2,...,d \ . \tag{6}$$

Intermediate normalization is assumed,

$$\langle\Psi^a|\Psi_0^a\rangle = \langle\Psi_0^a|\Psi_0^a\rangle = 1 \ . \tag{7}$$

The key equation in Lindgren's derivation[14] is the generalized Bloch equation

$$[\Omega,H_0]P = V\Omega P - \Omega PWP, \tag{8}$$

where W is the effective interaction

$$W = V\Omega \ . \tag{9}$$

An alternative form of (8) is

$$[\chi,H_0]P = QWP - \chi PWP \ , \tag{10}$$

where the correlation operator χ is defined by

$$\Omega = 1 + \chi \, . \tag{11}$$

The energies of interest are obtained by diagonalizing the effective Hamiltonian in the model space,

$$H_{eff}\Psi_0{}^a = E^a\Psi_0{}^a \, , \tag{12}$$

where

$$H_{eff} = PH\Omega P = P(H_0+W)P \, . \tag{13}$$

The correlation operator χ includes single, double, ... , virtual excitations and may be written as

$$\chi = C_1 + C_2 + \ldots = \sum_{ij} \{a_i^\dagger a_j\}\, t_j^i + \frac{1}{2}\sum_{ijkl} \{a_i^\dagger a_j^\dagger a_l a_k\}\, t_{kl}^{ij} + \ldots \tag{14}$$

t_j^i , t_{kl}^{ij} , ... , are excitation amplitudes, and the curly brackets denote normal order with respect to a reference (core) determinant. All terms, connected as well as disconnected, are included in (14). The operator used in CC is the excitation operator T, related to Ω by

$$\Omega = \{\exp(T)\} = 1 + T + \tfrac{1}{2}\{T^2\} + \ldots \quad . \tag{15}$$

T is given by summing the rhs of (14) over connected terms only. Perturbative or non-perturbative schemes for calculating the excitation operator and correlation energies may be derived from either of the following two equations, which include connected terms only[14]

$$[T,H_0] = (QV\Omega - \chi PV\Omega)_{conn} \, , \tag{16}$$

or

$$[T,H_0] = W_{op,conn} - (\chi W_{cl})_{conn} \, . \tag{17}$$

$W_{op,conn}$ describes all connected diagrams which have some open (non-valence) lines, corresponding to P→Q transitions. W_{cl} diagrams, with no external non-valence lines, describe P→P transitions. The latter also appear in the effective Hamiltonian, which may be written as

$$H_{eff} = PH_0P + W_{cl} \, . \tag{18}$$

The second term in equation (16) or (17) gives rise to the so-called folded diagrams.

The T operator for an open-shell system may be partitioned according to the number of valence orbitals excited,

$$T = T^{(0)} + T^{(1)} + T^{(2)} + \ldots \quad . \tag{19}$$

Haque and Mukherjee[21] have shown that partial decoupling of the equations is then possible, as the equations for $T^{(n)}$ involve only $T^{(m)}$ elements with $m \leq n$. This decoupling is helpful in reducing the computational effort required, and has been used in our calculations.[24-31]

The H_{eff} or W_{cl} diagrams (see Eq. (18)) may be separated into core and valence parts,

$$H_{eff} = H_{eff}^{core} + H_{eff}^{val} , \tag{20}$$

where the first term on the rhs consists of diagrams without any external lines. The eigenvalues of H_{eff}^{val} will then give directly the transition energies from the core, with correlation effects included for both core and valence electrons. The physical significance of these energies depends on the nature of the model space. Thus, electron affinities may be calculated by constructing a model space with valence particles only,[24,25,29] ionization potentials are given using valence holes,[26] and both types are included for the purpose of getting excitations out of a closed-shell system.[27,28,30] The expansion (19) is replaced in the latter case by

$$T = T^{(0,0)} + T^{(1,0)} + T^{(0,1)} + T^{(1,1)} + \dots \quad , \tag{21}$$

where the two superscripts give the numbers of valence holes and particles, respectively. The same decoupling holds as for Eq. (19).

Model Spaces

Model (P) spaces in OSCC or in many-body perturbation theory (MBPT) are usually constructed by first partitioning the orbitals into core, valence and particles, and including in P determinants with all possible combinations of valence orbitals (core orbitals are always occupied). This choice, called[32] a "complete" model space, is assumed in the well-known derivations of linked-diagram theorems.[33,14] The first MBPT method capable of handling more general model spaces was described and applied by Hose and Kaldor,[32,34] and a similar OSCC method was later given by Jeziorski and Monkhorst.[18] A complete model space presents a reasonable choice when all valence orbitals are close in energy and belong to the same type (holes or particles). Molecular excited states will usually involve both valence holes and particles with respect to a closed-shell reference determinant; complete model spaces would then be very large, with determinants describing multiple ionizations and additions of electrons. Such a model space will have a very broad energy span and involve many types of T elements, making the calculation unwieldy at best and unlikely to converge in most cases, due to the appearance of intruder states.[35] A reasonable model space for n-excited states would include only determinants with exactly n valence holes and n valence particles. This so-called "quasicomplete" space, proposed by Lindgren,[15] will lead in general to unlinked diagrams.[15,36,37] It has however been shown[37] that the usual coupled-cluster expansion remains linked for a quasicomplete space with one valence hole and one valence particle. Such spaces are used for singly-excited atomic and molecular states.[27,28,30] A more general method, applicable to any model spaces, has been derived recently by Mukherjee.[36] A key step in Mukherjee's derivation is abandoning the intermediate normalization (7). An arbitrary model space with m valence orbitals is defined, and operators of valence rank k ($0 \le k \le m$) are called m-open if they can lead to $P \to Q$ transitions in the m-valence sector. The resulting equations for Ω defined by (15) are[37]

$$[T, H_0]^{(k)}_{m\text{-op}} = \{V\Omega - \Omega W\}^{(k)}_{m\text{-op}} \tag{22}$$

$$\{\Omega W\}^{(k)}_{cl} = \{V\Omega\}^{(k)}_{cl} \, . \tag{23}$$

Equation (22) is identical with (16), except for the different structure of the model space. Equation (23) is somewhat more complicated than its complete-model-space counterpart (9), leading to coupled equations for the elements of the effective interaction W. These equations have low dimensionality and are simpler than the equations for the T elements, (16) or (22), and the added complications are not severe. An application of the method to several states of LiH is described below. A slightly different scheme has been recently applied to H_2 by Koch and Mukherjee.[38]

DIAGRAMS

The application of the coupled-cluster method is most easily described in terms of Goldstone-type diagrams. The atomic and molecular computations described below employ the CCSD approximation, including single and double excitations,

$$T \simeq T_1 + T_2 \, . \tag{24}$$

T_3 is calculated approximately[25] in some cases. T elements with no more than two valence lines (hole or particle) are calculated. This requires diagrams for $T_k^{(n,m)}$, with $k=1,2$ and $0 \le n+m \le 2$, where n and m are the numbers of incoming hole- and particle-valence lines, respectively, and k is the total number of incoming lines. The rules for drawing and calculating the diagrams are well-known (see e.g. Lindgren and Morrison,[14] Chap. 15). A modification of the program generating many-body perturbation theory diagrams[39] can be used. The Ω expansion under the approximation (24) is finite,

$$\Omega = 1 + T_1 + \tfrac{1}{2}\{T_1{}^2\} + T_2 + \tfrac{1}{3!}\{T_1{}^3\} + \{T_1 T_2\} + \tfrac{1}{4!}\{T_1{}^4\} + \tfrac{1}{2}\{T_1{}^2\, T_2\} + \tfrac{1}{2}\{T_2{}^2\} \, . \tag{25}$$

The closed-shell diagrams (n=m=0, k=1,2) have been listed by Cullen and Zerner.[7] More recent publications include full descriptions of one- and two-body diagrams with a single valence particle[29] (n=0, m=1, k=1,2), from which one-valence-hole diagrams are easily obtained by reversing all line directions, and of $T_k^{(1,1)}$ diagrams, with k=1,2.[30] Lowest-order $T_3^{(0,1)}$ diagrams have also been listed.[25]

APPLICATIONS

Several applications implementing the methods described above are presented in this section.

Electron Affinities and Ionization Potentials of the Alkali Atoms

A preliminary study[29] of Li and Na, with small basis sets, gave very good electron affinities (errors of 0.015eV or 3%). It should be noted that the Hartree-Fock EA's are negative (the energy of the anion is higher than that of the neutral atom), and correlation is crucial in getting correct values. The HF ionization potentials, on the other hand, are in fair agreement with experiment (the error is 0.2eV for Na), and very little improvement was obtained in the small-basis CCSD calculation.[29] Previous studies[40] using the configuration interaction (CI) method found that important contributions to the IP come from the core-valence correlation, i.e. the correlation between the ns and (n-1)p electrons. A serious problem encountered in the CI work is that inclusion of core-core correlation, in addition to core-valence, gives much poorer IP's. This effect results from the size-inconsistency of the CI method,[41] manifested by the appearance of unlinked terms in the expansion.

The CCSD method was applied to the alkali atoms using moderately large bases, from a set of 5 s, 5 p and one d contracted Gaussian-type orbitals for Li, to 11 s, 9 p and 5 d orbitals for Cs.[31] The M^+ ion served as the reference closed-shell system. Two sets of calculations were performed, correlating the ns electrons with or without the (n-1)sp shell. The main contribution to the electron affinities comes from the $(ns)^2$ correlation, which by itself brings the results to within 0.02eV of experiment. The inclusion of (n-1)s and p correlation has small effect (<0.03eV) and gives some improvement, cutting the error to 0.01eV. The correlated ^2S ionization potentials are within 0.02eV (Li) to 0.22eV (Cs) of experiment; when relativistic corrections, estimated from numerical Hartree-Fock and Dirac-Fock calculations, are added, the agreement with experiment is better than 0.1eV for all atoms. The ^2P ionization potentials are even better, with errors of 0.02-0.04eV.[31] It should be noted that the inclusion of both core-core and core-valence correlation causes no problems.

Excitation Energies of Closed-Shell Systems

Excitation energies and ionization potentials were calculated for the closed-shell atoms Be,[27] Ne,[27] Mg[28] and Ar,[28] and for the water molecule.[30] About 10 transition energies of each system were studied, and T_3 effects were included to lowest order.[25] All the results were within 0.1-0.2eV of experiment. This includes rather sensitive levels, such as the lowest D states of Mg, which appeared in reverse order (singlet below triplet) as they should. T_3 effects proved very important for Ne and isoelectronic H_2O, contributing 0.2-0.4eV to excitation energies and 0.5-0.7eV to ionization potentials. Be and Mg, essentially two-electron systems, showed as expected small effects of triple excitations, up to 0.15eV. Somewhat surprising were the small ($\leq$0.1eV) T_3 corrections for Ar. A possible explanation is the greater spatial extension of the Ar valence shell, as compared with Ne. Another interesting result involves the use of Hartree-Fock orbitals for the singly-ionized system rather than the neutral atom or molecule. Contrary to expectations, such orbitals gave poorer CCSD results; these were largely offset by bigger T_3 corrections, making the final CCSD+T values of the two orbital sets very close.

Incomplete Model Spaces: LiH

Convergence problems appear very often in multi-reference many-body perturbation theory (MBPT) and coupled-cluster calculations. They usually result from the use of "complete"[32] model spaces, comprising determinants constructed from all possible combinations of valence orbitals. The LiH molecule is a case in point. We tried to construct potential functions for the ground $X^1\Sigma^+$ state, with the configuration $1\sigma^2 2\sigma^2$, and the four lowest excited states, the $A^1\Sigma^+$ and $a^3\Sigma^+$ obtained by the $2\sigma \rightarrow 3\sigma$ transition and the $B^1\Pi$ and $b^3\Pi$ resulting from $2\sigma \rightarrow 1\pi$. Starting from the LiH^{++} determinant and adding the 2σ, 3σ and 1π orbitals as valence particles, the CCSD iterations could not be converged. The Hartree-Fock orbital energies are (in Hartree atomic units) -0.82, -0.39, -0.38, -0.30, and -0.26 for the 2σ, 1π, 3σ, 4σ and 5σ orbitals, respectively. A complete model space including the determinants $2\sigma^2$ and $3\sigma^2$, among others, will obviously have many Q-space determinants (e.g. $2\sigma 4\sigma$ and $2\sigma 5\sigma$) with energies inside the P-space range, a recipe for divergence. Freed and his coworkers[42] solved the problem within the MBPT framework by assigning equal energies to all valence orbitals, and including these energy shifts in the perturbation. The additional perturbation shows up in higher order only, and good results have been obtained in this way for a variety of atoms and molecules by second and third-order MBPT.[42] This stratagem cannot be used in CC work, as the infinite-order summation inherent in the method will bring the orbital energies back to their original values. Divergence can be avoided only by using an incomplete P space without the $1\sigma^2 3\sigma^2$ and $1\sigma^2 1\pi^2$ determinants, which are anyhow not expected to make major contributions to the molecular states of interest.

The concept of incomplete model spaces was first proposed by Hose and Kaldor,[32] who derived a general-model-space MBPT. Both the Hose-Kaldor formalism and the Jeziorski-Monkhorst[18] CC method include disconnected diagrams, which however do not spoil the size-consistence of the method, as no individual part of the disconnected diagram is a legitimate term by itself. Mukherjee[36] has recently described an incomplete-model-space CC scheme without disconnected diagrams, and this method is applied to LiH, with the P space including the determinants $1\sigma^2 2\sigma^2$, $1\sigma^2 2\sigma 3\sigma$, and $1\sigma^2 2\sigma 1\pi$. Preliminary calculations using a moderate-size basis give good excitation and dissociation energies. The T_e values (energy of the potential minimum relative to the ground-state minimum) are 3.25eV for the $A^1\Sigma^+$ state and 4.30eV for the $B^1\Pi$ state, compared with the experimental 3.29 and 4.33eV, respectively. The calculated dissociation energies are 2.42eV for the ground state, 1.05eV for the $A^1\Sigma^+$ state, and 0.01eV for the $B^1\Pi$ state. The corresponding experimental values are 2.52, 1.08, and 0.03eV. Computations using a larger basis set, particularly in the π sector, are now under way, and will be reported in due course.[43]

SUMMARY AND CONCLUSION

The open-shell coupled cluster method has been applied to the direct calculation of electronic transition energies, including excitation energies, electron affinities and ionization potentials, of several atomic and molecular systems. Virtual single and double excitations are summed to infinite order (the CCSD approximation); triple excitations are included in the lowest order they appear (CCSD+T) when there are more than two electrons in the valence shell. Good agreement with experiment is achieved. The errors in the calculated results are 0.01eV for the alkali atoms electron affinities, 0.02-0.1eV for their ionization potentials (when corrected for relativity), and 0.1-0.2eV for the excitation energies of Ne, Mg, Ar and H_2O. These errors are due to deficiencies in the basis sets and to approximations made in the excitation operator T. Of the latter, we believe the inclusion of T_3 to lowest-order only is more serious than the neglect of connected four- and higher-electron excitations.

REFERENCES

1. J. Hubbard, Proc. Roy. Soc. A240:539 (1957); ibid. A243:336 (1958).
2. F. Coester, Nucl. Phys. 7:421 (1958); F. Coester and H. Kümmel, Nucl. Phys. 17:477 (1960); H. Kümmel, K. H. Lührmann and J. G. Zabolitzky, Phys. Rept. 36:1 (1978).
3. O. Sinanoglu, Adv. Chem. Phys. 6:315 (1964).
4. J. Cizek, J. Chem. Phys. 45:4256 (1966); Adv. Chem. Phys. 14:35 (1969).
5. J. Paldus, J. Cizek and I. Shavitt, Phys. Rev. A 5:50 (1972); J. Paldus, J. Chem. Phys. 67:303 (1977); B. G. Adams and J. Paldus, Phys. Rev. A 20:1 (1979).
6. For a review see R. J. Bartlett, Ann. Rev. Phys. Chem. 32:359 (1981).
7. G. D. Purvis and R. J. Bartlett, J. Chem. Phys. 76:1910 (1982); J. M. Cullen and M. C. Zerner, J. Chem. Phys. 77:4088 (1982). Both references give the explicit CCSD equations, and the second one also shows the CCSD diagrams.
8. Y. S. Lee and R. J. Bartlett, J. Chem. Phys. 80:4371 (1984); Y. S. Lee, S. A.

Kucharsky, and R. J. Bartlett, J. Chem. Phys. 81:5906 (1984); M. Urban, J. Noga, S. J. Cole, and R. J. Bartlett, J. Chem. Phys. 83:4041 (1985); J. Noga and R. J. Bartlett, J. Chem. Phys. 86:7041 (1987).
9. F. E. Harris, Intern. J. Quantum Chem. S11:403 (1977).
10. H. J. Monkhorst, Intern. J. Quantum Chem. S11:421 (1977).
11. J. Paldus, J. Cizek, M. Saute and A. Laforgue, Phys. Rev. A 17:805 (1978); M. Saute, J. Paldus and J. Cizek, Intern. J. Quantum Chem. 15:463 (1979).
12. D. Mukherjee, R. K. Moitra and A. Mukhopadhyay, Pramana 4:247 (1975); Mol. Phys. 30:1861 (1975); A. Mukhopadhyay, R. K. Moitra and D. Mukherjee, J. Phys. B 12:1 (1979); D. Mukherjee and P. K. Mukherjee, Chem. Phys. 39:325 (1979); S. S. Adnan, S. Bhattacharyya and D. Mukherjee, Mol. Phys. 39:519 (1980); Chem. Phys. Lett. 85:204 (1981).
13. R. Offerman, W. Ey and H. Kümmel, Nucl. Phys. A273:349 (1976); R. Offerman, Nucl. Phys. A273:368 (1976); W. Ey, Nucl. Phys. A296:189 (1978).
14. I. Lindgren, Intern. J. Quantum Chem. S12:33 (1978); S. Salomonson, I. Lindgren and A. M. Martensson, Phys. Scr. 21:351 (1980); I. Lindgren and J. Morrison, "Atomic Many-Body Theory", Springer, Berlin, (1982).
15. I. Lindgren, Phys. Scr. 32:291, 32:611 (1985).
16. H. Nakatsuji, Chem. Phys. Lett. 59:362 (1978); ibid. 67:329 (1979); Chem. Phys. 75:425 (1983); ibid. 76:283 (1983); J. Chem. Phys. 80:3703 (1984).
17. H. Reitz and W. Kutzelnigg, Chem. Phys. Lett. 66:111 (1979); W. Kutzelnigg, J. Chem. Phys. 77:3081 (1981): ibid. 80:822 (1984).
18. B. Jeziorski and H. J. Monkhorst, Phys. Rev. A 24:1668 (1981); L. Z. Stolarczyk and H. J. Monkhorst, Phys. Rev. A 32:725, 32:743 (1985).
19. A. Banerjee and J. Simons, Intern. J. Quantum Chem. 19:207 (1981).
20. V. Kvasnicka, Chem. Phys. Lett. 79:89 (1981).
21. A. Haque and D. Mukherjee, J. Chem. Phys. 80:5058 (1984); Pramana 23:651 (1984).
22. J. Arponen, Ann. Phys. (NY) 151:311 (1983).
23. K. Tanaka and H. Terashima, Chem. Phys. Lett. 106:558 (1984).
24. A. Haque and U. Kaldor, Chem. Phys. Lett. 117:347 (1985).
25. A. Haque and U. Kaldor, Chem. Phys. Lett. 120:261 (1985).
26. A. Haque and U. Kaldor, Intern. J. Quantum Chem. 29:425 (1986).
27. U. Kaldor and A. Haque, Chem. Phys. Lett. 128:45 (1986).
28. U. Kaldor, Intern. J. Quantum Chem. S20:445 (1986).
29. U. Kaldor, J. Comput. Chem., in press.
30. U. Kaldor, J. Chem. Phys., in press.
31. U. Kaldor, submitted to J. Chem. Phys.
32. G. Hose and U. Kaldor, J. Phys. B 12:3827 (1979).
33. B. H. Brandow, Rev. Mod. Phys. 39:771 (1967).
34. G. Hose and U. Kaldor, Phys. Scr. 21:357 (1980); Chem. Phys. 63:165 (1981); J. Phys. Chem. 86:2133 (1982); Phys. Rev. A 30:2932 (1984); U. Kaldor, J. Chem. Phys. 81:2406 (1984).
35. T. H. Schucan and H. A. Weidenmuller, Ann. Phys. (NY) 73:108 (1972); ibid. 76:483 (1973).

36. D. Mukherjee, Chem. Phys. Lett. 125:207 (1986); Intern. J. Quantum Chem. S20:409 (1986).
37. D. Sinha, S. Mukhopadhyay, and D. Mukherjee, Chem. Phys. Lett. 129:369 (1986).
38. S. Koch and D. Mukherjee, preprint (1987).
39. U. Kaldor, J. Comput. Phys. 20:432 (1976).
40. G. H. Jeung, J. P. Daudey, and J. P. Malrieu, J. Chem. Phys. 77:3571 (1982); Chem. Phys. Lett. 94:300 (1983); J. Phys. B 16:699 (1983); S. P. Walch, C. W. Bauschlicher, P. E. M. Siegbahn, and H. Partridge, Chem. Phys. Lett. 92:54 (1982); H. Partridge, D. A. Dixon, S. P. Walch, C. W. Bauschlicher, and J. L. Gole, J. Chem. Phys. 79:1859 (1983); H. Partridge, C. W. Bauschlicher, S. P. Walch, and B. Liu, J. Chem. Phys. 79:1866 (1983); W. Müller, J. Flesch, and W. Meyer, J. Chem. Phys. 80:3297 (1984).
41. H. Primas, in "Modern Quantum Chemistry", O. Sinanoglu, ed., Academic, New-York (1965), Vol 2.
42. S. Iwata and K. F. Freed, J. Chem. Phys. 61:1500 (1974); K. F. Freed, in "Modern Thoretical Chemistry", G. A. Segal, ed., Plenum, New York (1977); H. Sung, K. F. Freed, M. F. Herman, and D. L. Yeager, J. Chem. Phys. 72:4158 (1980); M. G. Sheppard and K. F. Freed, J. Chem. Phys. 75:4525 (1981).
43. S. Ben-Shlomo and U. Kaldor, to be published.

VARIATIONAL APPROXIMATION TO THE NON-HERMITIAN DYSON BOSON HAMILTONIAN

M.C.Cambiaggio+ and J.Dukelsky
Departamento de Física, Comisión Nacional de Energía Atómica
Av. del Libertador 8250 - 1429 - Buenos Aires, Argentina

Boson expansion theories have recently generated a great deal of interest[1-5] as a convenient means for describing collective nuclear properties. Both unitary and nonunitary mappings have been used. The latter, referred to as the Dyson boson theory, has two advantages. It is exact and provides a finite boson expansion whereas the Hermitian theories are characterized by infinite boson expansions. However, some problems also arise. Due to the nonunitary nature of the Dyson mapping the normalization condition is not sufficient to normalize bra and ket vectors separately[5-7].

In addition, there is another problem that is always present when working with boson expansions. The boson Hilbert space is always larger than the original fermion Hilbert space. A physical subspace, spanned by the so-called physical boson basis states, may in principle be identified that is directly related to the fermion space. The rest of the ideal space, called the unphysical subspace, does not take into account Pauli effects. Consequently, spurious states may appear when using the mapped hamiltonian[5,8,9].

On the other hand, a complete diagonalization of the boson hamiltonian is not always possible in realistic calculations. Therefore, adequate approximations must be performed and the basis has to be truncated in a convenient and meaningful way. At present it is not clear how to address these questions, particularly in the case of the Dyson boson mapping. In this paper we propose to use a mean-field approximation and show that a Hartree-Bose approach that takes into account the non-Hermitian nature of the mapped hamiltonian gives good ground state

---
+ Member of the Consejo Nacional de Investigaciones Científicas y Técnicas, Argentina.

results when applied to a simplified shell model defined by a monopole pairing hamiltonian. A subsequent Tamm-Dancoff approximation (TDA) provides good results for the first excited state in the same case. Furthermore, and of particular importance, one gets a clear-cut way of truncating to a particular set of collective bosons which may be systematically improved. Finally, we present a way of correcting for the presence of spurious states within the mean-field approximation.

The Dyson boson expansion maps bifermion operators in terms of ideal boson creation and annihilation operators, B^+_{ij} and B_{ij}

$$
\begin{aligned}
(a^+_i a^+_j)_D &\rightarrow B^+_{ij} - \sum_{kl} B^+_{ik} B^+_{jl} B_{kl} \\
(a_j a_i)_D &\rightarrow B_{ij} \\
(a^+_i a_j)_D &\rightarrow \sum_{k} B^+_{ik} B_{jk}
\end{aligned}
\tag{1}
$$

where the index i completely specifies a single fermion state. The ideal boson operators satisfy the boson algebra

$$
\begin{aligned}
[B^+_{ij}, B^+_{kl}] &= [B_{ij}, B_{kl}] = 0 \\
[B_{ij}, B^+_{kl}] &= \delta_{ik}\delta_{jl} - \delta_{jk}\delta_{il}
\end{aligned}
\tag{2}
$$

and also that $B^+_{ij} = -B^+_{ji}$, $B_{ij} = -B_{ji}$

It is immediately seen that the transformation (1) is nonunitary and finite.

Then, collective boson operators Γ^+_n are introduced

$$\Gamma^+_n = \frac{1}{2}\sum_{ij} X^n_{ij} B^+_{ij} \tag{3}$$

Taking into account the nonunitary nature of the mapping (1) we propose to use eq.(3) for defining only the creation collective boson operators, γ_n instead

$$\gamma_n = \frac{1}{2}\sum_{ij} Y^n_{ij} B_{ij} \tag{4}$$

To ensure that the transformation given by (3) and (4) is canonical it is necessary that the following relations between the amplitudes X^n_{ij}

and Y^n_{ij} hold

$$\sum_{ij} X^n_{ij} Y^m_{ij} = 2\,\delta_{nm}$$
$$\sum_{m} X^m_{ij} Y^m_{kl} = \delta_{ik}\delta_{jl} - \delta_{il}\delta_{jk} \qquad (5)$$

Using (5) one gets the inverse transformation

$$B^+_{ij} = \sum_m Y^m_{ij}\, \Gamma^+_m$$
$$B_{ij} = \sum_m X^m_{ij}\, \gamma_m \qquad (6)$$

The Hartree approach consists in assuming that the ground state wave function is that of a condensate of only one kind of bosons (n=0) and minimizing the expectation value of the hamiltonian with respect to the transformation coefficients[10]. For a non-Hermitian hamiltonian such as the one obtained with the Dyson boson mapping (1) one has to consider a trial bra $\langle\phi|$ that is not equal to the hermitian conjugate of the trial ket $|\psi\rangle$,

$$|\psi\rangle = \frac{1}{\sqrt{N!}} \left(\Gamma^+_0\right)^N |0\rangle$$
$$\langle\phi| = \frac{1}{\sqrt{N!}} \langle 0| \left(\gamma_0\right)^N \qquad (7)$$

where $|0\rangle$ is the boson vacuum and N the number of bosons.

We now test this mean-field approximation in a simplified shell model defined by a monopole pairing hamiltonian,

$$\hat{H}_F = \sum_i \varepsilon_i a^+_i a_i - \frac{G}{4} \hat{P}^+ \hat{P} \qquad (8)$$

with

$$\hat{P}^+ = \sum_i a^+_i a^+_{\bar{i}} \quad , \quad \hat{P} = (\hat{P}^+)^+ \qquad (9)$$

ε_i is the single particle energy for the i orbit with angular momentum j_i and pair degeneracy $\Omega_i = j_i + ½$, i represents the set of quantum numbers j_i, m_i, $\bar{i}$ identifies the time-reversed state and G is the strength of the pairing interaction. The exact solution may be obtained by diagonalizing the hamiltonian (8) in the basis provided by the number of monopole pairs in each orbit, $|n_1, n_2, \ldots n_K\rangle$ with K being the number of single particle orbits and $0 \leq n_i \leq \Omega_i ; \sum_i n_i = N$. In particular we have used a set of parameters appropriate for the tin isotopes, K=5; ε_i=0,0.22, 1.90, 2.20, 2.80 MeV; G=0.187 MeV; Ω_i=3,4,1,2,6.

We apply the boson mapping (1) to the hamiltonian (8) and, furthermore, truncate the expansion by considering only zero angular momentum bosons, i.e. $B^+_{i\bar{i}}$. Using then the inverse collective transformation (6) one finally gets the mapped hamiltonian

$$\hat{H}_B = \sum_{inm} \varepsilon_i Y_i^n X_i^m \Gamma_n^+ \gamma_m -$$

$$- \frac{G}{4}\left[\sum_{in} Y_i^n \Gamma_m^+ - \sum_{inmp} Y_i^n Y_i^m X_i^p \Gamma_n^+ \Gamma_m^+ \gamma_p\right] \sum_{jq} X_j^q \gamma_q \tag{10}$$

where we have put $X^n_{i\bar{i}}=X^n_i$ and $Y^n_{i\bar{i}}=Y^n_i$ for simplicity. The expectation value of $\hat{H}_B$ with respect to the trial bra and ket defined in (7) gives

$$E = \langle \phi | \hat{H}_B | \psi \rangle = N \sum_i \varepsilon_i Y_i^0 X_i^0 -$$

$$- \frac{G}{4} N \sum_i \left[Y_i^0 - (N-1)(Y_i^0)^2 X_i^0\right] \sum_j X_j^0 \tag{11}$$

and the Hartree equations turn out to be

$$(\varepsilon_i - \lambda) Y_i^0 - \frac{G}{4} \sum_j Y_j^0 + \frac{G}{4}(N-1)\left[(Y_i^0)^2 \sum_j X_j^0 + \sum_j (Y_j^0)^2 X_j^0\right] = 0$$

$$(\varepsilon_i - \lambda) X_i^0 - \frac{G}{4} \sum_j X_j^0 + \frac{G}{2}(N-1) Y_i^0 X_i^0 \sum_j X_j^0 = 0 \tag{12}$$

$$\sum_i X_i^0 Y_i^0 = 2$$

where λ is the Lagrange multiplier related to the constraint given in the last equation.

The Hartree ground state energies for the tin isotopes, i.e., for the particular set of parameters given above, are shown in table 1. The exact results are included for comparison. The approximate energies agree very well with the exact ones except at the middle of the shell (N=8). The lower part of the table (N=8-14) was obtained considering holes instead of particles, i.e., replacing a^+_i and a_i by the corresponding hole operators $b_{\bar{i}}$ and $b^+_{\bar{i}}$ in the hamiltonian (8) before mapping it.

Table 1

Ground state energies for the tin isotopes

N	exact	Hartree	2b-diag.	Proj.Hartree
2	-2.624	-2.626	-2.625	-2.626
3	-3.258	-3.265	-3.261	-3.264
4	-3.419	-3.435	-3.427	-3.425
5	-3.084	-3.113	-3.100	-3.093
6	-2.209	-2.261	-2.239	-2.230
7	-0.701	-0.807	-0.685	-0.751
8	2.161	1.421	5.037	1.601
8	2.161	1.973	2.155	2.169
9	5.702	5.590	5.691	5.686
10	9.829	9.767	9.817	9.773
11	14.497	14.465	14.488	14.468
12	19.668	19.654	19.664	19.655
13	25.318	25.312	25.316	25.313
14	31.427	31.425	31.427	31.426

The next step is to get an approximation for excited states and with this aim we consider the TDA. First we write the Hartree approach in matrix form,

$$\sum_{j} h_{ij} X_j = \lambda X_i$$
$$\sum_{i} Y_i h_{ij} = \lambda Y_j \qquad (13)$$

with

$$h_{ij} = \varepsilon_i \delta_{ij} - \frac{G}{4} + \frac{G}{4}(N-1)\left[Y_i^o X_i^o + \delta_{ij} Y_i^o \sum_K X_K^o\right] \qquad (14)$$

where X_i^o and Y_i^o are the amplitudes obtained previously.

Diagonalizing the non-Hermitian Hartree matrix h_{ij} one gets for the lowest eigenstate the particular set X_i^o, Y_i^o (compare eqs. (13) and (12)). The rest of the solutions X_i^P, Y_i^P, $(p \neq 0)$ provide an adequate basis, orthogonal to the ground state, for calculating the TDA matrix. It is to be noted that for the particular hamiltonian (8) considered here one may work in a coupled, m-independent basis with i indicating only the quantum number j_i. In this case the Hartree matrix is of order K and the TDA matrix is of order (K-1).

The new basis states are

$$|\psi_p\rangle = \frac{1}{\sqrt{(N-1)!}}\,\Gamma_p^+ \left(\Gamma_o^+\right)^{N-1} |0\rangle$$

$$\langle\phi_p| = \frac{1}{\sqrt{(N-1)!}}\,\langle 0|(\gamma_0)^{N-1}\,\gamma_p \tag{15}$$

Diagonalizing the mapped hamiltonian (10) in this basis one gets the TDA results.

One may also write the TDA in matrix form, as in (13)

$$t_{ij} = \varepsilon_i\delta_{ij} - \frac{G}{4} + \frac{G}{2}(N-1)\left[Y_i^o X_i^o + \delta_{ij} Y_i^o \sum_\ell X_\ell^o\right] \tag{16}$$

and then diagonalize t_{ij}.

Table 2

Excitation energy of the first excited 0^+ state for the tin isotopes

N	exact	TDA	2b-diag.
2	1.755	1.763	1.754
3	1.796	1.796	1.793
4	1.880	1.849	1.874
5	2.008	1.930	1.997
6	2.200	2.053	2.182
7	2.104	2.086	1.942
8	1.672	1.129	3.707
8	1.672	1.946	1.747
9	1.939	2.023	2.125
10	1.981	2.092	1.963
11	2.088	2.194	2.077
12	2.245	2.319	2.239
13	2.410	2.454	2.408
14	2.574	2.593	2.573

The lowest TDA eigenvalue is the excitation energy of the first excited state and is shown in table 2 for the different tin isotopes. The exact results are included for comparison. As in the ground state case, the approximate energies agree very well with the exact ones except at the middle of the shell. Once again, the lower part of table 2 was obtained considering holes instead of particles.

The method described up to now provides not only a good approximation for the ground state and the first excited state but also an adequate collective boson basis generated from the Hartree boson and the TDA ones. These are ordered in energy and so one gets a clear-cut way of truncating to a particular set of collective bosons which may be systematically improved.

The results obtained from the diagonalization of the hamiltonian (10) truncated to only two collective bosons are shown in the third columns of tables 1 and 2. In general, the results are an improvement over the mean-field results.

Finally we study the problem of the existence of spurious components in the Hartree ground state wave function. First we notice that it may be expanded as

$$|\psi\rangle = \frac{1}{\sqrt{N!}}(\Gamma_0^+)^N|0\rangle = \frac{1}{2^N\sqrt{N!}}\sum_{n_1 \dots n_K}^{N} \frac{N!}{n_1! \dots n_K!}(x_1^0 B_1^+)^{n_1} \dots (x_K^0 B_K^+)^{n_K}|0\rangle \tag{17}$$

with $\sum_{i=1}^{K} n_i = N$ and similarly for the bra $\langle\phi|$.

It is clearly seen from (17) that spurious components appear when one or more of the occupation numbers n_i becomes greater than the corresponding degeneracy of the orbit Ω_i. Consequently, for projecting the wave function into the physical subspace it is necessary to restrict the sum in such a way that every n_i is always smaller than or equal to Ω_i. Calculating the expectation value of the hamiltonian with respect to this projected wave function one gets the projected energy

$$E_P = \frac{\langle\phi|P\hat{H}_B P|\psi\rangle}{\langle\phi|P|\psi\rangle} \tag{18}$$

The projected Hartree ground state energies obtained for the hamiltonian considered in this work are shown in the fourth column of table 1. These are a better approximation to the exact results than the Hartree energies. Nevertheless, this is only an approximate projection and problems to be solved still remain as for example the one related with the ambiguity due to the fact that the normalization condition is not enough to normalize the bra and the ket separately.

In summary, we have shown how to apply a mean-field approximation to the non-Hermitian Dyson boson hamiltonian. In the particular case of a monopole pairing hamiltonian the results obtained are very good. For the ground state the present results are as good as the ones obtained previously with different bosonic approximations[9,11]. The fact that the approximation is worse at the middle of the shell is most probably due to the presence of spurious components. The situation is the following. When one does a complete diagonalization in the boson space, the physical subspace is not mixed with the unphysical one but this always happens when using an approximation. As a consequence, spurious components appear in the approximate states which are more important when Pauli effects are stronger, i.e. at the middle of the shell. This is clear from the corrections obtained with the approximate projection procedure proposed in this work. How to improve this projection and to extend it to excited states is a matter for future studies.

The Hartree plus TDA approximation provides a method for selecting a collective subspace within which one may truncate the hamiltonian in an adequate way. The diagonalization of the hamiltonian in this collective subspace takes into account anharmonic effects neglected in TDA and improves the results.

References

1) E.R.Marshalek, Nucl.Phys. A347 (1980) 253
2) A.Klein and M.Vallieres, Phys.Lett. 98B (1981) 5
3) J.Dobaczewski, Nucl.Phys. A369 (1981) 237
4) T.Tamura, Phys.Rev. C28 (1983) 2840
5) Y.K.Gambhir, J.A.Sheikh, P.Ring and P.Schuck, Phys.Rev. C31 (1985) 1519
6) C.T.Li, Nucl.Phys. A417 (1984) 37
7) K.Takada, Phys.Rev. C34 (1986) 750
8) H.B.Geyer, C.A.Engelbrecht and F.J.W.Hahne, Phys.Rev. C33 (1986) 1041
9) C.T.Li, Phys.Rev. C29 (1984) 2309
10) J.Dukelsky, G.G.Dussel, R.P.J.Perazzo, S.L.Reich and H.M.Sofia, Nucl.Phys. A425 (1984) 93
11) A.Klein, C.T.Li, T.D.Cohen and M.Vallieres, Progress in Particle and Nuclear Physics, Ed.D.Wilkinson, Pergamon Press, Oxford (1983) Vol.9, p.183.

ON THE CONVERGENCE OF CLUSTER EXPANSIONS IN FINITE NUCLEI

M.C.Boscá, E.Buendía and R.Guardiola
Departamento de Física Moderna
Universidad de Granada
Granada (Spain)

1. Introduction

In spite of its deep interest, the Jastrow variational method has been scarcely applied to the description of finite nuclear systems.

The algorithms to compute the expectation values with Jastrow correlated wave functions are known since many years. Clark and Westhaus /1/ have developed several techniques appropriate to finite systems and known with the names of Iwamoto-Yamada (IY), Aviles-Hartog-Tolhoek (AHT), which is an averaged version of the former, as well as two multiplicative, van Kampen-like expansions FIY and FAHT. All that cluster expansions are extrapolation algorithms which attempt to evaluate A-body matrix elements in terms of few- (as few as possible) body matrix elements. Previous studies with these cluster expansions /2-3/ have shown the superiority of the multiplicative forms over the additive forms so that we will concentrate here in the FAHT cluster expansions.

It should also be mentioned the work of Fantoni and Rosati /4/ who have extended the FHNC formalism to finite systems. Hovewer, the lack of traslation invariance of the two-body distribution function in a finite and localized system makes the computational work very complicated and, up to our knowledge, this formalism has not yet been applied.

One of the troubles with standard cluster expansions is the abnormal behaviour of the resulting series. Unless the correlation volume is very small the various orders do not show clearly a definite behaviour: second and fourth order results are very close one to each other, whereas third order seems to be anomalous (see later).

Recently, Carlson and Kalos /5/ have shown the way of applying the Motecarlo random walk involving spin and isospin degrees of freedom to finite nuclear systems. The calculation turns out to be quite time consuming when state dependent correlations are used in the description of the wave function. As a consequence it is difficult to determine accurately the variational minimum and the final estimate of the binding energy has a quite large error.

Having in mind all these facts we have planned the study of the convergence of the FAHT cluster expansion by comparing the results at various orders with the exact (within statistical errors) results of Motecarlo method. In order to be able to carry the calculations up to a high order and with small statistical errors a simple interaction (Brink-Boeker /6/) has been considered. For the same reason the correlation factor has been taken of the state independent form. Because of these facts it has no sense to compare our results with the experimental values, but we believe the consequences regarding the convergence to have a general scope.

2. The FAHT cluster expansion in finite systems

Consider the correlation operator for the n-particle subsystem

$$F^{(n)} = \prod_{i<j}^{n} f_{ij} \tag{1}$$

and the corresponding n-particle hamiltonian

$$H^{(n)} = \sum_{i=1}^{n} T_i + \sum_{i<j}^{n} V_{ij} \tag{2}$$

where T and V represent the kinetic energy an the two-body potential energy operators. One defines the generalized subnormalization integral

$$I^{(n)}(z) = \langle \Phi | F^{(n)} \exp(zH^{(n)}) F^{(n)} | \Phi \rangle \qquad (3)$$

where Φ is the wave function representing the uncorrelated A-body system, and z is and auxiliary parameter. The word uncorrelated means that there are not short range correlations inside this wave function, but it may include long range correlations, such as deformations and/or configuration mixing. By means of equation (3) we intend to define A quantities, corresponding to the values of n=1,2,...,A.

The energy expectation value is obtained from eq.(3) by computing the logarithmic derivative with respect to the auxiliar paramenter z at z=0.

The cluster expansion results from the following chain of equations which define the cluster integrals $Y^{(n)}$:

$$\begin{aligned}
I^{(1)} &= Y_1 \\
I^{(2)} &= Y_1^2 \, Y_2 \\
I^{(3)} &= Y_1^3 \, Y_2^3 \, Y_3 \\
&\ldots\ldots\ldots\ldots\ldots\ldots \\
I^{(n)} &= Y_1^{\binom{n}{1}} \, Y_2^{\binom{n}{2}} \, Y_3^{\binom{n}{3}} \ldots Y_{n-1}^{\binom{n}{n-1}} \, Y_n \\
&\ldots\ldots\ldots\ldots\ldots\ldots
\end{aligned} \qquad (4)$$

up to n=A. The first equation defines Y_1, the second equation defines Y_2 and so on. Note that Y_n is a function of z.

In a low density system one expects that high order clusters have a negligible role, i.e., that $Y_p=1$ for p sufficiently large. Then, the chain of definitions of eq.(4) has a predictive character, allowing to compute I_A in terms of $Y_1, Y_2 \ldots Y_{p-1}$. In other words, the A-body matrix element I_A is approximately computed in terms of few body matrix elements.

The energy expectation value is given by

$$E \simeq \sum_{n=1}^{p} \binom{A}{n} \frac{Y'_n(z=0)}{Y_n(z=0)} \qquad (5)$$

and explicitely eq.(5) reads

$$E = A \frac{\langle \Phi | T_1 | \Phi \rangle}{\langle \Phi | \Phi \rangle} + \binom{A}{2} \left\{ \frac{\langle \Phi | f_{12}(T_1+T_2+V_{12}) f_{12} | \Phi \rangle}{\langle \Phi | f_{12}^2 | \Phi \rangle} - 2 \frac{\langle \Phi | T_1 | \Phi \rangle}{\langle \Phi | \Phi \rangle} \right\}$$

$$+ \binom{A}{3} \left\{ \frac{\langle \Phi | f_{12} f_{13} f_{23} (T_1+T_2+T_3+V_{12}+V_{13}+V_{23}) f_{12} f_{13} f_{23} | \Phi \rangle}{\langle \Phi | f_{12} f_{13} f_{23} \; f_{12} f_{13} f_{23} | \Phi \rangle} \right.$$

$$\left. - 3 \frac{\langle \Phi | f_{12}(T_1+T_2+V_{12}) f_{12} | \Phi \rangle}{\langle \Phi | f_{12} \; f_{12} | \Phi \rangle} + 3 \frac{\langle \Phi | T_1 | \Phi \rangle}{\langle \Phi | \Phi \rangle} \right\} \tag{6}$$

$$+ \binom{A}{4} \quad \ldots .$$

The calculation of the energy expectation value by means of eq.(6) presents some troubles. In the large A limit each term behaves like A, including the combinatorial factor in front of them. The basic contributions to each term, namely the quotients $\langle \Phi | F^{(n)} H^{(n)} F^{(n)} | \Phi \rangle / \langle \Phi | F^{(n)} F^{(n)} | \Phi \rangle$ are all of them O(1) so there must be strong cancellations between the various pieces of each contribution, i.e.,between the terms inside the pairs of braces. This requires high precision calculations in finite systems, or a further elaboration of the expansion in the large A limit.

We may resume the characteristics of the cluster expansion

i. The computation is simple at low orders. Actually second order requires standard Shell Model machinery.

ii. The extension to deal with excited states is straightforward (CBF theory).

iii. Long range correlations may be easily incorporated.

iv. There may be computational troubles when considering heavy nuclei, and

v. The expansions become more and more involved when high orders are required.

In addition, there is still pending the question of convergence.

3. Some calculations

In this section we will consider the four nuclei ^4He, ^8Be, ^{12}C and ^{16}O. The model wave function used is a Slater determinant with single particle states from the harmonic oscillator potential, corresponding to the configurations

$$
\begin{array}{lll}
{}^{4}\mathrm{He} & \ldots\ldots\ldots\ldots\ldots\ldots & (1s)^4 \\
{}^{8}\mathrm{Be} & \ldots\ldots\ldots\ldots\ldots\ldots & (1s)^4\,(1p_z)^4 \\
{}^{12}\mathrm{C} & \ldots\ldots\ldots\ldots\ldots\ldots & (1s)^4\,(1p_x)^4\,(1p_y)^4 \\
{}^{16}\mathrm{O} & \ldots\ldots\ldots\ldots\ldots\ldots & (1s)^4\,(1p_x)^4\,(1p_y)^4\,(1p_z)^4
\end{array}
\qquad (7)
$$

so that the nuclei are spin/isospin saturated, their wave functions corresponding to the SU(3) representation. Note the use of a cartesian representation of the single particle orbitals, which will simplify the computational work /2/ and, on the other hand, will facilitate the consideration of deformations. The model wave function depends only on one paramenter $\alpha = (m\omega/\hbar)^{1/2}$, or two parameters when deformations are considered. The deformation will be defined by means of the dimensionless quantity $d = \alpha_z / \alpha_x$, where α_z and α_x are the harmonic oscillator parameters characterizing the axially deformed system.

The interaction used is the B1 Brink-Boeker potential /6/ which has a Wigner-Majorana mixture. The radial functions of this interaction are gaussians, and this permits the exact computation of the required matrix elements. Finally the correlation factor has been taken with the simple form

$$f(r) = 1 + a\exp(-r_{12}^2/b^2) \qquad (8)$$

where a controls the depth of the correlation at very short distances, and b is the length of the correlation.

Table 1. Variational minima at second, third and fourth orders for several nuclei, and the corresponding contributions in MeV of the Cluster expansion

		$\alpha(fm^{-1})$	a	b(fm)	$E^{(2)}$	$\Delta E^{(3)}$	$\Delta E^{(4)}$
^{8}Be	2	.648	-0.48	.763	-45.02	- 0.54	3.37
	3	.680	-0.50	.845	-44.60	- 1.84	6.18
	4	.636	-0.46	.707	-44.98	- 0.83	2.29
^{12}C	2	.633	-0.49	.781	-82.93	- 4.93	5.75
	3	.690	-0.52	.913	-78.12	-12.64	15.30
	4	.633	-0.48	.754	-82.81	- 4.50	4.99
^{16}O	2	.640	-0.49	.819	-146.90	- 9.98	8.52
	3	.720	-0.55	.953	-137.60	-23.77	23.37
	4	.648	-0.51	.811	-146.62	-10.99	9.02

The results corresponding to this computation are shown in table 1. This table is organized into three blocks of three lines, each block corresponding to Be, C and O, respectively. The first line of each block shows the determination of the minimum at second order, the second line corresponds to the determination at third order and finally the third line corresponds to the minimum as determined at fourth order of the cluster expansion. There are two facts which can be analyzed by means of the numbers shown in table 1:

i. The stability of the minimum: There is a significative change in the value of the parameters between second and fourth order, from one side, and third order from the other side. The placement of the minimum is almost the same in second and fourth order.

ii. The third order contribution $\Delta E^{(3)}$ is almost cancelled by the fourth order contribution $\Delta E^{(4)}$. It seems that third order is anomalous. However, a definite statement will be put forward once the exact results are known.

Of course, one can also think that the cluster expansion is not convergent. Just to show that this is not the case we show in table 2 the results corresponding to the last row of table 1 for the various pieces of the hamiltonian, i.e., kinetic energy term, Wigner part of the potential and Majorana part of the potential. The abnormal behaviour of the cluster expansion is the result of strong cancellations between clearly convergent expansions.

Table 2. Separate contributions to the energy (Mev) related to the last row of table 1.

	$E^{(2)}$	$\Delta E^{(3)}$	$\Delta E^{(4)}$
KINETIC	317.18	- 3.70	0.24
WIGNER	- 82.14	31.75	- 9.17
MAJORANA	-381.66	-39.04	17.95
TOTAL	-146.62	-10.99	9.02

4. Variational Montecarlo in finite nuclei with spin and isospin degrees of freedom

From the Montecarlo variational point of view it is convenient to rewrite the energy expectation value in the form

$$E = \frac{\sum_N \int dR \; \Psi^*(R,N)\, \Psi(R,N) \left\{ \frac{1}{\Psi(R,N)} \sum_{N'} H_{N,N'} \; \Psi(R,N') \right\}}{\sum_N \int dR \; \Psi^*(R,N)\, \Psi(R,N)} \tag{9}$$

where R is a 3A-dimensional vector representing the instantaneous position of all nucleons, and N is a A-dimensional vector representing the third component of spin and isospin. The quantity (R,N) is the quantum mechanical wave function

$$\Psi(R,N) = \langle r_1 \upsilon_1 ;\; r_2 \upsilon_2 ;\; \dots ;\; r_A \upsilon_A \mid \Psi \rangle \tag{10}$$

and $H_{N,N'}$ is a shorthand way of representing the matrix element

$$H_{N,N'} = \langle \upsilon_1, \upsilon_2, \dots \upsilon_A \mid H \mid \upsilon_1', \upsilon_2', \dots \upsilon_A' \rangle \tag{11}$$

The interpretation of equation (9) is straightforward: we must compute the average value of the quantity in braces $1/\Psi \; H\Psi$ accordingly with the probability distribution

$$| \Psi(R,N) |^2 \;/\; \sum_N \int dR \; | \Psi(R,N) |^2 \tag{12}$$

so that equation (12) controls the sampling procedure. The way of generating R's ·and N's distributed accordingly with eq.(12) is a simple modification of Metropolis random walk procedure /5,8/. From a given point $\{R,N\}$ one tries to move to another point $\{R',N'\}$ by moving sequentially the particles and exchanging at random the spin/isospin quantum numbers of pairs of particles. The last sentence of this statement reeds some comments: among the 4^A quantities defined in equation (10) only few of them are different from zero, namely those which have the same value for the third component of total spin and isospin. The way of "moving" the spin of a given particle with this restriction is to exchange its value with the value of the spin of any other particle, itself included.

From the computational point of view, this sampling is very complex when the correlations are spin/isospin dependent /5/, but the calculation is fairly simple when correlations are state independent. It is convenient to realize the following facts:

i. When the hamiltonian depends on spin or isospin, one must also sample the spin variables N, even for state independent correlations.

ii. Metropolis moves of spin or isospin variables lower significantly the ratio of acceptances to total moves. A spin exchange is a big move, and thermalization is mandatory.

iii. In the case of a Wigner plus Majorana interaction it is not necessary to move the discrete degrees of freedom.

Table 3. Comparison between fourth order cluster expansion and Motecarlo variational calculation. The wave function of Be and C are spherically symmetric.

	KINETIC	WIGNER	MAJORANA	TOTAL
^{4}He	62.61 62.54 ±0.13	24.94 24.84 ± 0.20	-124.17 -123.85 ± 0.22	-36.62 -36.47 ± 0.12
^{8}Be	126.14 126.36 ±0.15	8.58 8.38 ± 0.20	-177.94 -178.24 ± 0.25	-43.23 -43.51 ± 0.16
^{12}C	211.33 211.82 ±0.20	-15.73 -14.92 ± 0.28	-277.91 -279.77 ± 0.34	-82.31 -82.87 ± 0.20
^{16}O	313.72 314.04 ±0.16	-63.56 -62.31 ± 0.50	-398.75 -402.59 ± 0.39	-148.59 -150.87 ± 0.33

Table 4. Same as table 3 for deformed Be (d=0.71) and C (d=1.275).

	KINETIC	WIGNER	MAJORANA	TOTAL
^{8}Be	130.48 130.66 ± 0.15	20.10 20.40 ±0.30	-205.01 -206.21 ± 0.33	-54.43 -55.15 ± 0.16
^{12}C	217.08 217.34 ± 0.20	- 9.18 - 8.66 ±0.47	-298.82 -300.94 ± 0.36	-90.92 -92.26 ± 0.34

We have not found significant differences in our Montecarlo calculations when moving or maintaining the values of the spin and isospin of the individual nucleons. In the calculation, the random walk has been continued following Malvin Kalos recommendation /10/, "until either computer time is exhausted or the statistics are judged acceptable".

Detailed results corresponding to the parameters at the variational minimum at fourth order are compared in tables 3 and 4 with the previously described cluster expansion at fourth order. Table 3 deals with spherically symmetric wave functions for all nuclei, whereas Table 4 takes into account also a deformation, oblate and prolate for ^{12}C and ^{8}Be, respectively.

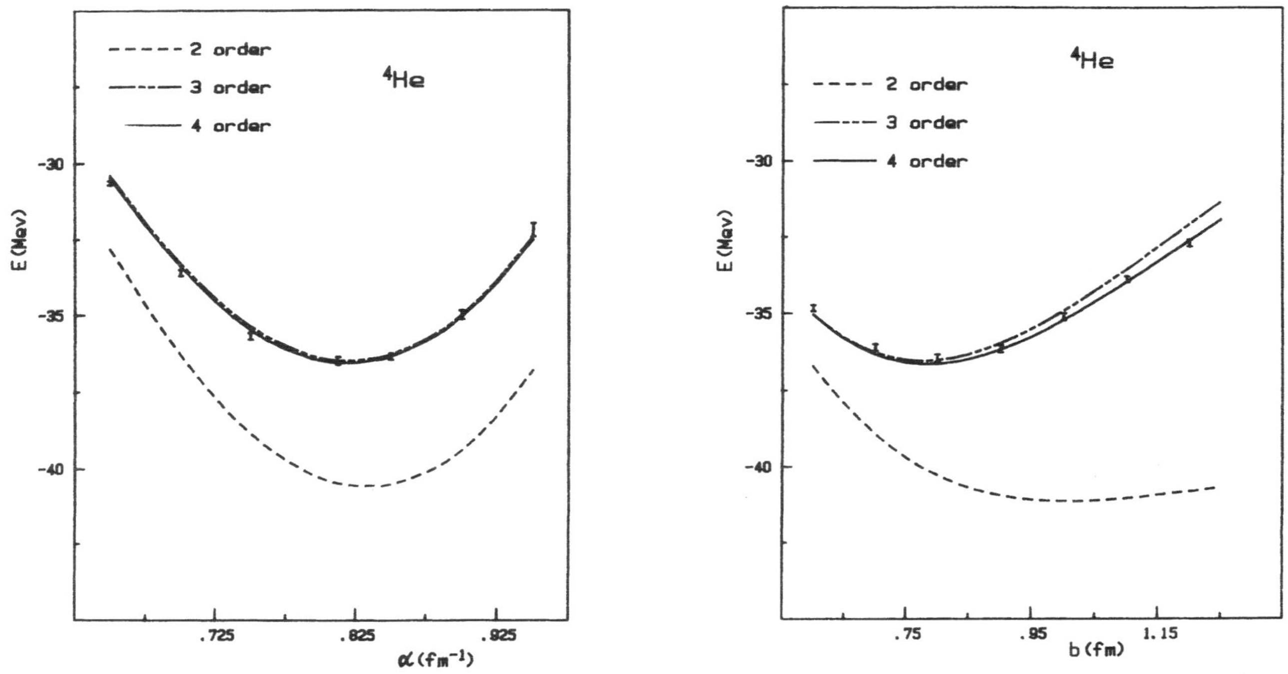

Figure 1. Comparison of Montecarlo calculation with various orders of the cluster expansion in ^{4}He. Left, at fixed correlation (a=-0.48, b=0.8 fm), right at fixed nuclear wave function (a=-0.48, α =0.8124 fm^{-1})

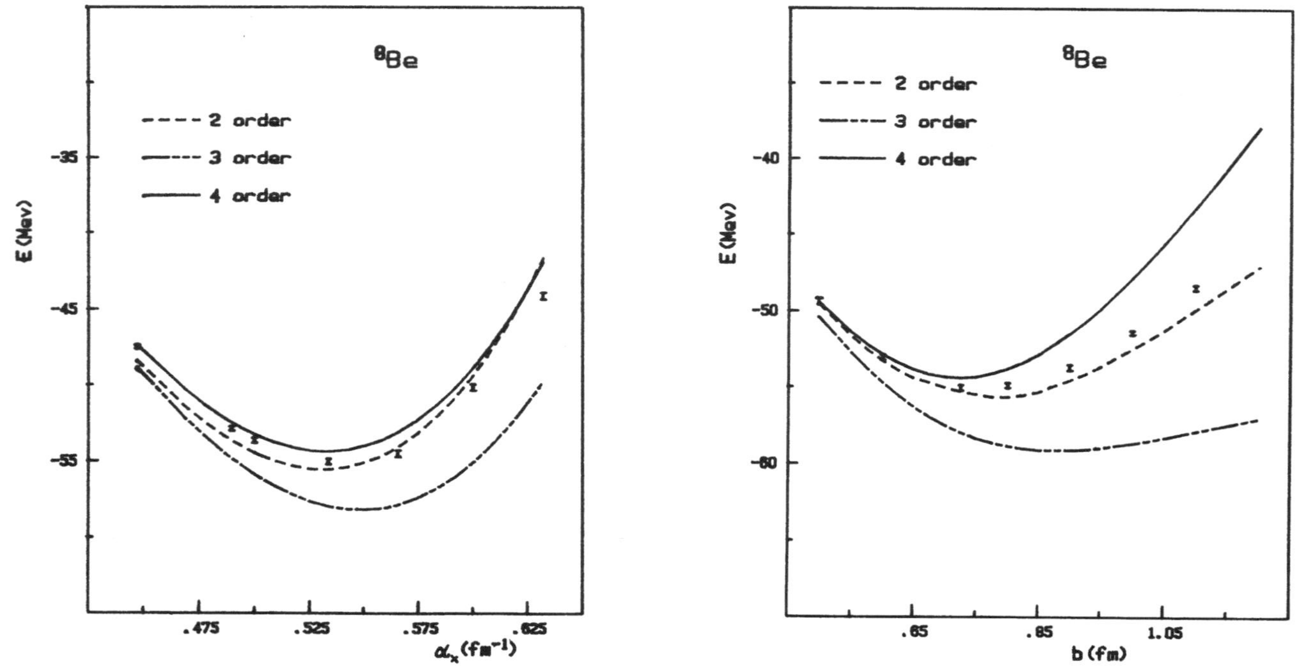

Figure 2. Comparison of Montecarlo calculation with various orders of the cluster expansion in ^{8}Be. Left, at fixed correlation (a=-0.46, b=0.7255 fm), right at fixed nuclear wave function (a=-0.46, d=0.71, α=0.7517 fm^{-1})

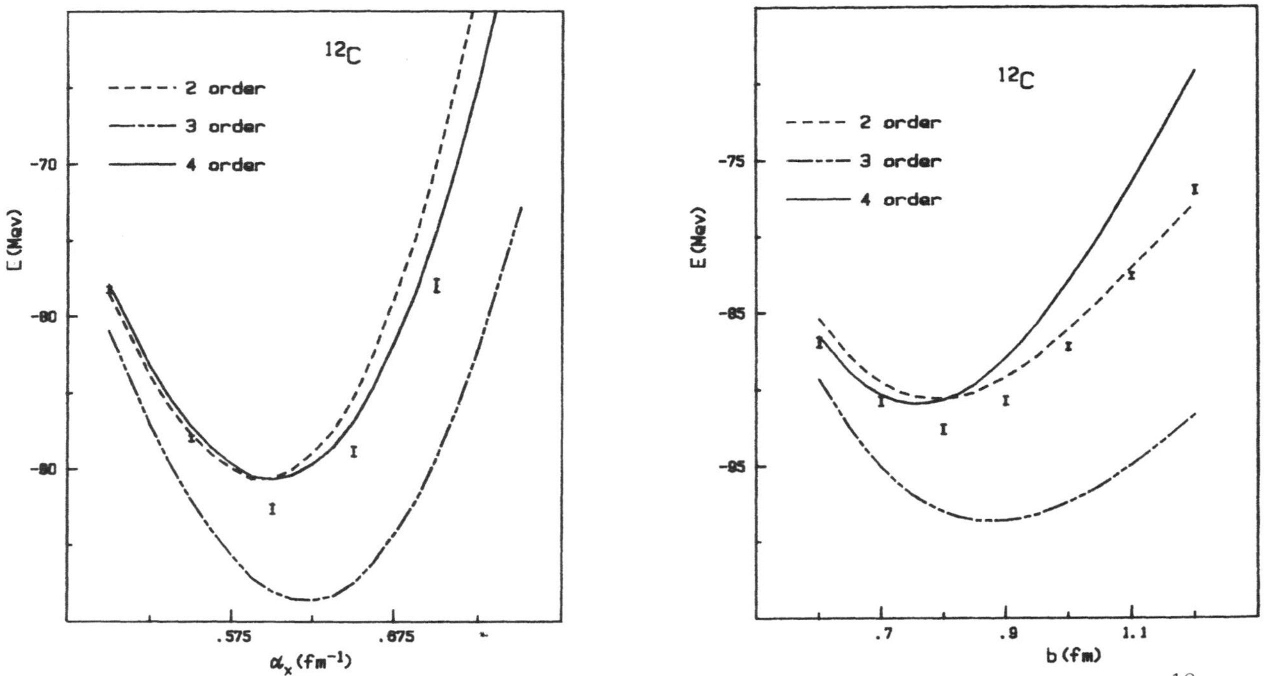

Figure 3. Comparison of Montecarlo calculation with various orders of the cluster expansion in ^{12}C. Left, at fixed correlation (a=-0.49, b=0.8 fm, d=1.275), right at fixed nuclear wave function (a=-0.49, d=1.275, α=0.6 fm^{-1})

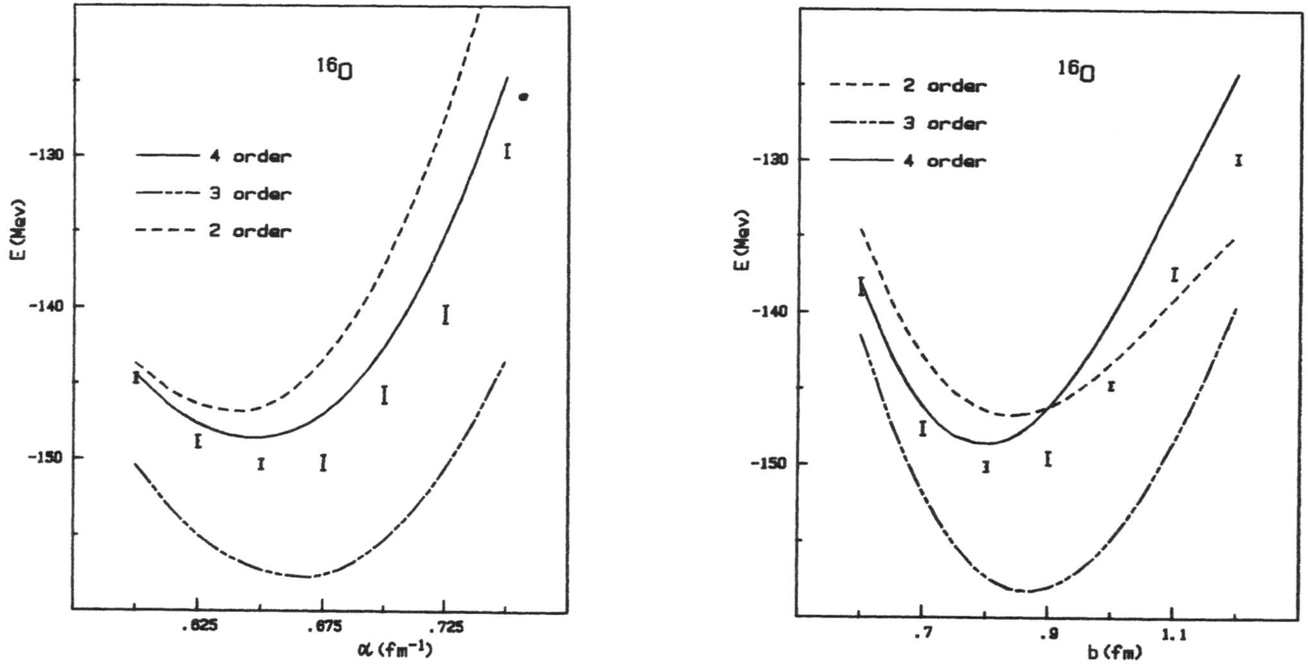

Figure 4. Comparison of Montecarlo calculation with various orders of the cluster expansion in ^{16}O. Left, at fixed correlation (a=-0.51, b=0.8 fm), right at fixed nuclear wave function (a=-0.51, α=0.6481 fm^{-1})

In figures 1 to 4 we show the comparison of Montecarlo calculations with the various orders of the cluster expansion in all nuclei considered, both as a function of the harmonic oscillator parameter and as a function of the correlation length. Results corresponding to ^{4}He have not too much interest. In this case fourth order cluster expansion calculation is the exact result, so that we merely check that both types of calculations have been carried out properly.

A glance to these figures, as well as to Tables 3 and 4, shows that both second and fourth order of the cluster expansion give a reasonably good estimate of the total energy of nuclei. On the other hand, calculations at third order give systematically too much binding energy. Moreover, the placement of the minimum at both second and fourth order is again well determined. To find significant differences between these two approximations one must go to extreme cases, namely at a really long range correlation or to a very small uncorrelated nucleus. In any case, the value of the energy obtained near the minimum deviates only few MeV (as few as one or two MeV) from the exact value.

5. Final comments

Our comparisons give confidence to calculations at low order in the cluster expansions in finite nuclei, at least with regard to bulk properties of nuclei. We have been able to push forward the cluster expansion up to fourth order because of the simplicity of both the wave function and the interaction, which has allowed a FORTRAN coded algebraic-like algorithm. Of course this will not be the case when dealing with realistic interactions or more physical single particle wave functions, like those of a Woods-Saxon potential. Equation (6) is not very practical when the cluster integrals have to be computed numerically. Actually, the strong cancellations present in each term of this equation may amplify unreasonably the error of the successive terms of the expansion. We beleive that future work should be devoted to find a rearrangement of the cluster contributions in eq. (6) which permits a Montecarlo calculation of the cluster integrals free of numerical uncertainties, so as to combine the simplicity of Montecarlo algorithm with the significant saving of computing time related to the cluster expansion.

Acknowledgements

This work has been supported bay CICyT, formerly CAICyT, under contract Nb. 1234/84.

References

1. J.W. Clark and P. Westhaus, Jour. Math. Phys. 9 (1968) 131 and 149.
2. R. Guardiola, Nucl. Phys. A328 (1979) 490.
3. R. Guardiola, A. Polls and J. Ros, Nuovo Cim. 59 (1980) 419.
4. S. Fantoni and S. Rosati, Nucl. Phys. A328 (1979) 478.
5. J. Carlson and M.H. Kalos, Phys. Rev. C32 (1985) 2105.
6. D.H. Brink and E. Boeker, Nucl. Phys. A91 (1967) 1.
7. E. Buendía and R. Guardiola, in "Recent Progress in Many Body Theories", Lecture Notes in Physics 198 (1983) 102. Springer (Berlin, 1983)
8. N. Metropolis, A.W. Rosenbluth, M.N. Rosenbluth, A.H. Teller and E. Teller, Jour. Chem. Phys. 21 (1953) 1087.
9. M.H. Kalos in "Montecarlo Methods in Statistical Physics", K. Binder Editor. Springer Verlag (Berlin, 1979)

SEMICLASSICAL MOLECULAR DYNAMICS OF WAVEPACKETS IN ONE-DIMENSIONAL PHASE SPACE

Azizul Haque and Thomas F. George

Departments of Physics and Chemistry
239 Fronczak Hall
State University of New York at Buffalo
Buffalo, New York 14260 USA

ABSTRACT

A semiclassical method for solving the quantum Liouville equation in one-dimensional phase-space is described. The development is based on constructing a Gaussian density matrix and is applicable to systems in pure and in mixed states having nonlinear interaction potentials. The density matrix is constructed using a set of dynamic variables whose expectation values are considered to be relevant for the dynamics. The self-consistent equations of motion are then derived for these expectations from the quantum Liouville equation using a projection scheme. The solution of these self-consistent equations provides the time evolution of the density matrix. The present method can yield, in principle, exact values for these expectations for all times. A model calculation is carried out to describe the vibrational motion of an arbitrary diatomic molecule on an anharmonic potential surface. However, the potentiality of this method lies in describing the time evolution of systems in mixed states and hence in describing the dynamics of molecular processes in condensed phases.

I. INTRODUCTION

Recent advances in the experimental study of the various molecular dynamical processes in condensed phase, such as energy transfer, molecular dissociation reactions, spectral line shapes, etc., require theoretical models for the quantitative understanding of the dynamical processes involved in condensed phases. There has been progress in studying equilibrium properties using classical [1], semiclassical [2], fully quantum mechanical [3] and quantum field theoretic methods [4,5]. Methods are also available for treating time-dependent processes within the classical framework [6]. However, very few theoretical developments are available for treating time-dependent processes incorporating quantum effects. These are the quantum corrections to the classically computed time-correlation functions [7], the exp(S) approach of Arponen and co-workers [4] and the semiclassical Gaussian wavepacket dynamics (GWD) approach developed notably by Heller [8]. The semiclassical GWD approach describes a self-consistent solution of the time-dependent Schrödinger equation and thus is restricted in its application to systems in pure states. Extension of this GWD method to the simulation of time-dependent properties of N-particle systems interacting through realistic pair potentials within the variational and nonvariational framework are also available in the literature [9]. Such application requires tedious thermal averaging, which arises from the fact that we have no knowledge about initial conditions of the N-particle system.

Our objective is to develop a similar GWD approach which as such is applicable to systems in pure and in mixed states. That is, when treating systems in mixed states, we do not need to perform tedious thermal averaging. Our development satisfies the maximum entropy principle [10] when treating equilibrium or nonequilibrium systems. However, we no not make the assumption that the exact nonequilibrium statistical density matrix is approximately equal to the local equilibrium one [11-13]. For an N-particle statistical system it is practically impossible to construct a density matrix which contain all information about the system. However, with the advent of projection operator techniques [11,14], it has been possible to construct density matrices which contain information sufficient for the calculation of various physical quantities of interest.

In this paper we are interested in a reduced description of the exact N-particle system, which is the time evolution of the N single-particle density matrices in a mixed state. We define our reduced density matrix, $\rho_{re}(X,X',t)$, as a product of N single-particle density functions

$$\rho_{re}(X,X';t) = \prod_{j=1}^{N} \phi_j(x_j,x_j';t) \quad , \tag{1}$$

where X is a vector with N coordinate components $x_1 \ldots x_N$. The time evolution of these density functions, ϕ_j are then obtained from the quantum Liouville equation using a projection operator scheme [11,14]. We define each single-particle density function $\phi_j(x_j,x_j';t)$ from the perspective of nonequilibrium statistical mechanics as [10,11,14]

$$\begin{aligned}\phi_j(x_j,x_j';t) &= \langle x_j|\hat{\phi}_j(t)|x_j'\rangle \\ &= \langle x_j|\exp[\sum_{\alpha=0} \lambda_{j\alpha}(t)\, A_{j\alpha}]|x_j'\rangle \quad ,\end{aligned} \tag{2}$$

which contains all information about the single particle system. The $\lambda_{j\alpha}(t)$'s are Lagrange multipliers, and the $A_{j\alpha}$'s are the dynamical variables. Since we are not interested in all the information contained in the ϕ_j's, we construct our ϕ_j's with respect to the six dynamical quantities

$$\begin{aligned} &A_{j0} = \hat{1}, \quad A_{j1} = \hat{x}_j, \quad A_{j2} = \hat{p}_j, \quad A_{j3} = \hat{x}_j^2, \\ &\qquad A_{j4} = \hat{p}_j^2, \quad A_{j5} = \hat{x}_j\hat{p}_j + \hat{p}_j\hat{x}_j \quad , \end{aligned} \tag{3}$$

where p_k is the momentum associated with the k-th particle and the hat designates an operator. As we shall see later, the choice of these dynamical quantities allows us to describe the time evolution of each single-particle density function incorporating quantum fluctuations. The time evolution of the expectations of these dynamical quantities, $\langle A_{j\alpha}\rangle$, are then obtained from the quantum Liouville equation using the projection operator scheme [11,14]. The choice of the single-particle density operator as given by Eq. (2) is by no means unique. Our choice is motivated by the physical consideration which is the maximum entropy principle [10,11].

We confine our development to one-dimensional phase space. In the next section we derive the equations of motion for the expectations, $\langle A_{j\alpha}\rangle$, in closed form and construct the corresponding density function for mixed states. In Sec. III we show that under certain conditions the density function for mixed states reduces to the density function for pure states. To describe the time evolution of the pure state density function, we then derive the equations of motion for the corresponding dynamical quantities. In Sec. IV we show that our maximum entropy-based density function can also describe the time evolution of a harmonic system in thermal equilibrium [15]. In Sec. V we solve the equations of motion for the pure state to describe the vibrational motion of an arbitrary diatomic molecule on a Morse potential surface. We then compare our results with those obtained using the existing Gaussian wavepacket dynamics method [8,9], and a discussion is provided in Sec. VI.

II. CONSTRUCTION OF THE DENSITY FUNCTION AND DERIVATION OF THE EQUATIONS OF MOTIONS FOR SYSTEMS IN MIXED STATES

We characterize our N-particle system by a Hamiltonian

$$H = \sum_{k=1}^{N} \frac{p_k^2}{2m_k} + V(x_1 \ldots x_n) \tag{4}$$

and a density matrix $\rho(X,X';t)$ which satisfies the quantum Liouville eqution

$$\frac{d\rho}{dt} = -iL\rho \equiv -(i/\hbar)[H,\rho] \quad , \tag{5}$$

where m_k is the mass of the k-th particle and V is the interaction potential. Since we are interested only in the time evolution of the N single-particle density functions, we partition our total density matrix as

$$\rho(t) = \rho_{re}(t) + \rho_{ir}(t) \quad , \tag{6}$$

where $\rho_{re}(t)$ is a reduced description of the N-particle interacting system and is represented by a product of N single-particle density functions as described in Eq. (1). $\rho_{ir}(t)$ represents the irrelevant degrees of freedom, since it does not contain any dynamical degrees of freedom of any single particle in the coupled N-particle system, but rather the correlations between single particle systems produced by their interaction.

We associate entropy S with our system by using the relation [10]

$$S = - k \, Tr\rho_{re}(t) \ln\rho_{re}(t) \tag{7}$$

where k is Boltzmann's constant. We maximize entropy subject to the constraints

$$Tr\rho_{re}(t) = 1 \tag{8a}$$

and

$$a_{j\alpha}(t) = \langle A_{j\alpha}(t)\rangle = TrA_{j\alpha}\rho_{re}(t) \equiv TrA_{j\alpha}\rho(t) \quad , \tag{8b}$$

where the $A_{j\alpha}$'s are the 6N dynamical variables of interest to us.

We now derive explicit expressions for the time evolution of the expectations, $a_{j\alpha}(t)$, using the time-dependent projection operator scheme [14] followed in constructing the maximum entropy distribution of the reduced density operator $\rho_{re}(t)$ in one-dimensional phase space. From now on we shall refer to these equations of motion as <u>reduced</u> equations of motion since they describe the time evolution of the reduced density operator $\rho_{re}(t)$. We shall use the projection operator technique in Liouville space [11,14]. In this space $\hat{H}$ and $\rho(t)$ can be written as $|H\rangle\rangle$ and $\rho(t)\rangle\rangle$. In this notation Eq. (8b) becomes

$$A_{j\alpha}(t) = \langle\langle\rho_{re}(t)|A_{j\alpha}\rangle\rangle \equiv \langle\langle\rho(t)|A_{j\alpha}\rangle\rangle \quad . \tag{8c}$$

For each degree of freedom j, we now define a 6 × 6 matrix with elements

$$D^j_{\alpha\beta}(t) = \langle\langle A_{j\alpha}|\rho_{re}(t)A_{j\beta}\rangle\rangle \equiv \mathrm{Tr}[A^\dagger_{j\alpha}\rho_{re}(t)A_{j\beta}]$$
$$\alpha,\beta = 0,1,\ldots 5 \tag{9}$$

and the Liouville space projection operator

$$P(t) = \sum_{j=1}^{N} \sum_{\alpha,\beta=0}^{5} |\rho_{re}(t)A_{j\alpha}\rangle\rangle[D^j(t)]^{-1}_{\alpha\beta} \langle\langle A_{j\beta}| \tag{10}$$

having the following properties:

a) $P(t)\,P(t') = P(t);\quad t > t'$
$$P^2(t) = P(t) \tag{11a}$$

b) $$P(t)|\rho(t)\rangle\rangle = |\rho_{re}(t)\rangle\rangle \tag{11b}$$

c) $\langle\langle A_{j\alpha}|P(t)\dot{\rho}(t)\rangle\rangle = \langle\langle A_{j\alpha}|\dot{\rho}_{re}(t)\rangle\rangle$

$$\text{where } \dot{\rho}(t) = \frac{d\rho(t)}{dt} \text{ and } \dot{\rho}_{re}(t) = \frac{d\rho_{re}(t)}{dt} \tag{11c}$$

d) $$P(t)|\rho(t)A_{j\beta}\rangle\rangle = |\rho_{re}(t)A_{j\beta}\rangle\rangle \quad . \tag{11d}$$

It has been shown in a separate communication [16] that the properties (11) can easily be derived using the definitions (9) and (10). P(t), therefore, is the projection operator, since it reduces the exact density matrix $\rho(t)$ to the simpler distribution $\rho_{re}(t)$.

Let us now assume that at some time t = t'

$$\rho(t') = \rho_{re}(t') \quad . \tag{12}$$

Using this assumption and introducing the complementary projection

$$Q(t) = 1 - P(t) \tag{13}$$

as shown in Ref. 16, we can write the exact reduced equations of motion (REM) for the $a_{j\alpha}(t)$'s from the quantum Liouville equation (5) as

$$\dot{a}_{j\alpha}(t) = - i\langle\langle A_{j\alpha}|L|\rho_{re}(t)\rangle\rangle + \sum_{\beta} M^j_{\alpha\beta}(t,t')a_{j\beta}(t) \quad , \tag{14a}$$

where we have introduced the 6 × 6 matrices

$$W^j_{\alpha\beta}(t,t') = -i\langle\langle A_{j\alpha}|LQ(t)U(t,t')|\rho_{re}(t')A_{j\beta}\rangle\rangle \quad , \tag{14b}$$

$$R^j_{\alpha\beta}(t,t') = \langle\langle A_{j\alpha}|U(t,t')|\rho_{re}(t')A_{j\beta}\rangle\rangle \quad , \tag{14c}$$

and

$$M^j_{\alpha\beta}(t,t') = \sum_{\gamma=0}^{5} W^j_{\alpha\gamma}(t,t')[R^j(t,t')]^{-1}_{\gamma\beta} \tag{14d}$$

Here U(t,t') is the time evolution operator

$$U(t,t') = \exp[-iL(t-t')] \quad . \tag{14e}$$

Equation (14d) can be recast in matrix notation:

$$M^j(t,t') = W^j(t,t')[R^j(t,t')]^{-1} \quad . \tag{14f}$$

Equation (14) describe the time evolution of the 5N dynamical quantities $a_{j\alpha}$ ($j = 1,\ldots N$; $\alpha = 1,2,\ldots 5$) and are exact. There are 5N nonlinear coupled differential equations for 5N unknown $a_{j\alpha}(t)$. In these equations the time derivative of $a_{j\alpha}$ at time t depends on all $a_{j\beta}$ at the same time. Note that we assume A_{j0} to be the unit operator, and normalization requires its expectation value to be independent of time, $a_{j0} = 1$. An alternative derivation of Eqs. (14) is also possible [14,16], where the time derivative of $a_{j\beta}$ at time t depends on all $a_{j\beta}$ at previous time $t' < s < t$, and the resulting equations are

$$\dot{a}_{j\alpha}(t) = -i\langle\langle A_{j\alpha}|L|\rho_{re}(t)\rangle\rangle - \int_{t'}^{t} ds \, \langle\langle A_{j\alpha}|L\bar{M}(t,s)Q(s)L|\rho_{re}(t)\rangle\rangle \tag{15a}$$

where

$$\bar{M}(t,s) = \exp[-i\int_0^t d\tau \; Q(\tau)L] \tag{15b}$$

is a time-ordered exponential. Now if we assume that condition (12) holds for all times, then $Q(t)\rho(t) = 0$, and we are left with the first term on the right-hand side of both Eqs. (14a) and (15a), which represents a mean field time evolution of the N-particle system. The second terms are the correlation terms and arise from the fact that $\rho(t) \neq \rho_{re}(t)$ for all times. If we retain up to a given order in the correlation terms in Eqs. (14a) and (15a), then they yield different approximations. However, in this paper we are interested only in the mean field time evolution of the N-particle system, where the time evoution of the expectations of the dynamical quantities, $A_{j\alpha}$, are given by

$$\dot{a}_{j\alpha}(t) = i/\hbar \, \mathrm{Tr}\{A_{j\alpha}[H,\rho_{re}(t)]\} \quad . \tag{16}$$

For our convenience, however, we evaluate explicitly the time evolution of the dynamical quantities

$$\sigma_{j1} = \langle x_j\rangle, \quad \sigma_{j2} = \langle p_j\rangle \tag{17a,b}$$

$$\sigma_{j3} = \langle x_j^2\rangle - \langle x_j\rangle^2, \quad \sigma_{j4} = \langle p_j^2\rangle - \langle p_j\rangle^2 \tag{17c,d}$$

$$\sigma_{j5} = [\langle x_jp_j + p_jx_j\rangle - 2\langle x_j\rangle\langle p_j\rangle] \quad , \tag{17e}$$

given by [16]

$$\dot{\sigma}_{j1} = \sigma_{j2}/m_j \quad , \tag{18a}$$

$$\dot{\sigma}_{j2} = -\langle V_j'(X)\rangle \quad , \tag{18b}$$

$$\dot{\sigma}_{j3} = \sigma_{j5}/m_j \quad , \tag{18c}$$

$$\dot{\sigma}_{j4} = -\langle V_j''(X)\rangle\sigma_{j5} \quad , \tag{18d}$$

$$\dot{\sigma}_{j5} = 2\{\frac{\sigma_{j4}}{m_j} - \langle V_j''(X)\rangle\sigma_{j3}\} \quad , \tag{18e}$$

where

$$V_j'(X) = \frac{\partial V}{\partial x_j} \quad , \tag{19a}$$

$$V_j''(X) = \frac{\partial^2 V}{\partial x_j^2} \quad , \tag{19b}$$

$$\langle V_j'(X)\rangle = \int_{-\infty}^{+\infty} dX \; V_j'(X)\rho_{re}(X,X;t) \quad , \tag{19c}$$

$$\rho_{re}(X,X;t) = \prod_{j=1}^{N} \phi_j(x_j,x_j;t) \tag{19d}$$

$$\phi_j(x_j,x_j;t) = \frac{1}{\sqrt{2\pi\sigma_{j3}}} \exp[-\frac{(x_j-\sigma_{j1})^2}{2\sigma_{j3}}] \quad . \tag{19e}$$

$\langle V_j''(X)\rangle$ is expressed in a similar way to Eq. (19c) by replacing $V_j'(X)$ with $V_j''(X)$. Equations (18) are the time-dependent self-consistent field (TDSCF) equations. For each particle j we obtain a closed set of five equations, which show correct self-consistent behavior in any potential and are coupled to each other. The first two equations in (18) express Ehrenfest's Theorem [17], and the third and fourth give a measure of the uncertainty in position and momentum measurement in the system. The fifth equation appears only when we are treating systems in a mixed state. For systems in a pure-state,

$$\sigma_{j5}^2 = (4\sigma_{j3}\sigma_{j4} - \hbar^2) \quad . \tag{20}$$

From Eq. (19) we find that for a successful application of the REM, the choice of the form of ϕ_j is crucial. Our particular choice, as described by Eqs. (2) and (19e), is by no means unique. We are motivated by the physical consideration which is the maximum entropy principle [10]. Such choice for ϕ_j connects the present semiclassical procedure with the more general problem of the derivation of REM in nonequilibrium statistical mechanics [18,20].

In the following we show that the choice of the dynamical quantities as described by Eq. (3) produces a Gaussian distribution for each ϕ_j in one-dimensional phase space. We now derive explicitly the phase-space representation (q,p) for one degree of freedom. The proof holds, however, for any N since we represent our reduced density function $\rho_{re}(X,X';t)$ by a product of N single- particle density functions $\phi_j(x_j,x_j';t)$ (Eq. (1)). We therefore, from now on, choose to drop the subscript j and replace Eq. (1) by

$$\sigma(x,x',t) \equiv \langle x|\sigma(t)|x'\rangle \tag{21a}$$

with

$$\sigma(t) = \exp[\sum_{\alpha=0}^{5} \lambda_\alpha(t) A_\alpha] \quad . \tag{21b}$$

As shown in Ref. 16, the Wigner representation [19] of the density operator, $\sigma(t)$, may be written in the form

$$\sigma_W(q,p;t) = \frac{1}{\pi}[\alpha\beta-\gamma^2/4]^{\frac{1}{2}}\exp[(\delta^2\beta+\phi^2\alpha-\gamma\delta\phi)/(4\alpha\beta-\gamma^2)]$$
$$\times \exp[\alpha q^2+\beta p^2+\gamma pq+\delta q+\phi p]$$

with

$$\iint dqdp\ \sigma_W(q,p,t) = 1 \quad , \tag{22}$$

and the corresponding coordinate representation is obtained from the transformation

$$\sigma(q+s,q-s;t) = \int_{-\infty}^{\infty} dp\ \sigma_W(q,p;t)\exp[2ips/\hbar] \quad . \tag{23}$$

Using the substitutions

$$q = (x+x')/2, \quad s = (x-x')/2 \tag{24}$$

in Eq. (23), we obtain

$$\begin{aligned}\sigma(x,x';t) = {} & \frac{1}{2\sqrt{\pi\beta}}(\gamma^2-4\alpha\beta)^{\frac{1}{2}}\exp[(2\beta\delta-\gamma\phi)^2/4\beta(4\alpha\beta-\gamma^2)] \\ & \times \exp[\tfrac{1}{4}(\alpha-\tfrac{\gamma^2}{4\beta})(x+x')^2 + \tfrac{1}{2}(\delta-\tfrac{\gamma\phi}{2\beta})(x+x') \\ & + \frac{1}{4\hbar^2\beta}(x-x')^2 - \frac{i\gamma}{4\beta\hbar}(x^2-x'^2) - \frac{i\phi}{2\beta\hbar}(x-x')] \quad ,\end{aligned} \tag{25}$$

with α, β, γ, δ and ϕ being real parameters, which may be expressed in terms of $\lambda_1,\ldots,\lambda_5$. Equations (22) and (25) accomplish our goal of expressing the maximum entropy distribution [Eq. (21b)] in phase space (q,p) and in the coordinate representation (x,x'). However, to obtain the TDSCF set of equations (18), we have used a different form of representation of these distribution functions, which were obtained by expressing Eqs. (22) and (25) in terms of the expectations of the dynamical quantities described by Eq. (17). They are related to the parameters by

$$\sigma_1(t) = \langle x(t)\rangle = (2\beta\delta-\gamma\phi)(\gamma^2-4\alpha\beta)^{-1} \tag{26a}$$

$$\sigma_2(t) = \langle p(t)\rangle = (2\alpha\phi-\gamma\delta)(\gamma^2-4\alpha\beta)^{-1} \tag{26b}$$

$$\sigma_3(t) = \langle x^2(t)\rangle - \langle x(t)\rangle^2 = 2\beta(\gamma^2-4\alpha\beta)^{-1} \tag{26c}$$

$$\sigma_4(t) = \langle p^2(t)\rangle - \langle p(t)\rangle^2 = 2\alpha(\gamma^2-4\alpha\beta)^{-1} \tag{26d}$$

$$\sigma_5(t) = \langle xp+px\rangle - 2\langle x\rangle\langle p\rangle = -2\gamma(\gamma^2-4\alpha\beta)^{-1} \tag{26e}$$

or

$$\alpha = c\sigma_4, \quad \beta = c\sigma_3, \quad \gamma = -c\sigma_5$$

$$\delta = c(\sigma_2\sigma_6 - 2\sigma_1\sigma_4), \quad \phi = c(\sigma_1\sigma_5 - 2\sigma_2\sigma_3) \quad ,$$

where

$$c = \tfrac{1}{2}(\gamma^2-4\alpha\beta) = \frac{2}{\sigma_5^2-4\sigma_3\sigma_4} . \tag{27}$$

Expressing the phase-space density function $\sigma_w(q,p,t)$ of Eq. (22) in terms of the $\sigma_j(t)$'s, we have

$$\sigma_w(q,p;t) = \frac{1}{\pi(4\sigma_3\sigma_4-\sigma_5^2)^{\frac{1}{2}}} \exp\{-\frac{2}{4\sigma_3\sigma_4-\sigma_5^2} [\sigma_4(q-\sigma_1)^2 + \sigma_3(p-\sigma_2)^2 - \sigma_5(q-\sigma_1)(p-\sigma_2)]\} , \tag{28}$$

and the corresponding coordinate representation (Eq. (25)) becomes

$$\sigma(x,x';t) = \frac{1}{\sqrt{2\pi\sigma_3}} \exp(-\sigma_1^2/2\sigma_3) \times \exp[-\frac{1}{8\sigma_3}(x+x')^2 + \frac{\sigma_1}{2\sigma_3}(x+x') + \frac{1}{4\hbar^2\sigma_3 c}(x-x')^2 + \frac{i\sigma_5}{4\sigma_3\hbar}(x^2-x'^2) - \frac{i}{2\sigma_3\hbar}(\sigma_1\sigma_5-2\sigma_2\sigma_3)(x-x')] . \tag{29}$$

Thus, the particular choice of the dynamical quantities, as depicted in Eq. (3), generates a Gaussian form for the representation of the corresponding single-particle distribution functions. In the following section we shall show that condition (20) reduces these mixed-state density functions [Eqs. (28) and (29)] to that of pure states. We shall also derive the REM for pure states.

III. REDUCED EQUATIONS OF MOTION (REM) FOR THE PURE-STATE DENSITY FUNCTION

Using the same projection scheme as in Sec. II, a self-consistent description for the time evolution of the pure-state density function can also be obtained. Following Heller [8], we define the reduced density function for pure states in the coordinate representation as

$$\sigma_H(x,x';t) = (\frac{2\alpha_1}{\pi\hbar})^{\frac{1}{2}} \exp[-\frac{\alpha_1}{\hbar}\{(x-x_t)^2 + (x'-x_t)^2\} + \frac{i}{\hbar}\alpha_2\{(x-x_t)^2 - (x'-x_t)^2\} + \frac{i}{\hbar}p_t(x-x')] , \tag{30}$$

where the parameters α_1, α_2, x_t and p_t are related to the $\sigma_i(t)$'s as follows:

$$\sigma_1(t) = x_t, \quad \sigma_2(t) = p_t , \tag{31a,b}$$

$$\sigma_3(t) = \frac{\hbar}{4\alpha_1}, \quad \sigma_4(t) = \frac{\hbar|\alpha|^2}{\alpha_1}; \qquad \alpha = i\alpha_1 + \alpha_2 \tag{31c,d}$$

$$\sigma_5(t) = \frac{\hbar\alpha_2}{\alpha_1} = [4\sigma_3(t)\sigma_4(t)-\hbar^2]^{\frac{1}{2}} . \tag{31e}$$

Therefore, $\sigma_H(x,x';t) = \frac{1}{\sqrt{2\pi\sigma_3}} \exp(-\frac{\sigma_1^2}{2\sigma_3})\exp[-\frac{1}{4\sigma_3}(x^2+x'^2) + \frac{\sigma_1}{2\sigma_3}(x+x')$

$$+ \frac{i\sigma_5}{4\hbar\sigma_3}(x^2-x'^2) - \frac{i}{2\sigma_3\hbar}(\sigma_1\sigma_5-2\sigma_2\sigma_3)(x-x')] \quad , \qquad (32)$$

and the corresponding phase-space representation is given by

$$\sigma_{HW}(q,p,t) = \frac{1}{\pi\hbar}\exp[-\frac{2}{\hbar^2}\{\sigma_4(q-\sigma_1)^2$$

$$+ \sigma_3(p-\sigma_2)^2 - \sigma_5(q-\sigma_1)(p-\sigma_2)\}] \quad , \qquad (33)$$

where σ_5 is given by (31e), which is the same as condition (20). These pure-state density functions can also be obtained directly from the mixed-state density functions (Eqs. (28) and (29)) using condition (20).

We now assume that the time evolution of the pure-state system is described by the approximate density functions, (32) and (33), for all times. This assumption then allows us to construct the SCF set of equations for the expectations of the corresponding dynamical quantities using Eq. (16). They are

$$\dot{\sigma}_1(t) = \frac{\sigma_2(t)}{m} \qquad (34a)$$

$$\dot{\sigma}_2(t) = -\langle V'(x)\rangle \qquad (34b)$$

$$\dot{\sigma}_3(t) = \frac{1}{m}(4\sigma_3(t)\sigma_4(t)-\hbar^2)^{\frac{1}{2}} \qquad (34c)$$

$$\dot{\sigma}_4(t) = -(4\sigma_3(t)\sigma_4(t)-\hbar^2)^{\frac{1}{2}}\langle V''(x)\rangle \quad . \qquad (34d)$$

This is a closed set of four equations which differ from the first four equations in the mixed case [Eqs. (18a) through (18d)] due to the fact that σ_5 is no longer an independent variable [Eq. (20)], and for the same reason we do not have any REM for σ_5 in the pure case.

The REM of Eq. (34) are obtained from Eq. (16) under the exact potential of the system. These equations, as shown below, are different from those obtained by solving the quantum Liouville equation (5) for pure states under the locally quadratic potential approximation [8]. They are

$$\dot{\sigma}_1(t) = \frac{\sigma_2(t)}{m} \qquad (35a)$$

$$\dot{\sigma}_2(t) = -V'(x)\big|_{x=\sigma_1} \qquad (35b)$$

$$\dot{\sigma}_3(t) = (4\sigma_3(t)\sigma_4(t)-\hbar^2)^{\frac{1}{2}}/m \qquad (35c)$$

$$\dot{\sigma}_4(t) = -(4\sigma_3(t)\sigma_4(t)-\hbar^2)^{\frac{1}{2}}V''(x)\big|_{x=\sigma_1} \quad . \qquad (35d)$$

Here the first two equations describe the classical motion for a system in a pure state and are not coupled with the other two equations [(35c) and (35d)], which describe the time evolution of the variances (Eqs. (17c), (17d)). Therefore, the present set of equations describe the trajectory of a particle whose position and momentum at time t are known from the center of the wavepacket. However, the trajectory of a particle is described by the wavepacket [Eqs. (32) and (33)] as a whole, which inevitably has certain spacial extension.

On the other hand, if we look at the SCF set of equations (34), we find that the first two equations (a and b) are coupled with the other two equations. Again, the right-hand side of Eq. (34b) is equal to the average of the force over the whole wavepacket and thus differs from Eq. (35b) due to the fact

$$\langle V'(x)\rangle \neq V'(x)\big|_{x=\sigma_1} \quad . \tag{36}$$

In Sec. V we shall analyze the relative meris of these approaches by studying the vibrational motion of an arbitrary diatomic molecule on a Morse potential surface. It is important to note here that for the mixed case, even if we start with a minimum uncertainty wavepacket the variances σ_3 and σ_4 are not constants of motion as the system evolves, which is evident from Eqs. (18). In deriving expressions (32) and (33) we assumed that the N-particle density function may be written as [20]

$$\rho(X,X';t) = \Psi(X,t)\ \Psi^*(X',t) \quad , \tag{37}$$

where $\Psi(X,t)$ is the exact wave function of the N-particle interacting system. We then introduced the approximation

$$\Psi(X,t) = \prod_{j=1}^{N} \psi_j(x_j,t) \quad , \tag{38}$$

where the $\psi_j\ (x_j,t)$'s are the single-particle wave functions and contain all information about the single-particle systems, including their phase. A reduced description of these single-particle wave functions was first introduced by Heller [8], which in terms of the σ_i's may be written as

$$\psi(x,t) = (2\pi\sigma_3)^{-\frac{1}{4}}\ \exp\{(\frac{1}{4\sigma_3} + \frac{i\sigma_5}{4\hbar\sigma_3})(x-\sigma_1)^2 + \frac{i\sigma_2}{\hbar}(x-\sigma_1) + \frac{i\gamma}{\hbar}\} \quad , \tag{39}$$

where for notational convenience we have dropped the subscript j. σ_5 is given by condition (20), and γ is the phase-factor. The density function corresponding to this wave function is given by expression (32), which we obtained from the maximum entropy distribution (29) using condition (20). Therefore, if we assume this Gaussian wave function (39) to approximate the exact single particle wave function for all times, the time evolution of this reduced wave function under the exact potential of the system can be obtained by solving the SCF set of REM given by Eqs. (34), along with the equation for the phase factor

$$\dot{\gamma}(t) = -\frac{\hbar}{4m\sigma_3} + \frac{V_2\sigma_3}{2} + \frac{\sigma_2^2}{2m} - V_0 \tag{40}$$

which is obtained from the Schrödinger equation

$$\langle E\rangle = \langle\frac{p^2}{2m} + V(x)\rangle = i\hbar\langle\psi|\dot{\psi}\rangle \tag{41}$$

The quantities V_0, V_2 in Eq. (40) are given by

$$V_0 = \langle V\rangle, \quad V_2 = \langle\frac{\partial^2 V}{\partial x^2}\rangle \quad .$$

Heller first evaluated this propogator (39) under the locally-quadratic potential approximation [8].

IV. CANONICAL DENSITY FUNCTION FOR A HARMONIC SYSTEM

In this section we show that the TDSCF set of equations (18), which describe the time evolution of any irreversible process under the exact potential of the system by using the reduced density matrix expressions (28) and (29), can be used to describe the time evolution of a harmonic system in thermal equilibrium [15]. When a system is in thermal equilbrium, we have the density matrix satisfying maximum entropy principle as [10]

$$\hat{\sigma}_T(t) = \exp(-\beta H)/\mathrm{Tr}[\exp(-\beta H)] \quad , \tag{42}$$

where $\beta = (kT)^{-1}$ and H is the Hamiltonian of the system. Under the quadratic potential approximation, where

$$H(q,p) = p^2/2m + \tfrac{1}{2}m\omega^2 q^2 \quad , \tag{43}$$

a Gaussian form of representation of the density operator (42) can be obtained, which in the coordinate representation (x,x') is given by

$$\sigma_T(x,x';t) = \left[\frac{m\omega \tanh(\beta\hbar\omega/2)}{\pi\hbar}\right]^{\frac{1}{2}} \times \exp\left\{\frac{-m\omega}{2\hbar \sinh(\beta\hbar\omega)}[(x^2+x'^2)\cosh(\beta\hbar\omega)-2xx']\right\} \quad . \tag{44}$$

Expectation values of the dynamical quantities (17), with respect to this density matrix, are

$$\sigma_1 = 0 \quad , \quad \sigma_2 = 0$$
$$\sigma_3 = \frac{\hbar}{2m\omega}\coth(\tfrac{1}{2}\beta\hbar\omega) \quad , \quad \sigma_4 = \tfrac{1}{2}m\omega\hbar\coth(\tfrac{1}{2}\beta\hbar\omega) \quad , \quad \sigma_5 = 0 \quad , \tag{45}$$

where for convenience we have dropped the j-subscript. Now expressing the thermal density function σ_T in terms of the $\sigma_i(t)$'s, we obtain

$$\sigma_T(x,x';t) = \frac{1}{\sqrt{2\pi\sigma_3}}\exp\left\{-\frac{1}{8\sigma_3}(x+x')^2 - \frac{\sigma_4}{2\hbar^2}(x-x')^2\right\} \quad , \tag{46}$$

and the corresponding phase-space density function becomes

$$\sigma_{TW}(q,p;t) = \frac{1}{2\pi\sqrt{\sigma_3\sigma_4}}\exp\left[-\frac{1}{2\sigma_3}q^2 - \frac{1}{2\sigma_4}p^2\right] \quad . \tag{47}$$

The time evolution of these density functions are found by solving the set of equations (18) with initial conditions given by Eq. (45) and the interaction potential given by expression (43). This is because Eqs. (46) and (47) do not satisfy condition (37). Our development, as described in Sec. II, however, is more general since it can be used for studying the relaxation of a system to thermal equilibrium with a thermal bath under the exact potential of the system.

V. VIBRATIONAL MOTION OF AN ARBITRARY DIATOMIC MOLECULE ON A MORSE POTENTIAL SURFACE

In this section we solve the TDSCF set of REM (34) to describe the vibrational motion of a diatomic molecule. We consider a diatomic molecule with two electronic states, a ground state $|g\rangle$ and an excited state $|e\rangle$. Its Hamiltonian is

$$H = |g\rangle H_g \langle g| + |e\rangle (W_{g,e} + H_e) \langle e| \quad . \tag{48}$$

We assume the ground state potential to be harmonic and the excited state potential to be given by a Morse oscillator. We then have

$$H_g = \frac{p^2}{2\mu} + \tfrac{1}{2}\mu\omega_g^2 (x-x_g)^2 \tag{49a}$$

$$H_e = \frac{p^2}{2\mu} + W_{g,e} + D_e(1-e^{-\beta(x-x_e)})^2 \quad , \tag{49b}$$

where μ is the reduced mass of the molecule, ω_g is the vibrational frequency on the lower potential surface, $W_{g,e}$ is the excitation energy from lower to the upper surface, D_e is the equilibrium dissociation energy of the upper potential surface, and β is a constant given by [21]

$$\beta = \{\frac{2\pi^2 c\mu}{\bar{D}_e h}\}^{\frac{1}{2}} \omega_e \quad , \tag{49c}$$

where c is the velocity of light, h is Planck's constant, and ω_e is the vibrational frequency that the anharmonic oscillator would have classically for an infinitesimal amplitude. For our purposes we assume

$$\omega_e = 4395 \text{ cm}^{-1}, \quad \mu = 0.5 \text{ a.u.}, \quad \bar{D}_e = 38{,}310 \text{ cm}^{-1}, \quad D_e = hc\bar{D}_e$$

$$\beta = 1.93 \text{ Å}^{-1}, \quad x_g = 0.504 \text{ Å}, \quad W_{g,e} = 26{,}230 \text{ cm}^{-1}, \quad x_e = 0.6325 \text{ Å} \quad . \tag{50}$$

We consider the molecule initially to be in its ground vibrational state satisfying the minimum uncertainty condition

$$\sigma_3\sigma_4 = \frac{\hbar^2}{4} \quad , \tag{51}$$

and we set initially $\sigma_1 = 0.604$ Å and $\sigma_2 = 0.0$ gm cm/s. We now assume that at time t = 0 there is a Franck-Condon transition from the ground to the excited potential surface. After this transition the molecule will start executing vibrational motion about the excited state equilbrium position X_e. To study this vibrational motion, we solve the TDSCF set of REM (34) in dimensionless form, where the dimensionless quantities $\hat{\sigma}_i$'s are related to the σ_i's as

$$\hat{\sigma}_1 = \{\frac{m\omega}{\hbar}\}^{\frac{1}{2}} \sigma_1, \quad \hat{\sigma}_2 = \frac{1}{\sqrt{m\hbar\omega}} \sigma_2, \quad \hat{\sigma}_3 = \frac{m\omega}{\hbar} \sigma_3, \quad \hat{\sigma}_4 = \frac{1}{m\hbar\omega} \sigma_4 \tag{52}$$

where $\omega = 2\pi c\omega_e$ with initial conditions (50), $\hat{\sigma}_3 = 5.0$, $\hat{\sigma}_4 = 0.05$ and with up to 200 time steps on the order of $\sim 0.3\times10^{-15}$ s. Variations of $\hat{\sigma}_1(t)$ and $\hat{\sigma}_2(t)$ with time are shown in Figs. 1 and 2, respectively. We have used the ordinary differential equation solver technique of Gear [22] to solve the TDSCF set of equations (34). In Fig. 3 we elaborate further on the performance of our SCF approach by tracing the path of σ_1 over the excited state potential surface. As evident from Figs. 1 and 2, given the initial σ_1 and σ_2 on the potential surface, which for the present case is $\sigma_1 = 0.604$ Å and $\hat{\sigma}_2 = 0.0$ gm cm/s, our TDSCF method describes anharmonic vibrational motion of the diatomic molecule over this surface from $\sigma_1 = 0.604$ Å to 1.08 Å.

For the sake of comparison, we also solve in dimensionless form the TDSCF set of REM (35), which describe the variations of $\sigma_i(t)$'s with time under the quadratic potential approximation. We use the same set of initial conditions as above. Time variations of $\hat{\sigma}_1(t)$ and $\hat{\sigma}_2(t)$ for the present case are shown in Figs. 4 and 5, respectively. In Fig. 3, we trace the path of $\sigma(t)$ obtained under the quadratic potential approximation using a dashed line to illustrate the performance of Heller's method compared to our TDSCF method.

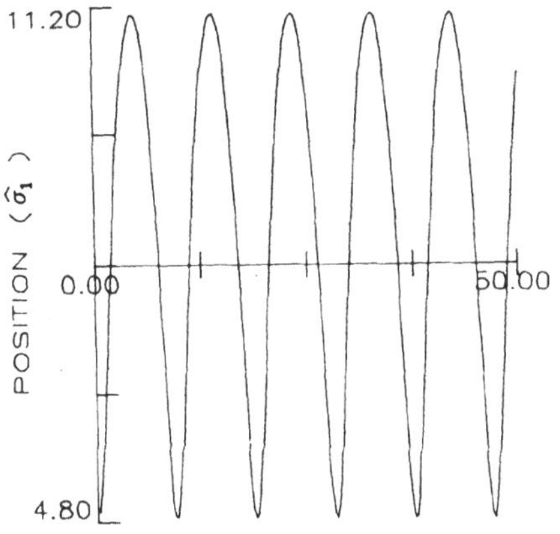

1

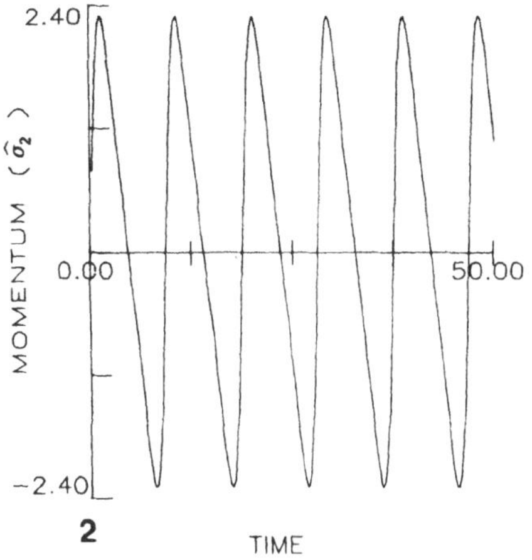

2

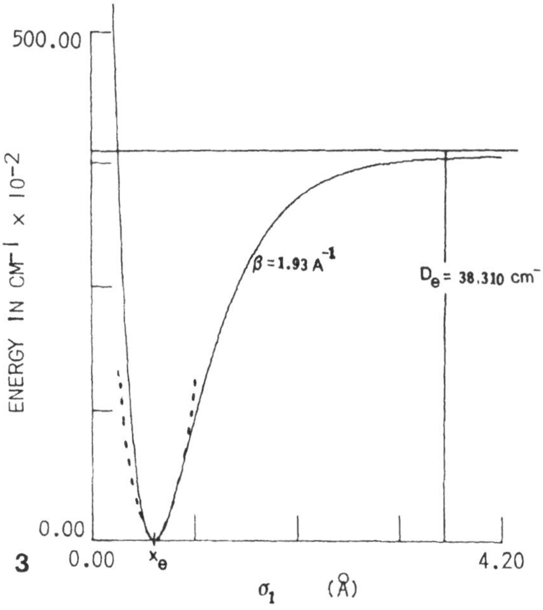

3

Fig. 1. The dimensionless mean displacement $\hat{\sigma}_1$ vs time for a Gaussian wavepacket propogated on a Morse potential.

Fig. 2. The dimensionless mean momentum $\hat{\sigma}_2$ vs time for a Gaussian wavepacket on a Morse potential.

Fig. 3. The Morse (solid line) potential function of an arbitrary diatomic molecule. The dashed line is obtained by considering up to the quadratic terms in $\sigma_1(t)$ in the Morse function.

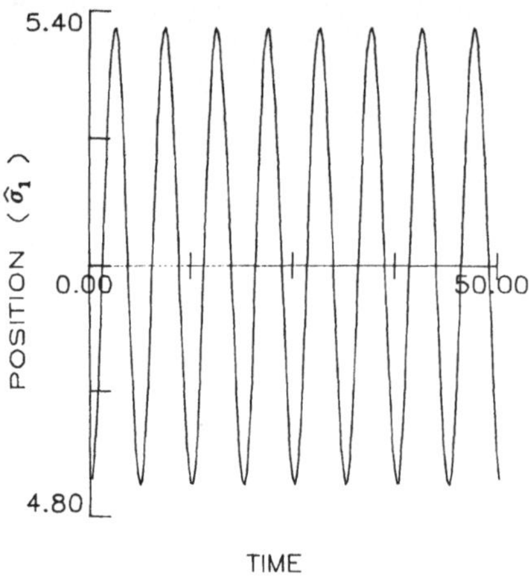

Fig. 4. The dimensionless mean displacement $\hat{\sigma}_1$ vs time for a Gaussian wavepacket propogated on a locally quadratic form of the Morse potential.

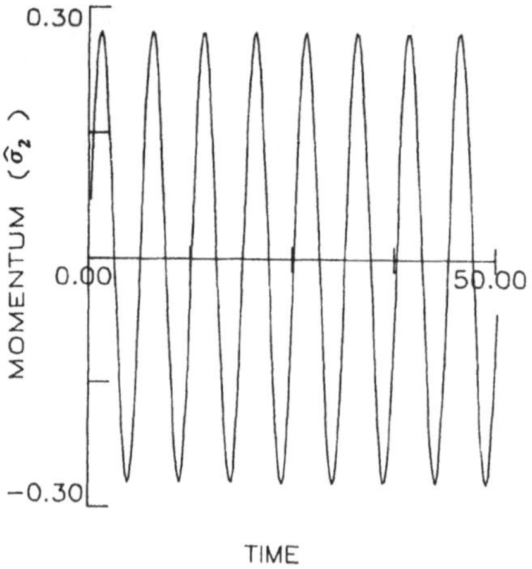

Fig. 5. The dimensionless mean momentum $\hat{\sigma}_2$ vs time for a Gaussian wavepacket propogated on a locally quadratic form of the Morse potential.

VI. DISCUSSION

We have described a method for studying the dynamical properties of irreversible statistical systems. Irreversibility is introduced into our system through quantum measurements [20], and this enables us to make use of the maximum entropy-based formulation (MEF). Use of MEF in constructing the reduced density matrix (2) eliminates the necessity of performing tedious thermal averaging [9]. Therefore, the present TDSCF method will be particularly suitable for studying the various dynamical processes in condensed phases. The present development resembles the derivation of thermodynamic theorems from statistical mechanics due to the fact that the construction of the density matrix and the corresponding REM are independent of the specific nature of the Hamiltonian. For this reason, we find that the present method can be used to describe the relaxation of a system to thermal equilibrium with a thermal bath under the exact and the quadratic potential approximations. Under certain conditions (20), the present method can also be used to describe the time evolution of systems in pure states. The derivation of the REM are based on a projection scheme, and the projection operators are defined in terms of the MEF density matrix. The TDSCF set of equations (35) and (40), which describe the time evolution of pure states, have been shown to be quite useful for describing a variety of molecular dynamical processes, including molecular scattering, electronic spectra, dissociation of clusters and thermal desorption from surfaces [9,23]. The present phase space TDSCF method enjoys all these advantages.

In deriving the TDSCF set of equations, we have not had to make the assumption that the exact nonequilibrium statistical density is in some sense approximately equal to the local equilibrium one, and thus the present method is much more general than the local equilbirium formulations. A close look at our TDSCF set of equations (18) shows that they do not contain $\hbar$. That is, even though we started our development using the quantum Liouville equation, the time evolution of our MEF-based density functions (28) and (29) is described by a classical TDSCF set of REM. Therefore, the present MEF-based TDSCF method is completely classical. This in turn suggests that the present procedure may be repeated for classical mechanics by replacing L in Eq. (5) with the classical Liouville operator. Each single-particle density function $\phi_j(x_j,x_j';t)$ should then be replaced by a phase space distribution which is Gaussian in x_j and p_j (28). We can then repeat the present procedure to obtain the TDSCF set of REM (18), and hence to confirm their classical nature. Our TDSCF method represents the lowest order of a systematic expansion, (14) and (15), and may therefore be improved by incorporating correlation terms order by order. Inclusion of the correlation terms will cause our REM to contain $\hbar$, and hence will depart from the classical picture. Therefore, the correlation terms may be considered as quantum corrections to our classical description [19]. However, for harmonic systems with normal mode x_j's, the TDSCF set of REM are exact. An alternative way to improve our TDSCF description would be to include cubic and higher moments to construct each single-particle density matrix $\phi_j(x_j,x_j';t)$. This would then be a departure from the Gaussian picture.

Although the inclusion of the correlation terms, (14) and (15), and the higher moments, (2), would improve our TDSCF description, the product ansatz, (1), for the N-particle density function implies neglect of exchange effects and an incomplete account of quantum mechanical correlations. This is one of the limitations of our single-particle description of an interacting N-particle statistical system. Implementation of the exchange effects for equilibrium Bose and Fermi systems are available in the literature [4,24]. Again, the present development is restricted to one dimensional phase space. Extension to the simulation of equilibrium and nonequilibrium statistical systems in three-dimensional phase space will be reported in the future.

ACKNOWLEDGMENTS

This research was supported by the Office of Naval Research, the Air Force Office of Scientific Research (AFSC), United States Air Force, under Contract F49620-86-C-0009, and the National Science Foundation under Grant CHE-8620274. The United States Government is authorized to reproduce and redistribute reprints for governmental purposes notwithstanding any copyright notatation hereon.

REFERENCES

[1] Various applications using classical methods are reviewed by H. L. Friedman, "A Course in Statistical Mechanics," Prentice-Hall, Englewood Cliffs, New Jersey (1985), and by J. P. Hansen and I. R. McDonald, "Theory of Simple Liquids," Academic Press, New York (1976).

[2] R. M. Stratt and W. H. Miller, J. Chem. Phys. 67:5894 (1977); J. G. Powles and G. Rickayzen, Mol. Phys. 38:1875 (1979).

[3] Several methods are reviewed by D. M. Ceperley and M. H. Kalos in: "Monte Carlo Methods in Statistical Physics," K. Binder, ed., Springer-Verlag, New York (1984), Chapt. 4.

[4] J. Arponen, Ann. Phys. (NY) 151:311 (1983), and references therein; J. Arponen, R. F. Bishop and E. Pajanne, preprint, 1987.

[5] Y. Takahasi and H. Umezawa, in: "Collective Phenomena," Vol. 2, Gordon and Breach, London (1975), pp. 55-80; H. Matsumoto, Y. Nakano, H. Umezawa, F. Mancini and M. Marinaro, Prog. Theor. Phys. 70:599 (83); I. Ojima, Ann. Phys. (NY) 137: 1 (1981); H. Matsumoto, I. Ojima and H. Umezawa, Ann. Phys. (NY) 152:348 (1984); H. Umezawa, H. Matsumoto and M. Tachiki, "Thermo Field Dynamics and Condensed States," North-Holland, Amsterdam (1982).

[6] A. Rahman, Phys. Rev. A 136:405 (1964); A. Rahman and F. H. Stillinger, J. Chem. Phys. 55:3336 (1971); L. Verlet, Phys. Rev. A 159:98 (1967); 165:201 (1968).
[7] P. A. Egelstaff, "An Introduction to the Liquid State," Academic Press, New York (1967), Chapt. 9; J. T. Hynes, Ph.D Thesis, Princeton, 1969. Quantum correction to classically-simulated I. R. and Raman Spectra are discussed by P. H. Bernes, S. R. White and K. R. Wilson, J. Chem. Phys. 74:4872 (1981); 75:515 (1981).
[8] E. J. Heller, J. Chem. Phys. 62:1544 (1975); 64:63 (1976); R. P. Feynman and A. R. Hibbs, "Quantum Mechanics and Path Integrals," McGraw-Hill, New York (1965).
[9] N. Corbin and K. Singer, Mol. Phys. 46:671 (1982); K. Singer and W. Smith, Mol. Phys. 57:761 (1986); R. B. Gerber, V. Buch and M. A. Ratner, J. Chem. Phys. 77:3022 (1982); R. D. Coalson and M. Karplus, Chem. Phys. Lett. 90: 301 (1982); J. Chem. Phys. 79:6150 (1983); D. Thirumalai and B. J. Berne, J. Chem. Phys. 79:5029 (1983); D. Thirumalai, E. J. Bruskin and B. J. Berne, ibid. 79:5063 (1983).
[10] E. T. Jaynes, Phys. Rev. 106:620 (1957); 108:171 (1957). For a detailed review, see "The Maximum Entropy Formalism," R. D. Levine and M. Tribus, eds., MIT Press, Cambridge (1978).
[11] B. Robertson, in: "The Maximum Entropy Formalism," R. D. Levine and M. Tribus, eds., MIT Press, Cambridge, (1978).
[12] H. Mori, J. Phys. Soc. Jpn. 11:1029 (1956); Phys. Rev. 112:1829 (1958); 115:298 (1959); H. Mori, I. Oppenheim and J. Ross, in "Studies in Statistical Mechancics," J. deBoer and G. E. Uhlenbeck, eds., Vol. I, North-Holland, Amsterdam (1962), p. 271 ff.
[13] J. A. McLennan, Phys. Fluids 4:1319 (1961); Adv. Chem. Phys. 5:261 (1963).
[14] B. Robertson, Phys. Rev. 144:151 (1966); 160:175 (1967); C. R. Willis and R. H. Picard, Phys. Rev. A 9:1343 (1974), and references therein; S. Mukamel, Phys. Rep. 93:1 (1982).
[15] R. P. Feynman, "Statistical Mechanics," Benjamin, New York (1972).
[16] A. Haque and T. F. George, Mol. Phys., submitted.
[17] See, for example, A. Messiah, "Quantum Mechanics," Vol. I, North-Holland, Amsterdam (1961), Chapt. 6.
[18] L. Onsager, Phys. Rev. 37:405 (1931); 38:2265 (1931); L. Onsager and S. Machlup, Phys. Rev. 91:1505 (1953); R. Zwanzig, Suppl. Prog. Theor. Phys. 64:74 (1978).
[19] E. P. Wigner, Phys. Rev. 40:749 (1932); M. Hillery, R. F. O'Connel, M. O. Scully and E. P. Wigner, Phys. Rep. 106:121 (1984).
[20] I. Prigogine, "From Being to Becoming," Freeman, New York (1980).
[21] G. Herzberg, "Spectra of Diatomic Molecules," van Nostrand Reinhold, New York (1950).
[22] A. C. Hindmarsh, "Gear: Ordinary Differential Equation System Solver," Lawrence Livermore Laboratory, Report UCID-30001, Revision 3 (December, 1974); C. W. Gear, "Numerical Initial Value Problems in Ordinary Differential Equations," Prentice-Hall, Englewood Cliffs, New Jersey (1971).
[23] S. Sawada, R. Heather, B. Jackson and H. Metiu, J. Chem. Phys. 83:3009 (1985); R. T. Skodje and D. G. Truhlar, J. Chem. Phys. 80:3123 (1984).
[24] E. Pollack and D. M. Ceperley, Phys. Rev. B 30:2555 (1984).

THEORY OF NEUTRON SCATTERING EXPERIMENTS ON MOMENTUM DISTRIBUTIONS IN QUANTUM FLUIDS

Richard N. Silver

MS B262, Theoretical Division
Los Alamos Neutron Scattering Center
Los Alamos National Laboratory
Los Alamos, NM 87545

I. INTRODUCTION

Momentum distributions are of fundamental interest to our understanding of the many body physics of quantum solids and fluids. Since the original suggestion by Hohenberg and Platzman,[1] there have been many experiments[2] with the goal of determining momentum distributions by scattering neutrons at momentum transfer high enough to invoke the impulse approximation (IA). The IA is questionable for helium, because the He-He potential is steeply repulsive at short distances leading to significant final state corrections. These must be understood in order to extract from experiment the parameters of interest, such as the Bose condensate fraction in ^{4}He or the Fermi surface discontinuity in ^{3}He. Most theories for final state corrections[1,3-5,6] have predicted a quasi-Lorentzian broadening of the IA. However, Gersch, et al.[7] argued, via a complex many-body cumulant derivation, that real space correlations result in a non-Lorentzian final state broadening. A simple quasiclassical theory for this prediction was given by Silver and Reiter,[8] who expressed the corrections in terms of the radial distribution function, g(r), and the He-He cross action. Until now a fully quantum theory, which included the correct physics for the quantitative correction of experiment, has been lacking.

In this paper, I present the first perturbative derivation of the final state corrections to the impulse approximation for deep inelastic neutron scattering experiments. The final state broadening is found to depend on g(r) and the He-He phase shifts. The theory satisfies the f-sum rule, the ω^2 sum rule ("kinetic energy") valid at high Q, and the ω^3 sum rule. In the structure of the theory, the self-energy terms alone would lead to quasi-Lorentzian broadening. However, these are exactly canceled by a part of the vertex terms which introduce g(r). Numerical results are presented for superfluid ^{4}He.

II. MOTIVATION

Let us review the physical picture first discussed by Gersch, et al.[7] and derived in a quasiclassical approximation by Silver and Reiter.[8]

A neutron scattering from a helium quantum fluid instantaneously imparts a momentum transfer Q and an energy transfer ω to an atom. The impulse approximation (IA) is obtained if Q and ω are large compared to any of the momentum and energy scales characterizing the fluid, so that the helium atom can be assumed to recoil freely. Then Q S(Q,ω) is a function only of a "scaling" variable, $Y = M(\omega - \hbar Q^2/2M)/\hbar Q$, and it is simply related to the initial momentum distribution n(p) according to

$$Q\,S_{IA}(Q,\omega) \equiv F_{IA}(Y) = \frac{Q}{\rho}\int \frac{d^3\vec{p}}{(2\pi)^3}\, n(\vec{p})\,\delta\!\left(\omega - \frac{\hbar Q^2}{2M} - \hbar\,\frac{\vec{Q}\cdot\vec{p}}{M}\right) \tag{1}$$

This equation has been extensively used in the analysis of experiment.[2]

However, the He-He potential is steeply repulsive at short distances violating the conditions for the IA. The final state scattering of the He atom by its neighbors should broaden the IA, according to

$$Q\,S(Q,\omega) \equiv F(Y) = \int_{-\infty}^{\infty} dY'\,R_{FS}(Y - Y')\,F_{IA}(Y') \tag{2}$$

$R_{FS}(Y)$ would be a Lorentzian[1] in the simple approximation that the He atom scatters at a constant rate $1/2\,\rho v\,\sigma_{He\text{-}He}$ where ρ is density, $v \equiv \hbar Q/M$ is velocity, and $\sigma_{He\text{-}He}$ is the cross section.

In reality, the final state scattering rate is not a constant because the initial positions of the helium atoms are strongly correlated. Figure 1 shows the He-He potential[9] and the radial distribution[10] function, g(r), for ^{4}He. The atoms sit in the attractive part of the potential (r > 2.7 Å) at some distance from the steeply repulsive core (r ≤ 2.3 Å) responsible for the final state scattering at high Q. In Wigner's quasiclassical approximation,[11] the variable Y is conjugate to the distance which the recoiling atom travels before reaching the core. As there are few collisions at very small collision distance, $R_{FS}(Y)$ should be narrower than the Lorentzian prediction and lack Lorentzian tails.

To obtain this physics in a fully quantum theory, I must retain the full correlations in the ground state, expressed through g(r), in a calculation of the dynamical scattering law, S(Q,ω), given by

$$S(Q,\omega) = \frac{Re}{\pi N}\int_0^{\infty} dt\,e^{i\omega t - \varepsilon t} < \hat{\rho}_{-Q}(0)\,e^{i\hat{H}t/\hbar}\,\hat{\rho}_Q(0)\,e^{-i\hat{H}t/\hbar} > \tag{3}$$

Here, $\hat{\rho}_Q(0) = \Sigma_k\, \hat{a}^+_{k+Q}\,\hat{a}_k$ is the density operator, N is the number of particles, <> denotes expectation values in the ground state of the full Hamiltonian, $\hat{H}$, including the interactions responsible for the strongly correlated g(r) and n(p). I do not wish to calculate g(r) and n(p). Rather, I take them as a given property of the ground state $|\psi_0>$ such as obtained by a variational or Monte Carlo calculation, or from a neutron diffraction experiment.

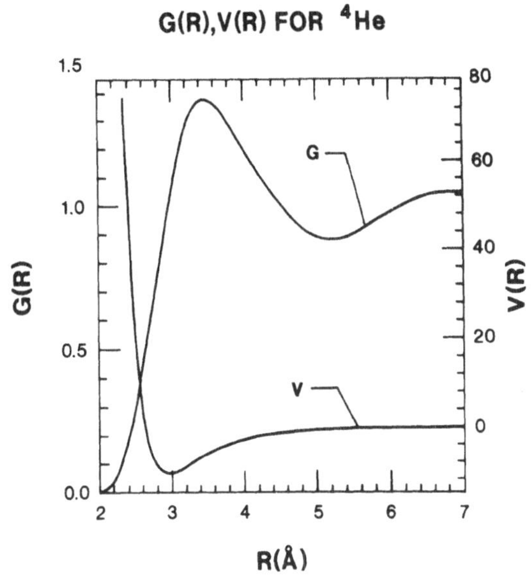

Fig. 1 He-He potential, V(r), and the radial distribution function, g(r), for ^{4}He at T = 0°K.

If I naively proceed to perturbatively expand only the time dependent part of (3), I find an infinite number of terms which diverge as inverse powers of $\omega - \hbar Q^2/2M + i\varepsilon$. A method for infinite order resummation is required. The Kubo formula for the frequency dependent electrical conductivity in a metal, $\sigma(\omega)$, has a form similar to Eq. (3). A similar problem occurs in which perturbation expansion results in an infinite number of divergent terms in inverse powers of $\omega + i\varepsilon$. One solution of this problem is to use Liouville perturbation theory and a diagonal projection operator method to resum all the singular terms.[12] I adopt a similar procedure to evaluate S(Q,ω), except that I use an off-diagonal projection operator appropriate for $Q \neq 0$. As is true for the perturbative derivation of Boltzmann equation answers for the resistivity starting from the Kubo formula, I find that vertex terms must be retained and these introduce the correlations, g(r).

In terms of Feynman diagrams, the perturbative expansion of Eq. (3) should yield the Dyson equation shown in Fig. 2. I assume that Q is sufficiently high that the dynamics of low momentum holes, created by $\hat{a}_k$ and represented by —◀— , occurs over a much longer time scale than the dynamics of high momentum particles, created by $\hat{a}^+_{k+Q}$ and represented by —▶— . The wiggly lines represent T-matrices. So the dynamics of holes can be ignored, but their instantaneous spatial correlations (defined by g(r,t) for t = 0) are important. If only the bare and self energy terms in Fig. 2 are included, the result would be quasi-Lorentzian broadening of the IA.[3] The vertex term in Fig. 2 includes a hole four point function, represented by ⊠, which is related to g(r) − 1. Inclusion of the vertex term yields a non-Lorentzian final state broadening of the IA. A precise meaning to the diagrams in Fig. 2 will be given in Secs. III and IV.

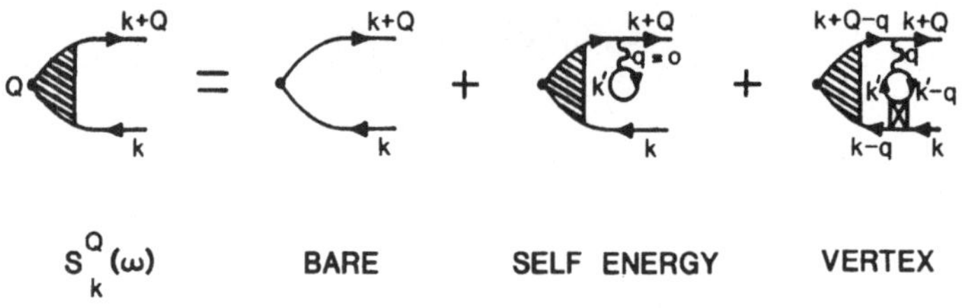

Fig. 2 Diagrammatic representation of the Dyson equation for deep inelastic neutron scattering.

III. HARD CORE PERTURBATION THEORY FOR DYNAMICS

In this section, I present a general framework for the Liouville perturbative expansion of S(Q,ω) in terms off-diagonal projection operator methods and ground state expectation values of products of creation and annhilation operators. In the following section I specialize to the particular case of "deep inelastic neutron scattering" (DINS) at high Q.

I define a "superoperator" as an operator $\tilde{S}$ which acts on the ordinary operator to its right, say $\hat{O}$, to create a new ordinary operator $\hat{O}'$, according to $\tilde{S}\hat{O} = \hat{O}'$. For example the Liouville superoperator $\tilde{L}$ is defined by

$$\tilde{L}\hat{O} = -[\hat{H},\hat{O}] \tag{4}$$

Here ˆ denotes an ordinary operator constructed out of sums of products of scalars and creation and annhilation operators, and ˜denotes a superoperator. If $\hat{H}$ can be written as the sum of kinetic $\hat{K}$, and potential, $\hat{V}$, terms, then similarly $\tilde{L} = \tilde{K} + \tilde{V}$.

Consider then the quantity which occurs in S(Q,ω), Eq. (3)

$$\hat{S}^Q(\omega) = \int_0^\infty dt e^{i\omega t - \varepsilon t} e^{i\hat{H}t/\hbar} \hat{\rho}_Q(0) e^{-i\hat{H}t/\hbar} \tag{5}$$

In terms of superoperators

$$\hat{S}^Q(\omega) = \frac{i\hbar}{\hbar\omega - \tilde{L}} \hat{\rho}_Q(0) \tag{6}$$

Eq. (6) can be expanded as a Dyson equation

$$\hat{S}^Q(\omega) = \frac{i\hbar}{\hbar\omega - \tilde{K} + i\varepsilon} \hat{\rho}_Q(0) + \frac{1}{\hbar\omega - \tilde{K} + i\varepsilon} \tilde{V} \hat{S}^Q(\omega) \tag{7}$$

Since $\tilde{K}\, \hat{a}^+_{k+Q}\hat{a}_k = (\varepsilon_{k+Q}-\varepsilon_k)\hat{a}^+_{k+Q}\hat{a}_k$, the singular terms in the expansion of (7) as inverse powers of $\omega + i\varepsilon - \hbar Q^2/2M$ occur as $(\hbar\omega^+ - \tilde{K})^{-1}$ operates on terms of the form $\hat{a}^+_{k+Q}\hat{a}_k$.

In general, $\hat{S}^Q(\omega)$ has the form of an infinite summation of terms consisting of a scalar times products of creation and annhilation operators. Following the treatment[12] of the singular terms in the Kubo formula developed by Argyres and Sigel, I seek a projection superoperator $\tilde{\Delta}$ which projects out only those components of an operator, $\hat{O}$, which create single particle excitations of momentum Q out of the ground state, i.e.

$$\tilde{\Delta}\hat{O} = \sum_k O_k \hat{a}^+_{k+Q}\hat{a}_k \tag{8}$$

Argyres and Sigel[12] used a diagonal projection superoperator, Q = 0, whereas I use an off-diagonal one. Δ must also satisfy $\tilde{\Delta}\tilde{\Delta} = \Delta$ and $\tilde{\Delta}\hat{a}^+_{k+Q}\hat{a}_k = \hat{a}^+_{k+Q}\hat{a}_k$.

Defining $\tilde{\Delta}' = 1 - \tilde{\Delta}$, straightforward manipulations then yield from the Dyson equation (7) for $\hat{S}^Q(\omega)$

$$\tilde{\Delta}\hat{S}^Q(\omega) = \frac{i\hbar}{\hbar\omega^+ - \tilde{K}}\hat{\rho}_Q(0) + \frac{1}{\hbar\omega^+ - \tilde{K}}\tilde{\Delta}\tilde{T}\tilde{\Delta}\tilde{\Delta}\hat{S}^Q(\omega) \tag{9}$$

where the $\tilde{T}$ superoperator is

$$\tilde{\Delta}\tilde{T}\tilde{\Delta} \equiv \tilde{\Delta}\tilde{V}\tilde{\Delta} + \tilde{\Delta}\tilde{V}\tilde{\Delta}'\frac{1}{\hbar\omega^+ - \tilde{K} - \tilde{\Delta}'\tilde{V}\tilde{\Delta}'}\tilde{\Delta}'\tilde{V}\tilde{\Delta} \tag{10}$$

Note that (10) is the superoperator analogue of the Hamiltonian $\hat{T}$ matrix equation.

A two body approximation would be to replace $\tilde{T}$ by

$$\tilde{T}\hat{O} = -[\hat{T},\hat{O}] \tag{11}$$

where the two body $\hat{T}$ matrix operator is

$$\hat{T} = \frac{1}{2\Omega}\sum_{k_1k_2Q} T_{k_1k_2Q}\hat{a}^+_{k_1+Q}\hat{a}^+_{k_2-Q}\hat{a}_{k_2}\hat{a}_{k_1} \tag{12}$$

Here $T_{k_1k_2Q}$ are the scalar components of the T-matrix, which can be expressed in terms of the phase shifts and scattering angles when ω is on-energy-shell. Note that exactly this two-body approximation[12] is used in the perturbative derivation of the resistivity starting from the Kubo formula.

I claim the following is an explicit construction of the projection superoperator required

$$\tilde{\Delta}\hat{O} = \sum_k \frac{\hat{a}^+_{k+Q}\hat{a}_k \langle[(\hat{a}^+_{k+Q}\hat{a}_k)^+, \hat{O}]\rangle}{n_k - n_{k+Q}} \tag{13}$$

Here [A,B] is a commutator, $n_k = \langle\hat{a}^+_k\hat{a}_k\rangle$, and $\langle\ \rangle$ denotes ground state average. Remarkably, the same formula works for Bosons and Fermions.

Within the two-body T-matrix approximation, I find that Eqs. (3), (9) and (13) constitute a closed system of equations for $S(Q,\omega)$. These depend on properties of the ground state through the n_k and through a four-point function

$$\Phi(k_1,k_2,Q) \equiv \langle \hat{a}^+_{k_1+Q}\hat{a}^+_{k_2-Q}\hat{a}_{k_2}\hat{a}_{k_1}\rangle \tag{14}$$

Usually I do not have complete information on $\Phi(k_1,k_2,Q)$. I do know its symmetry properties such as $\Phi(k_1,k_2,Q) = \Phi(k_1,k_2,K)$ where $K = k_2-Q-k_1$. I also know a sum rule

$$\frac{1}{N}\sum_{k_1k_2}\Phi(k_1,k_2,Q) = N\delta_{Q=0} + \rho\int d^3r\, e^{iQ\cdot r}(g(r)-1) \tag{15}$$

which follows from the definition of S(Q). Here, g(r) is the radial distribution function and $\rho \equiv N/\Omega$ is the density. As we shall see, the first term on the right hand side gives rise to self energy terms in the Dyson equations and the second term to vertex terms. Note that the -1 component of the second term exactly cancels the first term.

IV. S(Q,ω) AT HIGH Q

I now specialize to the problem of deep inelastic neutron scattering (DINS),[13] which is the behavior of S(Q,ω) for very high Q. For clarity, I restrict the calculation to ^{4}He. I will use the concept of high, capital "Q", and low, small "q", momenta to select the important terms at high Q. To define "high" and "low" momenta operationally, a high momentum Q is where to an excellent approximation $n_Q \simeq 0$ and $\rho\int d^3r\, e^{iQ\cdot r}(g(r) - 1) \simeq 0$ (or $\hat{a}_Q|\psi_0\rangle \simeq 0$). That is high momenta greatly exceed the typical momenta characterizing the condensed phase.

At high Q, I expect that two-body collisions dominate the final state broadening, so that the approximation Eq. (11) is valid. I also expect that the $T_{k_1k_2Q}$ can be taken to be the free particle T-matrix because the n_Q are negligible at high Q. A tedious, but straightforward, expansion of Eqs. (9), (11)-(13) would then yield a complicated set of equations for the components S^Q_k of $\tilde{\Delta}\hat{S}^Q(\omega)$ defined by

$$\tilde{\Delta}\hat{S}^Q(\omega) \equiv \sum_k S^Q_k \hat{a}^+_{k+Q}\hat{a}_k \tag{16}$$

At high Q, I find the following simplifications:

1) The projection operator can be reduced to

$$\tilde{\Delta}\hat{O} \simeq {\sum_k}' \frac{1}{n_k}\left\{\hat{a}^+_{k+Q}\hat{a}_k < \hat{a}^+_k\hat{a}_{k+Q}\hat{O} > + \hat{a}^+_k\hat{a}_{k-Q} < \hat{O}\hat{a}^+_{k-Q}\hat{a}_k >\right\} \tag{17}$$

where the prime on the summation means it is restricted to low momenta.

2) Terms generated of the form $\hat{a}^+_k\hat{a}_{k-Q}$ always have negligible coefficients $\Phi(k_1,k_2,Q)$, and so the second term in (17) can be dropped.

3) Terms generated from $\tilde{T}\hat{a}_k$ are negligible compared with terms generated from $\tilde{T}\hat{a}^+_{k+Q}$, because the corresponding $\hat{T}$ matrices are larger for high momenta due to the steeply repulsive He potential.

4) At high Q, the forward and backward scattering Bose symmetrized T-matrix

$$T^{sym}_{k-q+Q,k',q} \equiv T_{k-q+Q,k',q} + T_{k-q+Q,k',k'-k-Q} \tag{18}$$

is to a good approximation a function only of Q and q, i.e.

$$\lim_{high\ Q} T^{sym}_{k-q+Q,k',q} \simeq \bar{T}(Q,q) \tag{19}$$

5) I can take $\bar{T}(Q,q)$ to be on-energy-shell since I am interested in $\hbar\omega$ very close to the recoil energy $\hbar^2Q^2/2M$. (I may wish to consider off-energy-shell $\bar{T}(Q,q)$ for smaller Q, which would introduce[3] an asymmetry weighted toward high ω. This could also be important in the problem of electron scattering in nuclear physics.)

Following this line of reasoning, I arrive at a relatively simple equation for the S^Q_k

$$S^Q_k(\omega) = \frac{1}{\hbar\omega - \varepsilon_{k+Q} + i\varepsilon}\left[i\hbar + \frac{1}{\Omega n_k}\sum_q{}' S^Q_{k-q}(\omega)\,\overline{T}(Q,q)\sum_{k'}\Phi(k-q,k',q)\right] \tag{20}$$

Consider first the limit of a non-interacting system where $g(r) \to 1$. Then

$$\lim_{g(r)\to 1}\frac{1}{n_k}\sum_{k'}\Phi(k-q,k',q) = \delta_{q=0}N \tag{21}$$

would satisfy the sum rule, Eq. (15), and would be the correct non-interacting result. Then (20) is readily solved to yield

$$\lim_{g(r)\to 1} S^Q_k(\omega) = \frac{1}{\hbar\omega^+ - \varepsilon_{k+Q} - \rho\overline{T}(Q,0)} \tag{22}$$

This is the usual self-energy corrected propagator for a high Q particle in terms of the forward scattering T-matrix. This result would yield quasi-Lorentzian final state broadening of the impulse approximation. (Going off-energy-shell would introduce some asymmetry toward the high ω side of the recoil peak[3]).

However, $g(r)$ is very different from 1. While I don't exactly know $n^{-1}_k \Sigma_k \Phi(k-q,k',q)$, I do know the n_k weighted average of it from the sum rule, Eq. (15). Therefore, I approximate it by its average

$$\frac{1}{n_k}\sum_{k'}\Phi(k-q,k',q) \simeq N\delta_{q=0} + \rho\int d^3r\, e^{iq\cdot r}\{g(r)-1\} \tag{23}$$

This will lead to a convolution form of the final state broadening, Eq. (2). When Eq. (23) is substituted into Eq. (20) there are three terms on the right hand side. These terms may be represented by the Feynman diagrams shown in Fig. 2. A solution of Eqs. (20) and (23) can be obtained as follows.

$S^Q_k(\omega)$ is a function only of $k_\parallel$, where parallel ($\parallel$) is defined with respect to the direction of Q. Then one can sum over $q_\perp$. Define

$$\Gamma(q_\parallel) \equiv \frac{-1}{\Omega^{2/3}}\sum_{q_\perp}\overline{T}(Q,q)\,\phi(q)\,\frac{M}{\hbar^2 Q} \tag{24}$$

$$I(k_\parallel) \equiv \frac{\hbar^2 Q}{M} S^Q_{k_\parallel}(\omega) \tag{25}$$

and

$$\phi(q) \equiv \rho\int d^3r\, e^{iq\cdot r} g(r) \tag{26}$$

Using the scaling variable,[13] $Y \equiv M(\omega - \hbar Q^2/2M)/\hbar Q$, I obtain

$$(Y - k_\parallel)I(k_\parallel) = i\hbar + \frac{1}{\Omega^{1/3}}\sum_{q_\parallel}{}' I(k_\parallel - q_\parallel)\Gamma(q_\parallel) \tag{27}$$

The second term of (27) has the form of a convolution, and so the equations can be solved by Fourier transform from momentum space to real space

$$S^Q_{k_\parallel} = \frac{M}{\hbar Q}\int_0^\infty dx\, e^{ik_\parallel x} \exp\left\{i\int_0^x dx'(Y+\Gamma(x'))\right\} \tag{28}$$

where Γ(x) is the Fourier transform of $\overline{\Gamma}(q_\parallel)$, Eq. (24).

I use semiclassical methods,[14] which are certainly accurate at high Q, to solve for Γ(x). I start with the standard expression for $\overline{T}$ in terms of phase shifts.

$$T(Q,q) = -\frac{4\pi\hbar^2}{\mu i Q}\sum_{\ell\ even}(2\ell+1)[e^{2i\delta_\ell}-1]P_\ell(\cos\theta) \tag{29}$$

At high Q, a large number of ℓ contribute to (29). I can therefore replace the sum by an integral using the Poisson summation formula. The scattering angle, $\theta \simeq 2q_\perp/Q$, is small. I can therefore use the large ℓ/small angle representation of Legendre polynomials in terms of Bessel functions

$$P_\ell(\cos\theta) \simeq J_0((\ell+1/2)\theta) = \frac{1}{2\pi}\int_0^{2\pi} \exp\left[i\left(\ell+\frac{1}{2}\right)\frac{2}{Q}\vec{q}_\perp\cdot\vec{n}(\phi)\right]d\phi \tag{30}$$

where $\vec{n}(\phi)$ is a unit vector perpendicular to Q. The summation over $q_\perp$ in (24) simply yields a δ-function involving $r_\perp$ in Eq. (23). I replace the angular momentum ℓ by the impact parameter, $b = (\ell+1/2)2/Q$. Then $|r_\perp| = b$. I evaluate the phase shifts, δ(b), using the JWKB approximation.

V. RESULTS FOR DEEP INELASTIC NEUTRON SCATTERING

Following the steps outlined in Sec. IV yields the following final results for final state broadening function as defined in Eq. (2)

$$R_{FS}(Y) = \frac{1}{\pi} Re\int_0^\infty dx \exp\left\{i\int_0^x dx'(Y+\Gamma(x'))\right\} \tag{31}$$

where

$$\Gamma(x) = \frac{2\pi\rho}{i}\int_0^\infty bdb\, f_b\, g(\sqrt{x^2+b^2}) \tag{32}$$

$$f_b = e^{2i\delta(b)} - 1 + \sum_{M\neq 0} e^{2i\delta(b) + i\frac{M\pi Q}{2}b} \tag{33}$$

Γ(∞) is related by a constant to the He-He T-matrix. Only the M = -1 term is significant in the summation in Eq. (33), and it leads to the hard sphere glory oscillations of the He-He cross section. I will refer to Eq. (31) as the "hard core perturbation theory" result (HCPT).

The final results, Eqs. (31)-(33), meet all the requirements for a quantum theory discussed by Silver and Reiter.[8] Comparing the present results with the quasiclassical theory (QC), I find that the mathematical form from HCPT is remarkably similar. However, in the present theory: 1) forward diffractive scattering is properly taken into account, so that the second term produces glory oscillations (absent in QC) and the large x scattering rate is $1/2\rho v\, 2\pi r_s^2$ (2 x QC); 2) the db integrals now involve the phase shifts (rather than the "effective" hard sphere radius in QC), so that steeply repulsive potentials can also be handled; 3) the argument of g(r) is simpler, which allows HCPT to satisfy the ω^2 and ω^3 sum rules[15] for much smaller Q (not true in QC for $Q < 20$ Å^{-1}), and which eliminates the strong dependence on the detailed form of g(r) obtained in the QC theory; 4) the integral in the argument of the exponential extends to x (it was x/2 in QC); 5) there is a shift in the peak position due to the real part of Γ(x) (absent in QC). I have no explanation for the differences between these two theories.

The HCPT results also differ quantitatively from the work of Gersch, et al.[7] while they agree regarding the importance of spatial correlations. The decoupling approximation is different, HCPT expresses the results in terms of g(r) where Ref. 7 does not, HCPT properly handles the forward diffractive scattering where Ref. 7 does not, etc.

The HCPT results can be derived by an alternate procedure in which I approximate the Hamiltonian to retain only the high momenta and short distance (i.e. r such that $V(r) \sim O(\hbar^2Q^2/2M)$) components in the dynamics. The static expectation values are evaluated in a ground state determined from the low momenta and long range parts of the Hamiltonian. This was, in fact, the original route to Eqs. (31)-(33).

Results similar to (31)-(33) can be derived for Fermion systems (e.g. ^{3}He), except that the summation in (33) must be changed to account for the different statistics.

The final state broadening, Eq. (31), has been evaluated numerically for superfluid ^{4}He. Figure 3 compares the HCPT resolution function to a

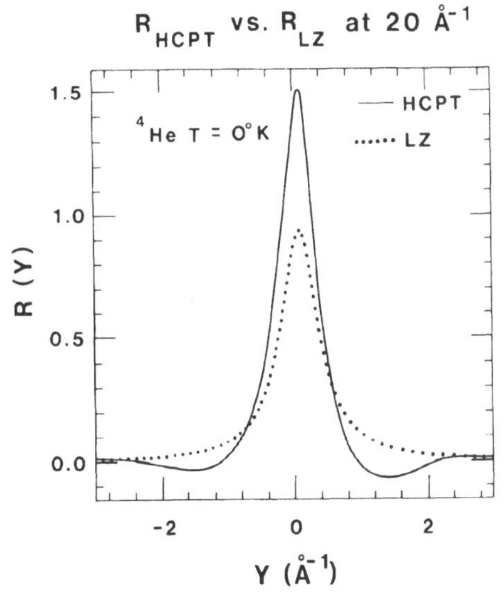

Fig. 3 Final state resolution functions, R(Y), calculated for ^{4}He at $Q = 20$ Å^{-1} in the present theory (HCPT) and in the quasi-Lorentzian approximation (LZ).

quasi-Lorentzian (LZ) obtained by taking $g(r) \rightarrow 1$ in Eq. (32). The $R_{IICPT}(Y)$ has a narrower FWHM, a zero second moment satisfying the kinetic energy (ω^2) sum rule, and no high frequency wings.

Figure 4 shows calculations of $QS(Q,\omega)$ for the HCPT, LZ, and IA models using a theoretical momentum distribution calculated by Lam, et al.,[16] which has an 11.9% Bose condensate fraction. For HCPT, the linewidth of the non-condensed atoms is comparable to the IA, but the Bose condensate peak is not clearly resolved. The LZ lineshape is much wider than the HCPT and the IA, and the glory oscillations (not shown) are much larger in LZ than in HCPT. It is remarkable that $QS(Q,\omega)$ turned out to be positive in this calculation as required, even though $R_{IICPT}(Y)$ is both positive and negative. This required a close relationship between $g(r)$ and $n(p)$.

Figure 5 shows the change in $QS(Q,\omega)$ between 20 Å^{-1} and 200 Å^{-1}. The Bose condensate peak only slightly sharpens at 200 Å^{-1}, but it is still not clearly resolved. The approach to the IA is very slow for He (logarithmic in Q), and the IA is never reached for a hard sphere potential no matter how high the Q. Final state corrections are important at any experimentally feasible Q.

Detailed numerical predictions for ^{4}He, ^{3}He and the hard sphere Bose liquid will be presented elsewhere.[17]

The convolution form for the final state broadening, Eq. (2), can fail for a variety of reasons: the k-dependence of the left hand side of Eq. (23) may be significant; Q may not be high enough to justify the on-energy-shell approximation for the T-matrix, etc. A detailed discussion of the corrections to Eqs. (2) and (31) at

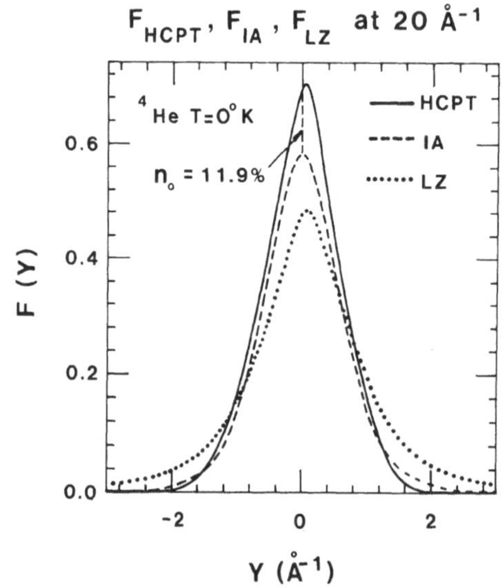

Fig. 4 Calculations of $QS(Q,\omega)$ in the present theory (HCPT), quasi-Lorentzian (LZ), and the impulse approximation (IA) for ^{4}He at $Q = 20$Å^{-1}.

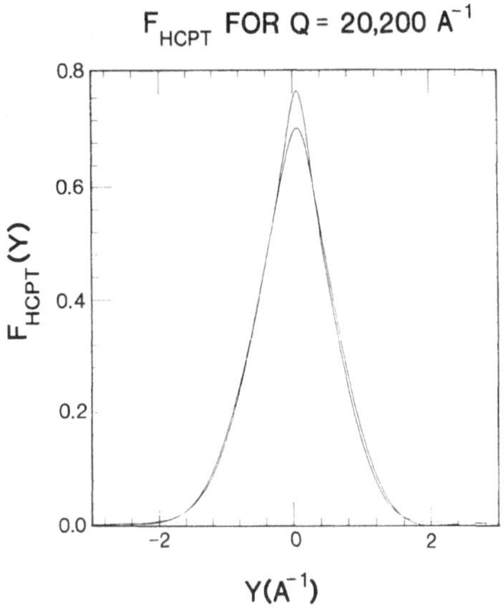

Fig. 5. Calculations of QS(Q,ω) in the present theory (HCPT) for ^{4}He at Q = 20 Å^{-1} and Q = 200 Å^{-1}.

lower Q is beyond the scope of this article, although the Q achieved in present generation experiments may not be high enough to apply the Eqs. (2) and (31) blindly.

VI. CONCLUSION

The hard core perturbation theory of deep inelastic neutron scattering experiments qualitatively confirms the earlier many-body cumulant theory of Gersch, et al.[7] and the quasiclassical theory of Silver and Reiter.[8] The quantitative predictions and the structure of the theory are new. I have shown how vertex corrections give rise to a non-Lorentzian, zero second moment lineshape for final state corrections.

The good news for experimentalists is that, at high enough Q, the final state broadening takes the form of a convolution and is smaller then the Lorentzian broadening theories would predict. The bad news is that neither the Bose condensate peak in ^{4}He, nor the Fermi surface discontinuity in ^{3}He, will be clearly resolved in any feasible DINS experiment. However, provided the final state theory is known and instrumental corrections understood, a deconvolution procedure (such as maximum entropy) might be feasible to extract the singular structures and other features of momentum distributions. There must now be a detailed effort to reanalyze momentum distribution experiments on quantum solids and fluids.

For theorists, it is truly remarkable that a projection superoperator method, originally designed to solve transport problems in the limit of Q = 0, can be extended to solve scattering problems at very high Q. This suggests that the method may be applicable to a wide variety of problems involving the calculation of dynamics, S(Q,ω), from a knowledge of static correlations, S(Q). An immediate application will be to momentum distribution experiments in nuclear and particle physics, such as electron nucleus scattering.

ACKNOWLEDGMENTS

Special thanks to J. W. Clark for his advice during the course of this project. I also thank K. Bedell, G. Reiter, and P. Sokol for many helpful discussions. This work was supported by the Office of Basic Energy Sciences of the U.S. Dept. of Energy.

REFERENCES

1. P. C. Hohenberg, P. M. Platzman, Phys. Rev. 152, 198 (1966).
2. For reviews, see E. C. Svennson, Proceedings of the 1984 Workshop on High Energy Excitations in Condensed Matter, LA-10227, p. 456; P. Sokol in Proceedings of the 1986 Banff Conference on Quantum Liquids and Solids, Can. Journ. Physics 1987. H. R. Glyde, E. C. Svennson, Chapter 13 in Neutron Scattering in Condensed Matter Research, Ed. K. Sköld and D. L. Price, to be published by Academic Press.
3. P. M. Platzman, N. Tzoar, Phys. Rev. B 30, 6397 (1984).
4. T. R. Kirkpatrick, Phys. Rev. B 30, 1266 (1984).
5. G. Reiter, R. Becher, Phys. Rev. B 32, 4492 (1985)
6. For other approaches see J. J. Weinstein, J. W. Negele, Phys. Rev. Letters 49, 1016 (1982) and V. F. Sears, Phys. Rev. B 30, 44 (1984).
7. H. A. Gersch, L. J. Rodriguez, Phys. Rev. A8, 905 (1973); L. J. Rodiguez, H. A. Gersch. H. A. Mook, ibid., 9, 2085 (1974).
8. R. N. Silver, G. Reiter, Phys. Rev. B1, 35, 3647 (1987).
9. R. Feltgen, H. Kirst, K. A. Kohler, H. Pauly. F. Torello, J. Chem. Phys. 76 (5), 2360 (1982).
10. E. C. Svennson, V. F. Sears, A. D. Woods, P. Martel, Phys. Rev. B 21, 3638 (1980).
11. G. Reiter, R. N. Silver, Phys. Rev. Letters 54, 1047 (1985).
12. P. N. Argyres, J. L. Sigel, Phys. Rev. Letters 31, 1397 (1973); Phys. Rev. B 9, 3197 (1974).
13. See G. B. West, Phys. Reports 18, 263 (1975).
14. See, e.g., M. V. Berry and K. E. Mount, Reports on Progress in Physics, 35, 315 (1972); R. K. B. Helbing, Journal of Chemical Physics, 50, 493 (1969).
15. See Appendix A. in E. Feenberg, Theory of Quantum Fluids, (Academic Press, New York, 1969).
16. P. M. Lam. J. W. Clark, M. L. Ristig, Phys. Rev. B 16, 222 (1977).
17. R. N. Silver, to be published.

EXCITATIONS AT HIGH MOMENTUM IN QUANTUM FLUIDS

B. Tanatar, E. F. Talbot, and H. R. Glyde

Department of Physics
University of Delaware
Newark, DE 19716

Abstract: In this paper we evaluate the shape of the dynamic form factor $S(Q,\omega)$ for wave vectors Q in the interval $3 \leq Q \leq 12$ Å^{-1} in liquid ^{4}He and ^{3}He. The purpose is to see whether the observed oscillations in the width, $W(Q)$, and the peak position, E_Q, of the $S(Q,\omega)$ in liquid ^{4}He can be reproduced beginning from the pair potential, $V(r)$. We find these oscillations can be reproduced and they originate from oscillations in the ^{4}He-^{4}He atom scattering amplitude that is used in the present RPA model. Similar oscillations are found in the $W(Q)$ of liquid ^{3}He. However, their amplitude is much smaller and probably could not be observed.

1. INTRODUCTION

The fundamental excitations in the density and spin-density of the quantum liquids ^{4}He and ^{3}He may be studied using inelastic neutron scattering methods.[1] For ^{4}He, the observed scattering intensity is proportional to the coherent dynamic form factor,[2]

$$S_c(Q,\omega) = \frac{1}{2\pi}\int_{-\infty}^{\infty} dt\, e^{i\omega t}\langle \rho(Q,t)\rho(-Q,0)\rangle . \tag{1}$$

Here $\hbar Q$ and $\hbar\omega$ are the momentum and energy transferred from the neutron to the fluid. The scattering creates a density excitation having this momentum and energy where

$$\rho(Q,t) = \sum_{\mathbf{k}} a^{\dagger}_{\mathbf{k}}(t)a_{\mathbf{k}+\mathbf{Q}}(t)$$

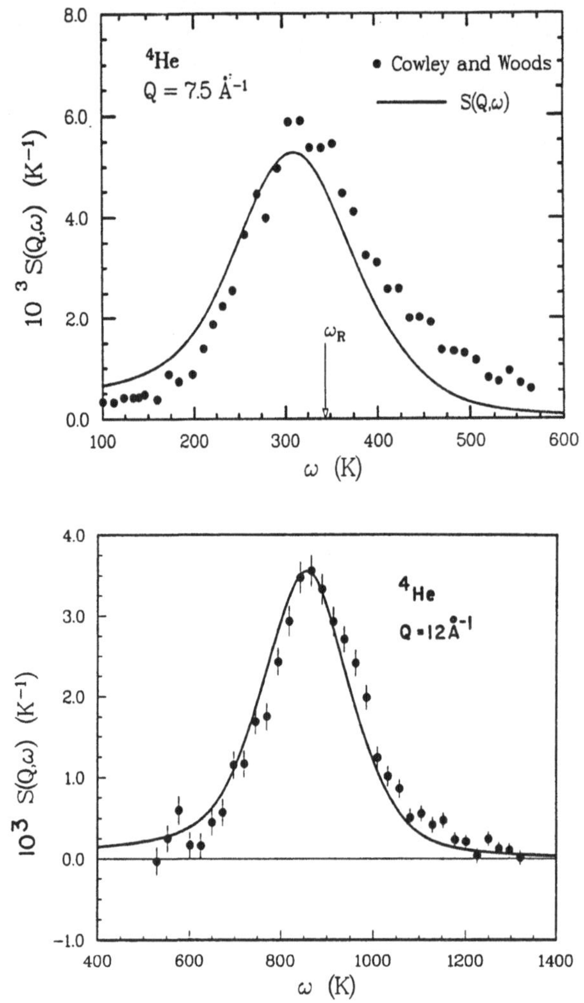

Figure 1: Scattered intensity from liquid ^{4}He observed by Cowley and Woods (Ref.3) and by Stirling et al. (Ref.19) at $Q = 7.5$ Å^{-1} and $Q = 12$ Å^{-1} respectively. The solid lines are the present calculations using model 1 and including resolution broadening.

is the Fourier transform of the number density of the fluid. Since the pioneering work of in the late 1950's, most neutron studies[1,3] have focussed on the low momentum transfer region, $Q \leq 3$ Å^{-1}. At low momentum, liquid ^{4}He supports collective excitations; the renowned phonon-roton mode proposed by Landau.[4] The response of liquid ^{4}He and $S(Q,\omega)$ at low Q has been studied in detail and this work is reviewed, for example, by Woods and Cowley, Price, and more recently by Glyde and Svensson.[1]

At higher Q ($3 \leq Q \leq 15$ Å^{-1}), the atoms in the liquid no longer respond collectively. Rather, the neutron transfers its momentum and energy to a single

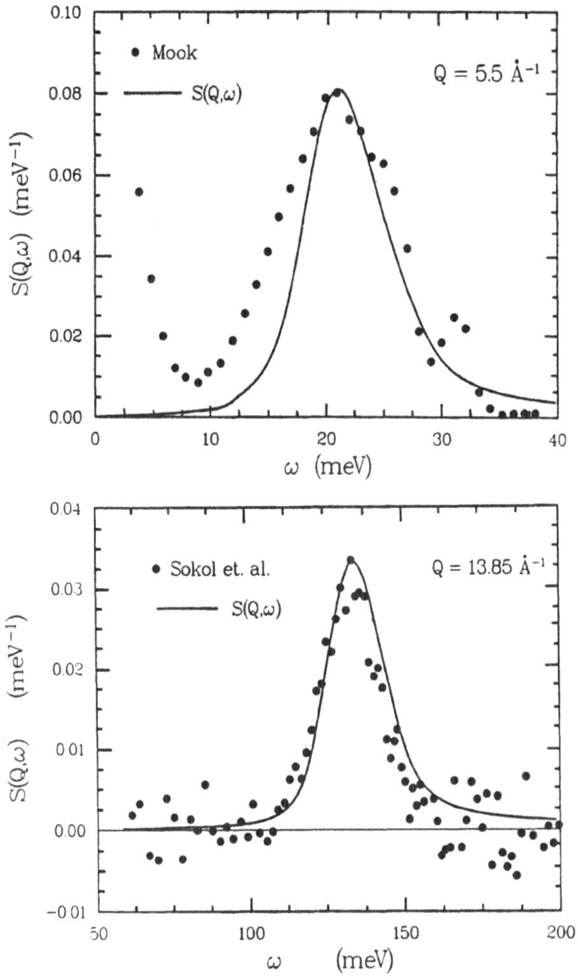

Figure 2: Scattered intensity from liquid ^{3}He observed by Mook (Ref.9) at $Q = 5.5$ Å^{-1} and by Sokol et al. (Ref.10) at $Q = 13.85$ Å^{-1}. The solid lines are the present calculations using model 2.

atom (nucleus) and this atom responds in a single particle manner. The $S(Q,\omega)$ is then a broad function of ω peaked near the free atom recoil energy, $\omega_R = \hbar Q^2/2m$ (see Fig.1). The width of $S(Q,\omega)$ is determined largely by the kinetic or Doppler broadening due to the zero point and thermal motion of the atoms.

In the range $3 \le Q \le 10$ Å^{-1}, Cowley and Woods[3] noted that the peak position, E_Q, and full width at half maximum, $W(Q)$, of $S(Q,\omega)$ oscillated with Q in liquid ^{4}He. The period of the oscillations was $Q \approx 3$ Å^{-1}. The amplitude of the oscillations in E_Q was small; $\sim 3-5\%$ of E_Q. However the amplitude of the oscillations in $W(Q)$ was large; $\sim$ 25% of $W(Q)$. The effect is small for E_Q but most important for $W(Q)$. The kinetic or Doppler contribution to $W(Q)$ is approximately proportional[5,6] to $\langle p^2\rangle^{1/2} Q/m$ where $\langle p^2\rangle^{1/2}$ is the rms value of the atomic momentum. Thus we expect the Doppler contribution to $W(Q)/Q$ to be constant. The oscillations observed in $W(Q)/Q$ therefore reflect departure from Doppler broadening and arise from interactions of the scattered atom with its neighbors after it is struck by the neutron. Martel et al.[6] proposed a simple model for $S(Q,\omega)$ in which these nearest neighbor interactions and the "interaction" component of $W(Q)$ was related to the observed ^{4}He-^{4}He atom scattering cross-section, $\sigma(Q)$. Since the observed[7] $\sigma(Q)$ oscillates with Q, the total $W(Q)/Q$ also oscillated with Q, in good agreement with experiment. Here we attempt a more fundamental explanation and description of this effect and of the shape of $S(Q,\omega)$ in general. For ^{4}He, as we shall see, our model is very simple.

Due to the large absorbtion cross-section of the ^{3}He nucleus for neutrons, neutron scattering studies[1] of excitations in liquid ^{3}He are very difficult. The most recent work at low Q is reported by Scherm et al.[8] The inelastic scattering cross-section is proportional to[1]

$$S(Q,\omega) = S_c(Q,\omega) + \frac{\sigma_i}{\sigma_c} S_I(Q,\omega) \tag{2}$$

where S_c is given by (1) and

$$S_I(Q,\omega) = \frac{1}{2\pi} \int dt\, e^{i\omega t} \langle \rho_z(Q,t)\rho_z(-Q,0)\rangle . \tag{3}$$

Here

$$\rho_z(Q,t) = \rho_\uparrow(Q,t) - \rho_\downarrow(Q,t) \tag{4}$$

is the Fourier component of the spin-density along the z-axis. Eq.(3) is valid for an isotropic fluid. In (1) $\rho(Q,t) = \rho_\uparrow(Q,t) + \rho_\downarrow(Q,t)$ is the total density including both spin states in ^{3}He. The ratio of the incoherent to coherent scattering cross-section is approximately $\sigma_i/\sigma_c = 0.25$. At higher Q, where $S_I(Q) = S_c(Q) = 1$, approximately 25% of the scattering is from the spin-density excitations in $S_I(Q,\omega)$ of (2).

As in ^{4}He, most neutron studies of ^{3}He have focussed on collective excitations at low Q. Collective excitations, however, cease to propagate in ^{3}He for $Q \geq 1.5$ Å^{-1}. Thus as in ^{4}He, we expect $S(Q,\omega)$ in ^{3}He to be characteristic of scattering from the interacting single particle excitations at $Q \geq 3$ Å^{-1}. In Fig.2 we show the scattering intensity observed by Mook[9] at $Q \approx 5.5$ Å^{-1} and by Sokol et al.[10] at $Q = 13.85$ Å^{-1}. The observed intensity is indeed a broad function of ω. In the interval $3 \leq Q \leq 7$ Å^{-1}, Mook[9] observed a substantial variation of $W(Q)$ with Q but no oscillation of $W(Q)/Q$ with Q. A goal here is therefore to evaluate $S(Q,\omega)$ in ^{3}He and especially its width $W(Q)$ to see whether observable oscillations are expected. We find only very small oscillations of $W(Q)/Q$ in ^{3}He; too small to be resolved with the current precision of measurements.

To calculate $S(Q,\omega)$ in the interval $3 \leq Q \leq 12$ Å^{-1} we use a simple Random Phase Approximation (RPA) model of the dynamic susceptibility. We assume that the most important interactions at high Q are collisions between pairs of atoms via the pair potential, $V(r)$. All interactions between pairs induced via collective excitations are assumed to be small in comparison. The interaction appearing in the RPA is therefore represented by a two-body T-matrix calculated[11] by summing ladder diagrams from the bare potential[12] $V(r)$. The only input is the density of the liquid and $V(r)$. In the next section we sketch how the full equation for $\chi(Q,\omega)$ may be approximated by the RPA at high Q. The models for the single particle energies and the T-matrix are discussed in Section 3. The results are presented in Section 4. A brief version of this work is to appear elsewhere and a full presentation is in progress.[12]

2. DYNAMIC SUSCEPTIBILITY

In this section we sketch how the RPA at high Q may be derived. The dynamic susceptibility $\chi(Q,t)$ corresponding to $S(Q,t)$ is

$$\chi(Q,t) = -\frac{i}{\Omega}\langle T\rho(Q,t)\rho(-Q,0)\rangle . \tag{5}$$

The Fourier transform $\chi(Q,\omega)$ of $\chi(Q,t)$ is related to $S(Q,\omega)$ by (at T=0 K),

$$S(Q,\omega) = -\frac{1}{n\pi}\chi''(Q,\omega). \tag{6}$$

In the case of ^{3}He, we can define a similar $\chi_I(Q,t)$ in terms of $\rho_z(Q,t)$ which is related to $S_I(Q,\omega)$ as in (6). The interaction appearing in $\chi(Q,\omega)$ is the spin-symmetric interaction, $2\Gamma^s = \Gamma_{\uparrow\uparrow} + \Gamma_{\uparrow\downarrow}$, while that appearing in χ_I is the spin-antisymmetric interaction, $2\Gamma^a = \Gamma_{\uparrow\uparrow} - \Gamma_{\uparrow\downarrow}$. Using the definition of $\rho(Q)$

$$\rho(Q) = \sum_{\mathbf{p}} a^\dagger_{\mathbf{p}} a_{\mathbf{p}+\mathbf{Q}} = \sum_{\mathbf{p}} a^\dagger_{\mathbf{p}-\mathbf{Q}/2} a_{\mathbf{p}+\mathbf{Q}/2}, \tag{7}$$

we may write

$$\chi(Q,t) = \frac{1}{\Omega^2}\sum_{\mathbf{p},\mathbf{p}'} \chi(\mathbf{p},\mathbf{p}';Q,t) \tag{8}$$

where

$$\chi(\mathbf{p},\mathbf{p}';Q,t) = -i\Omega\left\langle T a^\dagger_{\mathbf{p}-\mathbf{Q}/2}(t) a_{\mathbf{p}+\mathbf{Q}/2}(t) a^\dagger_{\mathbf{p}'+\mathbf{Q}/2}(0) a_{\mathbf{p}'-\mathbf{Q}/2}(0)\right\rangle . \tag{9}$$

The Fourier transform that we need in (6) is

$$\chi(Q,\omega) = \int \frac{d^4p}{(2\pi)^4}\int \frac{d^4p'}{(2\pi)^4}\chi(\mathbf{p},\mathbf{p}';Q,\omega) \equiv \chi(Q). \tag{10}$$

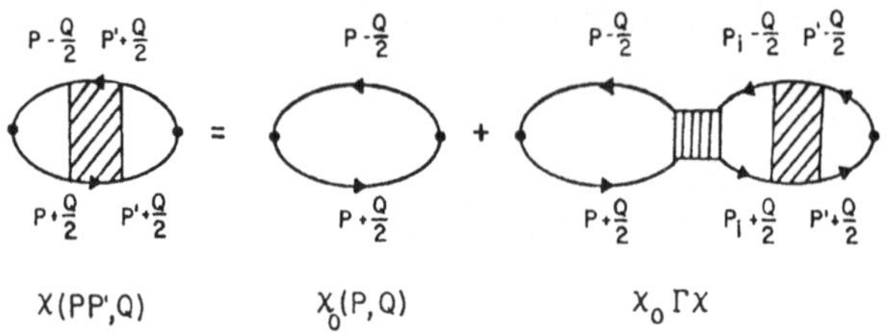

Figure 3: Diagrammatic representation of χ, χ_0 and Eq.(12).

Here we have introduced the four-vector notation, $p = (\mathbf{p}, \omega_p)$ and $Q = (\mathbf{Q}, \omega)$. The exact equation for $\chi(p, p'; Q)$ is

$$\begin{aligned}\chi(p, p'; Q) &= \chi_0(p, Q)\delta_{pp'}(2\pi)^4 \\ &+ \chi_0(p, Q) \int \frac{d^4p_i}{(2\pi)^4} \Gamma(p + Q/2, p_i - Q/2; p - Q - 2, p_i + Q/2)\chi(p_i, p'; Q)\end{aligned} \tag{11}$$

where

$$\chi_0(p, Q) = -2iG(p - Q/2)G(p + Q/2) \tag{12}$$

is the product of two single particle Green functions. Since p' is a parameter in (11), we may integrate over p' to remove it. To do this we define,

$$\begin{aligned}\chi(p, Q) &= \int \frac{d^4p'}{(2\pi)^4} \chi(p, p'; Q) \\ &= \chi(p, Q)\left[1 + \int \frac{d^4p_i}{(2\pi)^4} \Gamma(p + Q/2, p_i - Q/2; p - Q/2, p_i + Q/2)\chi(p_i, Q)\right].\end{aligned} \tag{13}$$

This is our basic equation which we will approximate to get RPA.

The $\chi(\mathbf{p}, \mathbf{p}'; Q, t)$ may be interpreted as the propagator of a particle-hole (p-h) pair excitation in the fluid. The product $a^{\dagger}_{\mathbf{p}'+\mathbf{Q}/2}(0)a_{\mathbf{p}'-\mathbf{Q}/2}(0)$ annihilates a particle from state $p - Q/2$ (creates a hole) and creates a particle in momentum state $p' + Q/2$ at $t = 0$. Clearly, for this to be possible $\mathbf{p} - \mathbf{Q}/2$ must lie within the Fermi sea ($|\mathbf{p} - \mathbf{Q}/2| \le p_F$ or $n_F(\mathbf{p} - \mathbf{Q}/2) \neq 0$) in the Fermi case or lie within the range of the filled momentum states $n_B(\mathbf{p})$ in the Bose case. In the Bose case $n_B(\mathbf{p})$ becomes very small for $|\mathbf{p}| \ge 1$ Å^{-1}. Between $t = 0$ and $t = t$, the additional pair propagates, interacts and scatters to other momentum states until it is annihilated from states $p + Q/2$, $p - Q/2$. The density excitation (10) can clearly be represented as a sum over all these p-h excitations. Following Abrikosov et al.,[14] we represent the $\chi(p, p; Q)$ by Fig.3.

Eq.(13) may be derived by a number of methods. Eq.(13) is Eq.(16) of Gottfried and Pičman[15] who derived it directly using Green function methods. Kadanoff and Baym[16] derived it using linear response theory and a recent clear discussion of this formalism is given by Green et al.[17] Abrikosov et al.[14] and Nozières[18] derive it using Green function methods beginning with the equivalent equation for the propagation of two particles added to the ground state. If we follow the last method we can see that the four-point interaction $\Gamma(1234) = \Gamma(p+Q/2, p_i - Q/2; p - Q/2, p_i + Q/2)$ contains all possible scattering processes between the p-h except those which have a p-h excitation as an intermediate state. The $\Gamma(1234)$ is clearly a very complicated interaction in general.

The basic assumption here is that the interaction between particles or a p-h pair at high momentum ($Q \geq 3$ Å^{-1}) is dominated by scattering between pairs via the interatomic potential $V(r)$. The steeply repulsive core of $V(r)$ will be especially important. Induced interactions involving "bubble" or "ring" diagrams will be assumed to be less important since collective modes no longer propagate for $Q \geq 3$ Å^{-1}. We assume, therefore, that $\Gamma(1234)$ can be well represented by summing ladder diagrams so that $\Gamma(1234)$ is given by the two-particle T-matrix.[11]

The T-matrix $\Gamma(1234)$ depends upon the relative incoming momentum, $\mathbf{k}$, the relative outgoing momentum, $\mathbf{k}'$, the total momentum, $\mathbf{K}$, and the total energy, E of the interacting pair,

$$\Gamma(p+Q/2, p_i - Q/2; p - Q/2, p_i + Q/2) = \Gamma(\mathbf{k}, \mathbf{k}'; \mathbf{K}, E) \tag{14}$$

where

$$\begin{aligned} \mathbf{k} &= \frac{1}{2}(\mathbf{p} - \mathbf{p}_i + \mathbf{Q}), & \mathbf{K} &= \mathbf{p} + \mathbf{p}_i, \\ \mathbf{k}' &= \frac{1}{2}(\mathbf{p} - \mathbf{p}_i - \mathbf{Q}), & E &= \omega_p + \omega_{p_i}. \end{aligned} \tag{15}$$

Using the properties of $G(p - Q/2)$ in $\chi_0(p, Q)$, it is possible to show[12] that in (11), the $|\mathbf{p} - \mathbf{Q}/2|$ and $|\mathbf{p}_i - \mathbf{Q}/2|$ must lie within the values for which $n(\mathbf{p} - \mathbf{Q}/2)$ is significant. For a Fermi system, this is $|\mathbf{p} - \mathbf{Q}/2| \leq p_F$. For $Q \gg p_F$, this requires $p \approx Q/2$. Similarly for ω_p and ω_{p_i} we must have $\omega_p + \omega/2 \leq \epsilon_F$ and $\omega_{p_i} - \omega/2 \leq \epsilon_F$. Thus at large ω near ω_R, $\omega_p \simeq \omega_{p_i} \simeq \omega/2$. And for large Q we have $k \simeq Q/2$, $k' \simeq -Q/2$ and $E \simeq \omega$, and

$$\Gamma(p+Q/2, p_i - Q/2; p - Q - 2, p_i + Q/2) \simeq \Gamma(Q, \omega). \tag{16}$$

In this approximation Γ is independent of the "legs" p and p_i and we may integrate over p and p_i in (13) to obtain

$$\begin{aligned} \chi(Q,\omega) &= \chi_0(Q,\omega) + \chi_0(Q,\omega)\Gamma(Q,\omega)\chi(Q,\omega) \\ &= \frac{\chi_0(Q,\omega)}{1 - \Gamma(Q,\omega)\chi_0(Q,\omega)} \end{aligned} \tag{17}$$

where

$$\begin{aligned} \chi_0(Q,\omega) &= 2\int \frac{d\omega_p}{2\pi} G(p - Q/2) G(p + Q/2) \\ &= \frac{2}{\Omega} \sum_{\mathbf{p}} \frac{n_{\mathbf{p}} - n_{\mathbf{p}+\mathbf{Q}}}{\omega - [\epsilon_{\mathbf{p}+\mathbf{Q}} - \epsilon_{\mathbf{p}}]} \end{aligned} \tag{18}$$

is the Lindhard function. We expect this approximation to be valid for $Q \gg \langle p^2 \rangle^{1/2}$ and for energy values $\omega \approx \omega_R$, near the center of the main peak of $S(Q,\omega)$.

3. MODELS OF LIQUID ^{3}He AND ^{4}He

We evaluate the RPA (17) for two models of χ_0 and of $\Gamma(Q,\omega)$. We discuss them below.

Model 1: Here χ_0 is the purely free Fermi or free Bose value, denoted by χ_0^0. This is obtained by setting $\epsilon_{\mathbf{p}} = \epsilon_{\mathbf{p}}^0 = p^2/2m$ and $n_{\mathbf{p}} = n_{\mathbf{p}}^0$ in (18). For ^{4}He we took $n_{\mathbf{p}}^0 = (e^{\epsilon_{\mathbf{p}}^0/kT} - 1)^{-1}$ with T=3.2 K, a temperature above the Bose-Einstein condensation temperature to avoid the condensate. For ^{3}He, $n_{\mathbf{p}}^0 = \theta(p - p_F)$. The $\Gamma(Q,\omega)$ in (17) was approximated by the scattering amplitude for a pair of helium atoms scattering in free space, $\Gamma(Q,\omega) \approx \Gamma_0(Q,\omega_R)$. The $\Gamma_0(Q,\omega_R)$ is therefore an "on-energy-shell" amplitude with $\hbar\omega_R = \hbar^2(p_1^2 + p_2^2)/2m$ set at the initial kinetic energy of the interacting pair. The Γ was expanded in angular momentum components, Γ_L. Since ^{4}He is a boson, only even L components of Γ contribute to the ^{4}He-^{4}He interaction,

$$\Gamma^{4-4}(Q,\omega) = 2a_e$$

where

$$a_e = \sum_{L \text{ even}} (2L+1)\Gamma_L(k,k)$$

is a sum over even Γ_L of $\Gamma(Q,\omega)$. In ^{3}He, the spin-symmetric interaction, Γ^s, appearing in $\chi_c(Q,\omega)$ and the spin-antisymmetric interaction appearing in χ_I are,

$$\Gamma_s = \frac{1}{2}(3a_o + a_e), \qquad \Gamma_a = \frac{1}{2}(a_o - a_e)$$

respectively. We use model 1 to describe the liquid ^{4}He.

Model 2: This is basically the Galitskii-Feynman-Hartree-Fock (GFHF) model developed by Glyde and Hernadi.[11] In (18) we use the GFHF energies

$$\epsilon_{\mathbf{p}} = \epsilon_{\mathbf{p}}^0 + \Sigma(\mathbf{p}, \epsilon_{\mathbf{p}}). \tag{19}$$

The $\Sigma(\mathbf{p}, \epsilon_{\mathbf{p}})$ is shown in Fig.4. The real part $\Sigma'(\mathbf{p}, \epsilon_{\mathbf{p}})$, of Σ is largely independent of p, since from Fig.4 we see that the difference $\epsilon_{\mathbf{p}} - \epsilon_{\mathbf{p}}^0$ is largely independent of p. Thus $\epsilon'_{\mathbf{p+Q}} - \epsilon'_{\mathbf{p}} \approx \epsilon^0_{\mathbf{p+Q}} - \epsilon^0_{\mathbf{p}}$ and the real part of Σ contributes little to χ_0. Also as seen in Fig.4, the imaginary part of Σ, $\epsilon''_{\mathbf{p}} = \Sigma''(\mathbf{p}, \epsilon_{\mathbf{p}})$ grows with p. Thus at large Q, $\Sigma''(\mathbf{p+Q}) \gg \Sigma''(\mathbf{p})$, and in (18) we have,

$$\epsilon_{\mathbf{p+Q}} - \epsilon_{\mathbf{p}} \approx \epsilon^0_{\mathbf{p+Q}} - \epsilon^0_{\mathbf{p}} + i\Sigma''(\mathbf{p+Q}). \tag{20}$$

The $\Sigma''(\mathbf{p+Q}) = \epsilon''_{\mathbf{p+Q}}$ oscillates with $|\mathbf{p+Q}|$ as seen in Fig.4. The resulting χ_0, denoted by χ_0^{HF}, contains the chief new feature of an imaginary part or "lifetime"

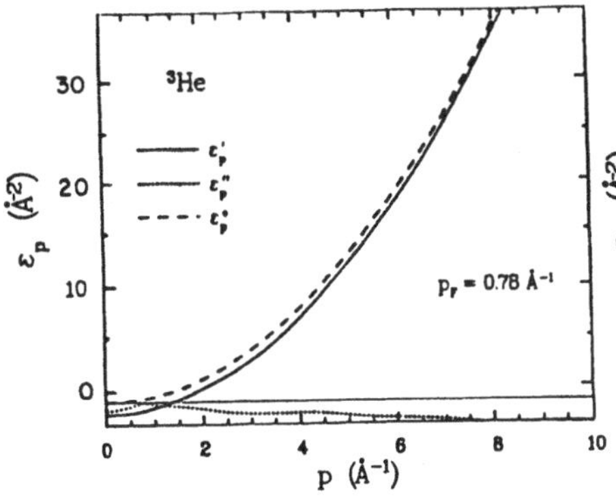

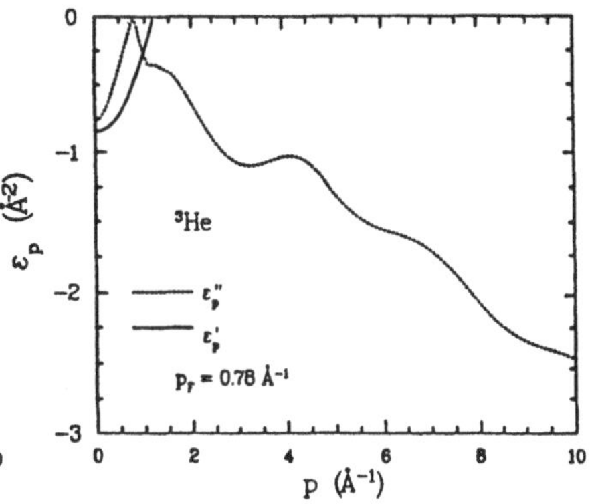

Figure 4: Single particle energy in ^{3}He in the GFHF approximation: ϵ_p'-real part, ϵ_p''-imaginary part, $\epsilon_p^0 = p^2/2m$. Right hand side shows an expanded scale of ϵ_p''.

in the single particle energies. The resulting $S_0^{HF} = -(n\pi)^{-1}\chi_0^{HF''}$ is very similar to the model constructed by Martel et al.[6] who added an oscillating imaginary part to the energy denominator of χ_0.

For $\Gamma(Q,\omega)$, we use the Galitskii-Feynman T-matrix. This is basically the scattering amplitude for two ^{3}He atoms in liquid ^{3}He. For $Q \geq 5$ Å^{-1}, we found $\Gamma(Q,\omega)$ was well approximated by $\Gamma_0(Q,\omega)$. Thus "Fermi liquid effects" contribute little for $Q \geq 5$ Å^{-1}. Models 1 and 2 were used to describe liquid ^{3}He.

4. RESULTS

In Fig.5 we show the scattering amplitude $\Gamma_0(Q,\omega_R)$ for two helium atoms scattering in free space as a function of Q ($\omega_R = \hbar Q^2/2m$). Both the Γ_0^{4-4}, describing the interaction of two ^{4}He atoms, and the spin-symmetric Γ_0^s for two ^{3}He atoms, oscillate with Q. The amplitude of the oscillations in Γ_0^{4-4} is approximately twice that for Γ_0^s. Aside from the oscillations, the imaginary part (ImΓ_0) is approximately proportinal to Q and the real part (ReΓ_0) is approximately independent of Q.

These scattering amplitudes can be used to calculate He-He atom cross-section $\sigma(Q)$, (e.g., $\sigma(Q) = -(2/Q)\Gamma_0''(Q,\omega_R)$). The resulting $\sigma(Q)$ agree well with the observed values[7] for both ^{3}He and ^{4}He.

The $S(Q,\omega)$ in ^{4}He calculated from (17) and model 1, which uses the interaction Γ_0^{4-4}, is shown in Fig.1. The $\chi_0^0(Q,\omega)$ in model 1 which employes a momentum distribution n_p^0 for non-interacting bosons is a narrow function of ω. The interaction Γ_0^{4-4} therefore makes an important contribution to the width, $W(Q)$, of $S(Q,\omega)$. From Fig.1 we see that the $S(Q,\omega)$ agrees quite well with the

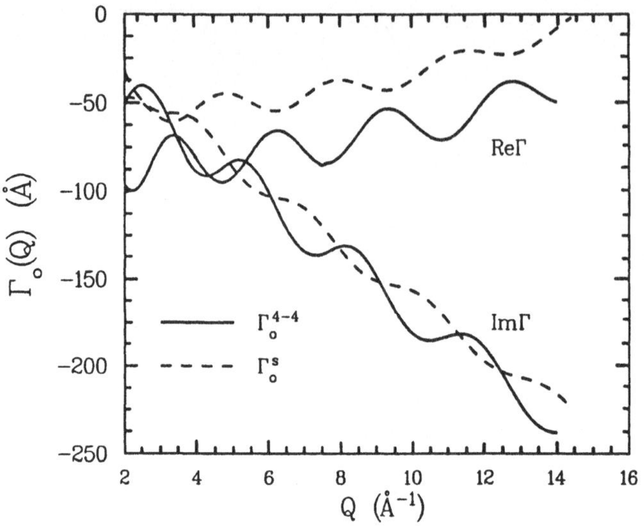

Figure 5: $\Gamma_0(Q,\omega_R)$ for two ^{4}He atoms (Γ_0^{4-4}) and two ^{3}He atoms (Γ_0^s).

observed scattering intensity near the peak where $\omega \approx \omega_R$. The approximations made to the full $\chi(Q,\omega)$ (13) to obtain the RPA (17) are valid for $\omega \approx \omega_R$. The approximations are poor in the wings of $S(Q,\omega)$ at large $\omega - \omega_R$. Model 1 appears to provide a resonable and simple description of $S(Q,\omega)$ at $Q \geq 4$ Å^{-1}.

In Fig.6 we show the width, $W(Q)$, and peak position, E_Q, obtained from our calculations of $S(Q,\omega)$ versus Q. These are compared with the recently observed values of Stirling et al.[19] and the model calculations of Martel et al.[6] For $Q \geq 4$ Å^{-1}, model 1 clearly predicts oscillations in both $W(Q)$ and E_Q with Q. Since there are no oscillations in $\chi_0^0(Q,\omega)$ with Q, these oscillations originate from the oscillations in Γ_0^{4-4} with Q displayes in Fig.5. For $Q \geq 4$ Å^{-1} the predicted oscillations in $W(Q)$ agree in phase and amplitude quite well with the observed values. For $Q \leq 4$ Å^{-1}, however, the present $W(Q)$ increases sharply and model 1 ceases to be valid. This is due especially to using a frequency independent interaction $\Gamma_0(Q,\omega_R)$ in the RPA. The full $\Gamma(Q,\omega)$ decreases with ω and is small at $\omega \to 0$. Thus at small Q, where ω is also small, the use of a constant interaction, $\Gamma_0(Q,\omega_R)$, becomes a poor approximation predicting too large a width, and too large a shift in peak position from E_Q. In Fig.6 the mean value of the observed $W(Q)/Q$ is approximately independent of Q, as expected in the impulse approximation. Our calculated $W(Q)/Q$ decreases with Q. We believe this is due again to using a constant $\Gamma_0(Q,\omega_R)$ which artificially increases the width $W(Q)$ at low Q.

In model 1, there is no self-energy of lifetime for the scattered atom in $\chi_0^0(Q,\omega)$. Including a lifetime $\epsilon''_{\mathbf{p}+\mathbf{Q}}$ which oscillates with $|\mathbf{p}+\mathbf{Q}|$ (see Fig.4) could also contribute to the oscillations of $W(Q)$ as proposed by Martel et al.[6]

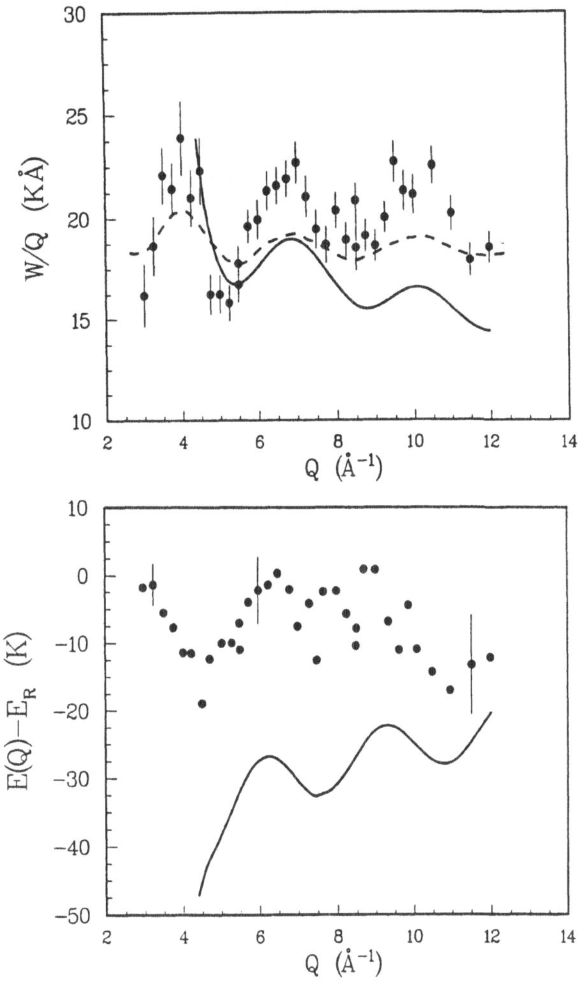

Figure 6: $W(Q)/Q$ and E_Q v.s. Q in liquid ^{4}He. The data is from Ref.19, solid line is the present calculations and dashed line is the Martel et al. (Ref.6) calculation.

Their model is based on a lifetime $\epsilon''_{\mathbf{p+Q}}$ which is assumed to be proportional to $\sigma(Q)$.

Turning to ^{3}He, we display the energy dependence of $\Gamma(Q,\omega)$ in liquid ^{3}He in Fig.7. There we see clearly that the imaginary part of $\Gamma(Q,\omega)$ is roughly proportional to ω in the range of interest. Thus, $\Gamma''(Q,0)$ is much smaller than $\Gamma(Q,\omega_R)$. Retaining the energy dependence is much more important at small Q than at large Q since ω_R grows as Q^2 and the width $W(Q)$ grows as Q. Hence at high Q, $S(Q,\omega)$ is more highly localized around ω_R. From Fig.7 we also see that $\Gamma(Q,\omega)$ is quite well approximated by $\Gamma_0(Q,\omega)$, the free atom scattering amplitude with frequency dependence retained.

The $S(Q,\omega)$ of liquid ^{3}He calculated using model 2 was shown in Fig.2. The width of $S(Q,\omega)$, calculated using both models 1 and 2, is shown in Fig.8. Also shown is the width calculated from χ_0^{HF} (Eqs. (18) and (20)), which has no interaction $\Gamma(Q,\omega)$ but contains the GFHF self-energy in the $\epsilon_{\mathbf{p}}$. The central observation is that models 1 and 2 predict approximately the same $W(Q)/Q$. This $W(Q)/Q$ agrees well with the values observed by Sokol et al.[10] None of the models predict oscillations of significant amplitude in $W(Q)/Q$ in ^{3}He. In particular, we are not able to reproduce the apparent variation of $W(Q)/Q$ with Q suggested by Mook's data. Our calculated $W(Q)/Q$ does agree well with the smaller values of $W(Q)/Q$ observed by Mook.[9] Thus in ^{3}He our calculated $W(Q)/Q$ does not oscillate in contrast to ^{4}He.

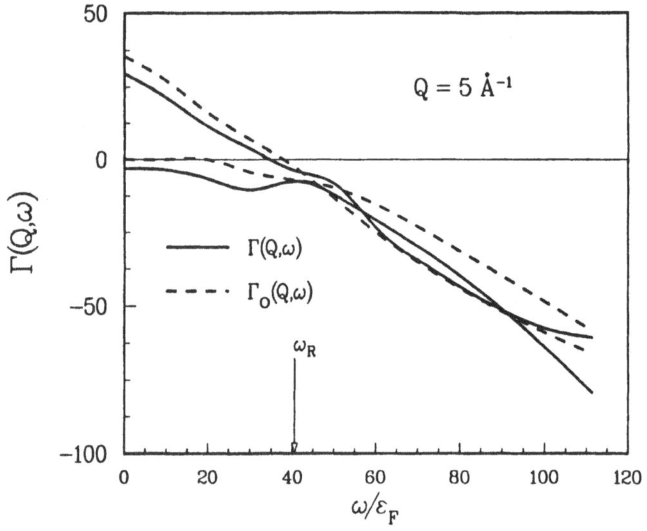

Figure 7: The energy dependence of $\Gamma(Q,\omega)$ at $Q = 5$ Å^{-1}: Γ-full Galitskii-Feynman T-matrix in liquid ^{3}He, Γ_0-free space T-matrix with energy dependence retained. Upper and lower curves show real and imaginary parts, respectively.

In liquid ^{3}He, the Doppler width of $\chi_0^0(Q,\omega)$ is large; $W(Q)/Q = \sqrt{2}\hbar v_F =$ 1.56 meVÅ. Thus in liquid ^{3}He a large fraction of the total width shown in Fig.8 comes from the kinetic (Doppler) width and rather little comes from the interaction $\Gamma(Q,\omega)$ in the RPA, or from the self-energy, Σ'', which appears in model 2. Therefore any oscillations in Γ or Σ'' will tend to be masked or buried in the Doppler width. For example, adding the interaction Γ_0 to the RPA increases the width from $W(Q)/Q = 1.56$ to $W(Q)/Q \approx 2.0$ meVÅ in Fig.8. From Fig.8 we also see that $W(Q)/Q$ for χ_0^{HF} is approximately 2.0 meVÅ. Adding the interaction $\Gamma(Q,\omega)$ to χ_0^{HF} in the RPA to complete model 2 actually decreases $W(Q)/Q$ somewhat. This is due to the frequency dependence of $\Gamma(Q,\omega)$.

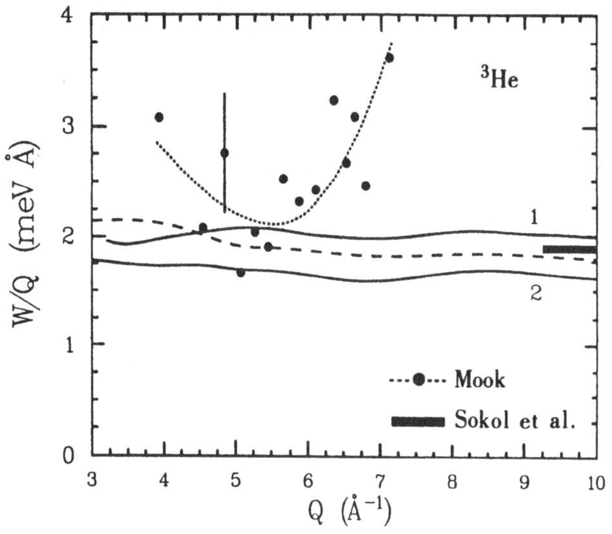

Figure 8: $W(Q)/Q$ v.s. Q in liquid ^{3}He. Solid curves are present calculations using models 1 and 2, dashed curve is S_0^{HF} without an interaction Γ in model 1. Data is by Mook (Ref.9) and Sokol et al.(Ref.10)

To summarize, we see from Figs.1 and 2 that the RPA with an interaction calculated from $V(r)$ by summing ladder diagrams provides a reasonable description of $S(Q,\omega)$ in the peak region at high Q. We emphasize that the approximations we made to get the RPA are valid only at high Q and $\omega \approx \omega_R$. Approximating Γ by a T-matrix is also valid only at large Q. The present calculations reproduce the observed oscillations in the width and peak position of $S(Q,\omega)$ in liquid ^{4}He, observed by Cowley and Woods[3], by Martel et al.[6] and most recently by Stirling et al.[19] In the present model, these oscillations originate from the oscillations in the T-matrix scattering amplitude which appears as the interaction in the RPA. The same model predicts only very small amplitude oscillations in $W(Q)/Q$ in ^{3}He; too small to be observed with present experimental accuracy. This is due to the larger kinetic width in ^{3}He and the smaller amplitude of the oscillations of the ^{3}He-^{3}He scattering amplitude. This prediction is in contrast to what would be expected if $W(Q)$ was assumed to be simply proportional to $\sigma(Q)$ in both ^{3}He and ^{4}He.

ACKNOWLEDGEMENTS

We gratefully acknowledge the support of the U.S. Department of Energy, Contract No. DE-FG02-84ER45083, and the U.S. Army Research Office which made this publication possible.

REFERENCES

1. A. D. B. Woods and R. A. Cowley, Rep. Prog. Phys. **36**, 1135 (1973); D. L. Price, in "The Physics of Liquid and Solid Helium", K. H. Bennemann and J. B. Ketterson, eds., Wiley-Interscience, New York (1976); H. R. Glyde and E. C. Svensson, Chap.13 in "Methods of Experimental Physics", K. Sköld and D. L. Price, eds., Academic Press, New York (1987); H. R. Glyde, in "Condensed Matter Research Using Neutrons", S. W. Lovesey and R. Scherm, eds., NATO ASI Series B**112**, Plenum, New York (1984).
2. M. Cohen and R. P. Feynman, Phys. Rev. **107**, 13 (1957); L. van Hove, Phys. Rev. **95**, 249 (1954); W. Marshall and S. W. Lovesey, "Theory of Thermal Neutron Scattering", Oxford University Press, Oxford (1971).
3. R. A. Cowley and A. D. B. Woods, Can. J. Phys. **49**, 177 (1971).
4. L. D. Landau, J. Phys. USSR **5**, 71 (1941); L. D. Landau, J. Phys. USSR **11**, 91 (1947).
5. P. C. Hohenberg and P. M. Platzman, Phys. Rev. **152**, 198 (1966).
6. P. Martel, E. C. Svensson, A. D. B. Woods, V. F. Sears, and R. A. Cowley, J. Low Temp. Phys. **23**, 285 (1976).
7. R. Feltgen, H. Kirst, K. A. Köhler, H. Pauly, and F. Torello, J. Chem. Phys. **76**, 2360 (1982); R. Feltgen, H. Pauly, F. Torello, and H. Vehmeyer, Phys. Rev. Lett. **30**, 820 (1973).
8. R. Scherm, K. Guckelsberger, B. Fåk, K. Sköld, A. J. Dianoux, H. Godfrin, and W. G. Stirling, Phys. Rev. Lett. **59**, 217 (1987).
9. H. A. Mook, Phys. Rev. Lett. **55**, 2452 (1985).
10. P. E. Sokol, K. Sköld, D. L. Price, and R. Kleb, Phys. Rev. Lett. **54**, 909 (1985).
11. H. R. Glyde and S. I. Hernadi, Phys. Rev. B**29**, 2873 (1984); H. R. Glyde and S. I. Hernadi, Phys. Rev. B**28**, 141 (1983).
12. B. Tanatar, E. F. Talbot, and H. R. Glyde, to be published; B. Tanatar, E. F. Talbot, and H. R. Glyde, in preparation.
13. R. A. Aziz, V. P. S. Nain, J. S. Carley, W. L. Taylor, and G. T. McConville, J. Chem. Phys. **70**, 4330 (1979).
14. A. A. Abrikosov, L. P. Gorkov, and I. E. Dzyaloshinski, "Methods of Quantum Field Theory in Statistical Physics", Dover, New York (1965).
15. K. Gottfried and L. Pičman, K. Dan. Vidensk. Selsk. Mat-Fys. Medd. **32**, 13 (1960).
16. G. Baym and L. P. Kadanoff, Phys. Rev. **124**, 287 (1961).
17. F. Green, D. Nielson, and J. Szymànski, Phys. Rev. B**31**, 2779 (1985).
18. P. Nozières, "Theory of Interacting Fermi Systems", Benjamin, New York (1964).
19. W. G. Stirling, B. Tanatar, E. F. Talbot, and H. R. Glyde, in preparation.

DYNAMIC STRUCTURE FUNCTION OF THE QUANTUM BOSON FLUID IN THE JASTROW THEORY

Mikko Saarela and Juhani Suominen

Department of Theoretical Physics
University of Oulu
Linnanmaa
SF-90570 Oulu 57, Finland

INTRODUCTION

The properties of inhomogeneous fluids form a chalenging field of investications in the many-body physics. The increase in the computing power has made it possible to get the Monte Carlo solutions for some ground state properties like a surface profile of a quantum fluid droplet[1]. At the same time variational techniques based on the Jastrow ansatz for the wave function have been developed to find the optimized solutions for various inhomogeneous systems[2,3,4].

The Jastrow wave function for boson fluids consists of the product of one- and two-particle correlation functions

$$\Psi(\vec{r}_1,\ldots,\vec{r}_A;t) = e^{\frac{1}{2}\sum_i u_i + \frac{1}{2}\sum_{i,j \atop i\neq j} v_{ij}} \tag{1}$$

where in the most general case u_i = u($\vec{r}_i$,$\vec{r}_j$;t) and v_{ij}=v($\vec{r}_i$,$\vec{r}_j$;t) are complex time-depentent functions. A systematic improvement to include also the three-partice correlation function is possible, but quite tedious. The presentation of the correlation functions in the coordinate space makes it possible to write optimizing equations in such a form that they directly apply to both homogeneous and inhomogeneous systems and thus the experimentally better known homogeneous systems can be used as a testing ground.

In this work we derive elementary excitation spectrum, dynamic linear response function, and the dynamic structure function for the boson fluid, and present results in the case of the homogeneous liquid ^4He. We start with the assumption that the physical system is in its optimized ground state which is then disturbed by an external time dependent field $U_{ext}(\vec{r}_i,t)$. This distubance causes changes in the correlation functions

$$u_i = \bar{u}(\vec{r}_i) + \delta u(\vec{r}_i;t) \tag{2}$$

$$v_{ij} = \bar{v}(\vec{r}_i,\vec{r}_j) + \delta v(\vec{r}_i,\vec{r}_j;t) \tag{3}$$

where $\bar{u}$ and $\bar{v}$ are the known ground state quantities. For the changes δu and δv we shall derive the optimizing equations[5] and they turn out to be the one- and two-particle continuity equations which then are solved numerically and compared with experiments.

A more traditional starting point for studing the dynamic properties of boson fluids is the Feynman-Cohen approach[6]. They introduced the following form for the wave function,

$$\Psi(\vec{r}_1,\ldots,\vec{r}_A;\vec{k}) = \Psi_0 \sum_i e^{i[\vec{k}\cdot\vec{r}_i + \sum_{j \neq i} g_{ij}]} \tag{4}$$

where Ψ_0 is the true ground state wave function and g_{ij} is the backflow factor

$$g_{ij} = \frac{\alpha \vec{k}\cdot\vec{r}_{ij}}{r_{ij}^3}. \tag{5}$$

The factor α is the variational parameter, for which a value α=-3.8 $\mathring{A}^3$ was obtained. It is worth pointing out that the function g_{ij} is an antisymmetric function under the exchange of particles i and j whereas the two particle correlation function in the Jastrow theory must be an even function. Thus there is no direct comparision between them. However, the basic idea that one- and two particle correlations should play the dominant role and the three-particle correlation could be neglected is the same. A lot of work has been put upon improving the results by Feynman and Cohen by adding further corrections into their wave function within the perturbation theory[7].

THE THEORY

In the variational approach we search for the extremum of the integral

$$\delta L = \delta \int_{t_0}^{t} dt \frac{\langle\vec{\Psi}(t)|H - i\hbar\partial/\partial t|\vec{\Psi}(t)\rangle}{\langle\vec{\Psi}(t)|\vec{\Psi}(t)\rangle} = 0 \tag{6}$$

where

$$\vec{\Psi}(t) = e^{-iA\mu t/\hbar}\Psi(\vec{r}_1,\ldots,\vec{r}_A;t) \tag{7}$$

and μ is the chemical potential of the system. The Jastrow form many-body wave function of eq.(1) includes now two unknown complex functions δu and δv. These functions will be solutions of the Euler-Lagrange equations

$$\begin{array}{ll} \frac{\delta L}{\delta(Re\delta u)} = 0 & ; \quad \frac{\delta L}{\delta(Im\delta u)} = 0 \\ \frac{\delta L}{\delta(Re\,\delta v)} = 0 & ; \quad \frac{\delta L}{\delta(Im\,\delta v)} = 0 \end{array}. \tag{8}$$

The Hamiltonian of the system

$$H = -\frac{\hbar^2}{2m}\sum_i \vec{\nabla}_i^2 + \sum_{\substack{i,j\\ i\neq j}} V(|\vec{r}_i - \vec{r}_j|) + \sum_i U_{ext}(\vec{r}_i;t) \tag{9}$$

includes besides the kinetic energy and the two-particle interaction $V(|\vec{r}_i - \vec{r}_j|)$ also an external single particle interaction $U_{ext}(\vec{r}_i;t)$. As stated earlier we assume that the undisturbed system, U_{ext}=0, has been optimized yieding $\bar{u}(\vec{r}_i)$ and $\bar{v}(\vec{r}_i,\vec{r}_j)$ and the chemical potential μ of the system. Thus from the expansion of the expectation value of H in terms of δu and δv we can drop all linear terms, and since we are interested only in small perturbations we need to keep only terms quadratic in δu and δv. The potential energy term, $V(|\vec{r}_i - \vec{r}_j|)$, commutes with δu and δv, and hence it gives no contribution. Then we are left with those ones coming from the kinetic energy, external potential, and the time differentiation.

$$
\begin{aligned}
\delta L = \delta \int_{t_0}^{t} dt \{ \frac{\hbar^2}{8m} [\int d^3r_1 \rho_1 |\vec{\nabla}_1 \delta u_1|^2 \\
+ \int d^3r_1 d^3r_2 \rho_{12} (\vec{\nabla}_1 \delta u_1^* \cdot \vec{\nabla}_1 \delta v_{12} + \vec{\nabla}_1 \delta u_1 \cdot \vec{\nabla}_1 \delta v_{12}^* + |\vec{\nabla}_1 \delta v_{12}|^2) \\
+ \int d^3r_1 d^3r_2 d^3r_3 \rho_{123} \vec{\nabla}_1 \delta v_{12}^* \cdot \vec{\nabla}_1 \delta v_{13}] \\
+ \int d^3r_1 \rho_1 U_{ext}(\vec{r}_1; t) + \frac{\hbar}{2} [\int d^3r_1 \rho_1 Im \delta \dot{u}_1 + \frac{1}{2} \int d^3r_1 d^3r_2 \rho_{12} Im \delta \dot{v}_{12}] \}
\end{aligned}
\quad . \quad (10)
$$

Here we have used the definitions of the one- and two-particle densities.

$$
\rho_1 \equiv \rho(\vec{r}_1; t) = \frac{A \int d^3r_2 \ldots d^3r_A |\Psi|^2}{\int d^3r_1 \ldots d^3r_A |\Psi|^2} = \bar{\rho}_1 + \delta\rho_1 \quad (11)
$$

$$
\begin{aligned}
\delta\rho_1 = \bar{\rho}_1 \delta u_1 + \int d^3r_2 [\bar{\rho}_{12} - \bar{\rho}_1 \bar{\rho}_2] \delta u_2 + \int d^3r_2 \bar{\rho}_{12} \delta v_{12} \\
+ \frac{1}{2} \int d^3r_2 d^3r_3 [\bar{\rho}_{123} - \bar{\rho}_1 \bar{\rho}_{23}] \delta v_{23}
\end{aligned}
\quad (12)
$$

$$
\rho_{12} \equiv \rho^{(2)}(\vec{r}_1, \vec{r}_2; t) = A(A-1) \frac{\int d^3r_3 \ldots d^3r_A |\Psi|^2}{\int d^3r_1 \ldots d^3r_A |\Psi|^2} = \bar{\rho}_{12} + \delta\rho_{12} \quad (13)
$$

$$
\begin{aligned}
\delta\rho_{12} = \bar{\rho}_{12} (\delta u_1 + \delta u_2 + \delta v_{12}) + \int d^3r_3 (\bar{\rho}_{123} - \bar{\rho}_3 \bar{\rho}_{12}) \delta u_3 \\
+ \int d^3r_3 \bar{\rho}_{123} (\delta v_{23} + \delta v_{13}) + \frac{1}{2} \int d^3r_3 d^3r_4 (\bar{\rho}_{1234} - \bar{\rho}_{12} \bar{\rho}_{34}) \delta v_{34}
\end{aligned}
\quad . \quad (14)
$$

The quantities $\bar{\rho}_1$ and $\bar{\rho}_{12}$ are again the densities of the undisturbed system. One should recognize that $\delta\rho_1$ and $\delta\rho_{12}$ depend on the three- and four-particle densities $\bar{\rho}_{123}$ and $\bar{\rho}_{1234}$ of the ground state, and these are the terms which will complicate the numerical work. For convenience we also introduce the two- and three particle distribution functions g_{12} and g_{123}, respectively.

$$
\rho_{12} = \bar{\rho}_1 \bar{\rho}_2 g_{12} \qquad ; \; \rho_{123} = \bar{\rho}_1 \bar{\rho}_2 \bar{\rho}_3 g_{123} \quad (15)
$$

The Euler-Lagrange equations (8) can now be written in the form of continuity equations.

$$
\vec{\nabla}_1 \cdot \vec{j}^{(1)}(\vec{r}_1; t) + i\delta\dot{\rho}_1 = D^{(1)}(\vec{r}_1, t) \quad (16)
$$

$$
\vec{\nabla}_1 \cdot \vec{j}^{(2)}(\vec{r}_1, \vec{r}_2; t) + i\delta\dot{\rho}_{12} = D^{(2)}(\vec{r}_1, \vec{r}_2, t) \quad (17)
$$

where both currents, $\vec{j}^{(1)}$ and $\vec{j}^{(2)}$, and the densities, $\delta\rho_1$ and $\delta\rho_{12}$, are complex. Since our main application in this work is the calculation of the linear response function of a homogeneous system, we set from now on $\bar{\rho}_1 = \bar{\rho}_2 = \rho_0$ which is the constant density of the system. Then we can write for the currents the following definitions

$$
\vec{j}^{(1)}(\vec{r}_1; t) = \frac{\hbar\rho_0}{2m} [\vec{\nabla}_1 \delta u_1 + \rho_0 \int d^3r_2 g_{12} \vec{\nabla}_1 \delta v_{12}] \quad (18)
$$

$$
\vec{j}^{(2)}(\vec{r}_1, \vec{r}_2; t) = \frac{\hbar\rho_0}{2m} [g_{12}(\vec{\nabla}_1 \delta u_1 + \vec{\nabla}_1 \delta v_{12}) + \rho_0 \int d^3r_3 g_{123} \vec{\nabla}_1 \delta v_{13}] + (1 \leftrightarrow 2) \quad (19)
$$

and the driving terms, $D^{(1)}$ and $D^{(2)}$, are

$$D^{(1)}(\vec{r}_1,t) = \frac{2}{\hbar}[\rho_0 U_{ext}(\vec{r}_1;t) + \rho_0^2 \int d^3r_2(g_{12}-1)U_{ext}(\vec{r}_2;t)] \tag{20}$$

$$D^{(2)}(\vec{r}_1,\vec{r}_2,t) = \frac{2}{\hbar}[\rho_0^2 g_{12} U_{ext}(\vec{r}_1;t) + \frac{1}{2}\rho_0^2 \int d^3r_3(g_{123}-g_{12})U_{ext}(\vec{r}_3;t)] \tag{21}$$

These equations are exact upto the level of the Jastrow ansatz for the wave function.

THE DYNAMIC STRUCTURE

In the study of the dynamic structure of quantum fluids the main quantity of interest is the dynamic linear response function, $\chi(\mathrm{k},\omega)$, defined as follows

$$\frac{\delta U(k,\omega)}{\delta\rho(k,\omega)} = \chi^{-1}(k,\omega) \tag{22}$$

The poles of $\chi(\mathrm{k},\omega)$ determine the elementary excitations $\omega=\omega(k)$ and its imaginary part is the dynamic structure function,

$$S(k,\omega) = \frac{\hbar}{2\pi\rho_0} Im\,\chi(k,\omega). \tag{23}$$

When ω=0 in eq.(22)one obtains the static response function $\chi(\mathrm{k})$.

These quantities can be calculated from the Fourier transforms of the continuity equations eqs.(16) and (17). In order to demonstrate the method let us make a simplifying approximation by setting δv_{12}=0. Then the one particle continuity equation can be written in the following form.

$$-\frac{\hbar^2 k^2 \rho_0}{4m}\delta u_{k,\omega} + \frac{1}{2}\hbar\omega\delta\rho_{k,\omega} = \rho_0 S_k \delta U(k,\omega) \tag{24}$$

where S_k is the static structure function and the functions $\delta u_{k,\omega}$, $\delta\rho_{k,\omega}$ and $\delta U_{k,\omega}$ are the Fourier transforms of the corresponding coordinate space functions $\delta u(\vec{r};t)$, $\delta\rho(\vec{r};t)$ and $U_{ext}(\vec{r};t)$. The relation between $\delta\rho_{k,\omega}$ and $\delta u_{k,\omega}$ is found from the definition for the one-particle density, eq.(12),

$$\delta\rho_{k,\omega} = \rho_0 S_k \delta u_{k,\omega}. \tag{25}$$

Using eq.(24) and(25) we can write the expression for the inverse of the response function

$$\rho_0\chi^{-1}(k,\omega) = -\frac{\hbar^2k^2}{4mS_k^2} + \frac{\hbar\omega}{2S_k}. \tag{26}$$

By setting $\chi^{-1}(k,\omega))=0$ we get the well known Feynman excitation spectrum

$$\varepsilon_k = \hbar\omega = \frac{\hbar^2k^2}{2mS_k} \tag{27}$$

leading to the one-pole approximation for the dynamic structure function

$$S(k,\omega) = S_k\delta(\omega - \frac{\varepsilon_k}{\hbar}). \tag{28}$$

Let us now return to the full set of continuity equations. From eq.(16) we can again deduce an expression for the linear response function

$$\chi(k,\omega) = \frac{2\rho_0 S_k}{\hbar\omega + \Pi(k,\omega) - \varepsilon_k}. \tag{29}$$

The new term, $\Pi(k,\omega)$, entering due to the change in the two particle correlation function can be identified as the proper energy which can be written in the form

$$\Pi(k,\omega) = \frac{\hbar^2 k^2}{4m} \int \frac{d^3q}{(2\pi)^3\rho_0} \Sigma_k(\vec{q}) \frac{\delta v_{k,\omega}(\vec{q})}{\delta\rho_{k,\omega}} \tag{30}$$

The term $\Sigma_k(\vec{q})$ inside the integral depends on the three-particle density. A simplest approximation which gives the small k limit corectly is the convolution approximation. That yields the following form for $\Sigma_k(\vec{q})$

$$\Sigma_k(\vec{q}) = S_+S_- - \frac{1}{2}(S_+ + S_-) + \frac{\vec{k}\cdot\vec{q}}{k^2}(S_+ - S_-) \tag{31}$$

where

$$S_\pm = S(|\frac{\vec{k}}{2} \pm \vec{q}|) \equiv S_{\frac{k}{2}\pm q}. \tag{32}$$

The excitation spectrum of the system will be determined now by the solutions of the equation

$$\begin{aligned} &\hbar\omega_p + \Pi(k,\omega_p) - \varepsilon_k = 0 \\ \Rightarrow \quad &\omega_p = \omega_p(k) \end{aligned} \tag{33}$$

They can be either real or complex. The lowest real solution determines the elementary excitation mode and the coplex solutions should correspond to the excited states which can decay. The real solutions yield a δ-function form for the dynamic structure function,

$$S_p(k,\omega) = \frac{S_k}{1 + \Pi'(k,\omega)}\delta(\omega - \omega_p). \tag{34}$$

with the derivative

$$\Pi'(k,\omega) \equiv \frac{d\Pi(k,\omega)}{d(\hbar\omega)}. \tag{35}$$

Whereas complex solutions

$$\begin{aligned} \omega_p &= \omega_0 + i\omega_1 \\ \Pi(k,\omega_p) &= \Pi_0(k,\omega_p) + i\Pi_1(k,\omega_p) \end{aligned} \tag{36}$$

will result a broad peak into $S(k,\omega)$. For the height of the peak we can write a simple expression

$$S_p(k,\omega = \omega_0) = \frac{S_k}{[1 + \Pi_0' + \frac{\Pi_1'^2}{1+\Pi_0'}]\pi\omega_1} \tag{37}$$

The full dynamic response function is then the sum of all these contributions,

$$S(k,\omega) = \sum_p S_p(k,\omega) \tag{38}$$

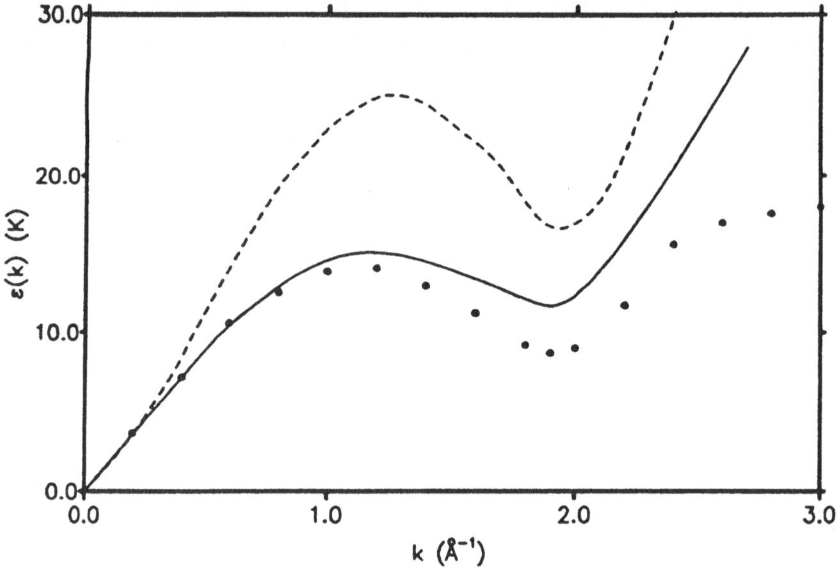

Fig. 1: The elementary excitation spectrum of liquid ^4He. The full line is the spectrum obtained by solving the eqs. (45), the black dots are the experimental results[10], and the dashed curve is the Feynman spectrum of eq. (27).

The static response function will also be modified by the proper energy term and it can be written in the following form,

$$\chi(k) = \frac{2S_k\rho_0}{\Pi(k,\omega=0) - \varepsilon_k} \tag{39}$$

THE RESULTS

Our first aim is calculate the excitation spectrum of the quantum boson system. That can be archieved by setting U_{ext}=0 in the continuity equations

$$\begin{aligned} \vec{\nabla}_1 \cdot \vec{j}^{(1)}(\vec{r}_1;t) + i\delta\dot{\rho}_1 = 0 \\ \vec{\nabla}_1 \cdot \vec{j}^{(2)}(\vec{r}_1,\vec{r}_2;t) + i\delta\dot{\rho}_{12} = 0 \end{aligned} \tag{40}$$

The main difficulty in solving these equations is to calculate $\delta\rho_{12}$ using its definition given in eq(14). There is, however, another exact form for which it is easier to invent approximations,

$$\delta\rho_{12} = \rho_0 g_{12}(\delta\rho_1 + \delta\rho_2) + \rho_0^3 \int d^3r_3 \frac{\delta g_{12}}{\delta(\delta u_3)}\delta u_3 + \rho_0^4 \int d^3r_3 d^3r_4 \frac{\delta g_{12}}{\delta(\delta v_{34})}\delta v_{34}. \tag{41}$$

For the first integral term the convolution approximation can be used again, but for the second one we invoke the hypernetted-chain equations for the inhomogeneous system[8]. The use of them leads to an integral equation

$$s_k(\vec{p},\vec{p}\,') - \int \frac{d^3q}{(2\pi)^3\rho_0} H_k(\vec{p},\vec{q}) s_k(\vec{q},\vec{p}\,') = y_k(\vec{p},\vec{p}\,') \tag{42}$$

where the unknown quantity $s_k(\vec{p},\vec{p}')$ is related to the Fourier transform of the functional derivative $\delta g_{12}/\delta(\delta v_{34})$. The functions H_k and y_k are known and depend only on the static structure function of the system,

$$H_k(\vec{p},\vec{q}) = (S_{p-q} - 1)(S_{\frac{k}{2}+q}S_{\frac{k}{2}-q} - 1) \tag{43}$$

$$\begin{aligned} y_k(\vec{p},\vec{p}') &= S_{\frac{k}{2}+p'}S_{\frac{k}{2}-p'}(S_{p+p'} - 1) \\ &+ \frac{1}{2}S_k(S_{\frac{k}{2}+p'}S_{\frac{k}{2}-p'} - 1)\int \frac{d^3q}{(2\pi)^3\rho_0}(S_{p-q} - 1)(S_{\frac{k}{2}+q} - 1)S_{\frac{k}{2}-q} - 1). \end{aligned} \tag{44}$$

Inserting these results into the continuity equations in the Fourier space we end up with the following set of integral equations

$$\left\{ \begin{array}{l} \frac{\hbar^2k^2}{4m}\int \frac{d^3q}{(2\pi)^3\rho_0}\sum_k(\vec{q})\delta v_{k,\omega}(\vec{q}) + (\hbar\omega - \varepsilon_k)\delta\rho_{k,\omega} = 0 \\ X_{k,\omega}(\vec{q})\delta v_{k,\omega}(\vec{q}) + \int \frac{d^3q'}{(2\pi)^3\rho_0}\delta v_{k,\omega}(\vec{q}')K_{k,\omega}(\vec{q},\vec{q}') = -\varepsilon_k\sum_k(\vec{q})\delta\rho_{k,\omega} \end{array} \right. \tag{45}$$

where

$$X_{k,\omega} \equiv \hbar\omega S_+S_- - \frac{\hbar^2}{2m}[(\frac{\vec{k}}{2} + \vec{q})^2S_- + (\frac{\vec{k}}{2} - \vec{q})^2S_+] \tag{46}$$

and

$$\begin{aligned} K_{k,\omega}(\vec{q},\vec{q}') &\equiv \hbar\omega S_+S_-s_k(\vec{q},\vec{q}') \\ &- \frac{\hbar^2}{2m}(S_{q-q'} - 1)[S_-S'_-(\frac{\vec{k}}{2} + \vec{q})\cdot(\frac{\vec{k}}{2} + \vec{q}') \\ &+ S_+S'_+(\frac{\vec{k}}{2} - \vec{q})\cdot(\frac{\vec{k}}{2} - \vec{q}')] - \frac{\hbar^2k^2}{4m}\Sigma_k(\vec{q})\Sigma_k(\vec{q}') \end{aligned} \tag{47}$$

The unknown function $\delta v_k(\vec{q})$ depends on $|\vec{q}|$ and the angle $\theta_{k,q}$ between $\vec{k}$ and $\vec{q}$. Thus we expand it in terms of Legendre polynomials

$$\delta v_k(\vec{q}) = \sum_l \nu_l(q)P_l(cos\theta_{k,q}) \tag{48}$$

and look for the solution of the multipliers ν_l(q). The second equation in (45) is actually the difference between the two continuity equations. That enables us to get rid of some uninteresting δ-functions present otherwise. It is also easy to check the validity of the sequential condition which states in this case that the second equation in (45) should be equal to zero when $\vec{q} = \frac{\vec{k}}{2}$.

As the numerical application we solve the set of equations (45) for liquid ^4He. The quantity which determines the homogeneous system is the static structure function which we choose to be in this application the experimentally measured function[9]. In the calculations we include Legendre polynomials upto l=6. There a good convergence of the Legendre expansion is already reached. The eigenvalue spectrum contains both real and imaginary solutions, but here we are only interested in the lowest real solution. The results are shown in Fig.1. The calculated curve follows nicely the experimental points[10] upto the maxon region, but the roton dip is not guite deep enough and the second plateau at k>2.5 K^{-1} is not seen at all. This may be due to the fact that the three particle density is approximated using the convolution approximation which is nesessary to get the small k behaviour correct, but it becomes worse at larger k where the short range correlations begin to play important role. One would also expect that at k>2.5 K^{-1} the three particle correlations begin to play some role. In Fig.2. we then show the one-phonon contribution to the total scattering. In the Feynman approach this is just the static structure function which will be equal to unity at large k whereas experiments tell that it should approach to zero. The proper

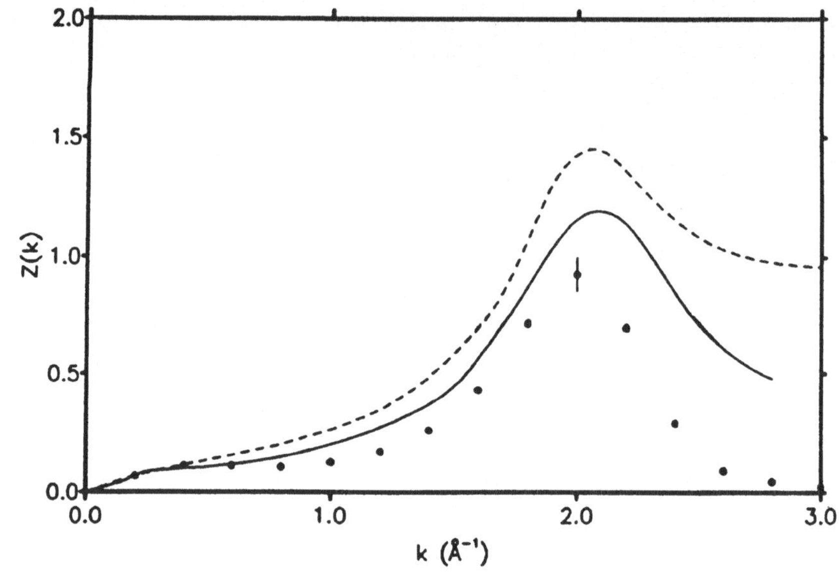

Fig. 2: The one-phonon contribution to the scattering defined in eq(34) . The identification of the curves is the same as in Fig.1.

energy term brings the theory into closer agreement with experiments but it is not quite large enough.

The calculation of the static response function does not require the knowledge of the $\delta\rho_{12}$ and is in that sense the most accurately determined quantity. In Fig.3. we present the results of the full calculation (see eq.(39). Comparison with experiments shows a good agreement even at large k.

THE CONCLUSIONS

In this work we have presented a variational procedure by which we can calculate the dynamic properties of quantum boson fluids. The starting point in this approach is the Jastrow ansatz for the wave function which contains complex, time-dependent one- and two-particle correlation factors. The continuity equations are derived and they are shown to give the opptimized result for the correlation functions. In the numerical calculations the convolution approximation is used in the calculation of the three particle density and the hypernetted approximation for the evaluation of the change in the radial distribution function, δg_{12}. The requirement for the approximations has been that they satify the sequetial condition between the one- and two-particle continuity equations. The results compare well with experiments at small k region but when k>1.5 K^{-1} improvements on the convolution approximation are required.

ACKNOWLEDGEMENTS

The authors wish to thank Eckhard Krotscheck, Mauri Puoskari and Pekka Pietiläinen for many valuable discussions. They also like to thank Alpo Kallio for stimulating working conditions. One of us (M.S.) would like to thank Hans Pauli for warm hospitality during his stay in Heidelberg where part of this work was done.

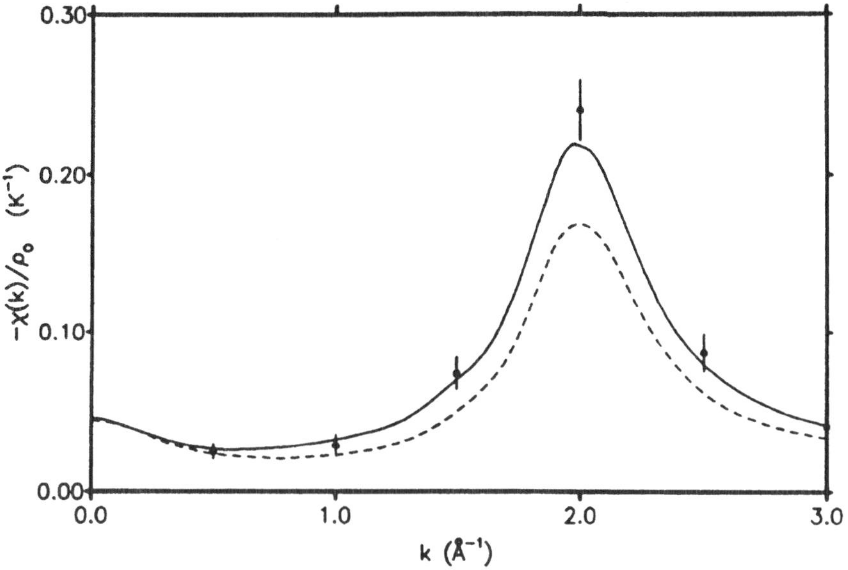

Fig. 3: The static response function of liquid ^4He. The full curve contains the proper energy correction of eq.(39), the vertical lines are the experimental results[10], and the dashed line is the Feynman result from eq(26).

REFERENCES

1. V.R. Pandharipande, J.G. Zabolitzky, S.C. Pieper, R.B. Wiringa, and U. Helmbrecht, Phys. Rev. Lett. 50, 1676 (1982).
2. M. Saarela, P. Pietiläinen and A. Kallio, Phys. Rev. B27, 231 (1983).
3. M.Saarela, Phys. Rev. B33, 4596 (1986).
4. E. Krotscheck, Phys. Rev. B32, 5713 (1985) and refereces in there.
5. E. Krotscheck, Phys. Rev. B31 4258 (1985).
6. R.P. Feynman and M. Cohen, Phys. Rev. 102, 1189 (1956).
7. E. Manousakis and V.R. Pandharipande, Phys. Rev. B30, 5062 (1984).
8. T. Morita, Prog. Theor. Phys. 20, 920 (1958).
9. H.N. Robkoff and R.B. Hallock, Phys. Rev. B24, 159 (1981). and Phys. Rev. B25, 1572 (1982).
10. R.A. Cowley and A.D.B. Woods, Can. J. Phys. 49, 177 (1971).

LENNARD-JONES BOSON MATTER

IN THE THERMODYNAMIC PERTURBATION SCHEME

V. C. Aguilera-Navarro*, Christina Keller and M. de Llano§

Physics Department, North Dakota State University
Fargo, ND 58105, USA

R. Guardiola

Departamento de Física Moderna, Facultad de Ciencias
Universidad de Granada, 18071 Granada, Spain

M. A. Solís

Instituto de Física, Universidad Nacional Antónoma de
México, 01000 México, D. F., México

ABSTRACT

The thermodynamic perturbation scheme for the very successful construction of classical liquid equations of state is applied in the quantum regime to the ground-state energy of a many-boson system interacting pairwise via a Lennard-Jones intermolecular potential. The ensuing density and attractive-coupling double expansion for the energy is series-analyzed with Padé techniques, as well as generalizations thereof. Results through sixth order compare with the Green Function Monte Carlo simulations for the same system at the one-percent level of accuracy.

INTRODUCTION

The original idea of van der Waals to describe a classical fluid by separating the pair-interaction potential into repulsive and attractive components has more recently developed into what is probably the best scheme to calculate equations-of-state for liquids. Different potential separations were proposed by McQuarrie & Katz[1], Barker & Henderson[2] and by Weeks, Chandler & Anderson[3], and became an integral part of "thermodynamic perturbation theory". The exactness of this theory in first order as the system density approaches close packing (for the case of hard-sphere-plus-attractive potentials) has been proved in one-[4], more recently in two-[5], and conjectured[5] to hold in three-dimensions.

---

* On leave from Instituto de Física Teórica, São Paulo, Brazil.

§ Support from U.S Army Research Office is gratefully acknowledged.

We present here a simple, quick and inexpensive quantum scheme for describing the ground-state energy of a moderately dense liquid like helium-four modeled by a system of bosons interacting via the Lennard-Jones (LJ) 6-12 intermolecular potential[6]. The results are within a percent or so of Green Function Monte Carlo (GFMC) calculations[7], which are generally accepted as giving the "exact" non-relativistic, Schroedinger ground-state energy of the many-particle system. The approach is inspired in the abovementioned thermodynamic perturbation theory, where, having split the pair potential into repulsive and attractive parts, one first describes the fluid with the repulsive interaction as accurately as possible, and incorporates the attractions as subsequent perturbations.

For the quantum (many-boson) system one begins with the well-known low-density expansion produced by quantum field perturbation theory[8] for the ground-state energy per particle

$$\frac{E}{N} \simeq \frac{2\pi\hbar^2}{m}\rho a\left[1 + C_1\sqrt{\rho a^3} + C_2\rho a^3 \ln(\rho a^3) + O(\rho a^3)\right] \tag{1}$$

$$C_1 = 128/15\pi^2, \quad C_2 = 8\left(\frac{4\pi}{3} - \sqrt{3}\right), \quad \frac{\hbar^2}{m} = 12.120904\ \mathrm{K}\text{-}\mathring{A}^2\ (\text{for } {}^4He),$$

where ρ is the particle density and a the S-wave scattering length associated with the assumed pair interaction. The series (1) can be considered as embodying the most rigorous information we possess of the fully-interacting many-boson system ground-state energy, in spite of the fact that it may diverge for all values of $\sqrt{\rho a^3}$, and be only an asymptotic series at best. In diagrammatic terms, the expression (1) is the result of summing to infinite order precisely those subclasses[9] of diagrams needed to determine the resulting non-power series exactly, through the order given. Optimized Jastrow variational techniques[10] as well as Brueckner ladder perturbative summations[11], will correctly reproduce up to the second term in (1), i.e., up to and including the $C_1\sqrt{\rho a^3}$ correction term. Unfortunately , it is not known to what extent further terms in (1) can be recovered from additional refinements of either the Jastrow or Brueckner schemes.

THERMODYNAMIC PERTURBATION SERIES

For zero scattering energy in a pair potential V(r) the S-wave Schroedinger equation is

$$u_0''(r) - \frac{m}{\hbar^2}V(r)u_0(r) = 0, \quad u_0(0) = 0, \quad \lim_{r\to\infty} u_0(r) = r - a \tag{2}$$

where the S-wave scattering length a can be written[12] as

$$a = \frac{m}{\hbar^2}\int_0^\infty dr\, r\, V(r)\, u_0(r). \tag{3}$$

For the Lennard-Jones 6-12 pair potential

$$V(r) = 4\epsilon\left[\left(\frac{\sigma}{r}\right)^{12} - \left(\frac{\sigma}{r}\right)^6\right] \tag{4}$$

a very accurate determination[13] of a using (3) directly leads to a=-176.3252906 Å. Being negative, this leads to imaginary terms in (1). A purely real expression for the energy can be obtained for (1) by simply expanding the RHS of (3) in powers of an appropriately defined "switching" parameter $0 \leqslant \lambda \leqslant 1$ to be associated with the attractive part, still to be specified, of (4). Thence

$$a = a_0 + a_1 \lambda + a_2 \lambda^2 + \cdots \tag{5}$$

substituted into (1) which is then rearranged into powers of λ leads to the double series expression for the energy per particle

$$\frac{E}{N} \simeq \frac{2\pi\hbar^2}{m a_0^2} x^2 \sum_{i=0}^{\infty} \frac{a_i}{a_0} e_i(x) \lambda^i, \qquad x \equiv \sqrt{\rho a_0^3}, \tag{6a}$$

$$e_i(x) \simeq 1 + C_{1i} x + C_{2i} x^2 \ln x^2 + O(x^2). \tag{6b}$$

Unlike (1) for $a<0$, (6a) is now real for all $0 \leqslant \lambda \leqslant 1$ since the a_i's of (5) (which can at worst be determined numerically) are all real. In (6) the dimensionless coefficients C_{1i}, C_{2i} have been deduced[14] for i=1,2,...,6 via the computer algebraic scheme MACSYMA[15] in terms of $C_1, C_2, a_1, a_2, \ldots, a_6$. Clearly, $C_{10} \equiv C_1$, $C_{20} \equiv C_2$.

The rearranged ("low-density" and "weak-attractive-coupling") double series (6) is the zero-temperature quantum analogue of the classical thermodynamic perturbation scheme. Finite-temperature generalizations can be constructed along the lines, e.g., of Huang's classic treatment[16] of the imperfect Bose gas employing Bogoliubov-Valatin[17] transformations; these would of course evolve into <u>triple</u> series. First, however, one must explore whether a "few-term-series" like (1) can yield, at zero temperature, sufficiently accurate results for liquid densities, and the extent to which higher-order coefficients are required for improvements.

Within the classical thermodynamic perturbation scheme Barker & Henderson (BH)[2] and Weeks, Chandler & Anderson (WCA)[3], respectively, split V(r), with $0 \leqslant \lambda \leqslant 1$, as

$$\theta(\sigma - r)V(r) + \lambda\, \theta(r-\sigma)V(r) \qquad \text{(BH)} \tag{7}$$

$$\theta(r_m - r)\left[V(r)+\varepsilon\right] + \lambda\left[\theta(r-r_m)V(r) - \varepsilon\,\theta(r_m - r)\right] \qquad \text{(WCA)} \tag{8}$$

In other words, the BH splitting is at the zero of the potential, namely σ, while the WCA splitting is at the zero of the corresponding <u>force</u> -dV(r)/dr, namely at $r_m = 2^{1/6}\sigma$, which is the minimum of the potential. It has been argued[18] that (7) is appropriate for describing dilute gases and (8) for dense liquids. For either splitting, λ-expansion coefficients of (5) have been calculated through 14th order in double-precision for the ^{4}He Lennard-Jones[19] as well as other potentials[20]. Table I lists the a_i values to sixth order, together with the resulting

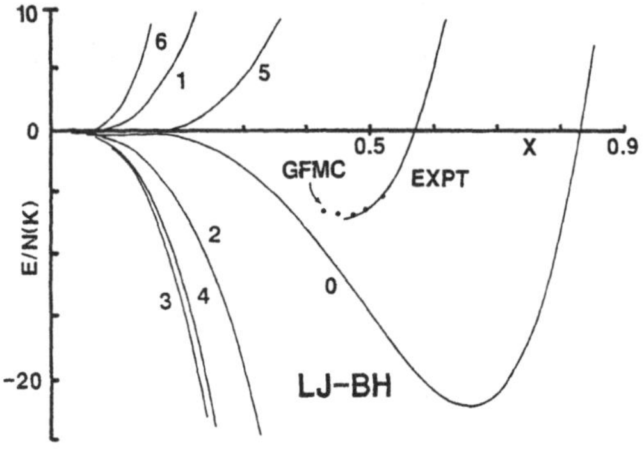

Figure 1. Plot of the first seven partial sums (0,1,2,...,6) of the thermodynamic perturbation expansion (6a) for the Barker-Henderson (BH) splitting (7) of the Lennard-Jones potential for bosons of ^{4}He mass, with $\lambda=1$. The curve labeled EXPT refers to a fitting[21] of the experimental results for liquid helium four while the dots refer to the Green function Monte Carlo (GFMC) data points for the ground state energy of a system of Lennard-Jones bosons. Energy units are in degrees Kelvin (K).

Table I. Expansion coefficients a_i, Eq. (5), for the LJ potential (4) split according to the BH and WCA methods, Eqs. (7) and (8) respectively. Also given are the "virial" coefficients C_{1i} and C_{2i} of Eq.(6b) which result.

	i	a_i (Å)	C_{1i}	C_{2i}
BH	0	2.092954	$128/15\sqrt{\pi} \simeq 4.814...$	$8(4\pi/3-\sqrt{3}) \simeq 19.65...$
	1	-2.799958	12.03605	78.61566
	2	-2.372997	-2.213165	-107.5271
	3	-2.300964	-9.596263	-75.56631
	4	-2.264692	-14.20699	2.055173
	5	-2.233809	-15.80247	98.14943
	6	-2.204089	-13.71285	184.0432
WCA	0	2.138328	$128/15\sqrt{\pi}$	$8(4\pi/3-\sqrt{3})$
	1	-2.721638	12.03605	78.61566
	2	-2.282284	-1.665258	-100.3695
	3	-2.208046	-8.711356	-74.68053
	4	-2.172995	-13.26767	-8.072665
	5	-2.144249	-15.19937	77.11551
	6	-2.116852	-13.97842	157.3693

i-th order low-density coefficients C_{1i}, C_{2i} appearing in (6). Figure 1 shows the behavior in $x \equiv \sqrt{\rho a_o^3}$ of the first seven partial sums (i=0,1,...,6) of (6) for the BH splitting. The WCA splitting gives qualitatively similar results. We compared both with the experimental results[21] as well as the GFMC data points[7] for the same pair potential used here. The equilibrium minimum of the GFMC calculation is only about 0.29K above the experimental value of -7.14K, or about 4% too high. These results are clearly meaningless at all densities, there being no physically reasonable pattern emerging from them. Even at vanishingly small x, where the $e_i(x)$ series in (6) should be adequate, there is no observed convergence in the attractive coupling λ series since from Table I it is clear that, for the LJ potential, the ratios $a_{i+1}\lambda/a_i \simeq O(1)$ for $\lambda = 1$.

SOFT- AND HARD-SPHERE REFERENCE FLUIDS

Extrapolation of (6) to non-zero x and λ is undoubtedly required. Padé[22] and other extrapolant methods have proved considerably successful in reproducing the correct behavior of functions for which only the beginning of an expansion series is available, and for which many (20 to 30) series coefficients are known, as in the critical phenomena associated with lattice-spin models[23], or in various simple quantum problems (like the quartic oscillator) of so-called "large-order perturbation theory"[24]. Little work appears to have been done to extract useful information from series bearing only a handful (3 to 4) of coefficients. As in the earlier classical studies[2,3] the first step is an accurate treatment of the purely repulsive fluid. In the low-density limit this is described by (1) with λ=0, namely,

$$\frac{E}{N} \simeq \frac{2\pi\hbar^2}{m a_o^2} x^2 e_o(x) \tag{9a}$$

$$e_o(x) \simeq 1 + C_1 x + C_2 x^2 \ln x^2 + O(x^2). \tag{9b}$$

For hard spheres, a_o is just the sphere diameter, and all possible Padé-like approximants $\epsilon_o(x)$ to the series $e_o(x)$ of (9b) have been studied[25]. The $\epsilon_o(x)$'s can be defined so that $\epsilon_o(x)-e_o(x) = o(x^2)$, if a value for the coefficient of the unknown term, say C_3x^2, is assumed. Of the twelve such extrapolants $\epsilon_o(x)$, only two present acceptable global fits to the boson hard sphere GFMC data[26] each with just one constant C_3. Both forms turn out to be almost coincident with each other from low up to moderate (physical, e.g., equilibrium liquid ^{4}He) densities, as well as with the GFMC data; they begin differing at higher densities. Of the two forms, the approximant having the smallest least-mean-squares-fit value with the four available GFMC data points, is

$$\epsilon_o(x) = \left[1 - \frac{\frac{1}{2}C_1 x}{1 - \frac{2C_2}{C_1} x\left(\ln x + [C_3 - \frac{3}{4}C_1^2]/2C_2\right)}\right]^{-2} \tag{10}$$

which gave C_3=25.110 ± 0.287 and a standard deviation of 0.0051 for the fit. The form (10) possesses a second-order (Bernal density) pole at x = 0.7245 ± 0.0045.

Since a_0 of (9b) scales together with the density ρ as $x=\sqrt{\rho a_0^3}$, at low enough densities a fluid of soft spheres with an S-wave scattering length a_0 should behave like a fluid of rigid spheres of diameter a_0. The question is whether this equivalence persists at higher densities, or nearly so. For reasons to be given below, the answer appears to be "yes", even to the extent of postulating that, for all practical purposes, a fluid of soft spheres described by potentials like (7) or (8) with $\lambda=0$ will possess a finite, ultimate density value at which the system energy-per-particle is essentially infinite. A real divergence definitely occurs for a classical hard sphere fluid, where the ultimate density is just the random-close-packing (or Bernal) density empirically found, in experiments[27] with ball-bearings for example, to occur at a packing fraction very close (to within four digits) to $2/\pi \approx 0.6366$, or at a density of about 86% of the ordered "primitive hexagonal" (closest) packing value of $\rho_0=\sqrt{2}/c^3$, where c is the rigid sphere diameter. The Bernal density for boson hard spheres as predicted by (10) is about 37% of ρ_0. This value is within the range of 30% to 86% expected a priori for boson particles; the lower bound value of 30% coming from the simple quantum estimate of allowing the classical Bernal density of $0.86\sqrt{2}/c^3$ to become $(0.86/2^{3/2})\sqrt{2}/c^3 \simeq 0.304\sqrt{2}/c^3$, since wave-mechanical diffraction makes rigid spheres of diameter c effectively possess a diameter of $\sqrt{2}c$ at high scattering energies. As mentioned, one can give several reasons in support of the soft- vs hard-sphere fluid equivalency, of which the simplest are the following two: a) The pair distribution function as extracted for a LJ fluid via molecular dynamics[28] is well-known to be virtually zero for separation distances $\lesssim 0.9\sigma \simeq 2.3$ Å. b) In numerically integrating the radial Schroedinger equation (2) at zero-scattering-energy for a pair of LJ particles, it was necessary[19] to assume that the reduced radial wave function $u(r)$ of (3) was virtually zero ($\lesssim 10^{-38}$ Å) for $r\lesssim 0.6\sigma \simeq 1.5$Å, in order to achieve multi-digit precision.

DENSITY-SERIES ANALYSIS

To incorporate the i-th order perturbation corrections ---needed to produce a self-bound liquid ground state---we look for all possible Padé-like approximants $\epsilon_i(x)$ such that $\epsilon_i(x)-e_i(x) = O(x^2)$, where the $e_i(x)$ (i=1,2,...) are the two-termed series in (6b). One arrives at four distinct forms, each of which upon expansion for small x give back (6b). These are: i) $1+C_{1i}/(1-C_{2i}x\ln x^2/C_{1i})$; ii) $(1-C_{1i}x-C_{2i}x^2\ln x^2)^{-1}$; iii) $(1+C_{2i}x^2\ln x^2)/(1-C_{1i}x)$; iv) $(1+C_{1i}x)/(1-C_{2i}x^2\ln x^2)$. The density behavior within the physical interval $0\leqslant\rho<\rho_B$ of each form was studied and the following conditions imposed: 1) Since the perturbation potential is negative, both first- and second-order contributions to the energy must be non-positive throughout the density interval. 2) The first-order contribution, moreover, must clearly increase monotonically in ρ. 3) In every order, the contribution must be finite; specifically, $\epsilon_i(x)$ forms with vanishing denominators in $0\leqslant x\leqslant x_B=\sqrt{\rho_B a_0^3}$ are immediately discarded. For orders 1 through 6, forms i, i, i, iii, ii and ii, respectively, survived this analysis for the BH case. They are graphed in Fig. 2 and compared against the repulsive-sphere, zero-order extrapolant (10) labeled by "0". Note from the Figure that $\epsilon_1(x)$ (curve labeled 1) increases in ρ. Hence, first-order energy-per-particle in the present (soft-sphere-fluid-based) perturbation scheme being $E_1/N\equiv(2\pi\hbar^2/m)\rho a_1\epsilon_1(\sqrt{\rho a_0^3})$ where $a_1<0$, will decrease faster in ρ than the familiar, exact

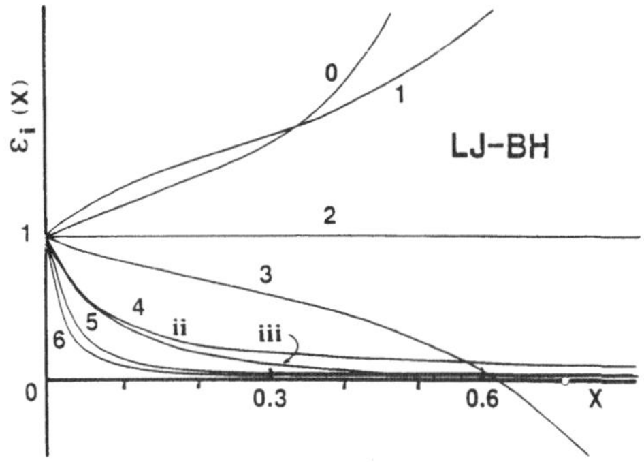

Figure 2. Approximants to the low-density series (6b) for orders i=0 through i=6 that survived the density series analysis, as explained in the text, for the BH splitting of the LJ potential. The open circle on the x-axis at 0.7245 marks the placement of the Bernal density second-order pole of $\epsilon_o(x)$ as given by (10).

first-order (ideal-gas-based) perturbation contribution $-2\pi\rho V_o R^3/3$ for purely attractive bosons interacting with a rectangular well of depth $-V_o$ and range R. In second order we have $E_2/N \equiv (2\pi\hbar^2/m)\rho a_2 \epsilon_2(\sqrt{\rho a_o^3})$ again with $a_2<0$. Curve 2 of Figure 2 shows an essentially constant $\epsilon_i(x)$ so that E_2/N decreases linearly in ρ, just as does the exact[29] second-order result $-(4\pi/15)\rho V_o^2 R^5$ for rectangular-well bosons in the ideal-gas-based perturbation theory. This comparison cannot be continued since the exact results are well-known[8] to diverge for bosons, order by order, as of third-order, because the unperturbed reference, zero-order state is the zero-momentum, plane-waves (ideal Bose gas) state. On the other hand, the soft-sphere-fluid-based perturbation results as of third-order are not only finite but also very close in behavior to that expected from the exactness[4,5] of 1st order thermodynamic perturbation theory as one approaches close packing, namely, $\epsilon_i(x_B) \simeq 0$. The only ambiguity that emerged in analyzing forms (i) through (iv) in each order occurred in 4th order. Here the analysis yields two acceptable forms (ii) and (iii); the latter one was retained for conforming best with the $\epsilon_4(x_B) \simeq 0$ criterion.

ATTRACTIVE-COUPLING-SERIES ANALYSIS

The energy-per-particle Eq.(6)---with the $e_i(x)$ replaced by the $\epsilon_i(x)$ just constructed---can now be computed through sixth order in λ for $\lambda = 1$. These results are illustrated in Figure 3 for the WCA case; qualitatively similar results hold for the BH potential splitting. Also displayed are the GFMC data for LJ

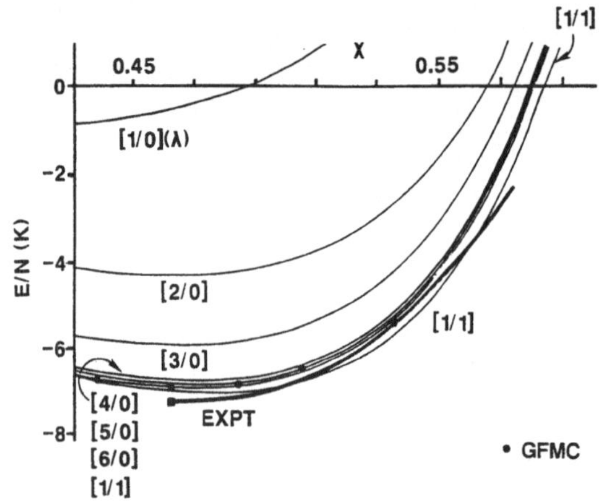

Figure 3. First few partial sums of (6a) with the $e_i(x)$ replaced by the $\epsilon_i(x)$ approximants displayed in Fig. 2, for the LJ boson system under the WCA potential splitting with $\lambda=1$. The curve labeled [1/1] is a lower bound to all n-th order partial sums of (6a) at least through n=6, as discussed in the text. The dots and the thick curve are as in Fig. 1.

bosons[7] (dots), where the reported errors are smaller than the dot size, as well as the experimental results for liquid helium four as summarized in ref.[21] (thick full curve). Binding occurs already in first order, denoted by [1/0](λ), and very rapid convergence is observed in succeeding orders. The curve labeled [1/1] is the one-one Padé approximant in λ, which emerges as an apparent lower bound to the perturbation calculations, at least through sixth order. The larger curvature observed in our converged results at higher densities ($x > 0.55$) compared with, say, the experimental curve, can undoubtedly be understood as a breakdown at higher densities of simulating the soft-spheres reference fluid by the hard spheres one, as we have done here.

Finally, to investigate the "smallness" of the attraction parameter $\lambda = 1$ at the predicted equilibrium densities we study the variation in the zero-pressure energy of all the Padé approximants [L/M](λ) to the λ-series (6) for which $2 \leqslant L+M \leqslant 6$, except the [0/N] approximants which being positive for all x, produce no binding. These are listed in Table II for both BH and WCA potential splittings; the final entry is the GFMC value. A

Table II. Minimum boson liquid energy per particle $(E/N)_{min}$ in degrees Kelvin and corresponding equilibrium (or saturation) density variable x_s which result from Padé analysis $[L/M](\lambda)$ of the rearranged, attractive-coupling series Eq. (6a), for both BH and WCA splittings of the LJ potential. The last entry refers to the corresponding Green Function Monte Carlo (GFMC) result of Ref. [7].

		BH		WCA	
L	M	x_s	$(E/N)_{min}(K)$	x_s	$(E/N)_{min}(K)$
2	0	0.4721	-5.148	0.4636	-4.266
1	1	0.4820	-8.128	0.4753	-6.957
3	0	0.4675	-6.480	0.4663	-5.865
2	1	0.4530	-7.152	0.4509	-7.078
1	2	0.4500	-7.113	0.4517	-7.074
4	0	0.4671	-6.651	0.4703	-6.685
3	1	0.4671	-6.676	0.4812	-7.582
2	2	pole in $0.44<x<0.46$		pole in $0.44<x<0.45$	
1	3	0.4722	-6.523	0.4747	-7.486
5	0	0.4677	-6.750	0.4710	-6.794
4	1	0.4709	-6.889	0.4711	-6.811
3	2	0.4672	-6.758	0.4638	-7.172
2	3	0.4673	-6.725	0.4640	-7.169
1	4	0.4673	-6.874	0.4731	-6.617
6	0	0.4680	-6.808	0.4713	-6.855
5	1	0.4685	-6.892	0.4718	-6.932
4	2	0.4690	-6.892	0.4714	-6.868
3	3	0.4680	-6.812	0.4677	-7.070
2	4	0.4667	-6.852	0.4723	-6.833
1	5	0.4675	-6.851	0.4716	-6.916
GFMC		0.448	-6.848 ± 0.018	0.463	-6.848 ± 0.018

clearer "picture" of the results (for the typical WCA case) can be seen in Figure 4 where we note that in the energy-density plane the minimum energies and densities predicted by all the Padé approximants in each fixed order L+M subtend well-defined rectangles. The smallness of these rectangles is a measure of the accuracy achieved in each perturbation order. Each rectangle center can roughly be taken as a "central" predicted value. These values are listed in Table III for the 6th order rectangles for both (BH and WCA) cases; the percentage error figures are with respect to the GFMC central values.

We conclude that the longstanding few-term, low-density expansion for a many-boson system can be rearranged into a meaningful perturbation expansion about a reference fluid of repulsive particles. The density behavior of both the reference system as well as of its perturbation corrections can be successfully series-analyzed. Finally, the resulting attraction coupling series converges rapidly and yields, through Padé approximation, an accuracy with respect to "exact" (Green Function Monte Carlo) results in the one-percent range.

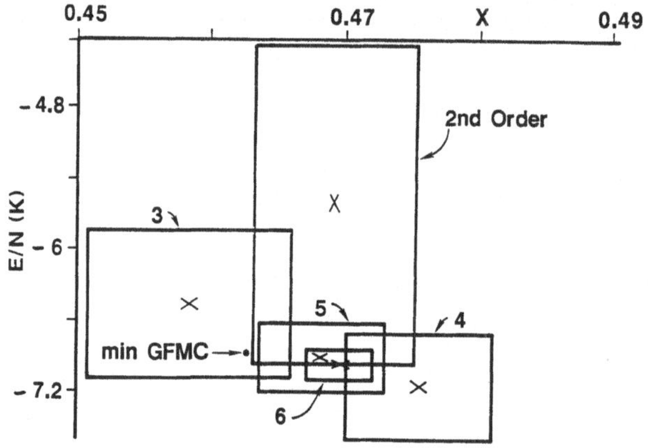

Figure 4. Convergence of the λ-Padé-approximants. Shown are the rectangles subtended (in the energy-per-particle and reduced density $x=\sqrt{\rho a_0^3}$ plane) by the equilibrium density energy minima predicted by all 2nd through 6th order Padé approximants $[L/M](\lambda)$ to the series (6a) with $e_i(x)$ replaced by the $\epsilon_i(x)$ of Fig. 2, for the LJ boson liquid under the WCA potential splitting. Crosses mark the geometric centers of each rectangle and the dot locates the GFMC energy minimum.

Table III. Coordinates in the energy-density plane of the geometrical center of the rectangles defined in Fig 4 for sixth order. Percentage figures refer to discrepancies with respect to the GFMC result. The uncertainties shown give the rectangle width and length.

	BH		WCA	
	6th order	error wrt GFMC	6th order	error wrt GFMC
$(E/N)_{min}(K)$	-6.850 ± 0.042	0.03%	-6.952 ± 0.118	1.5%
x_s	0.468 ± 0.001	4.5%	0.470 ± 0.002	1.5%

ACKNOWLEDGMENTS

Work of M. de Ll. was supported by the National Science Foundation (NSF) under grant DMR 8603038, of V.C.A.N. by Fundação de Amparo à Pesquisa do Estado de São Paulo (FAPESP), Brazil and of R.G. by Comisión Asesora de Investigación Científica y Técnica (CAICYT) Spain and of M.A.S. by Consejo Nacional de Ciencia y Tecnología (CONACyT), México.

REFERENCES

1. D.A. McQuarrie & J.L. Katz, J. Chem Phys. 44, 2393 (1966)
2. J.A. Barker & D. Henderson, Rev. Mod. Phys. 48, 597 (1976)
3. H.C. Andersen, D. Chandler & J.D. Weeks, Adv. Chem. Phys. 34, 105 (1976)
4. J.M. Kincaid, G. Stell & C.K. Hall, J. Chem. Phys. 65, 2161 (1976)
5. G. Stell & O. Penrose, Phys. Rev. Lett. 51, 1397 (1983)
6. J. de Boer & A. Michels, Physica 6, 945 (1938)
7. M.H. Kalos, M.A. Lee, P.A. Whitlock & G.V. Chester, Phys. Rev. B 24, 115 (1981)
8. A.L. Fetter & J.D. Walecka, Quantum Theory of Many-Particle Systems (McGraw-Hill, New York, 1971)
9. H.M. Hugenholtz & D. Pines, Phys. Rev. 116, 489 (1959)
10. K. Hiroike, Prog. Theor. Phys. 27, 342 (1962)
11. S.T. Beliav, Sov. Phys. J.E.T.P. 7, 289 (1958)
12. P. Roman, Advanced Quantum Theory (Addison-Wesley, NY, 1965)
13. G. Gutiérrez, M. de Llano & W.C. Stwalley, Phys. Rev. B 29, 5211 (1984)
14. V.C. Aguilera-Navarro, R. Guardiola, C. Keller, M. de Llano & M. Popovic, Phys. Rev. A 35, 3901 (1987)
15. R.H. Rand, Computer Algebra in Applied Mathematics: An Introduction to MACSYMA (Pitman, London, 1984)
16. K. Huang, in Studies in Statistical Mechanics, vol. II, ed. by G.E. Uhlenbeck (North-Holland, Amsterdam, 1963)
17. N.N. Bogoliubov, J. Phys. (USSR) 11, 23 (1947)
18. D. Chandler, J.D. Weeks & H.C. Andersen, Science 220, 787 (1983)
19. E. Buendía, R. Guardiola & M. de Llano, Phys. Rev. A 30, 941 (1984)
20. L.P. Benofy, E. Buendía, R. Guardiola & M. de Llano, Phys. Rev. A 33, 3749 (1986)
21. P.R. Roach, J.B. Ketterson & C.-W Woo, Phys. Rev. A 2, 543 (1970); R.A. Aziz & R.K. Pathria, Phys. Rev. A 7, 809 (1973)
22. G.A. Baker, Jr., & P. Graves - Morris, Padé Approximants, in Encyclopedia of Math. and its Applications, ed. by G.-C. Rota, vols. 13 & 14 (Addison-Wesley, New York, 1981)
23. D.S. Gaunt & A.J. Guttmann, in Phase Transitions and Critical Phenomena, vol. 3, ed. by C. Domb & M.S. Green (Academic, New York, 1974)
24. J. Cizek & E.R. Vrscay, Int. J. Quantum Chem. 21, 27 (1982)
25. G.A. Baker, Jr., M. de Llano & J. Pineda, Phys. Rev. B 24, 6304 (1981)
26. M.H. Kalos, D. Levesque & L. Verlet, Phys. Rev. A9, 2178 (1974)
27. G.D. Scott & D. M. Kilgour, J. Phys. D 2, 863 (1969); J.L. Finney, Proc. R. Soc. (Lond.) A 319, 479 (1970)
28. D. Levesque & L. Verlet, Phys. Rev. 182, 307 (1969)
29. V.J. Emery, J.L. Gammel & F.R.A. Hopgood, Phys. Rev. 132 10 (1963)

SEMIQUANTUM GASES: QUANTUM COLLECTIVE PHENOMENA IN CLASSICAL TEMPERATURE RANGE

Eugene P. Bashkin

Academy of Sciences
Institute for Physical Problems
117334, Moscow, USSR

1. INTRODUCTION

According to the traditional point of view any collective phenomena can exist either in systems of particles with a long-range interaction between them (e.g. the Langmuir oscillations in the Coulomb plasma etc.) or in the case of high enough density systems in which the mean interparticle distance is of the order of the interaction range (spin waves in condensed magnets etc.). If some low density system is considered the collective effects may appear in the system only at low enough temperatures compared with the quantum degeneracy temperature $\mathcal{E}_d$.

However it has recently become clear that fundamentally quantum-mechanical phenomena may exist even in low-density systems at high temperatures $T \gg \mathcal{E}_d$, i.e. in objects which usually have been described by means of classical statistical physics. The tenuous systems which are classical in view of statistics (the distribution function of particles is described by the Boltzmann-Maxwell formulae) but exhibit essentially quantum macroscopic properties will be designated as semiquantum gases.

From the quantitative point of view the phrase "semiquantum gases" implies that the temperature has been reduced so that we have entered the very interesting temperature region

$$\mathcal{E}_d \ll T \ll \hbar^2/mr_0^2 \qquad (1.1)$$

where m is the mass of the particle, r_0 the interaction range of the order of atomic size. The region (1.1) undoubtedly exists because of the availability of the small gas parameter $Nr_0^3 \ll 1$. Inequality (1.1) is equivalent to the following hierarchy of characteristic length in the system $N^{-1/3} \gg \Lambda \gg r_0$ when $\Lambda = \hbar/(mT)^{1/2}$ is the mean de Broglie wavelength of the particle. The fact that the de Broglie wavelength of the particle turns out to be larger than the own size of the molecule corresponds to the essentially quantum-mechanical situation. So one could expect the gas to exhibit quantum phenomena even in the classical temperature region $T \gg \mathcal{E}_d$ (In previous author's papers the tenuous systems under condition (1.1) were interpreted as "quantum gases". The word "semiquantum" seems to be more adequate since it allows to avoid confusing the case of degenerate gases at $T \ll \mathcal{E}_d$).

The gaseous isotopes of helium and hydregen ^3He, ^4He, H_2, D_2, HD, H$\uparrow$, the traditional objects investigated in low temperature physics, have appreciable saturated vapor pressures in the temperature region (1.1).

Another convenient object for experimental studying is the dilute ^3He - ^4He mixtures in which impurity excitations can form the semiquantum gas. Finally binary gases with a light electron component (a weakly ionized plasma, semiconductors with point defects) are the semiquantum gases for which the upper limit of the region(1.1) exceeds considerably even room temperatures $\hbar^2/m_e r_0^2 \sim 10^5 - 10^7 K$ (m_e is the bare or effective electron mass). The most interesting effects in considered systems are connected with the presence of spin polarization. That is why we shall basically discuss further properties of spin-polarized semiquantum gases. Here we shall enumerate briefly the most intriquing phenomena including both the effects which have already been observed and confirmed experimentally and the latest predictions of the theory.

2. THERMODYNAMICS

All the thermodynamical functions of a gas are determined by means of the virial expansion which is equivalent to the functional expansion in the power series of the statistical density matrix of ideal gas. So the contribution of interaction to the total free energy can be written in the form[1,2]

$$F_{int} = 4\sum_{p,p'} \{\Psi(\vec{p},\vec{p}\,') + \alpha^2 S(\vec{p},\vec{p}\,')\} n_0(\vec{p}) n_0(\vec{p}\,') \tag{2.1}$$

where $n_0(\vec{p})$ is the Maxwell distribution function for an ideal gas, $\alpha = (N_\uparrow - N_\downarrow)/N$ the degree of polarization. Functions Ψ and S are expressed in terms of the exact two-particles scattering amplitudes $f_\pm(\theta, q)$

$$\Psi(\vec{p},\vec{p}\,') = \frac{1}{4}[3A_+ + A_-], \quad S(\vec{p},\vec{p}\,') = \frac{1}{4}[A_+ - A_-],$$

$$A_\pm(\vec{q}) = -\frac{4\pi\hbar^2}{m}\{Re\, f_\pm(0,q) + \frac{mT}{\hbar}[Re\, f_\pm(\theta,q)\frac{\partial}{\partial q} Im\, f_\pm(\theta,q) - \tag{2.2}$$

$$- Im\, f_\pm(\theta,q)\frac{\partial}{\partial q} Re\, f_\pm(\theta,q)]\}, \quad 2\vec{q} = \vec{p} - \vec{p}\,'$$

Taking into account the quantum-mechanical identity of fermions, for the triplet (+) and singlet (-) scattering amplitudes under conditions (1.1) we have

$$f_- = -a(1 - iqa/\hbar), \quad f_+ = 0, \quad |a| \sim r_0 \tag{2.3}$$

Here a is the s-wave scattering length. Thus the dependence of due to the interaction corrections in the thermodynamics of a gas of particles with spin ½ on the degree of polarization reduces to the quadratic function at arbitrary (not certainly small!) values of α. It is connected with the circumstance that in the pair interaction approximation functions Ψ and S for a gas are not functionals of the distribution function (as it takes place in a dense Fermi-liquid). Therefore the dependence of F_{int} on α is fully determined by the dependence of occupation numbers $n_\uparrow(\vec{p})$, $n_\downarrow(\vec{p})$ on the degree of polarization in an ideal Maxwell gas.

3. MAGNETOTRANSPORT PHENOMENA

The most dramatic effects here are associated with the gigantic growth of the mean free path on polarizing the semiquantum Fermi-gas [3,4,1]. The physics of the phenomena in question is very simple. In the temperature range (1.1) the main contribution to the interaction is given by s-wave scattering for which relations (2.3) are valid because of the quantum-mechanical identity of particles. Therefore in the case of the complete polarization $\alpha \to 1$ when each pair of particles is in the triplet state the total cross section vanishes in the basic approximation (2.3) and the mean free path tends to infinity. The increase in the mean free path and consequently in all the transport coefficients is restricted by the contribution of p-wave scattering. It corresponds to the enhancement of the mean free path by the very large factor $(\Lambda/a)^4 \gg 1$ when the gas is completely polarized. The increase in the second sound absorption in dilute mixtures ^{3}He - ^{4}He placed in the external magnetic field has been detected by Greywall and Paalanen [5]. The Paris group [6] has observed the change of the thermal conductivity of spin-polarized gaseous ^{3}He.

4. SPIN WAVES AND LONG-RANGE CORRELATIONS

Collective oscillations of magnetization in tenuous gases in the classical temperature range is one of the most interesting and surprizing predictions of the theory [7,4,1]. Spin waves can propagate in the gas both in the low-frequency limit $\omega\tau \ll 1$ and in the case of $\omega\tau \gg 1$ where $\tau = (Na^2 v_T)^{-1}$ is the gas-kinetic relaxation time. The existense of spin waves (and hence long-range magnetic correlations) is the rather nontrivial property of a Boltzmann gas especially in the high-frequency case $\omega\tau \gg 1$.

Actually we got used to think that in a classical gas all changes in particle states occur only at the instant of collision while, in the interval between collisions, i.e. within the mean free path at the time less than τ, the gas particles move freely without any external disturbance. Therefore no high-frequency $\omega\tau \gg 1$ processes are possible in a classical gas. Consequently the phenomenon in question is the essentially quantal one, though it occurs in the classical temperature range. The physics of the effect here is similar to the situation in the neutron optics when the addition to the real part of the refractive index $\mathrm{Re}\, n - 1 \sim N \mathrm{Re} f(0,q)$ for the slow neutron beam, propagating through the tenuous system of scattering centers, exceeds considerably $\mathrm{Im}\, n \sim N \mathrm{Im} f(0,q) \sim N\sigma/\Lambda$ which describes the dissipation of the beam over the mean free path. The quantum-mechanical refraction of slow gas molecules by one another provided the condition (1.1) is satisfied leads to the appearance of the particular self-consistent field of the Fermi-liquid type even in the classical temperature region. Such a molecular field assures the possibility of propagation of weakly-damped spin waves with the spectrum

$$\omega = \Omega_H + (k v_T)^2/\Omega_{int}, \quad k v_T \ll |\Omega_{int}| = 4\pi |a| \hbar N \alpha / m, \quad \alpha \gg \frac{|a|}{\Lambda} \tag{4.1}$$

where $\Omega_H = 2\beta H/\hbar$ is the Larmor frequency. At small enough magnetic fields (e.g. when polarizing the gas by optical pumping) correlations between transverse spin fluctuations falls off with the distance very slowly, according to the r^{-1}-law, that is also rather extraordinary for a Maxwell gas with a short-range interaction between particles. The existence of nuclear spin waves in gaseous H$\uparrow$ was confirmed experimentally for the first time at Cornell University by means of nuclear magnetic resonance investigations [8,9]. At the same time spin waves in gaseous ^{3}He$\uparrow$ at $T \gtrsim \hbar^2/m r_0^2$ (the highly-damped diffusive regime) were detected in experiments of the Paris group [10,11]. At orbitrary temperatures the spectrum of transverse magnetic-moment fluctuations is defined by the equations [4,1]

$$\omega = \Omega_H + \frac{D_0 k^2}{1 + (D_0 N \alpha \gamma)^2}(-D_0 N \alpha \gamma - i), \quad \frac{1}{\gamma} = \frac{\hbar}{12}\int v^2 A(p) n_0(p) \frac{d^3 p}{(2\pi\hbar)^3}, \tag{4.2}$$

$$\frac{1}{3}\int \xi_1(p,p') A(p') n_0(p') \frac{v'}{v} \frac{d^3 p'}{(2\pi\hbar)^3} - A(p)\int \xi_0(p,p') n_0(p') \frac{d^3 p'}{(2\pi\hbar)^3} = 1$$

where ξ_0, ξ_1 are the first harmonics in the expansion of the function $\xi(\vec{p}, \vec{p}') \equiv \xi(\vec{q})$ in terms of the Legendre polynomials. The experimental evidence for existence of the effective self-consistent field in nondegenerate ^3He - ^4He solutions was given in the work [12] on the spin echo in mixtures.

5. GIGANTIC OPALESCENCE

Under certain conditions the interaction of particles with fluctuations of the transverse magnetization in spin-polarized gases may be considerably stronger than the interaction of the particles with each other. In the case all kinetic properties of a gas are determined by the scattering of paramagnetic molecules by fluctuations of the magnetization and the mean free path in a gas can decrease greatly (the situation is opposite to the case mentioned above in chapter 3). Transverse spin fluctuations may exhibit themselves as a homogeneous precession of the magnetization or slightly inhomogeneous spin waves. In the first case, for instance, the cross section for the inelastic scattering of the paramagnetic particle with the magnetic moment β_0 and the mass m_0 by transverse fluctuations at $\Omega_H \tau_s \gg 1$ (τ_s is the relaxation time for the longitudinal magnetization) has the form[13]

$$d\sigma = 2\pi\left(\frac{4\beta\beta_0 m_0}{\hbar^2}\right)^2 (1+\cos^2\theta)\alpha \frac{T}{\hbar\Omega_H}\frac{do'}{4\pi} \tag{5.1}$$

where θ is the angle between the vector ($\vec{\beta} - \vec{\beta}'$) and the direction of the spin-polarization vector. (Generally speaking the cross-section $d\sigma$ has a maximum at $\Omega_H \tau_s \approx 1$ and begins to fall off as the magnetic field H is reduced further).

As an illustration of the magnitude of the effect we consider the passage of the cold beam of Cs atoms through gaseous $^3He\uparrow$ (polarized by optical pumping) in the direction parallel to the spin-polarization vector. At $H \lesssim H_c$ = 1.2 kG the cross-section (5.1) begins exceeding the cross-section for the scattering of the atom by the He particle. (The magnitude of H_c for the scattering of the Cs beam by the hydrogen ($H\uparrow$)target is equal to 85 kG). For example, let us choose as a target the layer of gaseous $^3He\uparrow$ ($\alpha \approx 0.5; T \approx 6K$) with the density $N \sim 10^{14}$ cm^{-3} and thickness $d \sim 1$ cm which is absolutely transparent to the incident beam. Then after the not too strong magnetic field $H \lesssim 120\,G$ is applied the tenuous layer loses completely its transparency and becomes totally opaque since the mean free path of Cs atoms, connected now with the scattering by spin fluctuations, becomes less than 1 cm. Though the calculations found above are valid regorously only in the semiquantum range (1.1) there are nevertheless some reasons to hope the effect can be observed even at room temperature, say, for gaseous $Rb\uparrow$ target. The effects mentioned above influence also own transport properties of semiquantum gases.

6. MAGNETIC RESONANCE IN BINARY GASES

In the exchange approximation the paramagnetic resonance frequency of a system placed in an external magnetic field does not depend on parameters characterizing the interaction between particles of the system and is determined only by the gyromagnetic ratio. The situation is radically altered in the case of binary gases and the resonance frequencies depend essentially both on the interaction between particles belonging to the different components and on the thermodynamic state, i.e. they are functions of the temperature and the partial pressures of the two components. If the gyromagnetic ratios for particles of both components are the same and equal to γ , the spectrum of magnetic resonance frequencies is given by the relations [2,1]

$$\omega_1 = \gamma H, \quad \omega_2 = \gamma H\left[1 - \frac{N}{4T}(2\pi m_{12} T)^{-3/2}\int \xi(\vec{q}) e^{-q^2/2m_{12}T} d^3q\right.$$

$$N = N_1 + N_2, \quad m_{12} = m_1 m_2/(m_1+m_2), \quad \xi(\vec{q}) = A_+ - A_-, \tag{6.1}$$

where the functions $A_\pm(\vec{q})$ are determined by the expressions (2.2) with m replaced by $2m_{12}$. The solution of the first type corresponds to the usual precession of the total magnetization of the system around the applied magnetic field with the frequency γH. In the oscillations of the second type the total magnetization is conserved and does not oscillate although the separate magnetic moments of each of the components do execute the periodic precessions with the frequency ω_2.

7. SYSTEMS WITH AN ELECTRON COMPONENT [14,1]

In binary systems with a light electron component very many collective phenomena inherent to semiquantum gases (two branches of collective spin waves etc.) can also occur. It is important to emphasize that such systems (a weakly ionized gas, semiconductors with point defeets) may be expected to exhibit not only a considerable expansion of the temperature range, in which the conclusions of the theory of semiquantum gases are applicable, but an increase in the quantitative measure of collective effects. So, in the semimagnetic semiconductor with typical values of the electron density $N_e \sim 10^{16}-10^{17}$ cm^{-3} and of the paramagnetic impurity concentration $N_d \sim 10^{17}-10^{18}$ cm^{-3} at $T \sim 10^2 K$ the splitting of the magnetic resonance line turns out to be very large

$$\frac{|\omega_1-\omega_2|}{\omega_1} = \frac{2\pi\hbar}{m^*}|a_+-a_-|(N_e+N_d)\,\frac{th(\hbar\Omega_H/2T)}{\Omega_H} \sim 10^{-3}-10^{-1} \tag{7.1}$$

Besides, in the systems under consideration the appearance of qualitatively new phenomena is also possible. For example, in a weakly ionized gas virial corrections in the free energy connected with the interaction between electrons and neutrals lead to the creation of the thermodynamically equilibrium charge density wave existing within the temperature interval

$$T_{c1}^* < T < T_{c1}, \quad T_{c1} = \frac{\pi\hbar^2}{2m_e a_1^2}(N_e|a_1|^3)^{1/2}(N_n|a_1|^3)^{1/2}, \tag{7.2}$$
$$T_{c1}^* = T_{c1}(1+z^{-1})^{-1}$$

where $a_1 = 3a_+ + a_-$ is the ion charge in a plasma. At $T \lesssim T_{c2}$ where T_{c2} is defined by the formula (7.2) with a_1 replaced by $a_2 = a_+ - a_-$ the equilibrium ferro- or ferrimagnetic order appears in a Maxwell weakly-ionized gas. The analogous conclusion is also valid in the case of semimagnetic semiconductors. Unfortunately almost no quantum collective phenomena in nondegenerate electron systems have been studied yet in practice. Only in the system of electrons localized above the surface of liquid helium and of solid hydrogen the phenomenon of quantum-mechanical refraction by atoms of the saturated vapor has been already confirmed experimentally [14,15].

REFERENCES

1. E.P. Bashkin, Usp. Fiz. Nauk, 148, 393 (1986) Sov. Phys. Usp. 29(3), 238 (1986) ; Zh.Eksp. Teor. Fiz. 87, 1948 (1984) Sov. Phys.JETP 60, 1143 (1984); Pis'ma Zh. Eksp. Teor. Fiz. 40, 383 (1984) JETP Lett. 40, 1197 (1984) .
2. E.P.Bashkin. Zh. Eksp. Teor. Fiz. 86, 937 (1984) Sov. Phys.JETP 59, 547 (1984) .
3. E.P.Bashkin and A.E.Meyerovich. Pis'ma Zh. Eksp. Teor. Fiz. 26, 696 (1977) JETP Lett. 26, 534 (1977) ; Adv. Phys. 30, 11 (1981).
4. C.Lhuillier and F. Laloë. J. Phys.(Paris) 43, 197, 225, 833 (1982)
5. D.S. Greywall and M.A.Paalanen, Phys. Rev. Lett. 46, 1292 (1981)

6. M. Leduc, P.J. Nacher, D.S. Betts, J.M.Daniels, G.Tastevin, and F. Laloë, Preprint (1987)
7. E.P. Bashkin, Pis'ma Zh. Eksp. Teor. Fiz. 33, 11(1981) JETP Lett., 33, 8 (1981) .
8. B.R. Johnson, J.B.Denker, N. Bigelow, L.P. Levy, J.H. Freed, and D.M. Lee. Phys. Rev. Lett. 52, 1508 (1984); 53, 302 (1984).
9. L.P. Levy and A.F. Ruckenstein. Phys. Rev. Lett. 52, 1512 (1984); 53, 302 (1984).
10. P.J. Nacher, G. Tastevin, M. Leduc, S.B. Crampton, and F. Laloë. J. Phys. Lett. (Paris) 45, L-441 (1984)
11. G. Tastevin, P.J. Nacher, M. Leduc, and F. Laloë. J.Phys. Lett. (Paris) 45, L-249 (1985).
12. W.J. Gully and W.J. Mullin. Phys. Rev. Lett. 52, 1810 (1984)
13. E.P. Bashkin. Pis'ma Zh. Eksp. Teor. Fiz. 44, 322 (1986) JETP Lett. 44, 414 (1986).
14. E.P. Bashkin. Pis'ma Zh. Eksp. Teor. Fiz. 34, 86 (1981) JETP Lett. 34, 81 (1981) ; Zh. Eksp. Teor. Fiz. 82, 254, 1868 (1982) Sov. JETP 55, 152, 1076 (1982) .
15. V.V. Zav'yalov and I.I. Smol'yaninov. Zh. Eksp. Teor. Fiz. 92, 339 (1987)

SOME THOUGHTS ON THE NEW HIGH T_c SUPERCONDUCTORS

Richard A. Klemm

Department of Physics
University of California, San Diego
La Jolla, CA 92093

INTRODUCTION

Rarely in the history of physics has there been such excitement as there is right now. This field of high temperature superconductors is mov-so rapidly that the usual journals are barely capable of recording the developments; a number of workers have resorted to newspapers and press conferences in order to more rapidly inform others as to their new discoveries. The prospects for technological development appear to be outstanding: already the critical currents at liquid nitrogen temperature have been found to be more than adequate for most purposes, and the enormous critical field values at low temperatures are presently too large to measure. A wealth of experimental results on pressed powders are presently available, and some preliminary data on single crystals has been obtained. Some of these data have implications for the mechanism for superconductivity in these materials, and a number of models have been proposed. In this talk, I would like to discuss some of the experiments which have been performed to date, and their consequences for a theory. I will then discuss a number of models that have been presented, giving my biases regarding their strengths and weaknesses. Finally, I shall present a new model for one of the materials, which I believe is consistent with all of the data, and discuss its consequences.

SOME RELEVANT EXPERIMENTAL RESULTS

Ordinarily, a discussion of experimental results for superconducting materials is limited to the properties of the superconducting state itself; the normal state is generally that of a conventional metal, in which the individual electrons move essentially as free particles, scattering off impurities, phonons, and each other. However, in these exotic new materials, there is substantial evidence that the "normal" state is qualitatively different; hence it is also necessary to include a discussion of the experimental properties above T_c and in related non-superconducting materials. Hence, the experimental discussion will be divided into three parts: first a discussion of the materials, second a summary of some of the data regarding their superconducting properties, and finally a detailed description of their normal state properties. As it is impossible to list all of the experiments performed to date in the space and time allotted, I extend my apologies to the numerous workers whose data I have not included.

Materials Which are Superconducting at High Temperatures

Until 1986, the phrase "high temperature superconductivity" meant some material with a transition temperature T_c in the range 10-23K; the highest T_c of 23.2K was for Nb_3Ge. Within the last year, this record has been broken numerous times, and by large amounts. The pioneering advance was by Bednorz and Müller,[1] who found that $La_{2-x}Ba_xCuO_{4-y}$ was superconducting with $T_c \approx 30K$. It has since been found that substitution of Sr or Ca for Ba is possible; the highest T_c for a material of this type is with Sr doping of concentration $x \approx 0.15$ and negligible oxygen vacancy concentration y. Under pressure, it has been found to be as high as 52.5K.[2] In the absence of alkali earth doping, these materials undergo a structural phase transition at $T_O \approx 500K$, depending upon y.[3] Above T_O, La_2CuO_{4-y} is in a tetragonal crystal structure, consisting of layers of CuO_2 in a square configuration, separated by oxygens on either side of each Cu and a double La layer. Doping with alkali earths suppresses T_O and causes the material to become superconducting in the orthorhombic phase. It is possible, however, to suppress T_O below T_c by Sr doping concentration $x \approx 0.2$,[4] provided that y is large enough. Generally, though, the materials in this class are superconducting in the orthorhombic phase.

A second class of materials was first reported by Wu et al.[5] The first example of this material, originally referred to mysteriously as "the green powder", was subsequently found[6] to be $YBa_2Cu_3O_{7-\delta}$. This material is superconducting with a T_c in the range of 90-95K, depending upon y and perhaps mysterious preparation procedures. This material also undergoes a tetragonal to orthorhombic phase transition at $T_O \approx 600K$, but unlike the first class, the unit cell does not double at the transition temperature.[7] It has since been found[8] that Y may be replaced with practically all of the rare earths, without affecting T_c appreciably. These materials also consist of CuO_2 layers (which are corrugated in and out of the plane), but one of three Cu-O layers consists of parallel Cu-O chains. The Ba layers are between the Cu-O chain layer and the CuO_2 planar layers, and the Y or rare earth layers apparently isolate the triple layers of Cu-O-Ba from each other.

Recently, Ovshinsky et al.[9] reported that substitution of fluorine for some of the oxygen in the second class of compounds resulted in a compound that showed some appearances of superconductivity at $T_c \approx 155K$. This material was thought to be of the composition $YBa_2Cu_3F_2O_y$, but as of this date, neither the oxygen composition nor the crystal structure are known. A similar vanishing in the resistance of an as-yet unspecified material (presumably containing fluorine) has been reported verbally by Chu[10] to have a T_c of 225K. It remains to be seen if these materials do indeed turn out to be legitimate superconductors, but at least there now appears to be some indication that superconductivity at room temperature may not be too far off in the future.

Superconducting Properties of the New Materials

Next to the anomalously large values of T_c for these new materials, perhaps the most noteworthy aspect of the superconductivity in these materials is that while some experiments indicate entirely conventional superconductivity, others do not. Some of the discrepancies probably arise from the scarcity of high quality single crystals. Josephson tunneling[11] on Y-Ba-Cu-O indicated conventional superconductivity (i. e., the flux quantum is hc/2e, and the tunneling current is consistent with ordinary singlet superconductivity) in that system. Experiments on the La-Sr-Cu-O system, however, showed tunneling between grains of the material, but not between the La-Sr-Cu-O material and the Nb layer, which is inconsistent with conventional superconductivity.

Measurements of the isotope effect[12] in $YBa_2CuO_{7-\delta}$ by substitution of ^{16}O with ^{18}O indicated that there was no shift, although the Raman lines for the oxygen modes were shifted by the expected 4%. Thus, conventional electron-phonon coupling in that material is unlikely. Recent[13] unpublished results on La-Sr-Cu-O, however, are reported to exhibit an isotope effect, so that phonons are presumably involved at least to some degree in that material.

All of these materials are rather anisotropic in their superconducting properties, as evidenced from preliminary single crystal measurements.[14,15] The values of the upper critical field parallel to the layers can only be obtained with conventional equipment near to T_c. Extrapolating to lower temperatures from the slope at T_c, anisotropies of at least an order of magnitude are observed, and estimates of the parallel-field $H_{c2}(0)$ on the order of 1-2 Megagauss are obtained[16] for Y-Ba-Cu-O.

Although preliminary indications of three-dimensionality[17] from the fluctuation-induced conductivity have been reported, the region in temperature above T_c for which these fluctuations have been reported to be 3D-like is far above the dimensional-crossover temperature,[18] indicating that an experimental difficulty in subtacting the normal state background is likely to be present.

Early measurements of the critical current[19] indicated that the values obtained were too low to be of practical interest. However, more recent results on single crystals[20] have given acceptable values. Critical current values of greater than 10^6 A/cm^2 have been obtained, which is competitive with the best conventional materials. Naturally, the large anisotropy in these high T_c materials is important in the critical currents of single crystals, but in powders the main problem appears to be the granular nature of the material.

Normal State Properties

To date, the most experimental results in the normal state for any system is for the $La_{2-x}Sr_xCuO_{4-y}$ system. This system has been studied extensively for x=0. Magnetic susceptibility measurements on La_2CuO_{4-y} indicate that for $y \neq 0$, the ground state is an antiferromagnet.[21] This antiferromagnetism has been confirmed by unpolarized[22] and polarized[23] neutron diffraction experiments. Experiments on nearly stoichiometric single crystals also indicate antiferromagnetism, although at a much lower Néel temperature T_N.[24] Although for y=0.03, $T_N \approx 290$K, the dependence of T_N upon y is remarkable; for y=0, the antiferromagnetism vanishes![21] (This fact has not been appreciated by the vast majority of theorists working on spin models.) There have been reports of superconductivity in La_2CuO_4, but the materials appeared to be somewhat irreproducible[25] and hence likely to contain lanthanum vacancies, which would create the same effect as would strontium doping.

Transport measurements on La_2CuO_{4-y} indicate[21] that for y=0, the carriers are predominately holes, and for $T < 100$K, the resistivity is activated with an activation energy of ≈ 250K. Oxygen vacancies add electron carriers: for y=0.015, the resistivity is reduced from that for y=0, but the magnitude of the thermopower is also reduced. For y=0, the resistivity above 100K approaches a constant as T increases towards room temperature. Curiously, for y=0.03, the resistivity is larger than for y=0.015, which may be due to Anderson localization.[26] For doped materials, there are two types of resistivity behavior above T_c: in samples with a small amount of doping, the resistivity above T_c rises as the temperature is lowered, exhibiting a peak just before falling to zero at T_c.[27] In more doped samples, the resistivity in the normal states appears naively to be more metallic-like: it appears to be approximated by a constant plus a linear term.[28] Although in a few cases[29] this constant appears to be zero, in most of the samples with the sharpest transition temperatures, it is non-zero.[28]

Unfortunately, the crystal structure of $YBa_2Cu_3O_7$ is not preserved upon elimination of the barium, so that the yttrium compound analogous to La_2CuO_4 cannot readily be made. It has therefore been difficult so far to examine the magnetic behavior of the non-superconducting analogues. The resistivity and thermopower for those materials are similar to that for $La_{2-x}Sr_xCuO_4$, however.

THEORETICAL MODELS

To date, the theoretical models for the superconductivity in the new materials fall into the following classes: conventional electron-phonon, exciton, bipolaron, and spin. There have of course been band structure calculations;[30] but the bandwidths obtained from those calculations appear to be far too wide to describe the transport data. They certainly have nothing to say about the magnetism in these materials, as has been noted long ago regarding $LaMnO_3$ by Zener.[31]

The conventional electron-phonon picture has been presented by Weber,[32] who performed a detailed calculation of the phonon modes for the La_2CuO_4 structure. He concluded that breathing modes of the oxygen ions were important. He estimated T_c for this type of mechanism to be on the order of 40K, which could possibly be relevant for $La_{2-x}Sr_xCuO_{4-y}$. The recent isotope effect in that material[13] would tend to support the electron-phonon picture, although they would also be consistent with some other mechanism also involving phonons in an essential way. The Josephson tunneling data[11] do not appear to support this picture, however. This model also neglects the magnetic behavior of the normal state, and may have difficulty in explaining the normal state transport properties.

There are several types of exciton models. The models of Little[33] and Allender, Bray, and Bardeen[34] for quasi-one- and quasi-two-dimensional superconductors appear to have little likelihood of being applicable, for a number of reasons. The Little model depends upon the excitations of the oxygens normal to the Cu-O planes as enhancers for the electronic motion in the Cu-O plane. This would require a perfect match of the excitation frequencies with the in-plane electronic energies. In the Allender, Bray, and Bardeen model, the conduction electrons tunnel into the semiconducting layers, exciting excitons, which are then reabsorbed by other electrons. This model requires a precise matching of the Fermi energy in the metallic layers with the energy of excitation of the excitons in the semiconducting layers, which is not likely to happen in every case of Y being replaced by rare earths. The Little model also suffers from the same criticism.

The more interesting excitonic model is that of Varma, Schmitt-Rink, and Abrahams.[35] In this model, the excitons are localized about the Cu(II) ions. These Frenkel excitons are created by the polarization of the oxygen cloud about a Cu(III); the quanta of the excitation are Frenkel excitons. The attractive part of this localized picture is that the vertex corrections for the exciton exchange are not as destructive to pairing in this model as they are in the delocalized models of Little and Allender et al. However, as an electron (or a hole) moves from one copper site to another, the polarization cloud moves with it, so it seems to me to give merely a correction to the hole self-energy, rather than be a mechanism for the exchange of excitons. That is, I view this mechanism as being more related to the Fröhlich mechanism for conventional superconductivity, which does not result in superconductivity.

Hubbard Models

One model that has been used by a number of authors[36,37,38] to describe these materials is the two-dimensional extended Hubbard model,

$$H = \sum_{ij\sigma} t_{ij} a^{\dagger}_{i\sigma} a_{j\sigma} + (1/2) \sum_{ij\sigma\sigma'} U_{ij} n_{i\sigma} n_{j\sigma'} \quad , \tag{1}$$

where i,j label a copper or an oxygen site, σ,σ' label the spin, $a^{\dagger}_{i\sigma}$ creates a hole at site i with spin σ, t_{ij} is the effective hopping matrix element for holes between sites i and j, U_{ij} is the effective Coulomb potential between holes on sites i and j, and $n_{i\sigma}=a^{\dagger}_{i\sigma}a_{i\sigma}$. A half-filled band for this model is known to be an antiferromagnetic insulator, according to Monte Carlo calculations by Hirsch.[36] In the half-filled case, the number of holes per site is zero; for a non-half-filled band, such as would be obtained by doping La_2CuO_4 with Sr, there would be δ holes/site. Emery[37] has discussed this model, assuming the holes are primarily on the oxygen sublattice, as suggested by photoemission experiments.[39] He assumes that the stoichiometric La_2CuO_4 is an antiferromagnet, the long-range antiferromagnetic order appearing on the Cu^{++} S=1/2 sublattice. He then assumes the holes on the oxygen sublattice (which are also spin 1/2) to move about in the presence of the long-range antiferrmoagnetic order on the copper sublattice. This picture has been elaborated upon by Hirsch,[38] who considered the propagation of an oxygen hole in the presence of this antiferromagnetic background. He showed for an Ising model (obtainable from the above for large U relative to t, assuming near-neighbor hopping, on-site Coulomb repulsion only, and crystal field symmetry breaking), that an oxygen hole can only propagate outside of the domain of a Cu(II) spin (defined as the Cu(II) site and its four neighboring oxygen sites) if the spin on the copper is flipped. He then argued that if two oxygen holes propagate along together, one may leave a track of flipped spins, but the other flips them back, resulting in an undistorted spin order in the trail of the hole pair. Thus, Hirsch argues that the long-range antiferromagnetic order on the copper sites gives rise to a singlet pairing interaction between oxygen holes.

The above model is very interesting, and I believe has many redeeming features. However, it should be pointed out that it is based upon the assumption (which I believe to be false) that the ground state of La_2CuO_4 is an antiferromagnet. Since the magnetic susceptibility experiments[21] indicate that La_2CuO_{4-y} is an antiferromagnet only for $y \neq 0$ (and is non-magnetic for y=0), the ground state of the stoichiometric material is presumably something entirely different.

Superexchange Models

A model related to the Hubbard model is the superexchange model proposed by Anderson[40] et al. and Ruckenstein, Hirschfeld, and Appel.[41] The effective Hamiltonian for this model can be written as

$$H_{se} = J_1 \sum_{<ij>} \vec{S}_i \cdot \vec{S}_j + J_2 \sum_{(ij)} \vec{S}_i \cdot \vec{S}_j \quad , \qquad (2)$$

where <..> means nearest-neighbors and (..) means next-nearest-neighbors in the CuO_2 plane. The J_1 interaction can be derived from a Hubbard Hamiltonian with near-neighbor hopping and on-site Coulomb repulsion, $J_1 \sim t^2/U$. If the J_2 term were absent, the ground state of this Hamiltonian for a two-dimensional square lattice would be an antiferromagnet. The J_2 term is necessary to frustrate the Néel state for the square lattice. As Anderson pointed out,[40] for a two-dimensional triangular lattice with just the J_1 Hamiltonian is self-frustrating.

Anderson[40] argued that the ground state of La_2CuO_4 might be a resonating valence bond state. This state would arise from the superexchange Hamiltonian given by Eq. (2), provided that J_2 were sufficiently large relative to J_1. In mean-field theory, the resonating valence bond (RVB) state has a lower energy than does the Néel state for $J_2/J_1 > 1/8$. In the RVB state, the spins in the Hamiltonian are taken to be on the copper sites, and any holes in the valence band are taken to be holes on the copper sites. The RVB state consists of singlet pairs of Cu(II) spins. These singlet pairs resonate between degenerate configurations. For example, consider a square with copper ions on the corners. One configuration in an overall RVB state configuration would be for the spins on the left-hand side to be paired and the spins on the right-hand side to be paired, both in singlet states. Another (degenerate) configuration for the spins in the square would be for the two upper spins and the two lower spins to separately be paired in singlets. The actual state for a 2X2 lattice would thus be a linear combination of the two singlet pair configurations. One might think of this state as being analogous to benzene: one configuration involves an alternation of single and double bonds between the carbons in the six-membered ring, starting at the 1 position. Another configuration is identical, except that it starts at the 2 position. The ground state is a resonance between the two possible configurations, and is lower in energy than either of them individually.

Kivelson, Rokhsar, and Sethna[42] examined the excitations of the RVB state. They assumed that there were three types of excitations: fermion spin solitons (or spinons), boson holes with charge e, and the product of the charged bosons and spin solitons. Introduction of a hole into the RVB will give this latter type of excitation, as it creates an isolated spin on one site and an hole on another (initially adjacent) site. The hole is a boson because it represents an S=0 configuration of Cu(III). The spin soliton is created because a singlet pair has been broken by the introduction of a true hole. Kivelson, Rokhsar, and Sethna argue that the spin solitons will have a gap in their excitation spectrum, whereas the boson charges will be gapless. They then proposed that the boson holes would undergo boson condensation, giving rise to superconductivity with a flux quantum of hc/e. Unfortunately, the experiments appear to rule out this possibility.[11]

More recently, Anderson, Baskaran, Zou, and Hsu[43] made an interesting proposal: they suggested that the tetragonal to orthorhombic transition in $La_{2-x}Sr_xCuO_{4-y}$ is really a transition into the RVB state. They make the claim (which is contradicted by the experiments[21]) that the ground state for $x < 0.02$ is the Néel state, and that the mean-field transition temperature for the RVB state describes the tetragonal to orthorhombic distortion, except for $x > 0.15$ or so, where it describes the normal to superconducting transition. There are several things wrong with their picture. First, the experiments clearly indicate that as the oxygen vacancy level y is decreased to 0, the Néel temperature decreases to zero, so the stoichiometric compound is not an antiferromagnet. Second, it is difficult to understand where the broken symmetry enters into the RVB state. It is necessary to have a state with broken symmetry, as the orthorhombic state is a state of lower symmetry than is the tetragonal state. Anderson et al. argue that in one dimension the RVB state will not have a broken symmetry, but suggest that for two or three dimensions it might. However, as they pointed out in a note added, such a symmetry breaking can only occur for dimension greater than or equal to four. Hence, it seems that if the RVB state is at all appropriate to $La_{2-x}Sr_xCuO_{4-y}$, it could only describe the tetragonal phase. It may be relevant therefore for the case with sufficient oxygen vacancies and strontium doping ($x > 0.2$) in which the superconductivity occurs in the tetragonal phase. The other major objection to the RVB state in the orthorhombic phase is that it is inconsistent with the electron spin resonance data,[44] which I shall describe in detail below.

Evidence for Triplet Pairs in the Ground State of Orthorhombic $La_{2-x}Sr_xCuO_4$

Recently, Thomann et al.[44] presented electron spin resonance data for stoichiometric La_2CuO_4 and $La_{2-x}Sr_xCuO_{4-y}$. These data show clearly that the ground state of both systems contains triplet pairs, which they interpret as triplet pairs of spins on the Cu(II) ions. The triplet signal is characterized by the "half-field" transition: for an orthorhombic crystal field splitting much less than the magnetic field energy, the ESR signal at approximately one-half the magnetic field (i. e. twice the frequency) of the usual $\Delta m = \pm 1$ transition is allowed. In the absence of such a crystal field splitting, this would be a $\Delta m = \pm 2$ transition, and would be forbidden. In superconducting $La_{2-x}Sr_xCuO_{4-y}$, the triplet signal is also observed, provided that the material has an orthorhombic crystal structure. Since both the regular and the half-field transition occur for magnetic fields between H_{c1} and H_{c2}, the magnetic field can penetrate the superconductor in vortices, driving it normal within the vortex cores. Within these normal vortex cores, the triplet signal is observed.

As the temperature is decreased, the triplet signal intensity increases, indicating that the triplet pairs are a property of the ground state of La_2CuO_4, and of the normal state of $La_{2-x}Sr_xCuO_{4-y}$. The integrated intensity indicates that $\sim 2\times 10^{-3}$ of the Cu spins are flipped in the measurement, which may be due to the pairs forming a many-body state, and only the pairs near the grain or domain boundaries can be flipped by the field. It is interesting to note that the triplet signal is not seen in doped samples that remain tetragonal (i.e., for $x>0.2$ and y sufficiently large). This may be due to the half-field transition becoming forbidden, but it could also mean that the triplet pairs are only present in the ground state of the orthorhombic phase.

Phenomenological Model for the Ground State of La_2CuO_{4-y}

At this point, I would like to reiterate what I consider to be the most important features of the system La_2CuO_{4-y}. There are three experiments that I deem of primary importance in the construction of a phenomenological model: the magnetic susceptibility as a function of y, the neutron scattering experiments, and the electron spin resonance experiments. As I mentioned previously, the magnetic susceptibility experiments show that the Néel temperature is very strongly dependent upon the concentration y of oxygen vacancies. In fact, it is so strongly y-dependent that for y=0, T_N vanishes! In addition, the temperature dependence of the magnetic susceptibility for $y \neq 0$ is anomalous: for y=0.03, for example, as the temperature is decreased from above room temperature, the susceptibility decreases, and then begins to increase, as if the system were about to go ferromagnetic. Just before long-range ferromagnetic order sets in, however, the susceptibility peaks, and then decreases as T is lowered further towards T_N, where the long-range order sets in. In short, the magnetic susceptibility shows striking evidence of a competition between ferro- and antiferromagnetic order. The second important piece of imformation is the spin ordering as observed in the neutron diffraction experiments.[22] In the Néel state, the spins form ferromagnetic sheets in the orthorhombic bc plane, with an antiferromagnetic alternation in the a-axis direction. The spins point parallel to the orthorhombic c-axis direction. The fact that the spins point in the c-axis direction is no accident: it implies that the spin rotational invariance present in Eq. (2) is broken. Finally, the electron spin resonance experiments indicate that the ground state of the stoichiometric compound is non-magnetic, containing triplet pairs.

A model consistent with the above experimental observations is[45]

$$H = H_{se} + H_{re} + H_{s\text{-}ph} \tag{3}$$

where the superexchange part of the Hamiltonian, written as

$$H_{se} = J_1 \sum_{acnn} \vec{S}_i \cdot \vec{S}_j + J_2 \sum_{acnnnn} \vec{S}_i \cdot \vec{S}_j \tag{4}$$

is similar to Eq. (2), containing antiferromagnetic interactions between the Cu(II) spins which are nearest-neighbors and next-next-nearest-neighbors in the orthorhombic ac plane,

$$H_{re} = -J_3 \sum_{acnnn} \vec{S}_i \cdot \vec{S}_j \tag{5}$$

is the interaction between next-nearest-neighbors in the ac plane, which we take to be ferromagnetic due to a ring exchange process described below, and

$$H_{s\text{-}ph} = - \delta J_3 \sum_{cnn} S_i^c \, S_j^c \tag{6}$$

is the ferromagnetic interaction between nearest-neighbors in the c-axis direction, arising from the coupling of the spins to the orthorhombic crystal distortion.

Although it is not completely essential for this model that H_{re} be ferromagnetic, since the spin-phonon part of the Hamiltonian is sufficient to provide the symmetry-breaking and the competition between ferromagnetic and antiferromagnetic ordering, it does help to expain the strong dependence of the Néel temperature upon the oxygen vacancy concentration. In the ring exchange picture, one considers a ring of four copper and four oxygen ions. If the spins on the copper ions are all parallel, then the ring will be four-fold degenerate under rotations about the b-axis. This degeneracy will therefore lower the energy (in the absence of J_1) if all the spins on the ring are parallel, creating an effective attractive interaction J_3 between all of the spins on the ring. If we now turn on the superexchange between the near-neighbors on the ring with bare interaction J_{10}, the resulting interaction between the nearest-neighbors in the ac plane will be $J_1 \sim J_{10} - J_3$. Since oxygen vacancies on the ring remove the degeneracy responsible for the ring exchange interaction, they tend to suppress J_3 more than they do J_{10}, J_2, or δJ_3. Thus, oxygen vacancies tend to decrease J_3 and <u>increase</u> J_1, strongly favoring the Néel state.

The model presented has many possible ground states. For $y \neq 0$, we want it to describe the Néel state observed in the neutron diffraction experiments, which would be the ground state if J_2 were absent. However, for $y=0$, we want to have a ground state consisting of triplet pairs. To do so, we must have $J_2 \neq 0$, J_1 relatively small, and J_3 or δJ_3 relatively large. This state we call a resonating triplet pair (RTP) state. There are many such states, depending upon the details of the various spin-spin interactions. We want to have a state in which the overall spontaneous magnetization (both ferromagnetic and antiferromagnetic) is zero (or at least too small to measure).

This model suggests several possible scenarios for superconductivity in $La_{2-x}Sr_xCuO_{4-y}$. In the first scenario, the triplet pairs in the RTP ground state will condense into a triplet superconducting state. In case the holes are primarily on the oxygens, as suggested by photoemission experiments,[39] the oxygen holes will experience the spin "fluctuations" present in the RTP ground state. The anisotropy of the local spin order in the RTP ground state will give rise to an anisotropic effective pairing interaction between the oxygen holes, leading to superconductivity which could be either singlet or triplet, depending upon the details of the interactions. In the third scenario, the holes are predominately on the coppers, which would be somewhat related to the picture of Kivelson et al.[42] In this picture, I would imagine the holes would first pair and then condense, the pairing being at least partially due to the fact that it is easier to break up a m=0 triplet pair than a parallel spin pair.

It is tempting to consider a modification of our model for the case of $YBa_2Cu_3O_{7-\delta}$. Ignoring for the moment the quasi-one-dimensional chains of Cu-O, and focussing on the CuO_2 layers, we notice that one distinct difference between this material and $La_{2-x}Sr_xCuO_{4-y}$ is apparent. In the orthorhombic phase, the unit cell is doubled from its tetragonal phase size in the latter compound, but not in the former. Thus, the spin-phonon term in the Hamiltonian plays a very different role in the two materials. In the $YBa_2Cu_3O_{7-\delta}$ compound, the c-axis direction is parallel to the nearest-neighbor direction, so that J_1 becomes anisotropic due to the δJ_3 interaction: $J_1 = J_{10}$ in the a-axis direction, and $J_1=J_{10}-\delta J_3$ in the c-axis direction. Ignoring J_2 and J_3, we see that the dominant interaction in now the J_1 interaction in the a-axis direction, which does not depend upon the spin-phonon interaction. This resulting interaction is antiferromagnetic, which ought to give rise to a form of singlet superconductivity. In addition, we would expect no isotope effect from this mechanism, whereas for the $La_{2-x}Sr_xCuO_{4-y}$ material, we would expect an isotope effect for the orthorhombic phase.

Much work needs to be done to elucidate the details of the RTP states. More experiments, especially neutron diffraction experiments on single crystals with essentially no oxygen vacancies might be helpful. On the theoretical side, it would be useful to have Monte Carlo simulations to find the interesting possible RTP ground states.

ACKNOWLEDGEMENTS

The author would like to acknowledge many fruitful discussions with D. C. Johnston and H. Thomann. He would also like to express his thanks to the U. S. Army Research Office for a travel grant to enable him to attend this workshop.

REFERENCES

1. J. G. Bednorz and K. A. Müller, Z. Phys. B64, 169 (1986); J. G. Bednorz, K. A. Müller and M. Takashige, Science 236, 73 (1987).
2. C. W. Chu, P. H. Hor, R. L. Meng, L. Gao, and Z. J. Huang, Science, 235, 567 (1987).
3. P. Lehuédé and M. Daire, C. R. Acad. Sc. Paris, Ser. C 276, 1011 (1973); J. M. Longo and P. M. Raccah, J. Solid State Chem. 6, 526 (1973).
4. D. C. Johnston, private communication.
5. M. K. Wu, J. R. Ashburn, C. J. Tong, P. H. Hor, R. L. Wong, L. Gao, Z. J. Huang, Y. Q. Wang, and C. W. Chu, Phys. Rev. Lett. 58, 908 (1987).
6. R. J. Cava, B. Batlogg, R. B. van Dover, D. W. Murphy, S. Sunshine, T. Siegrist, J. P. Remeilka, E. A. Rietman, S. Zahurak, and G. P. Espinosa, Phys. Rev. Lett. 58, 1676 (1987).
7. A. J. Panson, A. I. Braginski, J. R. Gavaler, J. K. Hulm, M. A. Janocko, H. C. Pohl, A. M. Stewart, J. Talvacchio, and G. R. Wagner, Phys. Rev. B35, 8774 (1987).
8. P. H. Hor, R. L. Meng, Y. Q. Wang, L. Gao, Z. J. Huang, J. Bechtold, K. Forster, and C. W. Chu, Phys. Rev. Lett. 58, 1891 (1987).
9. S. R. Ovshinsky, R. T. Young, D. D. Allred, G. DeMaggio, and G. A. Van der Leeden, Phys. Rev. Lett. 58, 2579 (1987).
10. C. W. Chu, private communication.
11. J. S. Tsai, Y. Kubo, and J. Tabuchi, Phys. Rev. Lett. 58, 1979 (1987).
12. B. Batlogg, R. J. Cava, A. Jayaraman, R. B. van Dover, G. A. Kourouklis, S. Sunshine, D. W. Murphy, L. W. Rupp, H. S. Chen, A. White, K. T. Short, A. M. Mujsce, and E. A. Rietman, Phys. Rev. Lett. 58, 2333 (1987); L. C. Bourne, M. F. Crommie, A. Zettl, H.-C. zur Loye, S. W. Keller, K. L. Leary, A. M. Stacy, K. J. Chang, M. L. Cohen, and D. E. Morris, Phys. Rev. Lett. 58, 2337 (1987).
13. same authors as in Ref. 12, private communication.
14. T. R. Dinger, T. K. Worthington, W. J. Gallagher, and R. L. Sandstrom, Phys. Rev. Lett. 58, 2687 (1987).
15. Y. Hidaka, Y. Enomoto, M. Suzuki, M. Oda, and T. Murakami, Jpn. J. Appl. Phys. 26, L377 (1987).
16. T. P. Orlando, K. A. Delin, S. Foner, E. J. McNiff, Jr., J. M. Tarascon, L. H. Greene, W. R. McKinnon, and G. W. Hull, Phys. Rev. B35, 7249 (1987).
17. P. P. Freitas, C. C. Tsuei, and T. S. Plaskett, Phys. Rev. B36, to be published.
18. R. A. Klemm, J. Low Temp. Phys. 16, 381 (1974).
19. J. Z. Sun, D. J. Webb, M. Naito, K. Char, M. R. Hahn, J. W. P. Hsu, A. D. Kent, D. B. Mitzi, B. Oh, M. R. Beasley, T. H. Geballe, R. H. Hammond, and A. Kapitulnik, Phys. Rev. Lett. 58, 1574 (1987); Ref. 6.
20. G. W. Crabtree, J. Z. Liu, A. Umezawa, W. K. Kwok, C. H. Sowers, S. K. Malik, B. W. Veal, D. J. Lam, M. B. Brodsky, and D. W. Downey, preprint; Ref. 14.
21. D. C. Johnston, J. P. Stokes, D. P. Goshorn, and J. T. Lewandowski, preprint.
22. D. Vaknin, S. K. Sinha, D. E. Moncton, D. C. Johnston, J. M. Newsam, C. R. Safinya, and H. E. King, Jr., Phys. Rev. Lett. 58, 2802 (1987).
23. S. Mitsuda, G. Shirane, S. K. Sinha, D. C. Johnston, M. S. Alvarez, D. Vaknin, and D. E. Moncton, Phys. Rev. B36, to be published.
24. T. Freltoft, J. P. Remeilka, D. E. Moncton, A. S. Cooper, J. E. Fischer, P. Harshman, G. Shirane, S. K. Sinha, and D. Vaknin, Phys. Rev. B36, to be published.
25. P. M. Grant, S. S. P. Parkin, V. Y. Lee, E. M. Engler, M. L. Ramirez,

J. E. Vazquez, G. Lim, R. D. Jacowitz, and R. L. Greene, Phys. Rev. Lett. 58, 2482 (1987).
26. Ref. 21; R. A. Klemm, H. Thomann, and D. C. Johnston, preprint.
27. Ref. 25; R. B. van Dover, R. J. Cava, B. Batlogg, and E. A. Rietman, Phys. Rev. B35, 5337 (1987).
28. D. W. Murphy, S. Sunshine, R. B. van Dover, R. J. Cava, B. Batlogg, S. M. Zahurak, and L. F. Schneemeyer, Phys. Rev. Lett. 58, 1888 (1987).
29. Ref. 6.
30. L. F. Mattheiss, Phys. Rev. Lett. 58, 1028 (1987); J. Yu, A. J. Freeman, and J.-H. Xu, Phys. Rev. Lett. 58, 1035 (1987).
31. C. Zener, Phys. Rev. 81, 440 (1951); Phys. Rev. 82, 403 (1951).
32. W. Weber, Phys. Rev. Lett. 58, 1371 (1987).
33. W. A. Little, Phys. Rev. 134, A1416 (1964).
34. D. Allender, J. Bray, and J. Bardeen, Phys. Rev. B7, 1020 (1973).
35. C. M. Varma, S. Schmitt-Rink, and E. Abrahams, preprint.
36. J. Hirsch, Phys. Rev. B31, 4403 (1985).
37. V. J. Emery, Phys. Rev. Lett. 58, 2794 (1987).
38. J. E. Hirsch, Phys. Rev. Lett. 59, 228 (1987).
39. J. M. Tranquada, S. M. Heald, A. R. Moodenbaugh, and M. Suenaga, Phys. Rev. B35, 7187 (1987).
40. P. W. Anderson, Science 235, 1196 (1987); G. Baskaran, Z. Zou, and P. W. Anderson, Solid State Commun., to be published.
41. A. Ruckenstein, P. Hirschfeld, and J. Appel, preprint.
42. S. Kivelson, D. S. Rokhsar, and J. P. Sethna, Phys. Rev. B35, 8865 (1987).
43. P. W. Anderson, G. Baskaran, Z. Zou, and T. Hsu, Phys. Rev. Lett. 58, 2790 (1987).
44. H. Thomann, P. Tindall, D. C. Johnston, and R. A. Klemm, preprint.
45. R. A. Klemm, H. Thomann, and D. C. Johnston, preprint.

NATURE OF GIGANTIC RESISTANCE FLUCTUATIONS

IN METAL-OXIDE-SEMICONDUCTOR WIRES

Rajiv K. Kalia

Materials Science Division
Argonne National Laboratory
Argonne, IL 60439, USA

ABSTRACT

We summarize the experimental observations and the current theoretical understanding of resistance fluctuations in one-dimensional metal-oxide-semiconductor field-effect transistors.

INTRODUCTION

Recent advances in microfabrication have made it possible to create narrow electron channels in metal-oxide-semiconductor field-effect transistors (MOSFET).[1] Using shadowing techniques, one can create gates 10-20 nm wide or, alternatively, one can use an electric field to pinch 2D electron gas in an accumulation layer. Resistivity measurements[2] in these MOSFETs reveal a totally reproducible noise-like structure with varying gate voltage. The magnitude of these fluctuations grows considerably as the temperature is lowered and the temperature dependence of the resistivity is observed to be $\exp[A(T_0/T)^{1/2}]$.

THEORETICAL BACKGROUND

The essential features of these fluctuations seem to support the explanation put forward by Lee:[3] He shows that the nature of the structure seen experimentally is similar to the fluctuations inherent in Mott's variable-range hopping model. Since the typical length of experimental systems is 30-100 times the localization length and a Mott hop covers several localization lengths, it requires only a few hops to traverse the entire sample. In Mott hopping the resistance of the network is expected to be dominated by the resistance R_c of a single hop at the percolation threshold. Since each hop is exponentially activated and is expected to obey the lognormal distribution, there may exist substantial fluctuations in lnR and consequently orders-of-magnitude variations in the sample conductance.

FINITE-SIZE EFFECTS

Recently Serota et al. have investigated[4] the effect of varying the length of samples on conductance fluctuations, the crossover from Mott's $T^{-1/2}$ to activated behavior in sufficiently long wires, and also the change from 1D to 2D behavior. The average of the log resistance is found to increase as $(\ln 2\alpha L)^{1/2}$ while the fluctuations of the log resistance decrease according to the relation $\Delta \sim (\ln 2\alpha L)^{-1/2}$, α and L being the inverse localization length and the sample length, respectively. Thus, the resistance fluctuations are not just due to finite-size effects.

NATURE OF MAGNETORESISTANCE FLUCTUATIONS

In the variable-range hopping model, the sample resistance is due to a critical hop at the percolation threshold. Experimental support for this model suggests that conductance measurements in 1D MOSFETs in fact probe the nature of a single hop between a pair of localized states. This opens up several exciting possibilities, including the study of the effect of a magnetic field on a critical hop and how that alters the fluctuations.

The effect of a magnetic field enters through the Zeeman shift and the changes in the wave functions localized at different sites.[5] For Zeeman effect we use the variable-range model, allowing for double occupation at a site with an intrasite Coulomb repulsion U. Without spin-flip scattering, there are four hopping mechanisms: (1) from a singly-occupied site to an unoccupied site, (2) from a doubly-occupied site to an unoccupied site, (3) from a singly-occupied site to another singly-occupied site with an opposite spin, and (4) from a doubly-occupied site to a singly-occupied site. It should be noted that processes (2) and (3) are related by detailed balance. A magnetic field lowers the occupation probability of antiparallel-magnetic-moment states and consequently processes (2) and (3) are suppressed. So at sufficiently high magnetic fields, processes (1) and (4) are dominant and they have opposite effects on resistance fluctuations. For example, we have observed that the fluctuation arising from process (1) shift to lower values of the chemical potential when the magnetic field is increased. The reason for this shift is that an increase in the magnetic field lowers the occupation of antiparallel-magnetic-moment states and the hopping therefore involves parallel-magnetic-moment electrons with energy $E_i - 1/2\ g\mu_B H$. Since the hopping rate depends on the separation of these states from the chemical potential, an increase in H is equivalent to a reduction in the chemical potential and the fluctuations corresponding to process (1) therefore shift to lower values of the chemical potential. In a different range of chemical potentials, we observe an exactly opposite shift of resistance fluctuations; the fluctuations shift to a higher chemical potential when the magnetic field is increased. Such fluctuations reflect hopping process (4). In particular, it involves the hopping of an antiparallel-magnetic-moment electron from a doubly-occupied state to a singly-occupied state with a parallel magnetic-moment electron. In relation to the chemical potential, the energy of the hopping electron, $E_i + 1/2\ g\mu_B H$, increases with the magnetic field and to observe the same fluctuation, the chemical potential must therefore be increased. Both of these shifts have been observed experimentally.[6] The Zeeman effect increases the average resistance because the magnetic field suppresses the hopping rates corresponding to processes (2) and (3). The sample resistance saturates at sufficiently high magnetic fields.

The orbital effect is investigated by numerically solving for the localization lengths at different sites in the presence of a magnetic field.[5] Using these localization lengths, we obtain the percolation solution to the random-resistor network. The orbital effect is dramatically different from the Zeeman effect: There are no systematic shifts in the resistance fluctuations due to orbital effect, and the average resistance decreases with an increase in the magnetic field. The two effects can be easily separated experimentally through the orientation of the magnetic field: If the field is parallel to the sample, only the Zeeman effect survives. For magnetic fields perpendicular to the sample, both of these effects are present.

ACKNOWLEDGMENTS

This work was supported by the U.S. Dept. of Energy, Basic Energy Sciences-Materials Sciences, under Contract No. W-31-109-ENG-38 and a grant of time on the Energy Research Cray X-MP and Cray 2 computers at the Magnetic Fusion Energy Computing Center. The author would like to thank the U.S. Army Research Office for partial travel support.

REFERENCES

1. R. E. Howard, L. D. Jackel, P. M. Mankiewich, and W. J. Skocpol, Science 231, 346 (1986).
2. A. B. Fowler, A. Hartstein, and R. A. Webb, Phys. Rev. Lett. 48, 196 (1982); R. F. Kwasnik, M. A. Kastner, J. Melngailis, and P. A. Lee, Phys. Rev. Lett. 52, 224 (1984); R. A. Webb, A. Hartstein, J. J. Wainer, and A. B. Fowler, Phys. Rev. Lett. 54, 1577 (1985).
3. P. A. Lee, Phys. Rev. Lett. 52, 1641 (1984).
4. R. A. Serota, R. K. Kalia, and P. A. Lee, Phys. Rev. B 33, 8441 (1986).
5. R. K. Kalia, W. Xue, and P. A. Lee, Phys. Rev. Lett. 57, 1615 (1986).
6. R. A. Webb, J. J. Wainer, and A. B. Fowler, to be published.

THE FORMULATION OF DENSITY FUNCTIONAL THEORY

Jaime Keller and Carlos Amador

División de Ciencias Básicas, Fac. de Química
Universidad Nacional Autónoma de México
Apartado 70-528, 04510 México, D.F.

ABSTRACT

We have generalized the formulation of density functional theory in a form suitable for multiconfiguration calculations, and for the calculation of electric and magnetic response functions. The method allows the direct calculation of mixed valence states, spectra (and not just the energy eigenvalues and their differences) and in general, of the elementary excitations of the system.

INTRODUCTION

The ultimate goal of density functional theory (DFT) should be the formulation and development of a quantum mechanical problem in terms of the relevant particle densities $\rho_\alpha(r)$. The idea was born with the well known Thomas-Fermi theory and its succesive developments (see the paper by Kryachko[1] in the present volume), where the density $\rho(r)$ alone should be used and in principle obtained in a closed form from the theory itself. DFT should provide a full and faithful description of reality, a difficult task indeed if formulated in the simplest forms assumed until now. An example would be the use of a Thomas-Fermi density ρ_{TF} with only one energy parameter $\alpha\rho^{1/3}$, or "chemical potential", and the observed spectroscopy of a many particle system where a complete set of first ionization potentials is measured, corresponding to the N possible ways of extracting one particle from the system. Of course one of this first ionization potentials requires the least amount of removal energy and, in general, corresponds to the final state being the ionized N-1 particle system in its ground state (unless the ionization method employed can not bring, from physical symmetry considerations, the initial N particle system into the ground state N-1 particle system). Another well known example would be the existence of long lived stationary states of a system with energies larger than the true ground state (consider a system of a Na^+ and a Cl^- ions separated by a macroscopic distance, that will exist for a practically infinite long time). The simplest approaches to density functional theory will not account for this and other well known facts. This is not the case of more sofisticated formalisms, for example the Kohn-Sham[2] method (KS).

But if a theory is developed starting, at least, at the level of a density matrix $\gamma(1,1')$ (which contains all the information that in standard quantum mechanics is described as a set of configurations and the internal structure of each one of those configurations), the possibility of accounting both for the shell structure of each allowed configuration and for the configurational structure of every stationary state would be open. In fact, as we will see below an ansatz can be made (Keller and de Teresa[3]) that the one-matrix is to be expanded as

$$\gamma(1,1') = \sum_{ij} \beta^{ij}(\phi_i^*(1)\phi_j(1') + \text{h.c.}) \qquad (1)$$

in terms of orthonormal functions $|\phi_k(r)|^2 = \rho_k$ using an hermitian matrix of coefficicients β^{ij} which contain the possibility of accounting for the weight of configurations $|a^B|^2 = C^B$ and for the shell structure of each configuration B, as a summation over configurations A, B

$$\beta^{ij} = \sum_{AB} \Delta_{AB}^{ij}(a^{*A}a^B + \text{h.c.})/2 \qquad (2)$$

each configuration having a shell structure

$$s_A^i = \begin{cases} 1 \text{ if "state" } i \text{ is in configuration A} \\ 0 \text{ if state } i \text{ is not included in A} \end{cases} \qquad (3)$$

such that the coefficients

$$\Delta_{AB}^{ij} = \begin{cases} 1 \text{ if } \{\{s_A^k\} - s_A^i\} = \{\{s_B^k\} - s_B^j\} \\ 0 \quad \text{otherwise} \end{cases} \qquad (4)$$

from the orthonormality of the $\phi_i(r)$.

The KS method corresponding to $a^A = \delta^{AA'}$, a single configuration, where the $\beta^{ij} = \delta^{ij}$.

In the rest of the paper we describe, in the next section, the analysis of the problem made by Keller[4] and by Keller and Ludeña[5] and, in the following section, a practical procedure to use, and compute, the one-matrix (1).

THE ONE MATRIX AND THE DENSITY AS A MAPPING FROM THE MANY BODY WAVE FUNCTION TO THE DENSITY MATRIX

In standard quantum mechanics let the determinant (for fermions)

$$D_I = (1/(n!)^{1/2}) [\Phi_{I1}(1), \Phi_{I2}(2), \ldots] \tag{5}$$

represent a configuration in the n-electron Hilbert Space such that the total density matrix is

$$\rho_n = \sum_{IJ} C_I C_J^* |D_I\rangle\langle D_J| \tag{6}$$

and by $p = n-1$ reduction by integration and proper normalization obtain the one matrix $\rho_1(r_1, r_2)$, with the coefficients C_I optimized from the conditions for an stationary state (Keller and Ludeña and references therein) and the restrictions in orthonormality of the wave functions.

The idea of density functional theory DFT would be to find and expansion of the one matrix (by direct DFT procedures)

$$\gamma(1,1') = \rho_1(r_1, r_2) = \sum_{i=1}^{m} \lambda_i |\phi_i\rangle\langle\phi_i| = \sum_i \lambda_i \gamma_i(1,1') \tag{7}$$

which would correspond to the correct expression. For that purpose a functional should be defined

$$H[\rho_1] = \mathrm{Tr}_1 [(\alpha' - v(r))\rho_1] = \mathrm{Tr}_1[\hat{h}\rho_1] \tag{8}$$

such that

$$[1] \quad H_{\lambda_i}[\rho_1] \equiv \frac{\delta H[\rho_1]}{\delta\lambda_i} = \mathrm{Tr}_1 [\frac{\partial \hat{h}}{\partial\lambda_i} \rho_1 + \hat{h}\gamma_i] = 0 \tag{9}$$

and

$$[2] \quad (H_{\phi_i}[\rho_1](1) \equiv \frac{\delta H[\rho_1]}{\delta\phi_i(1)} = $$

$$= \mathrm{Tr}_1 (\frac{\delta\hat{h}}{\delta\phi_i(1)} \rho_1) + \lambda_i \int d1'\hat{h}(1,1')\phi_i^*(1') = 0 \tag{10}$$

allowing for [1] configuration optimization and [2] wave function optimization. But of course, both the (ensemble) n-representability for the ρ_1 and the orthonormality of the ϕ_i has to be introduced into [1] and [2] as auxiliary conditions: $0 \leq \lambda_i \leq 1$ and $\sum_{i=1}^{m} \lambda_i = n$; as well as $\langle\phi_i|\phi_j\rangle = \delta_{ij}$. Then we should minimize

$$\Omega[\rho_1] = H[\rho_1] - \mu(\sum_i \lambda_i - n) - \sum_{ij} \beta_{ij}(\langle\phi_i|\phi_j\rangle - \delta_{ij}) \tag{11}$$

Can this be done and

$$\frac{\partial\Omega}{\partial\lambda_i} = 0 \quad \text{as well as} \quad \frac{\partial\Omega}{\partial\phi_i^*} = 0 \tag{12}$$

be obtained? A practical scheme is sketched in the next section, based on the knowledge which has been developed about DFT in nearly 30 years, it has been intensively used in atomic, molecular, nuclear and solid state physics.

MULTICONFIGURATION IN DENSITY FUNCTIONAL THEORY. A CALCULATIONAL SCHEME.

Though density functional theory suggests the use of the electronic density itself in the calculation of all relevant properties, the most successful practical schemes (eg. Kohn & Sham), one way or another, rely on the use of the first order density matrix $\gamma(1,1')$ as mentioned above. Therefore, as γ is computed anyhow, and having abandoned the conceptual advantage of using just the density ρ , one is tempted to extract full profit of the first order density matrix. One practical calculational scheme is the following.

As a first step, the construction of an auxiliary N-particle "wave function"

$$\Psi_A(1,\ldots,\ell,\ldots,N) = \mathbf{S} \prod_{i\varepsilon M} \phi_i(r_\ell) \begin{cases} i \neq j & i,j \text{ for fermions} \\ i = j & \text{allowed for bosons} \end{cases} \tag{13}$$

where A refers to a particular configuration, **S** stands for a symmetrizing operator and the condition on the indices accounts for the Pauli principle. The mixing of several configurations is achieved through a linear combination of these "single configuration" N-particle "wave functions":

$$\Psi_c = \sum_A a_c^A \Psi_A \tag{14}$$

The auxiliary multiconfigurational wave function is then used to approximate the exact first order density matrix $\gamma^E(1,1')$

$$\gamma^E(1,1') \cong \gamma_c(1,1') = \int \Psi_c(1,2,\ldots,N)\Psi_c^*(1',2,\ldots,N)d2\ldots dN, \tag{15}$$

the well known Kohn-Sham procedure corresponds to $a_c^A = \delta^{AA'}$, that is, to a single configuration A'.

The local approximation energy functional acts on this density matrix

$$\hat{H}[\gamma_c(1,1')] = [-\frac{1}{2}\int \nabla_{1'}^2 \gamma_c(1,1')|_{1'\to 1} dr_1 + h.c.] + \int v_{eff}^c(1)\gamma_c(1,1)dr_1 \quad (16)$$

where v_{ff}^c is the configuration-dependent one particle effective potential. The energy functional is then minimized with respect to both the basis functions $\{\phi_i(r)\}$ and the linear combination coefficients $\{a_c^A\}$, subject to the orthonormalization and to the number conservation constraints. From the variation of the basis functions one obtains the usual Euler equation

$$v_{eff,i}^f + k_i^f - \mu\phi_i = 0 \quad (17)$$

where

$$v_{eff,i}^f = \frac{\partial \int v_{eff}\gamma_c(1,1)dr_1}{\partial \phi_i^*} \quad (18)$$

and

$$k_i^f = \frac{\partial (-\frac{1}{2}\int \nabla_{1'}^2 \gamma_c(1,1')|_{1'\to 1} dr_1)}{\partial \phi_i^*} \quad (19)$$

To illustrate the form of these functionals, we try the ansatz of the basis functions to be orthonormal

$$\int \phi_i^* \phi_j dr = \delta_{ij} \quad (20)$$

In this case, the first order density matrix, reduces to

$$\gamma(1,1') = \sum_{AA'} a^{*A} a^{A'} \sum_{\substack{i\varepsilon A \\ j\varepsilon A'}} \beta_{AA'}^{ij} \phi_i^*(1)\phi_j(1') \quad (21)$$

for fermions

$$\beta_{AA'}^{ij} = \begin{cases} 1 & \text{if } \{A'-j\} = \{A-i\} \\ 0 & \text{otherwise} \end{cases} \quad (22)$$

for bosons the value of $\beta_{AA'}^{ij}$ depends on the occurrence numbers of the multiply "occupied" states ϕ_j in A and A'.

The "kinetic energy" term becomes

$$-\sum_{AA'} \sum_{\substack{i\varepsilon A \\ j\varepsilon A'}} (a^A a^{*A'} \beta_{AA'}^{ij} \phi_i^*(1)\nabla^2\phi_j(1) + h.c.) \quad (23)$$

then some contributions to the kinetic energy term in the one-particle equation, that is, the derivative of (23) with respect to ϕ_i^*

$$- \sum_{A'} \sum_{j\epsilon A'} (a^A a^{*A'} \beta^{ij}_{AA'} \nabla^2 \phi_j(1) + h.c.) \tag{24}$$

will not necessarily be positive, depending on the relative signs of the a^A.

Through this term the differential equation, for the basis functions will be coupled. This coupling is one of the most important effects from the use of the full first order density matrix. It is here where the correlation corrections will appear. Unfortunately the non linearity (24) might render this approach unpractical if attempted to be solved in an analytical, exact form.

To test the feasibility of the scheme with approximate solutions, we are studying two possibilities:
In the first one we rely only on the coefficients $\{a_c^A\}$ to find the best energy. That is, we start by constructing a density matrix from the auxiliary orbitals of an usual Kohn-Sham calculation; with it, we perform a Kohn-Sham-like calculation including the "off-diagonal" terms and find the expectation value for the energy, then we vary the coefficients $\{ a_c^A \}$ used in the construction of the density matrix, subject to the condition of proper normalization, to find the minimum value of the energy; with these new coefficients we construct a new density matrix and the whole procedure is repeated until achieving self-consistency.

One can include the basis functions themselves as well, in the optimization procedure, though in an approximate manner. In this second approach the steps will be essentially the same, then the new basis functions obtained through a Kohn-Sham-like calculation, are modified accordingly by using a perturbational method. This is, in order to satisfy

$$\partial\hat{H}/\partial\phi_i^* = 0 \tag{25}$$

one can use the Kohn-Sham like solutions to

$$(H^o_{KS} - \varepsilon^o)\, \phi_i^o = 0 \tag{26}$$

to solve the Euler equation coming from the variation,

$$H_{KS}\phi_i^{opt} + \gamma^{jj} f = 0 \tag{27}$$

where $\gamma^{ij} f$ is the term produced by the coupling or off-diagonal terms mentioned above. The optimized function ϕ^{opt} .

$$\phi_i^{opt} = \phi_i^o + \delta\phi_i \qquad (28)$$

The corrections $\delta\phi_i$ are found treating the $\gamma^{ij}f$ terms as a perturbation to the original Kohn-Sham Hamiltonian.

We could consider an alternate formulation based on the density alone

$$\rho(r) = \gamma(1,1) \qquad (29)$$

where the shell structure of $\gamma(1,1')$ is kept in the auxiliary kinetic energy expression, as far as we can write (Keller[4])

$$\rho(r) = \sum_{ij}^{M} \beta^{ij}\rho_{ij} \in R \qquad (30)$$

with $\rho_{ij} = (\phi_i^*\phi_j$, $M \geq N$ and $(\beta^{ij}\rho_{ij} + \beta^{ji}\rho_{ji}) \in R'$ and the $(\phi_i|^2 = \rho_{ii}$ defining the auxiliary function ϕ_i up to a phase. And the total kinetic energy operator being then

$$\hat{k}\rho = \frac{1}{2}\sum_{ij}^{M} \beta^{ij} \int \phi_i^* \nabla^2 \phi_j \, d\tau \in R' \qquad (31)$$

The phases of each of the $\phi_i(r)$ auxiliary functions fixed by matching the appropriate boundary conditions.

We should remember that the Thomas-Fermi kinetic energy operator makes use of the "shell" structure of the free electron gas. This being a limiting case for "shell" structures.

In (30) and (31) we have included the configuration and shell internal symmetries of the many particle system both in the density function and in the kinetic energy operator. We can formally do it algebraically by symbolically introducing the "universal" kinetic energy operator

$$-(1/2) \sum_{j=1}^{\infty} \nabla^2_{<j|} \qquad (32)$$

where formally $\nabla^2_{<j|}$ is the Laplacian of the jth function ϕ_j, and $\rho(r)$ is assumed to have the form (30). The representation of (32) will change from system to system and, in principle, from stationary state to stationary state of a given system.

CONCLUSIONS.

We have presented both the basic ideas of a general DFT and a practical formulation of it. In previous volumes of this series[6, 7], we have discussed the response function of the system in terms of either the ground state density and the electronic density of states or of the Green's function of the system expanded in terms of the auxiliary functions ϕ_i, the extension here presented allows the introduction of the full effect of correlation into the formalism.

In refs. 4 and 5 we have then proposed that the formal structure of density functional theory should be stated in at least three theorems

(1) The total energy of the system, in any stationary state, is a unique functional of its particle density.

(2) There exists a variational principle for the energy functional of a stationary state of the system.

(3) the internal symmetry of the system (through a set of parameters $\{c_{i,N}\}$, for example) is a necessary part of the theory. The variational equations should read $\delta(E[\rho\ ;\ \{c_{i,N}\}] - \mu(\int\rho[\{c_{i,N}\}]d\tau-N) = 0$ and the parameters $c_{i,N}$ included either in the construction of the density function (as in the Kohn-Sham procedure) or in the energy functional itself.

The internal symmetry of the system is an important (hidden or explicit) part of the theory, and, when properly considered, the results have always compared reasonably well with experiment.

Acknowledgements

This work was partially supported by CONACTY, Mexico, project PCEXCNA-022702. The technical assistance of Mrs. Irma Aragon is also gratefully acknowledged.

References

1. E. Kryachko, present volume and see, for example, Density Functional Theory, J. Keller and J.L. Gázquez, Eds., Lecture Notes in Physics, Vol. 187 (Springer-Verlag, New York, 1983).
2. W. Kohn and L.J. Sham, Phys. Rev. A 140, 1133 (1965).
3. J. Keller and C. de Teresa, in La Química Teórica en México 1987, J. Keller Ed. (UNAM, México , 1987) 1180-1202.
4. J. Keller, Int. J. Quantum Chem., Quantum Chem. Symp. 20, 767 (1986).
   J. Keller, Journal of Molecular Structure (1988) in press.
5. J. Keller and E. Ludeña, Int. J. Quantum Chem., Quantum Chem. Symp. 21, 171-180 (1987).
6. J. Keller, C. Amador and C. de Teresa, in Condensed Matter Theories B. Malik, Ed. (Plenum, New York, 1986) Vol. 1, p. 195.
7. J. Keller, C. Amador, C. de Teresa and J.A. Flores, in Condensed Matter Theories, P. Vashista, Ed., (Plenum, New York 1986), Vol. 2, 131-140.

THE ENERGY DENSITY FUNCTIONAL THEORY: PRESENT STATUS-REFLECTIONS

Eugene S. Kryachko

Institute for Theoretical Physics
Kiev-130, USSR 252130

Verbaque praevisam rem
non invita sequentur

In this year all many-body theorists celebrate the 60th anniversary of the energy density functional theory. Sixty years ago, in 1927 Thomas[1] and Fermi[2] proposed independently the approximate, statistical model for a many-electron atom christened later by their names - the Thomas-Fermi model - and originated the modern rigorous theory.

The fundamental idea of the energy density functional theory for the ground-state many-electron systems is as follows. Let us consider an arbitrary stationary and nonrelativistic system of N electrons (N is finite) with the Hamiltonian

$$\hat{H} = -\frac{1}{2}\sum_{i=1}^{N} \nabla_i^2 + \sum_{1\leq i<j\leq N} |\vec{r}_i - \vec{r}_j|^{-1} \tag{1}$$

$$-\sum_{\alpha=1}^{M} Z_\alpha \sum_{i=1}^{N} |\vec{r}_i - \vec{R}_\alpha|^{-1} + \sum_{1\leq\alpha<\beta\leq M} Z_\alpha Z_\beta |\vec{R}_\alpha - \vec{R}_\beta|^{-1}$$

$$\equiv \hat{T} + \hat{U}_{ee} + \hat{U}_{en} + \hat{U}_{nn}$$

where $\vec{r}_i$ is the radius vector of ith electron, $\vec{r}_i \in R^3$; s_i is its spin projection on the zth axis of the spin space, $s_i \in \{\pm 1/2\}$; $\vec{R}_\alpha$ and Z_α are, respectively, the radius vector and charge of the α th nucleus. Within the Born-Oppenheimer approximation, the ground state of this N-electron system is determined via the quantum mechanical variational principle in terms of the energy functional

$$E[\Psi] \equiv <\Psi|H|\Psi>$$

$$= T[\Psi] + U_{ee}[\Psi] + U_{en}[\Psi] + U_{nn}[\Psi] \tag{2}$$

where $A[\Psi] \equiv <\Psi|\hat{A}|\Psi>$, and Ψ is assumed to be normalized to unity. The energy density functional theory suggests an existence of the equivalent ground-state variational principle, expressed in terms of the so called energy density functional

$$\mathcal{E}[\rho(x)] = \begin{cases} \mathcal{E}[\rho(\vec{r})] \\ \mathcal{E}[\rho_\uparrow(\vec{r}),\ \rho_\downarrow(\vec{r})] \end{cases} \tag{3}$$

which is universal as $E[\Psi]$ in view of Eq.(2) and which depends, in a functional manner, on the single variable of theory - one-electron density $\rho(\vec{r}) \equiv \rho_\uparrow(\vec{r}) + \rho_\downarrow(\vec{r})$, or its spin components, $\rho_\uparrow(\vec{r})$ and $\rho_\downarrow(\vec{r})$, within the spin-polarized formulation. Here $\rho_\uparrow(\vec{r})$ is the one-electron density of the spin-up electrons, and respectively, $\rho_\downarrow(\vec{r})$ is that of spin-down electrons. From the pragmatical point of view, for practical needs of quantum chemistry, such the "density" reformulation of the traditional variational principle appears rather attractive, since $\rho(\vec{r})$ is a function of three spatial variables independing on the total number of electrons of the system in question. Well, but there arises immediately the problem of validity of this rather strong suggestion. It is usually accepted that the familiar Hohenberg-Kohn theorem[3] is its formal justification. Naturally, formulating the basic thesis of the energy density functional theory for many-electron systems about the existence of the "energy mapping",

$$E[\Psi] \mapsto \begin{cases} \mathcal{E}[\rho(\vec{r})] \\ \mathcal{E}[\rho_\uparrow(\vec{r}),\ \rho_\uparrow(\vec{r})] \end{cases} \tag{4}$$

we assume implicitly the existence of such "mapping of variables",

$$\Psi \longleftrightarrow \rho\ . \tag{5}$$

It is clear that the mapping (5) is well-defined iff (if and only if) the classes of functions in its left- and right-hand sides are rigorously determined. We designate the class of "admissible" N-electron wave functions by $\mathcal{L}_N$. Obviously, $\Psi \in \mathcal{L}_N$ if the energy functional (2) is well-defined. In the other words, the following conditions are satisfied:

(Fi) $$< \Psi|\Psi >\equiv \sum_{s_1 \cdots s_N} \int d^3\vec{r}_1 d^3\vec{r}_2 \cdots |\Psi(\vec{r}_1, s_1; \cdots; \vec{r}_N, s_N)|^2 < \infty\ , \tag{6a}$$

i.e., $\Psi \in \mathcal{L}_N \subset L^2_\sigma(R^{3N})$ where $L^2_\sigma(R^{3N})$ is the Hilbert space of antisymmetric, square-integrable wave functions;

(Fii) $$T[\Psi] < \infty\ , \tag{6b}$$

i.e., $\Psi \in \mathcal{L}_n$ is a differentiable function of all spatial coordinates, and an each component of $\nabla_{\vec{r}_i}\Psi$ belongs to $L^2_\sigma(R^{3N})$.

One can demonstrate[4] that for the Coulomb quantum systems these conditions (6a-b) determine completely the class $\mathcal{L}_N$ of admissible functions on which $E[\Psi]$ is well-defined. Henceforth we assume that any $\Psi \in \mathcal{L}_N$ is normalized to unity: $< \Psi|\Psi >= 1$.

Now, defined above the class of admissible N-electron wave functions, one can determine the corresponding set $\mathcal{N}_N$ of admissible "one-electron densities" by means of the so called reduction mapping

$$\rho_\Psi(\vec{r}, s) \equiv N \sum_{s_2 \cdots s_N} \int d^3\vec{r}_2 \cdots d^3\vec{r}_N |\Psi(\vec{r}, s; \vec{r}_2, s_2; \cdots; \vec{r}_N, s_N)|^2 . \tag{7}$$

It is worthy of notice that from the given definition of $\mathcal{N}_N$ there follows that, firstly, any one-electron density as the element of $\mathcal{N}_N$ is N-representable, that is, for any $\rho(\vec{r}, s) \in \mathcal{N}_N$ the inverse mapping from ρ to Ψ does exist, and secondly, $\rho(\vec{r}, s) \in \mathcal{N}_N$ possesses the following properties:

(Di) nonnegativity of $\rho(\vec{r}, s)$ anywhere in R^3;

(Dii)
$$\sum_s \int d^3\vec{r}\rho(\vec{r}, s) \equiv \int d^3\vec{r}\, [\rho_\uparrow(\vec{r}) + \rho_\downarrow(\vec{r})] \tag{8a}$$

$$= N_\uparrow + N_\downarrow = N,$$

i.e., $[\rho(\vec{r})]^{1/2} \in L^2(R^3)$, N_s is the number of electrons with the spin s ;

(Diii) $\rho(\vec{r}) = \sum_s \rho(\vec{r}, s)$ is a differentiable function relative to any coordinate of $\vec{r} \in R^3$, and

$$\nabla\rho(\vec{r}) \in L^2(R^3) \tag{8b}$$

since by the Schwarz inequality (see Lieb[5]),

$$[\nabla\rho_\Psi(\vec{r})]^2 \leq$$

$$4N\rho_\Psi(\vec{r}) \cdot \sum_{s_1 s_2 \cdots s_N} \int d^3\vec{r}_2 \cdots d^3\vec{r}_N [\nabla_{\vec{r}}\Psi(\vec{r}, s_1; \vec{r}_2, s_2; \cdots; \vec{r}_N, s_N)]^2. \tag{8c}$$

Furthermore, one can prove fairly easy[5] that $\nabla[\rho_\Psi(\vec{r})]^{1/2} \in L^2(R^3)$, since from the inequality (8b) it follows directly

$$\int d^3\vec{r}(\nabla[\rho_\Psi(\vec{r})]^{1/2})^2 = \frac{1}{4}\int d^3\vec{r}\frac{[\nabla\rho_\Psi(\vec{r})]^2}{\rho_\Psi(\vec{r})} \tag{8d}$$

$$\equiv 2\int d^3\vec{r} t_W[\rho_\Psi(\vec{r})] \leq T[\Psi] ,$$

i.e., the term $t_W^{1/2}[\rho(\vec{r})]$ ($t_W[\rho(\vec{r})]$ is the so called von Weizsäcker term[6]) is a square-integrable one for any $\rho(\vec{r}) \in \mathcal{N}_N$. Therefore, the Thomas-Fermi atom cannot be described by any N-electron wave function from the class $\mathcal{L}_N$, i.e., $\rho_{TF}(\vec{r})$ is not N-representable, because $t_W^{1/2}[\rho_{TF}(\vec{r})]$ is not square-integrable for the Thomas-Fermi one-electron density[7,8].

Let us introduce the concept of a shape factor $s_\rho(\vec{r})$ for any "density" $\rho(\vec{r}) \in \mathcal{N}_N$:[9]

$$s_\rho(\vec{r}) \equiv \rho(\vec{r})/_N . \tag{9}$$

One can demonstrate [10] that due to this relation (9), the Thomas-Fermi model [7,11] "absorbs" the dependence on the total number of electrons. Equation (8a) tells us that $s_\rho(\vec{r})$ is normalized to unity and obeys (Di). Nevertheless, for any $\rho(\vec{r}) \in \mathcal{N}_N$ $s_\rho(\vec{r})$ is not one-electron density which belongs to $\mathcal{N}_1$, since in general, $s_\rho(\vec{r})$ cannot be interpreted as 1-representable ground-state density. It means mathematically that the conditions (Di) - (Diii) are necessary for determining $\mathcal{N}_N$, but not sufficient. In fact, the class of N-representable densities of N-electron systems is more extended and involves such densities which cannot be realized, through a shape factor, as those for one-electron systems. $\mathcal{N}_N$ must contain such the densities which equally with the familiar asymptotical behavior at infinity, far from nuclei, and the electron-nuclear cusp conditions[12] have the so called zero-flux surfaces $S_1, S_2, \cdots, S_{M-1} \subset R^3$ where

$$\nabla\rho(\vec{r}) \cdot \vec{n}_\alpha = 0 , \tag{10}$$

$\vec{n}_\alpha$ is the normal to S_α at the point $\vec{r} \in R^3$.[13] The existence of such surfaces at finite $\vec{r}$ for free atoms and ions with $M = 1$ is impossible due to the boundary conditions.

If $N > 2$ and $M \geq 2$ these zero-flux surfaces induce the partitioning of R^3 into the atomic-like fragments. In general, any real one-electron density can be represented by the complete set of closed equidensity curves,

$$\mathcal{L}_C(\vec{r}) = \{ \vec{r} \in R^3 | \rho(\vec{r}) = C = \text{const} \} ,$$

some of which are distributed inside an each atomic-like fragment with the centre at $\vec{R}_\alpha$, the postion of the α th nucleus with charge Z_α.

It becomes apparent that now we enable to perform the rigorous "jump" in the mapping (4) on basis of traditional variational principle in terms of $E[\Psi]$. But if really this mapping exists, it is of a great importance to formulate the "density" analogon of the variational principle in the independent and closed manner which suggests, firstly, an explicit form of $\mathcal{E}[\rho(\vec{r})]$ or $\mathcal{E}[\rho_\uparrow(\vec{r}), \rho_\downarrow(\vec{r})]$, and secondly, a priori definition of the class $\mathcal{N}_N$ of variable one-electron densities with the explicit form of the N-representable submapping $\rho \mapsto \Psi \in \mathcal{L}_N$. Otherwise, the "density" analogon of the variational principle is lack in view of possible applications. Naturally one can expect that both these conditions are not independent. In fact, the explicit form of the energy density functional suggests an existence of the rigorous procedure, based on the submapping $\Psi \mapsto \rho$, by means of which one can construct $\mathcal{E}[\rho(\vec{r})]$ or $\mathcal{E}[\rho_\uparrow(\vec{r}), \rho_\downarrow(\vec{r})]$ from $E[\Psi]$. Such procedure is called usually as "of-first-principles" one, and the corresponding energy density functional is named the "of-first-principles" functional. Then, assuming that the class $\mathcal{N}_N$ of densities is well-defined, one can formulate the stationary ground-state variational principle,

$$\delta\{ \mathcal{E}[\rho(\vec{r})] - \mu\Big[\int d^3\vec{r}\rho(\vec{r}) - N\Big] \} = 0 , \ \rho(\vec{r}) \in \mathcal{N}_N , \tag{11}$$

where μ is the so called chemical potential of the system in question[14,15], and further, solve the Euler-Lagrange-type equation of "motion" for the density,

$$\frac{\delta\mathcal{E}[\rho(\vec{r})]}{\delta\rho(\vec{r})} - \mu = 0 , \qquad \rho(\vec{r}) \in \mathcal{N}_N , \tag{12}$$

where the first term means the functional derivative. Equation (12) is equivalent to determining the optimal one-electron density $\tilde{\rho}_\circ(\vec{r}) \in \mathcal{N}_N$ such that [14]

$$\left[\frac{\delta\mathcal{E}[\rho(\vec{r})]}{\delta\rho(\vec{r})}\right]_{\tilde{\rho}_\circ(\vec{r})} = \mu . \tag{13}$$

Due to the functional $\mathcal{E}[\rho(\vec{r})]$ in Eq. (11) to be "of-first-principles" one, there exists, by definition, the N-representable submapping $\rho \mapsto \Psi$ such that the optimal density $\tilde{\rho}_\circ(\vec{r}) \in \mathcal{N}_N$ corresponds the optimal N-electron wave function $\Psi_\circ \in \mathcal{L}_N$ (a symbol "tilde" means that the fulfilment of the equalities $\Psi_\circ = \tilde{\Psi}_\circ$ and $\rho_\circ \equiv \rho_{\Psi_\circ} = \tilde{\rho}_\circ$ are not required in general), and

$$\mathcal{E}[\tilde{\rho}_\circ(\vec{r})] \ = \ E[\tilde{\Psi}_\circ] . \tag{14}$$

To advance in constructing the "of-first-principles" procedure, let us consider some simple examples, and start from the simplest one-electron system with the Hamiltonian

$$\hat{H} = -\frac{1}{2}\nabla^2 + v(\vec{r}) \tag{15}$$

where

$$v(\vec{r}) = -\sum_{\alpha=1}^{M} Z_\alpha |\vec{r} - \vec{R}_\alpha|^{-1} \tag{16}$$

plays the role of the "external" potential.† In that case the energy functional $E[\Psi]$ with $\Psi \in \mathcal{L}_1$ takes the form of the energy density functional,

$$E[\Psi] = \int d^3\vec{r}\{t_W[\rho_\Psi(\vec{r})] + v(\vec{r})\rho_\Psi(\vec{r})\} \equiv \mathcal{E}[\rho_\Psi(\vec{r})] \tag{17}$$

since

$$\rho(\vec{r}) = |\Psi(\vec{r})|^2 \, . \tag{18}$$

Therefore, for one-electron system the "of-first-principles" procedure becomes trivial, and the second condition satisfies due to the explicit form of the mapping of variables given by Eq. (18).

Under transition from one- to two-electron systems, one should emphasize that in quantum theory of many bodies, in particular, quantum chemistry, there exists the hierarchy of levels of accuracy which is generated by a choice of the proper subclass of trial functions in $\mathcal{L}_N$, on which the energy functional $E[\Psi]$ is extremalized [17]. Obviously, if $\mathcal{D}_1 \subseteq \mathcal{D}_2 \subseteq \mathcal{L}_N$, $E_\circ^{(2)} \leq E_\circ^{(1)}$ where

$$E_\circ^{(i)} \equiv \inf_{\Psi\in\mathcal{D}_i} \{E[\Psi]\} \, , \; i = 1, 2. \tag{19}$$

In the author's opinion, these "hierarchical rules of play", arising due to inaccessibility of the exact level of accuracy by seaching the extremum of $E[\Psi]$ over the whole class $\mathcal{L}_N$, constitute the art of quantum chemistry. Thus, with respect to energy, the uppermost level corresponds to the Hartree approach, although it cannot be realizable within $\mathcal{L}_N$. The next basic level occupies the restricted Hartree-Fock model. Intermediate sublevels correspond to diverse independent-particle models. The exact level is the lowest in this hierarchy of levels realizing within the class $\mathcal{L}_N$. A "distance", in units of energy, between the Hartree-Fock level and the exact one is, by definition, the correlation energy due to Löwdin [18] (see also the recent rewiews [19,20] on the correlation energy problem in the energy density functional theory). It is clear that for one-electron systems both the levels, Hartree-Fock and exact ones, coincide with each other. Sitting at the Hartree-Fock level, one can write for two-electron systems the relation similar to Eq.(18),

$$\rho(\vec{r}) = 2|\varphi(\vec{r})|^2 \tag{20}$$

Where $\varphi(\vec{r})$ is the orbital. Hence, at the Slater determinant point

$$\Psi(x_1, x_2) = 2^{-1/2}\varphi(\vec{r}_1)\varphi(\vec{r}_2)[\alpha(s_1)\beta(s_2) - \beta(s_1)\alpha(s_2)]$$

†The term "external" refers to the nuclear subsystem which is external to the electronic one. In all cases we assume that a given N-electron system has the stationary ground state, i.e., it is stable. This condition is satisfied under imposing definite constraints on M and Z_α with the given N.[16]

the Hartree-Fock energy functional $E_{HF}[\Psi]$ becomes also the energy density functional,

$$E[\Psi] = \int d^3\vec{r}\{t_W[\rho_\Psi(\vec{r})] + \frac{1}{4}\rho_\Psi(\vec{r})v_H(\vec{r}) + \rho_\Psi(\vec{r})v(\vec{r})\} \equiv \mathcal{E}_{HF}[\rho\Psi(\vec{r})] \tag{21}$$

where

$$v_H(\vec{r}) = \int d^3\vec{r}\,'\rho_\Psi(\vec{r}\,') \mid \vec{r} - \vec{r}\,' \mid^{-1} \tag{22}$$

is the Hartree potential. Therefore, one can conclude that at the Hartree-Fock level for two-electron systems, which differs completely from the exact level, we construct the "of-first-principles" energy density functional. Let us call it the Hartree-Fock energy density functional unlike the energy density functional which exists at the exact level due to the Hohenberg-Kohn theorem. The latter is defined as the Hohenberg-Kohn one. For one-electron systems it is given explicitly by Eq.(17). It is natural that the hierarachy of levels given determines our inability to reach the exact-level Everest which is uniquely existing one in Nature. From this standpoint one should consider the Hartree-Fock "of-first-principles" energy density functional $\mathcal{E}_{HF}$ $[\rho(\vec{r})]$, given by Eq.(21) for two-electron systems, as the reflection of narrow-mindedness inherent for us. In reality there must exist the Hohenberg-Kohn functional, $\mathcal{E}_{HK}[\rho(\vec{r})]$, only. Applying to it the stationary variational principle (11), one obtains the exact level in terms of $\rho_\circ(\vec{r})$ and $\Psi_\circ$, and also the exact ground-state energy $E_\circ = E[\Psi_\circ] = \mathcal{E}_{HK}[\rho_\circ(\vec{r})]$. Within the given treatment of two-electron systems, one should emphasize the following. The Hartree-Fock energy density functional $\mathcal{E}_{HF}$ $[\rho(\vec{r})]$, Eq.(21), is defined on the class of one-electron densities obtained from the subclass S_2 of all Slater determinants, belonging to $\mathcal{L}_2$, under the reduction mapping (7). On the other hand, if a certain $\rho(\vec{r})$, being viewed as candidate for a "density", obeys (Di)-(Diii),one can construct the orbital $\varphi(\vec{r}) = \sqrt{\frac{1}{2}\rho(\vec{r})}$ and further the corresponding Slater determinant which satisfies the conditions (Fi)-(Fii). Therefore, all one-electron densities from $\mathcal{N}_2$ are representable by Slater determinants. Then, what does represent themselves other one-electron densities which correspond to other levels of the given hierarchy system[2]. Since the conditions (Di)-(Diii) are necessary, $\mathcal{N}_2$ does not involve densities which are non-2-representable by Slater determinants but 2-representable. In other words, to all other levels of accurary there correspond also densities from $\mathcal{N}_2$ which are 2-representable by Slater determinants. Therefore, for two-electron systems the submapping $\rho \mapsto \Psi$ is multivalued, i.e., is not one-to-one. March and Young[21], and Harriman[22] (see also Refs.[5,23]) demonsrate that this one-to-many correspondence for $\rho \mapsto \Psi$ is preserved for arbitrary N-electron systems. It is clear this nature of the submapping of $\mathcal{N}_N$ onto $\mathcal{L}_N$ makes difficult the construction of the "of-first-principles" procedure which apparently has to involve, in a somewhat implicit way, the definite constraints providing a choice of the one-to-one branches of this submapping. Otherwise, we get into difficulties under solving the variational problem for the "of-first-principles" energy density functional and finding the optimal density, since the so called inverse problem of the energy density functional theory, consisting in reproduction of N-electron wave function from the optimal density, becomes ill-defined (the similar inverse problem arises in crystallography[24]).

The one-to-many nature of the submapping of $\mathcal{N}_N$ onto $\mathcal{L}_N$ should be also taken into account under deriving the familiar virial relations[17]. Following to Fock[25] (see also Gombás[7]), one can obtain these relations as follows. Any nonvanishing number λ induces a scaling transformation on R^3: for any radius vector $\vec{r} \in R^3$ there corresponds, in general, a new vector $\vec{R} = \lambda\vec{r} \in R^3$, which further induces the corresponding transformations on $\mathcal{L}_N$ and $\mathcal{N}_N$:

$$\Psi_\lambda(\vec{r}_1, s_1; \cdots; \vec{r}_N, s_N) \equiv \lambda^{3N/2}\Psi(\lambda\vec{r}_1, s_1; \cdots; \lambda\vec{r}_N, s_N) \tag{23a}$$

and

$$\rho_\lambda(\vec{r}, s) \equiv \lambda^3\rho(\lambda\vec{r}, s) \tag{23b}$$

where $\rho \in \mathcal{N}_N$ and $\Psi \in \mathcal{L}_N$. Evidently, the scaling transformations, defined on $\mathcal{L}_N$ and $\mathcal{N}_N$, are consistent in view of the reduction mapping (5), i.e., $\rho_\lambda(\vec{r}) \equiv \rho_{\Psi_\lambda}(\vec{r}) = \lambda^3\rho_\Psi(\vec{r})$. Notice that the factor λ^3 in (23b) is the Jacobian of scaling transformation to provide a fulfilment of (Dii). A similar term is introduced in (23a). Fock[25] proposed the way of obtaining the virial relations based on scaling-type variations of the optimal density. Since this procedure refers to the energy functional $E[\Psi]$, it suggests implicitly the submapping $\rho \mapsto \Psi$ to be given and fixed. If this submapping is not defined, or in general, does not exist as for the Thomas-Fermi model, the relations are obtained in such a way for a given energy density functional are named as virial-like ones. The example of virial-like relations for the Thomas-Fermi model was discussed by Gombás[7].

The one-to-many mapping of $\mathcal{N}_N$ onto $\mathcal{L}_N$ is also of inportance for investigating the energy density functionals, $\mathcal{E}[\rho(\vec{r})]$ or $\mathcal{E}[\rho_\uparrow(\vec{r}), \rho_\downarrow(\vec{r})]$ given the literature and having the same structure as $E[\Psi]$,

$$\begin{aligned}\mathcal{E}[\rho_\uparrow(\vec{r}), \rho_\downarrow(\vec{r})] = T[\rho_\uparrow(\vec{r}), \rho_\downarrow(\vec{r})] + E_{Coulomb}[\rho(\vec{r})] + \\ E_{xc}[\rho_\uparrow(\vec{r}), \rho_\downarrow(\vec{r})] + \int d^3\vec{r}\rho(\vec{r})v(\vec{r})\end{aligned} \tag{24}$$

where

$$\begin{aligned}U_{ee}[\rho_\uparrow(\vec{r}), \rho_\downarrow(\vec{r})] = E_{Coulomb}[\rho(\vec{r})] + \\ E_{xc}[\rho_\uparrow(\vec{r}), \rho_\downarrow(\vec{r})]\end{aligned} \tag{25}$$

and

$$E_{Coulomb}[\rho(\vec{r})] = \frac{1}{2}\int d^3\vec{r}\rho(\vec{r})v_H(\vec{r}) \tag{26a}$$

$$E_{xc}[\rho_\uparrow, \rho_\downarrow] = \int d^3\vec{r}\sum_s \mathcal{E}^{xc}_s[\rho_\uparrow, \rho_\downarrow]\rho_s(\vec{r}) \ . \tag{26b}$$

$\mathcal{E}^{xc}_s[\rho_\uparrow, \rho_\downarrow]$ is the exchange-correlation energy per particle of the sth spin electronic subsystem. The problem of the rigorous formulation of the inverse problem as the artifact of one-to-many nature of submapping $\rho \mapsto \Psi$ has been studied by some authors[26]. Speaking generally, this problem possesses negative solution for the functionals of the type (24). We refer all theories, based on the energy density functionals of the type (24)-(26) and for which the inverse problem cannot be rigorously formulated, to the class of non-representable ones. It is clear that to this class of theories one can refer those functionals which evaluated for any stable N-electron system with the ground-state energy $E_\circ$, give the minimum

$$\mathcal{E}[\tilde{\rho}_{\circ\uparrow}(\vec{r}), \tilde{\rho}_{\circ\downarrow}(\vec{r})] < E_\circ$$

where $\tilde{\rho}_{os}(\vec{r})$ is the optimal density of sth electrons obtained as a result of applying stationary variational principle to Eq.(24). The another criterion of non-representability of the given theory is based on its adequacy of describing the spurious self-interaction term in $E_{Coulomb}[\rho(\vec{r})]$[27]. One can demonstrate[28] that $\mathcal{E}[\rho_\uparrow, \rho_\downarrow]$ provides the adequate description of self-interaction, or as usually said, is self-interaction-free, if

$$\triangle_{SIC}[\rho_s] \equiv E_{Coulomb}[\rho_s] + E_{xc}[\rho_s, 0] = 0 \tag{27}$$

for s=$\uparrow, \downarrow$. If Eq.(27) is not satisfield, a given theory is non-representable. One can recorrect $\mathcal{E}[\rho_\uparrow, \rho_\downarrow]$ in such a way that the conditions (27) are fulfilled. The modified, or self-interaction corrected energy density functional $\mathcal{E}_{SIC}[\rho_\uparrow, \rho_\downarrow]$ can be constructed as follows[28,29]

$$\mathcal{E}_{SIC}[\rho_\uparrow, \rho_\downarrow] \equiv \mathcal{E}[\rho_\uparrow, \rho_\downarrow] - \sum_s \triangle_{SIC}[\rho_s] \, . \tag{28}$$

One can expect that the new functional (28) is representable Abut is not that case in general since in some cases the kinetic energy functional $T[\rho_\uparrow, \rho_\downarrow]$ appears to be ill-defined and is not purely the kinetic energy [30]. Furthermore, usually the new functional (28) reduces the energy of the given N-electron system, i.e.,

$$\mathcal{E}_{SIC}[\tilde{\tilde{\rho}}_{o\uparrow}, \tilde{\tilde{\rho}}_{o\downarrow}] < \mathcal{E}[\tilde{\rho}_{o\uparrow}, \rho_{o\downarrow}]$$

where $\tilde{\tilde{\rho}}_{os}(\vec{r})$ is the optimal density of sth electrons with respect to the self-interaction corrected functional (28). In particular, the traditional energy density functional $\mathcal{E}_{SIC}[\rho_\uparrow, \rho_\downarrow]$ with the exchange-correlation energy term of the Kohn-Sham type gives the optimal "ground-state" energy for the atoms He, Li, Be, B etc. lower the exact energy E_o (see Table 2 presented by Perdew [29]). By definition, there does not exist any $\mathcal{L}_N$ -representable level which lies below the exact one relative to the decreasing energy. Therefore, such the energy density functional refers to the class of non-representable theories.† The related problems of correctness of $\mathcal{E}_{SIC}[\rho_\uparrow, \rho_\downarrow]$ have been discussed elswhere[29,32]. Any non-representable energy density functional produces a change of the hierarchy system of levels of accuracy by adding non-representable levels, particularly, the lowest non-representable one which describes the artificial "ground-state". Naturally one can use this "ground-state" level as the reference one for evaluating "correlation-like" energy [19]. Therefore, a fulfilment of the self-interaction constraints (27) can be interpreted as the necessary condition for $\mathcal{E}[\rho_\uparrow, \rho_\downarrow]$ to be representable. The inverse statement is incorrect.

One can consider the discussion presented as a certain, rather simplified projection of the whole sixty-year energy density functional theory onto the direction which leads to its rigorous formalism. In fact, the basic points of this discussion are the following:

(Ti) The energy density functional should be rigorously derivable from the energy functional. This requirement provides this density functional to be representable and hence, by definition, to be self-interaction free.

(Tii) The mapping of $\mathcal{N}_N$ onto $\mathcal{L}_N$ is one-to-many.

(Tiii) The energy density functional of the Hohenberg-Kohn type is not unique in the sense that there exist a large number of the representable energy density functionals

†A similar situation has been arised earlier in the N-represetability problem[31].

each of which is well-defined within the appropriate one-to-one branch of the mapping of $\mathcal{N}_N$ onto $\mathcal{L}_N$. One can say that this is a payment of desire for working with the energy functional depending on density rather a wave function. Nevertheless, the Hohenberg-Kohn-type energy density functional is unique in the sense that it is defined on such the branch of the mapping which involves the exact ground state, under assumption that the latter is nondegenerate.

(Tiv) The correct description of the whole set of one-to-one branches of the mapping of $\mathcal{N}_N$ onto $\mathcal{L}_N$ requires such the procedure with the help of which one can "visualize" all densities as elements of $\mathcal{N}_N$. It is clear that the procedure based on scaling transformations is not suitable.

The rigorous density functional formalism, taking into account these "building-up" proposals (Ti)-(Tiv), has been recently proposed [33-35,24]. Its basic idea consists in the following (see(Tiv)). For a fixed direction $\vec{e}[\vec{r}] = \vec{r}/r$ in the three-dimensional space R^3, any pair of densities, $\rho_1(\vec{r})$ and $\rho_2(\vec{r})$, belonging to $\mathcal{N}_N$ can be treated as two continuously-differentiable one-dimensional curves, $\rho_1(r, \vec{e}[\vec{r}])$ and $\rho_2(r, \vec{e}[\vec{r}])$, which can be topologically deformed to each other. Analytically that deformation, or homotopy, represent itself the generalized scaling, or local-scaling transformation. All these transformations form the group whose action on $\mathcal{L}_N$ partitions the latter into the adjacent orbits, on which all one-to-one branches are realized. Such the formalism allows us

(i) to obtain the explicit form of the kinetic and exchange-correlation energy density functionals[34] ;

(ii) to formulate the spin-polarized version of the density functional theory;

(iii) to generalize the devoloped formalism on excited states, particularly, to give the rigorous envelope of the Slater's concept of the transition state[37];

(iv) to come beyond the Born-Oppenheimer approximation;

(v) to formulate rigorously the concept of self-consistent field, originated by Slater[37], and Kohn and Sham[38], which can be viewed as "jumping" from one orbit to the another within a single superorbit.

All these results are summarized in the book[36].

References

[1] L.H. Thomas, The calculation of atomic fields, Proc. Cambridge Phil. Soc. 23: 542 (1974)

[2] E. Fermi, Un metodo statistico per la determinazione di alcune prioretà dell'atome, Rend. Acad. Naz. Lincei 6: 602 (1927)

[3] P. Hohenberg and W. Kohn, Inhomogeneous electron gas, Phys. Rev. 136: B864 (1964)

[4] E.H. Lieb, Variational principle for many-fermion systems, Phys. Rev. Lett. 46: 457 (1981); Ibid. 47: 69(E) (1981)

[5] E.H. Lieb, Density functionals for Coulomb systems, Int. J. Quantum Chem. 24: 243 (1983)

[6] C.F. von Weizsäcker, Zum theorie der kernmassen, Z. Phys. 96: 431 (1935)

[7] P. Gombás, "Die statistische theorie des atoms und ihre anwedungen", Springer, Wien (1949)

[8] J. Katriel and M.R. Nyden, A comparison between hydrogenic and Thomas-Fermi expectation values, J. Chem. Phys. 74: 1221 (1981)

[9] R.G. Parr and L.J. Bartolotti, Some remarks on the density functional theory of few-electron systems, J. Phys. Chem. 87: 2810 (1983)

[10] T.T. Nguyen-Dang, R.F.W. Bader and H.Essén, Some properties of the Lagrange multiplier μ in density funcional theory, Int. J. Quantum Chem. 22: 1049 (1982)

[11] N.H. March, The Thomas-Fermi approximation in quantum mechanics, Adv. Phys. 6: 1 (1957)

[12] V.H. Smith,Jr. and I. Absar, Basic concepts of quantum chemistry for electron density studies, Isr. J. Chem. 16: 87 (1977)

[13] R.F.W. Bader and T.T. Nguyen-Dang, Quantum theory of atoms in molecules - Dalton revisited, in: "Adv. in Quantum Chem.", P.-O. Löwdin, d., Vol. 14, Academic, New York (1981)

[14] R.G. Parr, R.A. Donnelly, M. Levy and W.E. Palke, Electronegativity: the density functional viewpoint, J. Chem. Phys. 68: 3801 (1978)

[15] N.H. March and R.F.W. Bader, Relation between chemical and ionization potentials in atoms, Phys. Lett. 78A: 242 (1980)

[16] E.H. Lieb, Bound on the maximum negative ionization of atoms and molecules, Phys. Rev. A29: 3018 (1984)

[17] S.T. Epstein, "The variation method in quantum chemistry", Academic, New York (1974)

[18] P.-O. Löwdin, Some aspects on the correlation problem and possible extensions of the independent-particle model, in: "Correlation effects in atoms and molecules", R. Lefebvre and C. Moser, ds., Vol. XIV, Adv. Chem. Phys., I. Prigogine and S. Rice, Eds., Wiley-Interscience, London (1969)

[19] J.R. Sabin and S.B. Trickey, On the systematic assessment of correlation effects in local density models, in: "Local density approximations in quantum chemistry and solid state physics", J.P. Dahl and J. Avery, ds., Plenum, New York (1984)

[20] M. Cook and M. Karplus, Electron correlation and density-functional methods, J. Phys. Chem. 91: 31 (1987)

[21] N.H. March and W.H. Young, Variational methods based on the density matrix, Proc. Phys. Soc. 72: 182 (1958)

[22] J.E. Harriman, Orthonormal orbitals for the representation of an arbitary density, Phys. Rev. A24: 680 (1981); A kinetic energy density functional, J. Chem. Phys. 83: 6283 (1985)

[23] E.V. Ludeña, An approximate universal energy functional in density functional theory, J. Chem. Phys. 79: 6174 (1983). G. Zumbach and K. Maschke, New approach to the calculation of density functionals, Phys. Rev. A28: 544 (1983)

[24] E.S. Kryachko, I. Zh. Petkov and M.V. Stoitsov, Method of local-scaling transformations and density-functional theory in quantum chemistry. II, Int. J. Quantum Chem. 32: 467 (1987)

[25] V. Fock, Bemerkung zum virialsatz, Z. Phys. 63: 855 (1930)

[26] O. Gunnarsson, M. Jonson and B.I. Lundqvist, Descriptions of exhange and correlation effects in inhomogeneous electron systems, Phys. Rev. B20: 3136 (1979). O. Gunnarsson and R.O. Jones, Extensions of the LSD approximation in density functional calculations, J. Chem. Phys. 72: 5357 (1980); Density functional calculations for atoms, molecules and clusters, Phys. Scr. 21:394 (1980). G.E.W. Bauer, General operator ground-state expectation values in the Hohenberg-Kohn-Sham density-functional formalism, Phys. Rev. B27: 5912 (1983); Evidence for a fundamental inadequacy of the gradient expansion of the exchange-correlation energy functional in the Hohenberg-Kohn-Sham density-functional theory from Compton scattering experiments, Ibid. B30: 1010 (1984). S. Valone, A one-to-one mapping between one-particle densities and some n-particle ensembles, J. Chem. Phys. 73: 4653 (1980)

[27] C.A. Coulson and C.S. Sharma, Correction for self-interaction in the Thomas-Fermi potential with application to f electrons in atoms, Proc. Phys. Soc. 79: 920 (1962)

[28] J.P. Perdew and A. Zunger, Self-interaction correction to density-functional approximations for many-electron systems, Phys. Rev. B23: 5048 (1981)

[29] J.P. Perdew, Self-interaction correction, in: "Local density approximations in quantum chemistry and solid state physics", J.P. Dahl and J. Avery, eds., Plenum, New York (1984)

[30] J. Callaway and N.H. March, Density functional methods: theory and approximations, in: "Solid State Physics", vol. 38, H. Ehrenreich and D. Turnbull, eds., Academic, New York (1984). F.W. Averill and G.S. Painter, Virial theorem in the density-functional formalism: forces in H_2, Phys. Rev. B24: 6795 (1981). A.R. Williams and U. von Barth, Applications of density functional theory to atoms, molecules, and solids, in: "Theory of the inhomogeneous electron gas", S. Lundqvist and N.H. March, eds., Plenum, New York (1983)

[31] A.J. Coleman, Structure of fermion density matrices, Rev. Mod. Phys. 35: 668 (1963)

[32] J.G. Harrison, An improved self-interaction-corrected local spin density functional for atoms, J. Chem. Phys. 78: 4562 (1983). R.A. Heaton, J.G. Harrison and C.C. Lin, Self-interaction correction for density-functional theory of electronic energy bands of solids, Phys. Rev. B28: 5992 (1983)

[33] I.Zh. Petkov, M.V. Stoitsov and E.S. Kryachko, Method of local-scaling transformations and density-functional theory in quantum chemistry, Int. J. Quantum Chem. 29: 149 (1986)

[34] E.S. Kryachko, I.Zh. Petkov and M.V. Stoitsov, Method of local-scaling transformations and density-functional theory in quantum chemistry. III, Int. J. Quantum Chem. 32: 473 (1987)

[35] E.S. Kryachko and E.V. Ludeña, Many-electron energy density functional theory: point transformations and one-electron densities, Phys. Rev. A35: 957 (1987)

[36] E.S. Kryachko and E.V. Ludeña, "Energy density functional theory in quantum chemistry", Reidel, Dordrecht (1987)

[37] J.C. Slater, "Quantum theory of molecules and solids. Vol. 4. The self-consistent field for molecules and solids", McGraw-Hill, New York (1974)

[38] W. Kohn and L.J. Sham, Self-consistent equations including exchange and correlation effects, Phys. Rev. 140: A1133 (1965)

STRUCTURES OF SMALL METAL CLUSTERS

M. Manninen

Laboratory of Physics
Helsinki University of Technology
SF-02150 Espoo, Finland

1. INTRODUCTION

Ultrafine particles and small atomic clusters have been a subject of extensive study in the last years[1-3]. The development of experimental techniques to produce clusters of desired size has inspired theorists of both solid state physics and atomic and molecular physics to calculate the properties of smallest clusters. Practically all theoretical methods for calculating the electronic structure have been applied, from the simple jellium model of the solid state physics[4-7] to the highly accurate multiconfiguration Hartree-Fock calculations of the quantum chemistry[8-9]. All these methods have shown to be able to give some new information for better understanding of the properties of small clusters. The chemistry and physics of the clusters depend on the atoms it is build up. Thus the structures of metal clusters[8-11] are different of those of semiconductors[12] or insulators[13]. The metallic bond is due to free conduction electrons, whereas the semiconductors are bound by directed covalent bonds. Still another group of clusters are the rare gas clusters, where the bonding is due to weak and pairwise polarization potentials and the most stable clusters have icosahedral geometries[14-15]. Generally the geometries of the most stable clusters, as well as the magic numbers, are very different for clusters bound in different ways. In this paper we concentrate on metal clusters, or more specifically on alkali or noble metal clusters of less than 100 atoms.

The experiments have shown[16-17], that alkali metal clusters have magic numbers 2, 8, 20, 40, 58, 92; i.e. the abundances of these clusters are clearly higher than those of the neighboring clusters. While these results are obtained by growing the clusters in a rare gas environment, similar magic numbers have been observed for noble metals, where ionized clusters are formed by sputtering from a plane surface[18]. The magic numbers for the positive and negative ions are the numbers of *valence electrons* in these clusters. This indicates that in the simple metal clusters the properties of the ions or their number are not dominating in determining the stability of the cluster. This is in accordance with the idea of the simple jellium model, where the ions merely provide a positive background charge which holds together the valence electrons. If the electrons (noninteracting or interacting) are assumed to move in a spherical potential well, the resulting magic numbers coincide with the observed ones. Better calculations have only been performed for so small clusters that the magic numbers bigger than 8 could not have been studied. For these small clusters the calculated geometries are fascinating, very different from bulk lattice geometry, and far from being spherical (except for the magic cluster of 8 atoms). Thus the jellium model has been sometimes criticized in the view of the results of the *ab initio* calculations for smallest clusters. However, the jellium model explains the magic numbers, and as will be discussed later, gives probably more insight to the origin of the magic numbers in simple metal clusters than more complicated theories. Also, it has been shown that the geometries of the smallest alkali metal clusters can be qualitatively understood in terms of the jellium model[11] and that the geometries of the magic clusters may even be predicted using jellium based models[7].

The standard way to perform *ab initio* calculations for small clusters is to expand the single-electron wave functions in terms of atomic and molecular orbitals. Analytic Slater orbitals are usually used and in Hartree-Fock calculations these are often expanded in terms of Gaussian orbitals. While these methods are accurate and widely used they suffer of the use of a basis set. It is very difficult to choose the proper basis functions and the size of the basis set. In the present work we introduce an other way to calculate the single electron wave functions, i.e. direct numerical integration in the three-dimensional space. Since no advantage is taken of the symmetry the method is well suited for calculating the geometries of metal clusters. This method is, however, not without problems either: it requires a smooth pseudopotential and is limited in accuracy. Nevertheless, it is very effective in determining the geometries of small alkali metal clusters, and has the advantage that the geometry can be relaxed at the same time as the self-consistency of the electronic structure.

The stability against charging is an experimentally important property of the clusters. When more and more electrons are stripped out from the cluster, the Coulomb repulsion of the positive charge distribution eventually wins the binding energy of the cluster which then explodes spontaneously. This fragmentation is a complicated dynamical process which has been studied using molecular dynamics in the case of rare gas clusters which can be described with simple pair potentials[19]. In the case of metals this can not be done and only the static properties of the fragmentation have been studied. It has been found that for alkali metals the simple jellium model gives correctly the energetically most preferable fragmentation channels[20]. This suggests that the jellium model can be used to study qualitatively the fragmentation of larger clusters where other methods have not been used. The results show (the obvious thing) that the most preferable fragmentation channels are those where the products are magic clusters. However the distribution of fragments from magic clusters are different from those where the parent is not magic. This is a result of the fact that the magic clusters have less possible fragmentation channels than other clusters.

The plan of this paper is as follows. In Section 2 the basic equations of the density functional Kohn-Sham method are reminded and the jellium model for clusters is described. In Section 3 pseudopotential corrections are introduced to the mere jellium model for better understanding why the jellium model predicts the correct magic numbers for the alkali and noble metals. In Section 4 the relaxation method for solving the density functional equations is introduced and the results for geometries of pseudopotential clusters are given. Section 5 discusses Coulomb explosion in terms of the jellium model. The conclusions are given in Section 6.

2. JELLIUM MODEL

In the jellium model the electronic structure is calculated using the density functional Kohn-Sham method and local density approximation (LDA) for the exchange-correlation energy. The equations to be solved self-consistently are

$$n(\mathbf{r}) = \sum_i |\psi_i(\mathbf{r})|^2 \tag{1}$$

$$-\frac{1}{2}\nabla^2\psi_i(\mathbf{r}) + V_{eff}(\mathbf{r})\psi_i(\mathbf{r}) = \varepsilon_i\psi_i(\mathbf{r}) \tag{2}$$

$$V_{eff}(\mathbf{r}) = \phi(\mathbf{r}) + \mu_{xc}(n(\mathbf{r})) \tag{3}$$

$$\phi(\mathbf{r}) = -4\pi(n(\mathbf{r}) - n_+(\mathbf{r})) \tag{4}$$

where n is the electron density, ψ_i one-electron wave function, ε_i one-electron eigenvalue, ϕ the electrostatic potential, and n_+ the density of all positive charges in the system (in the case of jellium it is the positive background charge, in the case of pseudopotential cluster it is the charge corresponding the pseudopotentials). The sum in Eq.(1) goes over the occupied states, in the ground state over the N lowest energy states (N is the number of electrons in the system). For the local exchange-correlation potential the parametrized formula of Vosko *et al*[21] was used (this formula is based on the data of Ceperley and Alder[22]). Once the positive charge density distribution n_+ and the number of electrons N are fixed, the self-consistent solution of Eqs.(1)-(4) gives the ground state electronic structure of the system.

In the jellium model the positive charge density distribution is chosen to be a homogeneously charged sphere:

$$n_+(\mathbf{r}) = n_0\,\theta(R - r) \tag{5}$$

where n_0 is the density of the jellium, R the cluster radius, and θ a step function. The cluster radius is related to the number of electrons as

$$R = r_s \sqrt[3]{N} \tag{6}$$

where r_s is the usual electron density parameter:

$$n_0 = \frac{4\pi r_s^3}{3} \tag{7}$$

The electron density n_0 is set equal to the valence electron density of the metal and defines thus the metal in question.

The jellium model does not have any free parameters. The solution of the Eqs. (1)-(6) gives the properties of the clusters of jellium. The single electron energy eigenvalues are very similar to those of an independent particle in a spherical potential well. This leads to a shell model where the states are filled in the order: 1s, 1p, 1d, 2s, 1f, 2p, 1g, 2d, 1h, 3s, etc, and the corresponding magic numbers are 2, 8, 18, 20, 34, 40, 58, 68, 90, 92 etc. Since some levels have nearly an accidental degeneracy, only 2, 8, 20, 40, 58, and 92 are seen experimentally as dominant magic numbers in alkali and noble metal clusters. In the total energy the magic numbers are seen as discontinuities in the derivative of the total energy as a function of the cluster size. This means a jump in the chemical potential. Thus plotting the second derivative of the total energy of the cluster as a function of the number of atoms in the cluster the magic numbers are seen as peaks. This is illustrated in Fig. 1. The density functional formalism of equations (1)-(4) can be generalized to take into account the spin. In the spin-density functional formalism the single electron states are going to be filled according to the Hund rules. This leads to the result that also half-filled shells would show as magic numbers, as shown by the dashed lines in Fig. 1. In real metal clusters the crystal fields are going to reduce the large degeneracy of the angular quantum number and the effect of the Hund rules and spin disappears in determining the magic numbers.

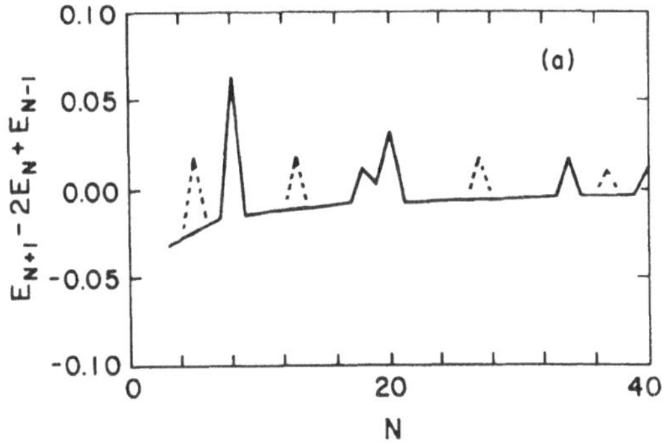

Fig. 1. *Second derivative of the total energy as a function of the cluster size for jellium clusters. The magic numbers show as peaks. The dashed line gives the result of the spin-density functional formalism showing the spurious magic numbers at half-filled shells.*

The results of the jellium model are qualitatively the same for all densities and thus explain why all alkali and noble metals have the same magic numbers. In the spherical model it is easy to calculate also other electronic properties of the cluster. In Fig. 2 the electronic polarizability of sodium clusters is shown as a function of the cluster size[6]. The result of the mere jellium model is qualitatively in a very good agreement with the experimental results. Quantitatively the theory gives about 20% too low polarizabilities, but this can be mainly due to the use of the local density approximation for the exchange-correlation energy[6]. This gives support in applying the jellium model also for calculating electronic excitations (dynamical polarizability)[5].

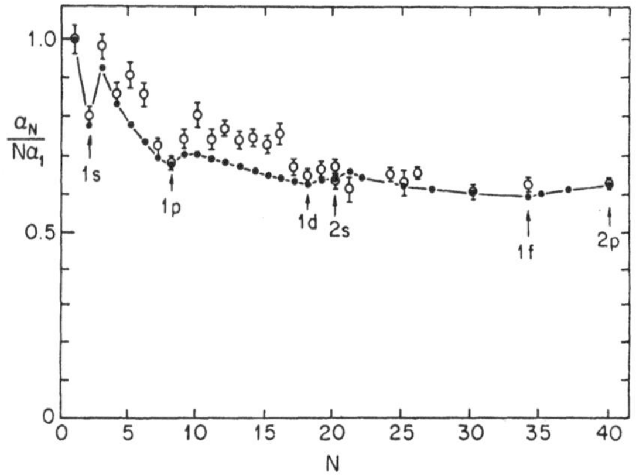

Fig. 2. *Static electronic polarizability as a function of the cluster size (ref 6). The open circles are experimental results (ref. 23) and the black dots the results of the jellium model.*

3. CORRECTIONS TO THE JELLIUM MODEL

The result that the jellium model predicts the correct magic numbers for simple metal clusters suggests that the variation of the total energy as a function of the cluster size is governed by the single electron eigenvalues which are determined in a nearly spherical potential. This means that the effective potential of the Kohn-sham equations has to be smooth and that the other contributions of the total energy are not as sensitive to the cluster size and geometry. For getting a first estimate of the energy variation in real clusters the ion pseudopotentials can be taken into account using perturbation theory in the similar way as the jellium model has been earlier used for studying surfaces[24] and vacancies[25] in simple metals. The expression for the total energy in the density functional theory is

$$E_{TOT} = T[n] + E_{xc}[n] + \frac{1}{2}\int d\mathbf{r}\int d\mathbf{r}' \frac{n(\mathbf{r})n(\mathbf{r}')}{|\mathbf{r}-\mathbf{r}'|}$$

$$+ \frac{1}{2}\sum_{i,j} \frac{1}{|\mathbf{R}_i-\mathbf{R}_j|} - \sum_i \int d\mathbf{r} V_{ps}(\mathbf{r}-\mathbf{R}_i)n(\mathbf{r}) \qquad (8)$$

Here T and E_{xc} are the kinetic and exchange-correlation energy functionals, $\mathbf{R}_i$ are the ion sites in the cluster, and V_{ps} is the electron-ion pseudopotential. The three first terms in Eq.(8) depend only on the electron density distribution whereas the two last terms are the self-interaction of the positive charges and the electrostatic interaction between the electrons and the ions. If now the electron density is calculated in the spherical jellium model and the ion pseudopotentials are treated by first order perturbation theory, the total energy can be written in the form[26]

$$E_{TOT} = E_{jellium} + E_M + \text{[small correction terms]} \qquad (9)$$

where $E_{jellium}$ is the total energy of the pure jellium model and E_M so-called Madelung energy:

$$E_M = \frac{1}{2}\int d\mathbf{r}\int d\mathbf{r}' \frac{n_+(\mathbf{r})n_+(\mathbf{r}')}{|\mathbf{r}-\mathbf{r}'|} + \frac{1}{2}\sum_{i,j} \frac{1}{|\mathbf{R}_i-\mathbf{R}_j|} + \sum_i \int d\mathbf{r} \frac{n_+(\mathbf{r})}{|\mathbf{r}-\mathbf{R}_i|} \qquad (10)$$

The idea to write the total energy in this form is that the Madelung energy depends only on the geometry (ion sites) of the cluster but not on the electronic structure or the form of the pseudopotential. The correction terms are smaller than the jellium energy and the Madelung energy and especially their dependence on the structure of the cluster is expected to be small[26]. If the correction terms are neglected, the geometry of the cluster can be solved simply by minimizing the Madelung energy with respect to the ion sites $\mathbf{R}_i$. This gives two important results: (i) the optimal geometries of the clusters are very different of any bulk lattice structure and (ii) the variation of the structural dependence of the total energy is about of the same order as the variation of the electronic part ($E_{jellium}$) as shown in Fig. 3. The inclusion of the correction terms do not change the geometries of the clusters, they change the total energies of the clusters slightly, but the variation of the structure dependent part of the total energy remains to be as large as that in Fig. 3. The spherical jellium model with pseudopotential corrections can thus not explain why the magic numbers are so clearly those of the pure jellium model. This means that in calculating the total energy the electronic response to the ions has to be taken into account (self-consistency). In the language of the perturbation theory it is clear that the total energy is reduced more in those cases where the electrostatic energy is large (the electron distribution of the spherical jellium model is a bad approximation). This will reduce the variation of the structural part of the total energy whereas it will not affect essentially to the electronic part as long as the single-electron levels remain nearly degenerate. In the next section it is going to be shown that at least in the case of smallest clusters this is indeed the case.

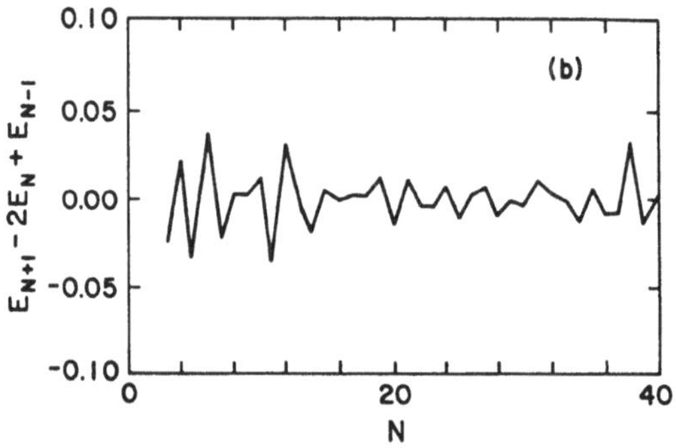

Fig. 3. *The second derivative of the madelung energy for spherical clusters with point ions in their optimal configuration (ref. 26). The variation of this curve should be compared with that of Fig.1.*

4. PSEUDOPOTENTIAL CLUSTERS

Several calculations have been reported on the electronic structures and geometries of smallest alkali metal clusters, containing from 1 to 7 atoms[8-10]. These calculations have been based on Hartree-Fock or LDA methods and the electronic structure is obtained using LCAO-MO-techniques. The geometry of the cluster is then obtained by minimizing the total energy. In the present report a different approach is taken in solving the Kohn-Sham-equations (1)-(4). Instead of using any basis functions, the Schrödinger equation and the Poisson equation are solved numerically in a three-dimensional grid using relaxation methods (for details see Ref. 11). The advantages of the relaxation methods are: (i) no problems in choosing the basis set, (ii) the self-consistency of the Kohn-Sham-equations and the optimal ion sites can be obtained simultaneously, (iii) the convergence is very fast, and (iv) all geometries are essentially equally easy to calculate. The disadvantage is that only very smooth pseudopotentials can be used for describing the ions. In this report results for clusters for up to eight electrons are reviewed.

The model pseudopotential chosen is as smooth as possible:

$$V_{ps}(r) = \begin{cases} -\dfrac{Z}{r} & \text{if } r > r_c \\ -\dfrac{3Z}{2r_c^3}\left(r_c^2 - \dfrac{r^2}{3}\right) & \text{if } r < r_c \end{cases} \tag{11}$$

where Z is the valency of the ion and r_c a cut-off radius. This potential is the electrostatic potential of a homogeneously charged sphere. The radius r_c corresponds roughly to the electron density parameter r_s of the metal in question. The results for monovalent clusters are qualitatively similar for all values of r_c between 2 and 6, corresponding to the density range of alkali and noble metals[11]. To further test the sensitivity of the properties of the clusters on the pseudopotentials calculations have been made where one of the monovalent pseudopotentials is replaced by a divalent pseudopotential. Results for the geometries of clusters containing 2 to 8 valence electrons are shown in Fig. 4. It is seen that clusters of up to 6 electrons are nearly planar, whereas the cluster with eight electrons is in all cases nearly spherical in shape. These structures can be understood by studying the filling of the single electron energy levels in the spherical jellium model. The first two electrons fill the spherical 1s-state. The next 6 electrons fill the p-shell. Due to the interplay between the geometry and the filling of this shell it is energetically favorable to fill the p-shell in the order, say p_x, p_y, p_z. This means that the electron density distribution of clusters with 3 and 4 electrons is cigar-shaped whereas for the clusters with 5 and 6 electrons it is disk-shaped. The minimization of the electrostatic energy requires the ions distribution to take the same shape as the electron distribution. In the magic eight electron cluster the electron shell is full and the electron density distribution is spherical.

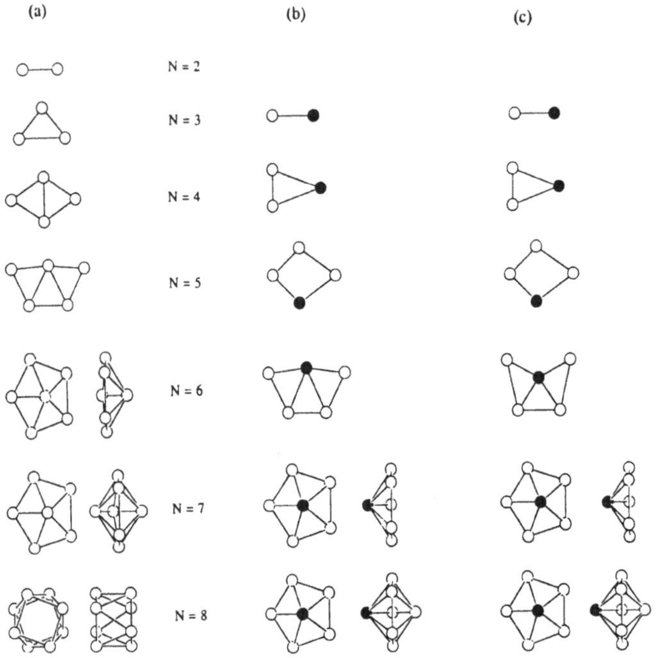

Fig. 4. *Geometries of the pseudopotential clusters. Open circles are monovalent pseudopotentials (r_c=4.0), filled circles divalent atoms (r_c=4.0 in (b) and r_c=3.0 in (c)). N is the number of electrons in the cluster.*

The simple interpretation of the cluster geometries implies that the electronic structure of the pseudopotential clusters is very similar to that of the spherical jellium. This is indeed the case as seen in Fig. 5 where the eigenvalues of the occupied energy levels are given as a function of the cluster size. The p-like states are nearly degenerate for all clusters and clearly separated from the doubly occupied s-like state. This means that the crystal field splitting of the levels is small compared to the level spacing of the jellium model. This is the reason for the success of the jellium model in explaining the magic numbers.

The calculated total energies and the average electronic polarizabilities of the pseudopotential model agree very well with the corresponding results of the mere jellium model[11]. Nevertheless, one should note that in calculating magnetic properties the effect of the crystal field splitting becomes important. In the case of the pseudopotential clusters it is seen for example that the Hund rules of the spherical model are not any more valid and the spurious magic numbers of half-filled shells disappear in agreement with the experimental result.

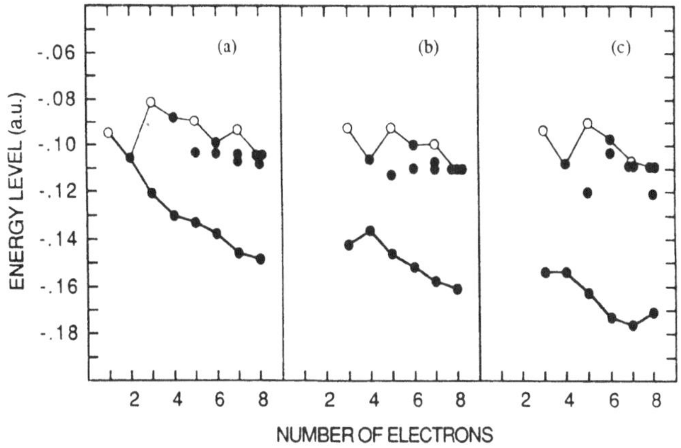

Fig. 5. *Energy eigenvalues of the occupiued states as a function of the cluster size for clusters shown in Fig. 4. Open circles denote levels with only one electron.*

5. COULOMB EXPLOSION OF THE JELLIUM CLUSTERS

Experimentally the clusters are often observed after ionization. If several electrons are stripped out from the cluster it may become energetically unstable due to the Coulomb repulsion. The possibility of fragmentation and the resulting fragmentation channels are then of great interest also in the experimental point of view. While in the rare gas clusters there seems to be a simple critical size which determines the stability of the charged cluster[19] in metals the situation seems to be more complicated. Again the idea of the shell model of the jellium can be used as a starting point. Of charged clusters the most stable are those where the number of electrons corresponds to a magic number. Then e.g. the positively charged clusters (charge +1) have magic numbers (number of atoms) 3, 9, 21, 41, etc. This is seen experimentally in charged noble metal clusters[18].

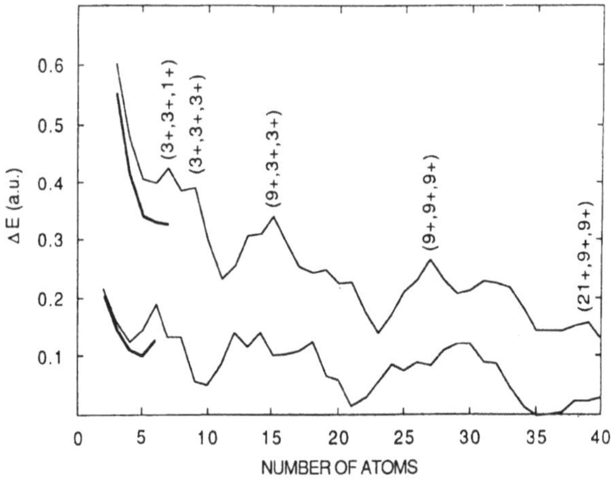

Fig. 6. *The energy released in fragmentation of triply (upper curve) and doubly charged (lower curve) jellium clusters (thin lines) as a function of the cluster size. Thick lines show the results of Hartree-Fock calculations (ref 20). The fragmentation products are shown in pharentheses for the maxima of the curve.*

The energy release in fragmentation of doubly and triply charged jellium clusters is shown in Fig. 6. Two obvious things can be seen: The energy release is at a minimum when the parent cluster is magic and at maximum when the daughter clusters are magic. The results are for jellium corresponding to the density of lithium. The absolute values of the energy are not very useful for determining if the fragmentation is possible, because the jellium can not describe properly the cohesion of lithium metal. Similar results have been reported for sodium by Iñiguez *et al*[27]. In sodium the jellium results might have also quantitative meaning.

More confidence for the results of the jellium model gives a comparison to *ab initio* all electron Hartree-Fock results. Rao *et al*[20] have shown that for the smallest clusters where the Hartree-Fock calculations were made the general trend of the energy release (see Fig. 6) and the energetically most preferred fragmentation channels were in perfect agreement with the results of the jellium model.

In addition to the energetically preferred fragmentation channel described in Fig. 6 there exists a large number of other possibilities. To demonstrate the richness of possible daughter clusters we show in Fig. 7 the distribution of fragmentation products of all channels where the fragmentation is energetically allowed (all channels are assumed to be as probable). Magic (37 atoms) and nonmagic (30 atoms) clusters lead to very different distributions. Figure 7 only shows the daughters of the primary fragmentation: doubly charged daughters can further fragment to smaller clusters.

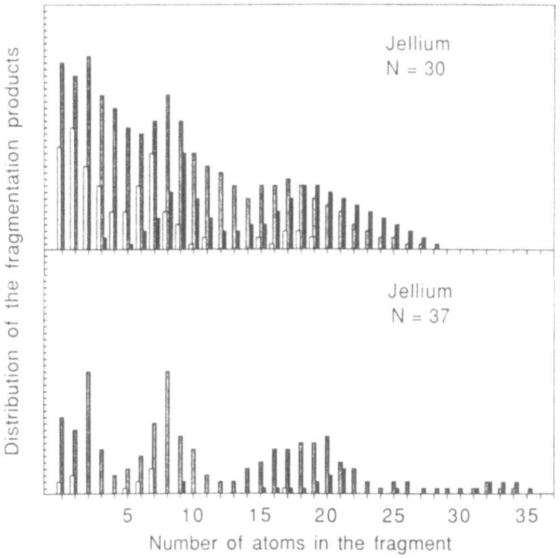

Fig. 7. *Distribution of daughter clusters in fragmentation of triply charged 30 and 37 atom clusters. All spontaneous fragmentation channels are assumed to be as probable. Open bars indicate neutral daughters, shaded bars singly charged and black bars doubly charged daughters.*

6. CONCLUSIONS

Electronic properties of alkali and noble metal clusters are dominated by a shell structure of the valence electrons which is similar to that of a spherical jellium model. The crystal field splitting is small compared to the spacing between the energy levels of different shells. This shell structure results to interesting geometries of small clusters and predicts that clusters up to hundreds of atoms may resemble more amorphous material than a piece of a crystal.

REFERENCES

1. T. D. Märk and A. W. Castleman, Jr., Adv. in Atomic and Molecular Phys. **20**, 65 (1985).
2. W. P. Halperin, Rev. Mod. Phys. **58**, 533 (1986).
3. R. Kubo, A. Kawabata, S. Kobayashi, Ann. Rev. Mater. Sci. **14**, 49 (1984).
4. W. Ekardt, Phys. Rev. Lett. **52**, 1925 (1984).
5. M. J. Puska, R. M. Nieminen, and M. Manninen, Phys. Rev. B. **31**, 3486 (1985).
6. M. Manninen, R. M. Nieminen, and M. J. Puska, Phys. Rev. B **33**, 4297 (1986).
7. M. Manninen, Solid State Commun. **59**, 281 (1986).
8. B. K. Rao and P. Jena, Phys. Rev. B **32**, 2058 (1985).
9. T. H. Upton, Phys. Rev. Lett. **56**, 2168 (1986).
10. J. Martins, J. Buttet, and R. Car, Phys. Rev. B **31**, 1804 (1985).
11. M. Manninen, Phys. Rev. B **34**, 6886 (1986).
12. D. Tománek and M. A. Schlüter, Phys. Rev. Lett. **56**, 1055 (1986).
13. T. P. Martin, Physica **127B**, 214 (1984).
14. J. G. Allpress, J. V. Sanders, Aust. J. Phys. **23**, 23 (1970).
15. O. Echt, K. Sattler, E. Regnagel, Phys. Rev. Lett. **47**, 1121 (1981).
16. W. D. Knight, K. Clemenger, W. A. de Heer, W. A. Saunders, M. Y. Chou, and M. Cohen, Phys. Rev. Lett. **52**, 2141 (1984).
17. W. D. Knight, W. A. de Heer, K. Clemenger, and W. A. Saunders, Solid State Commun. **53**, 445 (1985).
18. I. Katakuse, T. Ichihara, Y. Fujita, T. Matsuo, T. Sakurai, and H. Matsuda, Int J. Mass Spect. and Ion Processes **67**, 229 (1985).
19. J. G. Gay abd B. J. Berne, Phys. Rev. Lett. **49**, 194 (1982).
20. B. K. Rao, P. Jena, M. Manninen, and R. M. Nieminen, Phys. Rev. Lett. **58**, 1188 (1987).
21. S. H. Vosko, L. Wilk, and M. Nusair, Can. J. Phys. **58**, 1200 (1980).
22. D. M. Caperley and B. J. Alder, Phys. Rev. Lett **45**, 566 (1980).
23. W. D. Knight, K. Clemenger, W. A. de Haar, and W. A. Saunders, Phys. Rev. B **31**, 2539 (1985).
24. N. D. Lang and W. Kohn, Phys. Rev. B **1**, 4555 (1970).
25. M. Manninen and R. M. Nieminen, J. Phys. F **8**, 2243 (1978).
26. M. Manninen, Silid State Commun. **59**, 281 (1986).
27. M. P. Iñigues, J. A. Alonso, M. A. Aller, and L. C. Báldas, Phys. Rev. B **34**, 2152 (1986).

GENERALIZED GRADIENT APPROXIMATIONS FOR EXCHANGE AND CORRELATION: NUMERICAL TESTS AND PROSPECTS

John P. Perdew[†], Manoj K. Harbola[‡], and Viraht Sahni[‡]

[†]Department of Physics
Tulane University
New Orleans, LA
U.S.A. 70118

[‡]Department of Physics
Brooklyn College, CUNY
Brooklyn, NY
U.S.A. 11210

ABSTRACT

In the widely-used local spin density (LSD) approximation, the exchange-correlation energy of a many-electron system is approximated as $\int d^3r\ e(n_\uparrow,n_\downarrow)$. Recently, generalized gradient approximations of the form $\int d^3r\ f(n_\uparrow,n_\downarrow,\nabla n_\uparrow,\nabla n_\downarrow)$ have been proposed by Langreth, Mehl and Hu (LM) and by Perdew and Wang (GGA). Results of numerical tests of LSD, LM and GGA for atoms, molecules and jellium surfaces are presented here. Apparently LM is best for atomic ionization energies and jellium surface energies, while GGA is best for atomic total energies, dimer dissociation energies, and total and spin-magnetization densities deep inside atoms. Prospects are assessed for further refinements of the density functionals, leading to higher numerical accuracy.

INTRODUCTION

The ground-state energies and self-consistent structures of atoms, molecules and solids are routinely calculated within the Kohn-Sham density functional theory[1]. Many-body effects are incorporated via the exchange-correlation energy, which is usually treated within the local spin density (LSD) approximation[1]:

$$E_{xc}^{LSD}[n_\uparrow,n_\downarrow] = \int d^3r\ e_{xc}(n_\uparrow,n_\downarrow) \quad , \tag{1}$$

where $n_\sigma(\underset{\sim}{r})$ is the density of electrons of spin σ, and $e_{xc}(n_\uparrow,n_\downarrow)$ is the exchange-correlation energy per unit volume for an electron gas with uniform spin densities $n_\uparrow$ and $n_\downarrow$.

The LSD approximation is exact in the slowly-varying limit, where $n_\uparrow(\underset{\sim}{r})$ and $n_\downarrow(\underset{\sim}{r})$ vary slowly over space. The leading correction in this limit is the second-order density-gradient term[1-4]. The gradient expansion approximation (GEA) retains this correction:

$$E_{xc}^{GEA}[n_\uparrow,n_\downarrow] = E_{xc}^{LSD}[n_\uparrow,n_\downarrow] + \int d^3r \left\{ C_{\uparrow\uparrow}(n_\uparrow,n_\downarrow) \frac{|\nabla n_\uparrow|^2}{n_\uparrow^{4/3}} + C_{\downarrow\downarrow}(n_\uparrow,n_\downarrow) \frac{|\nabla n_\downarrow|^2}{n_\downarrow^{4/3}} + C_{\uparrow\downarrow}(n_\uparrow,n_\downarrow) \frac{\nabla n_\uparrow \cdot \nabla n_\downarrow}{(n_\uparrow n_\downarrow)^{2/3}} \right\} . \quad (2)$$

When applied to real systems of interest, LSD is often a decent first approximation, but the correction provided by GEA is usually disappointing or even disastrous. Real systems are not even close to the slowly-varying limit. This fact was realized by Ma and Brueckner[3] (MB), who first calculated the gradient coefficient for the correlation energy of a spin-unpolarized system. They also proposed the first generalized gradient approximation

$$E_C^{MB}[n] = \int d^3r \; e_C(n,|\nabla n|) , \quad (3)$$

which attempts to mimic a partial summation of the gradient series to all orders in $|\nabla n|$. Here $n = n_\uparrow + n_\downarrow$.

Gunnarsson and Lundqvist[5] have argued that LSD works decently, even beyond its formal domain of validity, because the LSD approximation for the density of the exchange hole surrounding an electron correctly integrates to -1, while the LSD approximation for the density of the correlation hole correctly integrates to 0.

Langreth and Perdew[6] showed that the GEA approximation for the density of the correlation hole <u>fails</u> to integrate correctly to 0. They Fourier-analyzed this hole and argued that the contribution from wavevectors $\underset{\sim}{K} \lesssim f|\nabla n|/n$ is spurious and should be cut off (set to zero). Subsequently, Langreth and Mehl[7] (LM) constructed an analytic functional for this generalized gradient approximation, which Hu and Langreth[8] extended to spin-polarized systems. The cut-off parameter $f = 0.15$ will be employed in all LM calculations to be reported below.

Perdew and Wang[9-12] analyzed the exchange hole in real space. The exact exchange hole density is everywhere negative and integrates to -1, but the GEA exchange hole contains an undamped long-range sinusoidal oscillation which integrates to no definite value. Imposition of the properties of the exact exchange hole via real-space cutoffs implies a generalized gradient approximation (GGA) for the exchange energy[10]:

$$E_x^{GGA}[n_\uparrow,n_\downarrow] = \frac{1}{2}E_x^{GGA}[2n_\uparrow] + \frac{1}{2}E_x^{GGA}[2n_\downarrow] \ , \tag{4}$$

where

$$E_x^{GGA}[n] = A_x \int d^3r\ n^{4/3}F(s) \ , \tag{5}$$

$$s = |\nabla n|/2k_F n \ , \tag{6}$$

$$F(s) = (1 + 1.296s^2 + 14s^4 + 0.2s^6)^{1/15} \ , \tag{7}$$

$A_x = -0.73856$ a.u., and $k_F = (3\pi^2 n)^{1/3}$. This GGA density functional reduces to GEA for the exchange energy[2,12] in the slowly-varying limit ($s<<1$), but predicts exchange energies for atomic systems that are much more accurate than those of LSD or GEA.

Alternative generalized gradient approximations for the exchange energy have been proposed: Becke[13] and De Pristo and Kress[14] fit F(s) to exact exchange energies of atoms. Vosko and Macdonald[15] fit F(s) to exact exchange energy-densities, and thereby obtained a significant improvement over LSD in the self-consistent electron density. Ghosh and Parr[16] combined a Gaussian approximation for the Wigner distribution function with the gradient expansion of the kinetic energy-density.

For the correlation energy, Perdew[17] constructed a new functional (GGA), based upon two modifications of the LM correlation functional: (1) The most natural separation between exchange and correlation was made. (2) Many-body effects beyond the random phase approximation (RPA) were incorporated. Accurate parametrizations of the LSD integrand[18] $e_{xc}(n_\uparrow,n_\downarrow)$ and of the gradient coefficient[19] C(n) are known beyond RPA. Thus

$$E_c^{GGA}[n_\uparrow,n_\downarrow] = \int d^3r\ e_c(n_\uparrow,n_\downarrow) + \int d^3r\ d^{-1}\ e^{-\Phi}\ C(n)\ |\nabla n|^2/n^{4/3} \ , \tag{8}$$

where

$$\Phi = 1.745\ \tilde{f}\ \frac{C(\infty)}{C(n)}\ \frac{|\nabla n|^2}{n^{7/6}} \ , \tag{9}$$

$$d = 2^{1/3}\left[\left(\frac{1+\zeta}{2}\right)^{5/3} + \left(\frac{1-\zeta}{2}\right)^{5/3}\right]^{1/2} . \tag{10}$$

$\zeta = (n_\uparrow - n_\downarrow)/n$, and $\tilde{f} = 0.11$. All equations have been written in atomic units (energies in hartrees, distances in bohrs). An alternative generalized gradient approximation for the correlation energy, based upon the Colle-Salvetti approach of quantum chemistry, has been suggested by Lee, Yang and Parr.[20]

Additional numerical tests and applications of LM or GGA have already been reported.[21-26] The aim of the present work is to compare LSD, LM, GGA and exact properties of atoms, molecules and jellium surfaces, in order to assess the progress that has been made and the prospects for further improvements in the density functionals.

Surfaces excepted, all calculations reported here are fully self-consistent. Formulas for the functional derivatives

$$v_{xc}^{\sigma}(\underset{\sim}{r}) = \delta E_{xc}/\delta n_{\sigma}(\underset{\sim}{r}) \ , \tag{11}$$

which serve as spin-dependent exchange-correlation potentials, have been reported elsewhere[8,10,17]. In Kohn-Sham theory[1], the spin densities are constructed as they would be in a spin-unrestricted Hartree-Fock calculation, but with the nonlocal Fock operator replaced by the local potential (multiplication operator) of Eq.(11). The total energies are similarly constructed.

Total energies of atoms in the exchange-only approximation ($E_c=0$) are reported in Table 1. The exact energies for comparison are those of the Optimized Potential Model of Talman *et al.*[27], which are marginally higher than Hartree-Fock energies because the optimized potential is constrained to be local. The LSD total energies are not sufficiently negative. LM corrects most of this error, and GGA is even better. The GGA exchange energies (not shown here) are in error by less than 1%[10].

Correlation energies of atoms are reported in Table 2. The exact correlation energies for comparison are the semi-empirical values of Veillard and Clementi,[28] with the sign of the Lamb-shift contribution corrected.[29] The LSD correlation energies are too negative by almost a factor of two, while the GEA correlation energies (not shown here) actually have the wrong sign.[3,17] LM corrects most of the error, and GGA is even more accurate. However, the remaining errors of LM and GGA are largest for partially spin-polarized atoms like P; this observation suggests that the spin-dependence (d^{-1}) of LM and GGA should be improved.

The numerical results in this section are consistent with those found earlier using Hartree-Fock densities.[10,17,23,24] The present results are however <u>fully</u> self-consistent: For example, each GGA energy in Table 2 was taken from a calculation in which the GGA exchange-correlation potential

Table 1. Total energies of atoms in the exchange only approximation (hartrees).

Atom	LSD	LM	GGA	Exact
H	-0.457	-0.494	-0.500	-0.500
He	-2.724	-2.846	-2.872	-2.862
Li	-7.193	-7.394	-7.443	-7.432
Be	-14.223	-14.501	-14.591	-14.572
N	-53.71	-54.23	-54.45	-54.40
Ne	-127.49	-128.26	-128.68	-128.55
Ar	-524.51	-526.02	-526.93	-526.81
Zn	-1773.89	-1776.50	-1778.18	-1777.83

Table 2. Correlation energies of atoms (hartrees).

Atom	LSD	LM	GGA	Expt.
H	-0.022	-0.009	-0.002	0.000
He	-0.111	-0.052	-0.043	-0.042
Li	-0.150	-0.060	-0.052	-0.046
Be	-0.224	-0.102	-0.093	-0.094
N	-0.42	-0.21	-0.20	-0.19
Ne	-0.74	-0.41	-0.38	-0.39
Na	-0.80	-0.43	-0.41	-0.40
Mg	-0.88	-0.48	-0.46	-0.44
P	-1.11	-0.61	-0.59	-0.55
Ar	-1.42	-0.81	-0.80	-0.79

was iterated to self-consistency, while each GGA energy in Table 1 was taken from a similar calculation without the correlation potential.

ATOMIC IONIZATION ENERGIES

First ionization energies of atoms are reported in Table 3. The systems chosen are those in which both the neutral atom and its positive ion have spherically-symmetric ground-state densities. The experimental values have been compiled by Moore.[30] LSD predicts reasonably accurate ionization energies, except for H, Cr and Cu. LM and GGA correct the error for H, but not for Cr and Cu. Although the results are mixed, it appears on average that LM is better than LSD, while GGA is worse.

DIMER DISSOCIATION ENERGIES

The dissociation energy is the energy needed to break up a molecule into its constituent atoms. Kutzler and Painter[25,26] have self-consistently

Table 3. Ionization energies of atoms (electron volts).

Atom	LSD	LM	GGA	Expt.
H	13.03	13.67	13.66	13.60
He	24.28	24.80	24.99	24.59
Li	5.47	5.56	5.65	5.39
Be	9.04	9.06	9.24	9.32
Na	5.36	5.20	5.51	5.14
Mg	7.74	7.57	7.93	7.65
K	4.52	4.37	4.65	4.34
Ca	6.22	6.07	6.38	6.11
Cr	7.45	7.50	7.52	6.77
Cu	8.36	8.24	8.53	7.73
Zn	9.70	9.41	9.84	9.39

Table 4. Dimer dissociation energies (electron volts). Values from Refs. (25) and (26).

Molecule	LSD	LM	GGA	Expt.
Li_2	1.01	0.60	0.96	1.03
B_2	3.84	3.38	3.22	2.9
C_2	7.22	6.18	6.18	6.2
N_2	11.34	10.10	10.48	9.91
O_2	7.49	6.30	5.94	5.2
F_2	3.39	2.39	2.19	1.65

calculated dissociation energies for several dimers (Table 4). For the generalized gradient approximations (LM and GGA), they found it important not to sphericalize the densities of nonspherical atoms; instead they used orbitals of cubic symmetry (e.g., $p_x^1 p_y^1 p_z^2$ for the O atom). Their calculations were performed at the energy-minimizing bond length for the observed ground-state symmetry.

Excluding the purely s-bonded systems like H_2 and Li_2, LSD badly overbinds molecules. Much of this error is eliminated in LM, and GGA is even better. Similar improvements may be expected in the cohesive energies of solids.

Kutzler and Painter have used parametrizations of the LSD integrands $e_{xc}(n_\uparrow, n_\downarrow)$ slightly different from those used in this paper. For their LSD and GGA calculations, they used Vosko, Wilk and Nusair's parametrization[31] of the true correlation energy, while this paper uses Perdew and Zunger's.[18] For their LM calculations, they used Vosko, Wilk and Nusair's parametrization[31] of the RPA correlation energy, while this paper uses von Barth and Hedin's[32].

The surface energy or surface tension of a metal is the energy required to create a unit area of fresh surface. The simplest metal surface is that of jellium, in which the charge on the positive ions is smeared out into a uniform semi-infinite background filling the half-space $x<0$. The electrons are confined inside the metal by the Kohn-Sham effective potential $v_{eff}(x)$, which may either be calculated self-consistently or modelled analytically. In the linear-potential model[33],

$$v_{eff}(x) = (x-a)\, F\, \theta(x-a) \;, \tag{12}$$

where the cutoff position a is determined by charge neutrality and where

$$F = \frac{1}{2}\, \bar{k}_F^{\,3} / y_F \;. \tag{13}$$

The bulk density is $\bar{n} = \bar{k}_F^{\,3}/3\pi^2 = 3/4\pi r_s^3$, and y_F is a dimensionless slope parameter which ranges from $y_F = 0$ (the infinite barrier model) to ∞.

The surface energy has kinetic, electrostatic and exchange-correlation components. The last component is typically several times bigger than the total, and will be considered here. Table 5 presents the exchange and correlation contributions to the surface energy in the infinite barrier model. The exact exchange contribution is known,[34] but the "exact" correlation contribution is only known within the RPA[35]. For this rapidly-varying density profile, the LM approximation is poor. GGA is slightly better than LSD for the exchange component, and vastly better for the correlation component. However, LSD gives the best account of exchange and correlation taken together, because of a cancellation of errors.

Under variation of the slope parameter y_F from 0 to ∞, the model passes continuously from the infinite-barrier to the slowly-varying limit.[34] Table 6 shows how the LSD, LM, GGA and exact surface exchange energies behave under this variation. In the range of physical surface profiles ($0.5 \lesssim y_F \lesssim 4$), LM and GGA are of comparable accuracy. Both are superior to LSD, but both underestimate the exact[34] surface exchange energy.

To find the physical value of y_F for a particular choice of bulk density parameter r_s, it is only necessary to minimize the total surface energy. Table 7 shows y_F values and surface exchange-correlation energies determined in this way over the range of metallic densities, $2 \lesssim r_s \lesssim 6$. The minimization was performed within LSD[22].

Table 5. Jellium surface energies σ in the infinite barrier (y_F=0) model (erg/cm^2).

Component	r_s	LSD	LM	GGA	"Exact"
x	2.07	1109	-1174	399	715
	4	154	-163	55	99
	6	46	-48	16	29
c	2.07	118	1319	538	673
	4	27	215	84	104
	6	10	71	26	34
xc	2.07	1227	145	937	1388
	4	181	52	139	203
	6	56	23	42	63

Table 6. Surface exchange energy σ_x (in units of $\bar{k}_F^3 \times 10^{-3}$a.u.) versus density-profile parameter y_F of the linear potential model (1 hartree/bohr^2= 1.557x10^6 ergs/cm^2).

y_F	LSD	LM	GGA	Exact
0.	0.894	0.946	0.322	0.576
0.5	1.078	0.381	0.490	0.736
1.	1.252	0.656	0.670	0.899
2.	1.609	1.128	1.077	1.268
4.	2.439	2.104	2.039	2.177
6.	3.438	3.189	3.148	3.274
8.	4.469	4.267	4.240	4.309
10.	5.513	5.341	5.323	5.382

Table 7. Surface exchange-correlation energies σ_{xc} for realistic jellium-surface density profiles (ergs/cm^2).

r_s	y_F	LSD	LM	GGA	WV
2.	3.39	3291	3431	3089	3454
2.5	2.67	1478	1561	1353	1561
3.	2.12	770	825	689	818
4.	1.37	280	306	240	300
5.	0.94	130	145	108	140
6.	0.67	71	81	58	77

The exact surface exchange correlation energies for jellium are not known. Prior to the development of the generalized gradient approximations, the best estimates were obtained by the wavevector interpolation method[35]

(WV), which predicts a correction to LSD approximately equal to (1300 ergs/cm^2)/r_s^3, independent of the profile parameter y_F. The WV values are also shown in Table 7. The LM values closely agree with the WV values; both predict a small positive correction to LSD.

Recent wavefunction - variational calculations[36,37] suggest that the positive correction to the LSD surface exchange-correlation energy could be much larger than the LM and WV methods predict. At any rate, there is little evidence for the small negative correction predicted by GGA. The GGA error arises at least in part from the GGA underestimation of the surface exchange energy (Table 6).

Studies of surface energies usefully complement studies of atomic exchange-correlation energies. The atomic energies sample the density functional mainly over the range of moderate density gradients ($0.2 \lesssim s \lesssim 1$), while the surface energies strongly sample the range of larger gradients ($s \gtrsim 1$). A tentative conclusion is that GGA applied to spin-polarized systems is correct for $s \lesssim 1$ and in error for $s \gtrsim 1$. This might have been expected, since the GGA functional retains the GEA description of the exchange-correlation hole very close to the electron it surrounds, cutting off only the long-range part of the GEA hole.[11] For very large density gradients ($s \gtrsim 1$), the GEA description of the exchange hole begins to fail even very close to the electron, as shown by the errors of the GEA kinetic energy for rapidly-varying densities.[38]

SELF-CONSISTENT ATOMIC DENSITIES IN THE EXCHANGE-ONLY APPROXIMATION

Density functional theory predicts total and spin-magnetization densities as well as energies, and consideration now passes from the latter to the former. Since "exact" correlated densities are often unavailable, the discussion here will be restricted to the exchange-only approximation.

The electron spin-magnetization density at the nucleus, $m(0) = n_\uparrow(0)-n_\downarrow(0)$, determines the hyperfine interaction. LSD predicts reasonable values for m(0) when m(0) arises directly via an excess of valence s-electrons of one spin. But in atoms like N, P or Mn, this density arises indirectly: an excess of valence p- or d- electrons of one spin polarizes the spin-paired s-electrons in the atom. Under these conditions, Wilk and Vosko[39] found that LSD often predicts the wrong sign for m(0). Table 8 displays m(0) for the N atom in the exchange-only approximation. The "exact" value is from a spin-unrestricted Hartree-Fock calculation.[39] Unlike LSD, both LM and GGA predict the correct sign for m(0), and GGA gives the best account of its numerical value.

Table 8. Spin-magnetization density m(0) at the nucleus of the nitrogen atom in the exchange-only approximation (atomic units).

LSD	LM	GGA	"Exact"
-0.052	0.003	0.086	0.188

Table 9. Highest occupied orbital energy and orbital-energy differences for atoms in the exchange-only approximation (hartrees).

Atom	Orbital	LSD	LM	GGA	Exact
Ne	2p	-0.443	-0.437	-0.457	-0.846
	2p-2s	0.82	0.84	0.84	0.87
	2p-1s	29.79	29.95	30.06	29.97
Zn	4s	-0.185	-0.179	-0.195	-0.308
	4s-3d	0.16	0.16	0.16	0.24
	4s-3p	2.78	2.80	2.80	2.91
	4s-3s	4.33	4.36	4.37	4.49
	4s-2p	36.40	36.43	36.44	36.51
	4s-2s	41.28	41.36	41.40	41.48
	4s-1s	344.69	345.19	345.56	345.51

The highest occupied orbital energy in an atom controls the decay of the density into the vacuum. As Table 9 shows, this Kohn-Sham eigenvalue is not nearly negative enough in LSD, leading to a density that is too diffuse in the outer part of the atom. The error is essentially uncorrected in LSD and GGA, as might have been expected: the generalized gradient approximations are continuously-differentiable functionals which cannot display the derivative discontinuity[40] present in the exact functional. Expressed differently, there is a residue of self-interaction error[18] in any generalized gradient approximation.

Kohn-Sham eigenvalue <u>differences</u>[7] (Table 9) gauge the accuracy of the density in the interior of the atom. The exact eigenvalue differences for comparison are [33] those of Talman's Optimized Potential Model,[27] and <u>not</u> the Hartree-Fock eigenvalues (which reflect the nonlocality of the Fock operator). Table 9 shows that LSD, LM and GGA reproduce these differences rather accurately in Ne. In Zn, all of the approximations place the 4s eigenvalue too close to the 3d and lower-lying eigenvalues. LM and GGA are superior to LSD for the deeper orbitals, reflecting a more accurate density deep inside the atom.

Table 10. Density moments $\langle r^k \rangle = \int d^3r \, r^k \, n(r)$ for atoms in the exchange-only approximation (atomic units). The exact values for Ne were calculated in Ref.(7) from the potential of Ref.(27).

Atom	Moment	LSD	LM	GGA	Exact
H	$\langle r^{-1} \rangle$	0.96	0.99	1.00	1.00
	$\langle r \rangle$	1.60	1.53	1.55	1.50
	$\langle r^2 \rangle$	3.50	3.16	3.27	3.00
Ne	$\langle r^{-1} \rangle$	30.95	31.05	31.09	31.10
	$\langle r \rangle$	8.07	8.01	8.06	7.90
	$\langle r^2 \rangle$	10.04	9.82	10.06	9.40

The density moments $\langle r^k \rangle$ in atoms tell a similar story. Table 10 shows that the density moment $\langle r^{-1} \rangle$ is too small in LSD. LM removes most of this error, while GGA removes essentially all of it. However, the density moments $\langle r \rangle$ and $\langle r^2 \rangle$ are less reliably improved. Thus GGA gives an improved account of the density deep inside an atom. This conclusion is consistent with the speculation of the preceding section that GGA is accurate in the range of moderate density gradients, since $s \lesssim 1$ in the interior of an atom[33].

CONCLUSIONS AND PROSPECTS

While the generalized gradient approximations (LM or GGA) are not yet perfected, they tantalize with the possibility to correct the local spin density (LSD) approximation simply and accurately for atoms, molecules and solids.

The GGA in particular seems to provide an accurate description of spin-unpolarized systems with moderate density gradients ($s \lesssim 1$). Refinements appear to be required in at least two directions: (1) a better account of the exchange energy for larger density gradients ($s \gtrsim 1$), and (2) a more accurate description of the spin dependence of the gradient term in the correlation energy. A possible third refinement would be (3) the use of a smooth wavevector cut-off of the gradient term in the correlation energy, as advocated by Langreth and Vosko,[41] in place of the artificially sharp cutoff used by LM and GGA.

This work was supported in part by the National Science Foundation under Grant No. DMR84-20964, and by the Research Foundation of the City University of New York (PSC-BHE).

REFERENCES

1. W. Kohn and L.J. Sham, Self-Consistent Equations Including Exchange and Correlation Effects, Phys. Rev. 140: A 1133 (1965).
2. L.J. Sham, Approximations of the Exchange and Correlation Potentials, in: "Computational Methods in Band Theory," P.M. Marcus, J.F. Janak and A.R. Williams, eds., Plenum, New York (1971).
3. S.-K. Ma and K.A. Brueckner, Correlation Energy of an Electron Gas with a Slowly Varying High Density, Phys. Rev. 165: 18 (1968).
4. M. Rasolt, Inhomogeneity Corrections to the Ground-State Properties of Itinerant Ferromagnets, Phys. Rev. B16: 3234 (1977).
5. O. Gunnarsson and B.I. Lundqvist, Exchange and Correlation in Atoms, Molecules, and Solids by the Spin-Density-Functional Formalism, Phys. Rev. B13: 4274 (1976).
6. D.C. Langreth and J.P. Perdew, Theory of Nonuniform Electronic Systems. I. Analysis of the Gradient Approximation and a Generalization that Works, Phys. Rev. B21: 5469 (1980).
7. D.C. Langreth and M.J. Mehl, Beyond the Local-Density Approximation in Calculations of Ground-State Electronic Properties, Phys. Rev. B 28: 1809 (1983); erratum ibid. 29: 2310 (1984).
8. C.D. Hu and D.C. Langreth, A Spin-Dependent Version of the Langreth-Mehl Exchange-Correlation Functional, Phys. Script. 32: 391 (1985); erratum of Ref. 17.
9. J.P. Perdew, Accurate Density Functional for the Energy: Real-Space Cutoff of the Gradient Expansion for the Exchange Hole, Phys. Rev. Lett. 55: 1665 (1985).
10. J.P. Perdew and Y. Wang, Accurate and Simple Density Functional for the Electronic Exchange Energy: Generalized Gradient Approximation, Phys. Rev. B 33: 8800 (1986).
11. J.P. Perdew, What's Right and What's Wrong with the Density-Gradient Expansions for the Exchange and Correlation Energies?, in "Condensed Matter Theories, Vol. 2," P. Vashishta, ed., Plenum, New York (1987).
12. J.P. Perdew and Y. Wang, Electron Density Functionals from the Gradient Expansion of the Density Matrix: The Trouble with Long-Range Interactions, in "Mathematics Applied to Science," J.A. Goldstein, S. Rosencrans and G. Sod, eds., Academic Press, New York (1987).
13. A.D. Becke, Density Functional Calculations of Molecular Bond Energies, J. Chem. Phys. 84: 4524 (1986).
14. A.E. De Pristo and J.D. Kress, Rational Function Representation for Accurate Exchange Energy Functionals, J. Chem. Phys. 86: 1425 (1987).
15. S.H. Vosko and L.D. Macdonald, Exchange-Only Energy Functionals from Atomic Exchange Energy Densities, in "Condensed Matter Theories, Vol. 2," P. Vashishta, ed., Plenum, New York (1987).
16. S.K. Ghosh and R.G. Parr, Phase-Space Approach to the Exchange-Energy Functional of Density Functional Theory, Phys. Rev. A 34: 785 (1986).
17. J.P. Perdew, Density-Functional Approximation for the Correlation Energy of the Inhomogeneous Electron Gas, Phys. Rev. B 33: 8822 (1986); erratum ibid. 34: 7406 (1986).
18. J.P. Perdew and A. Zunger, Self-Interaction Correction to Density-Functional Approximations for Many-Electron Systems, Phys. Rev. B23: 5048 (1981).
19. M. Rasolt and D.J.W. Geldart, Exchange and Correlation in the Non-uniform Electron Gas, unpublished. The form is presented in Ref. 17.
20. C. Lee, W. Yang and R.G. Parr, Development of the Colle-Salvetti Correlation Energy Formula into a Functional of the Electron Density, unpublished.

21. U. von Barth and A.C. Pedroza, The Cohesive Energy and Charge Density Form Factors of Beryllium as a Test of the Langreth-Perdew-Mehl Approximation, Phys. Script. 32: 353 (1985).
22. A.-R.E. Mohammed and V. Sahni, Density-Functional-Theory Studies of Correlation-Energy Effects at Metallic Surfaces, Phys. Rev. B 31: 4879 (1985).
23. A. Savin, H. Stoll and H. Preuss, An Application of Correlation Energy Density Functionals to Atoms and Molecules, Theor. Chem. Acta. 70: 407 (1986).
24. J.B. Lagowski and S.H. Vosko, An Analysis of Local and Gradient Corrected Correlation Energy Functionals Using Electron Removal Energies, unpublished.
25. F.W. Kutzler and G.S. Painter, Nonlocality in the Density Functional Description of Bonding in Li, N, O, and F Dimers, unpublished.
26. F.W. Kutzler and G.S. Painter, Effects of Nonsphericity in Atomic Energies Calculated with Nonlocal Density Functional Theory, unpublished.
27. K. Aashamar, T.M. Luke and J.D. Talman, Optimized Central Potentials for Atomic Ground-State Wavefunctions, At. Data Nucl. Data Tables 22: 443 (1978).
28. A. Veillard and E. Clementi, Correlation Energy in Atomic Systems. V. Degeneracy Effects for the Second-Row Atoms, J. Chem. Phys. 49: 2415 (1968).
29. H. Stoll and A. Savin, private communication (1983).
30. C.E. Moore, Ionization Potentials and Ionization Limits Derived from the Analyses of Optical Spectra, Nat. Stand. Ref. Data Ser., Nat. Bur. Stand. (U.S.) 34 (1970).
31. S.H. Vosko, L. Wilk and M. Nusair, Accurate Spin-Dependent Electron Liquid Correlation Energies for Local Spin Density Calculations: A Critical Analysis, Can. J. Phys. 58: 1200 (1980).
32. U. von Barth and L. Hedin, A Local Exchange-Correlation Potential for the Spin Polarized Case: I, J. Phys. C 5: 1629 (1972).
33. V. Sahni, J.B. Krieger and J. Gruenebaum, Metal Surface Properties in the Linear Potential Model, Phys. Rev. B 15: 1941 (1977).
34. V. Sahni, J. Gruenebaum and J.P. Perdew, Study of the Density-Gradient Expansion for the Exchange Energy, Phys. Rev. B 26: 4371 (1982).
35. D.C. Langreth and J.P. Perdew, Exchange-Correlation Energy of a Metallic Surface: Wave-Vector Analysis, Phys. Rev. B 15: 2884 (1977).
36. E. Krotscheck, W. Kohn and G.-X. Qian, Theory of Inhomogeneous Quantum Systems. IV. Variational Calculations for Metal Surfaces, Phys. Rev. B 32: 5693 (1985).
37. E. Krotscheck and W. Kohn, Nonlocal Screening in Metal Surfaces, Phys. Rev. Lett. 57: 862 (1986).
38. J.P. Perdew, V. Sahni, M.K. Harbola and R.K. Pathak, Fourth-Order Gradient Expansion of the Fermion Kinetic Energy: Extra Terms for Nonanalytic Densities, Phys. Rev. B 34: 686 (1986).
39. L. Wilk and S.H. Vosko, Investigation of the Spin-Density Functional Method for Calculating Spin-Magnetic-Moment Densities, Phys. Rev. A 15: 1839 (1977).
40. J.P. Perdew, R.G. Parr, M. Levy and J.L. Balduz, Density-Functional Theory for Fractional Particle Number: Derivative Discontinuities of the Energy, Phys. Rev. Lett. 49: 1691 (1982).
41. D.C. Langreth and S.H. Vosko, Exact Electron Gas Response Functions at High Density, Phys. Rev. Lett. 59: 497 (1987).

EFFECTIVE INTERACTIONS AND ELEMENTARY EXCITATIONS IN NUCLEAR MATTER*

J. Wambach, D. Pines and K. F. Quader

University of Illinois at Urbana-Champaign
Department of Physics
1110 West Green Street
Urbana, Illinois 61801

ABSTRACT

A polarization potential theory for nuclear matter is presented. Starting from a realistic two-body interaction we construct pseudopotentials which describe the effective quasiparticle interaction. Using these the linear response in various spin-isospin channels is evaluated.

INTRODUCTION

Polarization potential theory (PPT) was first developed and applied to ^{3}He and ^{4}He by Aldrich and Pines.[1] The basic idea is to describe the effects of strong short-range correlations in the liquid, produced by the nearly hard-core interatomic potential, by self-consistent renormalized fields called polarization potentials. The strengths of these potentials are determined from empirical inputs and conservation laws. Backflow correlations and high-frequency multipair (multiparticle) excitations are incorporated in a consistent way. The polarizations potentials describe the elementary excitation spectrum of the quantum liquid for finite wavevectors and frequencies. In this sense the polarization potential approach is a post-Landau theory.

Similar to ^{3}He the nucleon-nucleon potential is very repulsive at short distances resulting in strong positional correlations among the particles. Thus it is expected that a polarization potential theory for nuclear matter can be developed along the lines of ^{3}He. There are, however, several aspects which render the nuclear case more complicated. First of all, nucleons are not as "rigid" as He-atoms. Consider the ratio of the Fermi energy at saturation density to the first excited state of a constituent of the liquid. In ^{3}He this ratio is $\sim 10^{-3}$ while in nuclear matter it is only $\sim 10^{-1}$. The "softness" of the nucleons gives rise to three-body forces, which are important in the proper binding characteristics. Secondly, the nucleon-nucleon interaction has a strong tensor component, which is totally absent in the ^{3}He interatomic potential. This tensor interaction is crucial for binding.

Without the tensor interaction, nuclear matter would be unbound at all densities. Thus the saturation mechanism is quite different from ^3He.

POLARIZATION POTENTIAL THEORY

In PPT the consequences of strong interparticle interactions in a multicomponent quantum liquid are described by a set of self-consistent scalar and vector fields, ϕ_i and $\vec{A}_i$ which are both wavevector and frequency dependent. They are related to the average density and current fluctuations as

$$\phi^i_{pol}(q, \omega) = f^i_s(q) \langle\rho_i(q,\omega)\rangle \tag{2.1a}$$

$$\vec{A}^i_{pol}(q, \omega) = f^i_v(q)\langle\vec{j}_i(q, \omega)\rangle. \tag{2.1b}$$

where the index i specifies whether the various fluid components oscillate in or out of phase. The functions f^i_s and f^i_v denote the quasiparticle interactions (pseudopotentials) in the fluid which provide the restoring forces for the fluctuations. Given the potentials ϕ^i_{pol} and $\vec{A}^i_{pol}$ one immediately derives the elementary excitation spectrum.

We are mainly interested in the response of the system to external fields, ϕ^i_{ext}, which are weakly coupled to the density. They induce density fluctuation which are linearly proportional to ϕ^i_{ext}

$$\langle\rho_i(q, \omega)\rangle = \chi^i(q, \omega)\, \phi^i_{ext}(q, \omega) \tag{2.2}$$

and involve the density-density correlation function χ^i. The influence of the self-consistent fields ϕ^i_{pol} is taken into account by considering the system response to the sum $[\phi^i_{ext} + \phi^i_{pol}]$

$$\langle\rho_i(q, \omega)\rangle = \chi^i_{sc}(q, \omega)\, [\phi^i_{ext}(q, \omega) + \phi^i_{pol}(q, \omega)] \tag{2.3}$$

where χ^i_{sc} is the response to the effective field $[\phi^i_{ext} + \phi^i_{pol}]$.

The effect of backflow correlations can be obtained from the density-current relation

$$\omega\langle\rho_i(q, \omega)\rangle - \vec{q} \cdot \langle\vec{j}_i(q, \omega)\rangle = \delta_i. \tag{2.4}$$

If the currents are conserved individually, i.e. $\delta_i = 0$, we find

$$\phi^i_{pol}(q, \omega) = [f^i_s(q) + (\omega^2/q^2)f^i_v(q)]\, \langle\rho_i(q, \omega)\rangle \tag{2.5}$$

as a straightforward extension of (2.1a). In general, of course, δ_i will not be zero and (2.5) is incorrect. It rigorously applies, however, if particles of all species oscillate in phase, as a consequence of particle number conservation. Using (2.2), (2.3) and (2.5) we derive the PPT-expression of density-density correlation function

$$\chi^i(q, \omega) = \frac{\chi^i_{sc}(q, \omega)}{1 - [f^i_s(q) + (\omega^2/q^2)f^i_v(q)]\chi^i_{sc}(q, \omega)} \tag{2.6}$$

which is the central equation of the theory.

The effect of multipair modes on the excitation spectrum is taken into account by writing χ^i_{sc} as

$$\chi^i_{sc}(q, \omega) = \alpha^i(q)\chi^i_0(q, \omega) + (1 - \alpha^i(q))\chi^i_m(q, \omega) \tag{2.7}$$

where χ^i_0 is the single-pair response and χ^i_m is the multipair response. The functions $\alpha^i(q)$ take into account the reduction of the single-pair excitations which necessarily accompany the presence of multipair modes. In practice both α^i and χ^i_m contain adjustable parameters which are determined from experiment.

The scalar restoring forces f^i_s are related to suitably regularized position space bare potentials as indicated in Fig. 1a. They are characterized by a range, r^i_c, at which the interaction changes sign and a strength, a^i, of the effective repulsion in the liquid. The momentum space pseudopotentials (Fig. 1b) are obtained via Fourier transformation

$$f^i_s(q) = \int drr^2 j_0(qr)f^i_s(r). \tag{2.8}$$

In the $q \to 0$ limit these are related to empirical Fermi liquid parameters which are used to fix the core heights a^i.

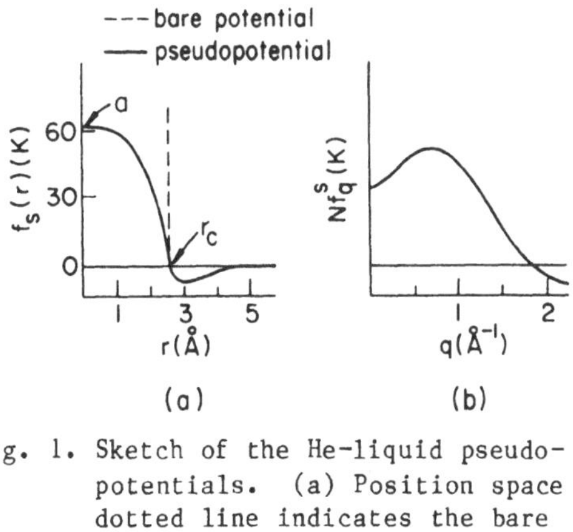

Fig. 1. Sketch of the He-liquid pseudopotentials. (a) Position space (the dotted line indicates the bare potential) (b) momentum space

The vector restoring forces $f^i_v(q)$ can be related to the quasipair effective masses $m^*_i(q)$ if the i-th current is conserved. Using sum rules and the high-frequency limit of χ^i and χ^i_{sc} one finds

$$m_i^*(q) = m + Nf_v^i(q). \tag{2.9}$$

where N is the number density. It can also be shown, that $m_i^*(q)$ in the long-wavelength limit reduces to m_o^*, the quasiparticle effective mass at the Fermi surface. In practice, as argued above, relation (2.9) only holds if all fluid components oscillate in phase and it shall only be used in this case.

PSEUDOPOTENTIALS IN NUCLEAR MATTER

"Bare Interaction"

In contrast to ^3He, where the bare potential is relatively unique, in the nuclear case there is a host of phase-shift equivalent potentials. Among those only the local ones are of interest here. A suitable choice is the Reid potential.[2] We restrict ourselves to the V_6-form

$$V_{Reid}(\vec{r}) = \sum_{ST} V_{ST}^c(r)O_S O_T + V_T^t(r)S_{12}(\hat{r})O_T \tag{3.1a}$$

$$O_S = 1,\ \vec{\sigma}\cdot\vec{\sigma}';\ S_{12}(\hat{r}) = 3\vec{\sigma}\cdot\hat{r}\vec{\sigma}'\cdot\hat{r} - \vec{\sigma}\vec{\sigma}';\ O_T = 1,\ \vec{\tau}\vec{\tau}' \tag{3.1b}$$

which leads to the most important correlations in the nuclear system. This potential is, however, not sufficient for constructing the pseudopotentials of interest.

First, in order to bind nuclear matter, V_{Reid} has to be supplemented by the second-order tensor interaction $V_{ten}^{(2)}$. $V_{ten}^{(2)}$ can be easily obtained from closure, since it involves high energy intermediate states.[3] Then

$$V_{ten}^{(2)}(\vec{r}) = -\frac{V_1^t(r)^2}{\overline{E}}(S_{12}(\hat{r})\vec{\tau}\cdot\vec{\tau}')^2$$

$$= -\frac{V_1^t(r)^2}{\overline{E}}(3 - 2\vec{\tau}\vec{\tau}')(6 + 2\vec{\sigma}\vec{\sigma}' - 2S_{12}(\hat{r})) \tag{3.2}$$

where $\overline{E} \approx 200$ MeV. The resulting "bare potentials" are displayed by the dashed lines in Fig. 2.

Second, in the nuclear system, three-body interactions (TBI) are important. They can be represented by density-dependent two-body interactions, which are suitable for constructing pseudopotentials. We choose the following form

$$V_{TBI}(\vec{r}) = g[\rho]\,\frac{e^{-\alpha r}}{\alpha r};\qquad \alpha = (2m_\pi)^{-1} \tag{3.3}$$

where $g[\rho]$ is a strength parameter and α is a typical range.[4] The strength parameter will be determined from the nuclear binding energy curve, as discussed in the next section. Since $V_{TBI}(r)$ has no spin- or isospin dependence, it adds the same amount to each of the first four channels in Fig. 2.

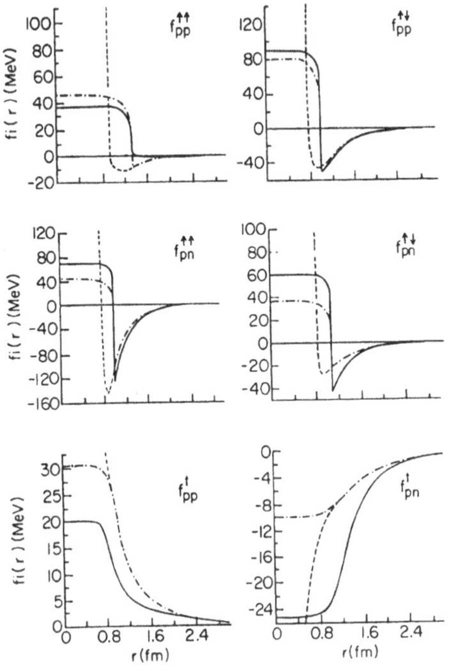

Fig. 2. Position space pseudopotentials in the (p,n,↑,↓)-channels. The full lines include zero-point motion as well as exchange and screening effects. In the dashed-dotted lines exchange and screening effects are neglected. The dashed lines indicate the bare potentials.

Empirical Zero-order Landau Parameters

To construct position space pseudopotentials we need to know the six lowest-order empirical Landau parameters F_0, F_0', G_0, G_0', H_0 and H_0' of nuclear matter. We briefly comment on their assignment.

F_0 is related to the nuclear compression modulus, K, by

$$K = 6\varepsilon_F(1 + F_0) \tag{3.4}$$

where ε_F is the Fermi energy (ε_F = 48 MeV for m_0^* = 0.8 m). The compression modulus can be extracted theoretically from empirical breathing mode energies in finite nuclei[5] and is found to be close to the free Fermi gas value K ≈ 230 MeV. According to our earlier discussion F_0 has to contain two parts, a two-body piece $F_0^{(2)}$ and a three-body piece $F_0^{(3)}$. The relative contribution of both at a given density ρ can be obtained from the nuclear saturation curve E/A(ρ). It is commonly accepted that two-body interactions give a curve close to that indicated by the dashed line in Fig. 3. To obtain proper binding one has to add a correction $E^c/A(\rho)$ which results in the full curve in

Fig. 3. This correction, in our treatment, comes from three-body interactions and hence fixes $F_0^{(3)}$

$$F_0^{(3)} = \frac{3m_0^*}{k_F^2} \frac{d\rho^2}{d\rho} \frac{dE^{(c)}/A}{d\rho}. \tag{3.5}$$

Knowing K, the two-body part $F_0^{(2)}$ is then determined. While $F_0^{(3)}$ determines the density-dependent strength parameter $g[\rho]$ in (3.3), $F_0^{(2)}$ has to be used in regularizing the bare potentials in Fig. 2.

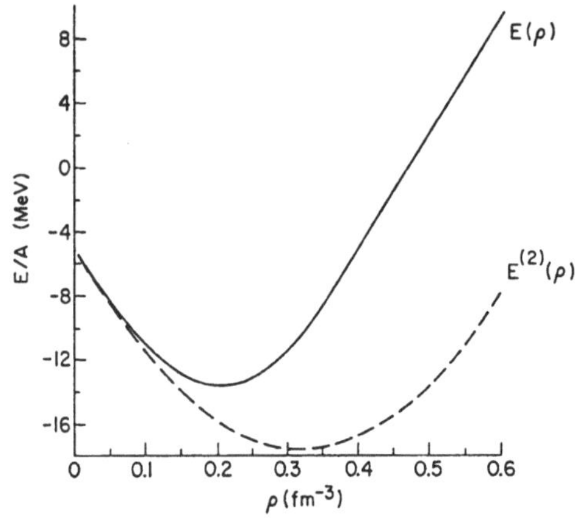

Fig. 3. Nuclear matter saturation curves. The full line includes two-body as well as three-body interactions. The dashed line gives the result with two body interactions only.[6]

The Landau parameter F_0' is related to the symmetry β as

$$\beta = \frac{\varepsilon_F}{3} (1 + F_0') \tag{3.6}$$

and is, most reliably, extracted from fits of the semiempirical mass formula. This yields $\beta \simeq 33$ MeV or $F_0' = 1.07$. The spin parameter G_0 is not well determined from nuclear phenomenology, but recent studies on isoscalar 1^+-states[7] indicate that it is close to 0.3. The spin-isospin parameter G_0' is fairly reliably determined from Giant Gamow-Teller resonances in nuclei and is found to be $\simeq 1.3$.[8] The tensor parameters H_0 and H_0' can be estimated from the free nucleon-nucleon T-matrix[9] which gives $H_0 = 0.45$ and $H_0' = -0.14$.

Scalar Pseudopotentials

The position space scalar pseudopotentials f_s^i are characterized by core radii r_c^i and core heights a^i. In the simplest approach r_c^i is defined as the radius at which the bare interaction changes sign (Fig. 2.) Such a choice does not, however, reflect the role played by zero-point motion of the particles in the liquid, which act to increase the range of the repulsive interaction. In ^{3}He this increase is typically of the order of 10%. The scale is set by the ratio of the zero point amplitude $\langle R^2\rangle^{1/2}$ to the mean interparticle spacing. In nuclear matter at saturation density this ratio is 3 times larger, such that we expect core radii to increase by 30%. The core heights a^i are then determined by relating the volume integrals of the short-range regularized interactions to the Fermi-Liquid parameters. The resulting pseudopotentials are given by the dashed-dotted curves in Fig. 2 and the corresponding momentum space potentials are displayed in the upper part of Fig. 4.

In the polarization potential theory of ^{3}He the quantum statistical correlations are taken into account by choosing the core radius for particles with parallel spin larger than that for particles with antiparallel spin. In nuclear matter we will treat Pauli effects somewhat differently. For distances $r > r_i$ we start with fully antisymmetrized two-body interactions

$$\tilde{V}(\vec{r}) = (1 - \mathcal{P})\, V(\vec{r}) \tag{3.7}$$

where $\mathcal{P} = -P_r P_\sigma P_\tau$ is the generalized Pauli exchange operator. Including exchange therefore modifies the long-range part of the interaction. In the representation (3.1) the effect of P_σ and P_τ is easily worked out. We find

$$\tilde{V}^c_{ST}(r) = \sum_{ST} (\delta_{SS'}\delta_{TT'} + A^c_{SS'TT'}P_r)V^c_{S'T'}(r) \tag{3.8a}$$

$$\tilde{V}^t_T(r) = \sum_{T'} (\delta_{TT'} + A^t_{TT'}P_r)V^t_{T'}(r). \tag{3.8b}$$

A^c and A^t are matrices containing spin-isospin statistical factors. To be represented as pseudopotentials the nonlocal interactions $P_r V^c_{ST}$ and $P_r V^t_T$ have to be approximated by local interactions. This can be done using the exchange-hole or Slater approximation.[10]

It is a well known phenomenon from the theory of the electron gas that exchange contributions are screened by medium polarization. The inclusion of this effect is absolutely crucial in obtaining the correct ground-state properties. In general, screening is expressed in terms of a dielectric constant which describes the medium polarization. This effect can be included in nuclear matter by evaluating generalized dielectric constants in the ring approximation. The position space pseudopotentials, which include exchange and screening effects are indicated by the solid curves in Fig. 2. The corresponding momentum space potentials are given in lower part of Fig. 4.

Vector Pseudopotentials

To model the vector pseudopotentials $f^i_v(q)$ we evaluate the quasipair effective mass $m^*_i(q)$. If the current is conserved f^i_v and m^*_i are uniquely related (2.9). This will be true only in S = 0, T = 0 channel. In the large q-limit the average quasiparticle energy and the quasipair energy will be given by the sum of the kinetic energy and the average potential energy $\Delta U (\Delta U \approx 50$ MeV). Thus

$$\frac{q^2}{2m^*(q)} = \frac{q^2}{2m} + \Delta U. \tag{3.9}$$

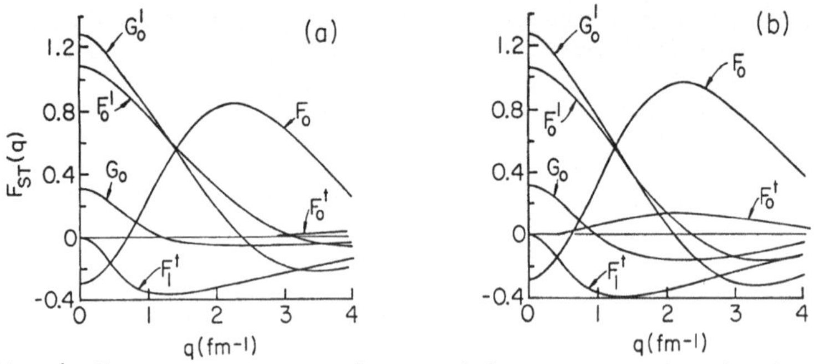

Fig. 4. Momentum space pseudopotentials at saturation density ($k_f = 1.36\ fm^{-1}$). The interactions are normalized such that in the $q \to 0$ limit the Fermi liquid parameters are obtained. In the upper part (a) exchange and screening have been neglected while in the lower part (b) both have been included.

In the $q \to 0$ limit on the other hand $m^*(q)$ becomes equal to m_o^*. We interpolate between these two limits as follows

$$m^*(q) = \frac{m}{1 + 2m\Delta U/(q^2 + q_o^2)} \tag{3.10}$$

where q_o is determined from the $q \to 0$ limit.

LINEAR RESPONSE

Dynamic Structure Functions

The imaginary part of the density-density correlation function (2.6) gives the dynamic structure function, which is measured in inelastic scattering experiments

$$S^i(q,\ \omega) = -\frac{Im}{\pi}\chi^i(q,\ \omega). \tag{4.1}$$

The index i denotes the various spin and isospin channels (S, T). The tensor interactions are involved in the spin-longitudinal response ($\vec{\sigma} \cdot \hat{q}$) and the spin-transverse response ($\vec{\sigma} \times \hat{q}$). As an example we display in Fig. 5 S^{01} which corresponds to protons oscillating against neutrons, while their spins remain in phase. This is the nuclear matter analog of nuclear giant dipole resonances. At small q we find a collective mode which is, however, quickly Landau-damped as q increases.

Static Polarizabilities and Liquid Structure Functions

The inverse energy moments of S^i determine the q-dependent polarizabilities κ^i

$$\kappa^i(q) = \int_0^\infty d\omega\ \frac{S^i(q,\ \omega)}{\omega}. \tag{4.2}$$

Equivalently these polarizabilities may be obtained from the $\omega \to 0$ limit of the dynamic structure functions. Hence, according to (2.5), there are

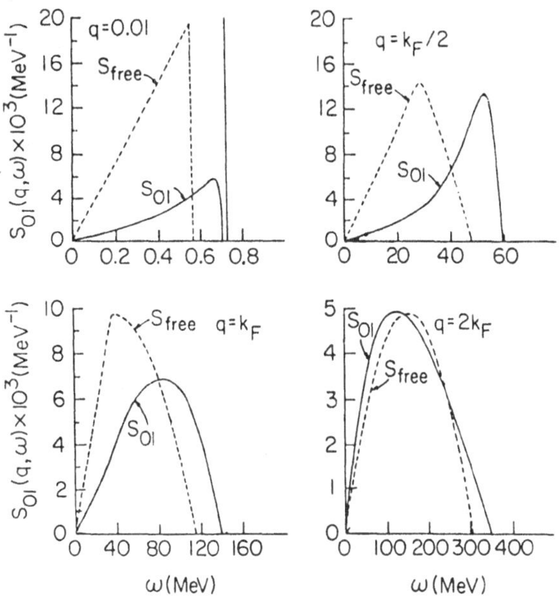

Fig. 5. Dynamic structure functions in the scalar-isovector channel (S = 0, T = 1) as a function of excitation energy ω and momentum transfer q. The full lines indicate the interacting case while the dashed lines give the free Fermi gas responses.

no backflow effects in the static limit and $\kappa^i(q)$ provided a measure of $f^i_S(q)$. Our results for the four central (S,T)-channels are given in Fig. 6.

The static structure functions are obtained as the zeroth moments of S^i

$$S^i(q) = \int_0^\infty d\omega S^i(q, \omega) \tag{4.3}$$

In TDA $S^i(q)$ is always equal to the free gas value. Hence the deviations measure the amount of ground-state correlations. Our results (Fig. 7) indicate that there is a strong suppression in the (0,0)-channel, as a consequence of the strong repulsion of $F_0(q)$ (Fig. 4) between $q \approx 1\ fm^{-1}$ and $q \approx 4\ fm^{-1}$.

SUMMARY AND CONCLUSIONS

We have extended the PPT of Aldrich and Pines to nuclear matter. Although similar in spirit important modifications are required. These are necessitated by the presence of tensor forces and three-body

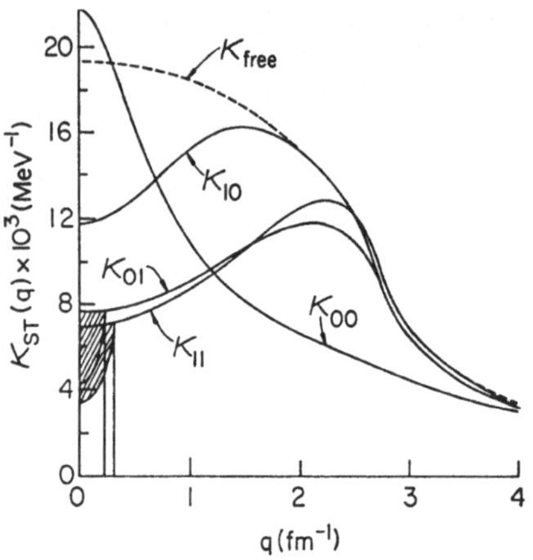

Fig. 6. Static polarizabilities in the four (S,T)-channels, driven by the central interactions, as a function of momentum transfer q. The Fermi gas polarizability is given by the dashed curve. The shaded areas indicate the contributions from the undamped collective modes and the vertical lines denote the momenta at which they merge with the particle-hole continuum.

interactions. The resulting position space and momentum space pseudopotentials appear physically reasonable and allow for a straightforward calculation of the dynamic and static response. Our results are, in general, in quite good agreement with those obtained by other methods.

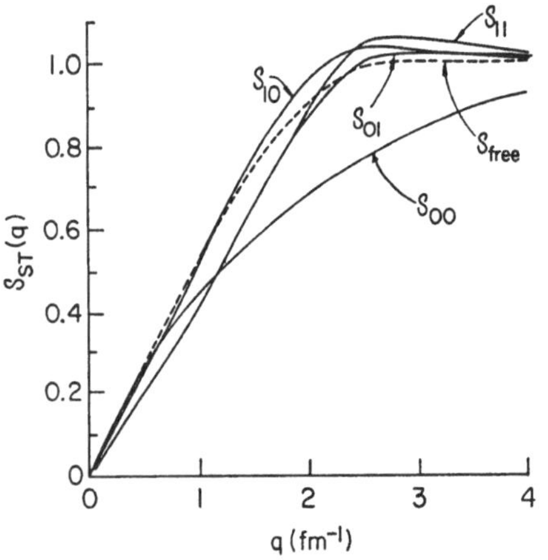

Fig. 7. Liquid structure functions in the four central channels (S,T) as a function of momentum transfer q. The Fermi gas result is indicated by the dashed line.

The pseudopotentials which we have obtained provide a natural basis for the construction of scattering amplitudes, which in turn make possible improved calculations of transport properties of nuclear matter (both cold and hot). We expect that it will not prove difficult to examine the role played by isospin polarization, and to develop the corresponding theory for neutron matter, a problem of considerable importance for neutron star formation and evolution, as well as for the behavior of pulsars.

REFERENCES

1. C. H. Aldrich III and D. Pines, J. Low Temp. Phys. 25:677 (1976); ibid., 32:698 (1978).
2. P. V. Reid, Ann. Phys. 50:411 (1968).
3. T. T. S. Kuo and G. E. Brown, Phys. Lett. 18:40 (1966).
4. V. R. Pandharipande, private communication.
5. J. P. Blaizot, Phys. Rep. 64:171 (1980).
6. B. D. Day and R. B. Wiringa, Phys. Rev. C32:1057 (1985).
7. R. Laszewski and J. Wambach, Comments Nucl. Part. Phys. 14:321 (1985).
8. G. F. Bertsch, D. Cha and H. Toki, Phys. Rev. C24:533 (1981).
9. M. A. Franey and W. G. Love, Phys. Rev. C31:488 (1985).
10. D. M. Clement, Nucl. Phys. A205:398 (1973).

* This work is partially supported by Grants NSF-PHY-84-15064 and NSF-PHY-86-00377.
A travel grant from the U.S. Army Research Office is thankfully acknowledged.

THE PARTICLE-HOLE INTERACTION IN FINITE NUCLEI*

W. H. Dickhoff

Department of Physics, Washington University

St. Louis, Missouri 63130, U.S.A.

Abstract The standard Random Phase Approximation (RPA) is extended by including the exchange of one– and the coupling to two–phonon excitations in the residual interaction. Within this scheme a microscopic understanding of the excitation spectra of doubly closed-shell nuclei, like e.g. ^{16}O, is possible. The mechanism to generate the many low-lying isoscalar natural parity states in ^{16}O is discussed and related to the attraction of the residual nucleon-nucleon (NN) interaction. The connection of this residual particle-hole (ph) interaction with the treatment of the single-particle (sp) self-energy is discussed.

I. INTRODUCTION

The microscopic description of nuclear excitations is only partly succesful in explaining observed experimental properties. The spectra of supposedly simple nuclei like ^{16}O and ^{40}Ca have defied a truly microscopic description up to now. Whereas in a simple picture one expects first one-particle one-hole (1p1h) negative parity excitations on top of the doubly closed-shell ground state, the first excited state in these nuclei is actually a 0^+ state and many more low-lying positive parity states are observed. Noting that all these states have isoscalar character, it is readily possible to interpret the four lowest T = 1 states in ^{16}O (0^-, 1^-, 2^-, 3^-) in terms of $2s_{1/2} - 1p_{1/2}^{-1}$ and $1d_{5/2} - 1p_{1/2}^{-1}$ 1p1h states. Various investigations have pointed out the need for many-particle many-hole correlations in describing the isoscalar positive parity states[1] as well as the importance of the lowest 3^- and 1^- T = 0 "bosons" as basic building blocks for generating these states.[2] Recent shell-model calculations[3] show, however, that even within an extended 2p2h basis the observed level structure is not reproduced.

RPA calculations for ^{16}O (see e.g. Ref. 4) have mainly focused on negative parity states. Essentially only the four above mentioned negative parity states can be obtained low in energy. This is a common feature of standard RPA, most clearly demonstrated in the schematic model[5] in which only one state (for each J^π,T) is either lowered in energy (for an attractive residual interaction) or pushed up (repulsive interaction). Calculations using realistic interactions of G matrix type[6,7] do not alter this but have the correct feature of moving down isoscalar natural parity states considerably and pushing up isovector states moderately. They fail, however, to produce the observed multiplicity of isoscalar negative parity states observed at low energies.

*This research was supported in part by the Condensed Matter Theory Program of the Division of Materials Research of the U.S. National Science Foundation under Grant No. DMR–8519077 and in part by the U.S. Army Research Office.

The situation for positive parity states is even worse. At most a reasonable description of the 2^+ giant quadrupole resonance can be given which in a discrete basis will be the *lowest* 2^+ state one can obtain. This situation is not essentially altered when a continuum calculation is performed. Obviously, the positive parity spectrum will be strongly influenced by 2p2h excitations which have unperturbed energies above 23 MeV. This is already roughly 7 MeV lower in energy than the $2\hbar\omega$ 1p1h states which can produce positive parity excitations. A straightforward extension of RPA to include 2p2h states,[8] however, leads to similar calculational problems as in the shell-model without leading to much new insight into this problem.

II. THEORETICAL FRAMEWORK

The information on the spectrum of a nucleus is contained in the ph propagator Π which depends on two times. In the energy representation it reads

$$\Pi(\alpha\beta^{-1},\gamma\delta^{-1};E) = \sum_{\lambda\neq 0} \frac{X^{\lambda *}_{\alpha\beta} X^{\lambda}_{\gamma\delta}}{E - E_\lambda + i\eta} - \frac{Y^{\lambda *}_{\delta\gamma} Y^{\lambda}_{\beta\alpha}}{E + E_\lambda - i\eta} \quad , \tag{1}$$

where E_λ corresponds to the excitation energy of the exact eigenstate $|\psi_\lambda\rangle$ with respect to the exact ground state $|\psi_0\rangle$ and X,Y are the usual a^+a ph transition amplitudes. The subscripts α, β, γ and δ refer to a convenient, but arbitrary sp basis. Experimental evidence from (e, e′) reactions[9] shows that most low-lying states in ^{16}O including those of positive parity are strongly excited. This implies that these states are connected by a one-body (a^+a) transition operator to the ground state and are therefore contained in Eq. (1).

The main purpose of this work is to extend the set of diagrams included in the usual RPA result for Π in order to describe the following physical picture. First it is observed that in RPA it is relatively simple to obtain e.g. a low-lying collective 3^-, $T = 0$ state[4,6,7] in accord with experiment. Such a phonon can then be used to construct two-phonon states of positve parity which can then be admixed with the $2\hbar\omega$ 1p1h positive parity states. This will then lead to positive parity phonons which come down in energy on the back of two such 3^- phonons. For the negative parity states in turn, one can combine the original negative parity phonon with such new 0^+, 2^+ phonons to mix with the $1\hbar\omega$ 1p1h excitations to produce modified phonons etcetera. Clearly this requires a self-consistency scheme which will be discussed below.

It is possible to separate off from the ph propagator Π a "free" part Π^f which represents two non-interacting but exact sp propagators.[10] The poles of Π^f correspond to "unperturbed" ph energies which can be related to the experimental energies in the $A \pm 1$ systems. All coupling of the particle and the hole with the medium is then already taken into account.

The irreducible (with respect to Π^f) ph interaction I can now be defined by

$$\Pi = \Pi^f + \Pi^f I \Pi \quad . \tag{2}$$

The corresponding reducible interaction R is given by $R\Pi^f = I\Pi$. Eq. (2) is displayed graphically in Fig. 1 where also the connection between R and I is shown. Since the ph interaction I as defined here, depends only on the energy E it is possible to derive from Eq. (2) an eigenvalue equation for the bound states.[11] The conventional RPA equation can be obtained by assuming Π^f to be diagonal in some sp basis (e.g. harmonic oscillator) and approximating I by an energy independent residual interaction. As an example one can think of the local G matrix type interactions discussed in Ref. 12 which take care of the short-range correlations between nucleons. This will now be the starting point here to extend the description of I to include also long-range correlations.

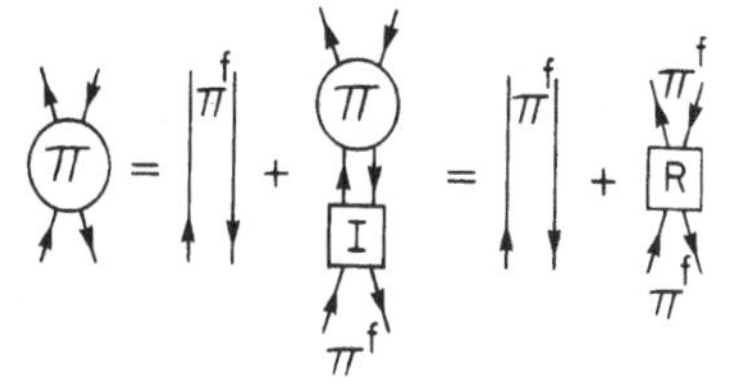

Figure 1. Diagrammatic representation of the ph propagator Π (Eq. (2)). The box labeled I corresponds to the irreducible ph interaction whereas the reducible one, R, contains all the iterations.

Inclusion of the diagrams shown in Fig.2 provides a systematic inclusion of long-range effects as will be shown below. It also contains no double counting. In other contexts similar schemes have been been discussed in the literature,[11,13,14] but only recently its importance for finite nuclei was demonstrated by treating the resulting energy dependence of I correctly.[10] Consider first the correlations which are included by approximating R by a G matrix in Fig. 2. This contribution was studied in Ref. 10 and is shown in Fig. 3. The idea is to allow the particle and the hole to exchange the phonons of the system, these have a low "mass" and will therefore provide a long-range component into the residual interaction.

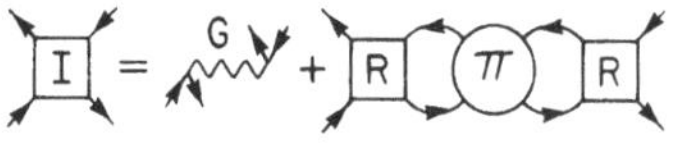

Figure 2. The wiggly line in this figure stands for a G matrix. This improvement of the residual ph interaction is proposed here in order to understand the ^{16}O excitation spectrum.

The non-linear formulation of the problem, I depends on the solution Π, has the advantage that there is an automatic screening included which prevents the occurrence of instabilities, which are signalled by imaginary eigenvalues.[10] This formulation also consistently includes exchange diagrams such that in infinite systems the Pauli principle sum rules are fulfilled.[15] A unique feature of finite systems is the presence of poles in the residual interaction I due to the presence of 2p2h propagation.[10] The energy dependent eigenvalue problem is solved with matrices which have the same dimension of the conventional RPA, due to the presence of poles in I one obtains however, many more eigenvalues (see Ref. 10 for calculational details). Another feature of finite systems is that phonons can have considerable lower energy than the unperturbed energies corresponding e.g. to 1p1h states. This shift is then translated in the occurrence of poles in I which are correspondingly lowered in energy and will now propagate as $(E - \varepsilon_{ph} - E_{phonon})^{-1}$ appropriate to Fig. 3.

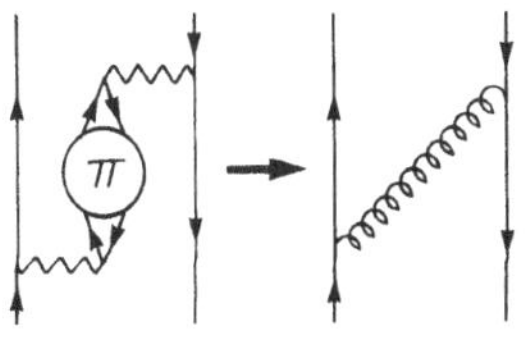

Figure 3. Microscopic one-phonon exchange contribution to the residual ph interaction and its phenomenological interpretation.

This discussion shows that one can lower the unperturbed energies for positive parity states in ^{16}O from 1p1h ($2\hbar\omega$) to 1p1h ($1\hbar\omega$) + phonon which implies a shift from 30 MeV to 18 MeV, respectively, using the experimental energy of the lowest 3^- state. One may therefore conclude that an important element is still missing to understand the ^{16}O excitation spectrum. To obtain this element one should observe first that Fig. 3 simply represents the dressing with full ph correlations of an unperturbed ph state which is exchanged between the original particle and hole. It is then natural to extend this dressing to the connecting interactions, replacing G by R. This then leads to the approximation to I displayed in Fig. 2.

It will now be shown that this extension leads naturally to a consistent microscopic description of ^{16}O. Consider the second contribution to I in Fig.2 with Π replaced by Π^f. Expanding R one obtains

$$R = G + G\Pi G + \cdots \quad . \tag{3}$$

Since the first term has already been considered, the next term in Eq. (3) must now be treated. The corresponding diagram is shown in Fig. 4. Its skeleton structure is obtained only in fourth order in the basic interaction. It is however, the key to understand the excitation spectrum of ^{16}O. Its contribution to Π reads

$$\begin{aligned}\Pi^J_{2ph}(12^{-1},34^{-1};E) = &-\sum_{J_1J_256}(2J_1+1)(2J_2+1)(-1)^{j_1+j_2+j_3+j_4+J}\begin{Bmatrix}J_1 & J_2 & J\\ j_4 & j_3 & j_6\end{Bmatrix}\begin{Bmatrix}J_1 & J_2 & J\\ j_2 & j_1 & j_5\end{Bmatrix}\\ &\int\frac{d\omega_1}{2\pi i}\int\frac{d\omega_2}{2\pi i}\int\frac{d\omega_3}{2\pi i}\, g_1(\omega_1+E)g_2(\omega_1)g_3(\omega_2+E)g_4(\omega_2)\\ &<15^{-1}|\, G^{J_1}\Pi^{J_1}(E-\omega_3+\omega_2)G^{J_1}|\, 36^{-1}>\, g_5(\omega_3+\omega_1-\omega_2)\, g_6(\omega_3-\omega_1+\omega_2)\\ &<52^{-1}|\, G^{J_2}\Pi^{J_2}(\omega_3-\omega_2)\, G^{J_2}|\, 64^{-1}> \quad ,\end{aligned} \tag{4}$$

where th sp propagators g are assumed to be diagonal, which is however, no restriction on the following arguments, and the proper angular momentum coupling has been performed. We note that J is also generic for isospin. After performing the internal energy integration over ω_3, one can show that the integration over ω_1 and ω_2 leads to a factorization of the form $\Pi^f I_{2ph}\Pi^f$, the subscript 2ph standing for 2 phonon. This contribution I_{2ph} to the residual ph interaction contains a number of terms[16] from which a typical example will be further discussed. This term is labeled by I_{2ph}^{+++} referring to the $i\eta$ structure of Π^{J_1},g_5,g_6 and Π^{J_2} respectively, from which it originates. Keeping only the important energy denominators its contribution to the RPA A–matrix reads

$$\begin{aligned}<p_1h_2|\, I_{2ph}^{+++}(E)\,|\, p_3h_4> = &\frac{1}{E-E_{n_1}-(\varepsilon_5-\varepsilon_2)+i\eta}\\ &\frac{1}{E-E_{n_1}-E_{n_2}+i\eta}\,\frac{1}{E-E_{n_2}-(\varepsilon_3-\varepsilon_6)+i\eta}\ ,\end{aligned} \tag{5}$$

where E_{n_i} correspond to the poles in Π^{J_i} and ε_i refers to the poles of the sp propagators. Eq. (5) shows that this contribution contains intermediate two-phonon propagation corresponding to an unperturbed energy $E_{n_1} + E_{n_2}$. Two phonon contributions to giant resonances have also been discussed in Ref. 17, there, however, the problem is formulated in a linear way which leads to bare phonons. Here the non-linear aspect is stressed which leads to a self-consistent formulation still within the RPA framework but with an extended residual interaction. It is important to note that two phonon energies in ^{16}O and ^{40}Ca are of similar magnitude as the unperturbed 1p1h energies. The phonon picture associated with such a contribution is illustrated in Fig. 4. Note that in the present formulation, as for the one-phonon exchange contribution discussed in Ref. 10 one solves the usual RPA eigenvalue equation albeit with an energy dependent residual interaction I(E).

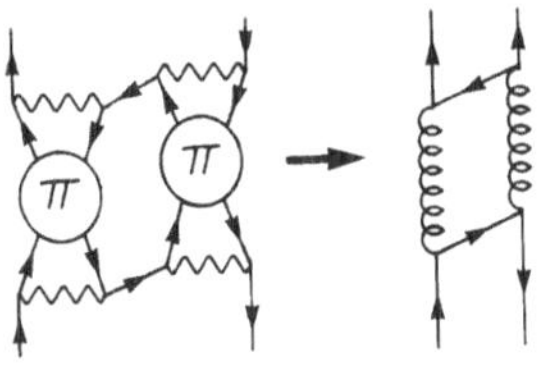

Figure 4. Similar illustration as in Fig. 3 for the two-phonon coupling contribution to the ph interaction which is discussed in the text.

To illustrate how this non-linear scheme can produce the spectrum of ^{16}O, a few iteration steps are discussed now qualitaively. First one starts with a conventional RPA calculation using e.g. the local G matrix of Ref. 12. Such an interaction is sufficiently attractive to bring down a 3^- and 1^- T = 0 state[6,7] to roughly the experimental energy. From these two bare RPA phonons one can then construct a few low-lying positive parity two-phonon states in the residual interaction I_{2ph}. In a next step one diagonalizes the new residual interaction I, this will bring down these 0^+, 2^+ and 4^+ states since the interaction is attractive in these channels[6] as it is for 3^- and 1^- states. The next iteration step contains then already enough phonons to generate the whole set of excited states which are observed in ^{16}O.

This can be explicitly demonstrated by using the experimental excitation energies of the four lowest phonons in ^{16}O (0^+, 3^-, 2^+,1^-) in the construction of unperturbed two-phonon energies. Gathering now the various unperturbed energies, 1p1h and two-phonon, one can compare these with the energies after the "final" diagonalization: the experimental ones. This comparison is made in Fig. 5 where the two-phonon unperturbed states (full lines) and the normal 1p1h unperturbed energies (dashed lines) are compared with experiment. Excitation energies up to 14 MeV were considered and the experimental information was taken from Ref. 18.

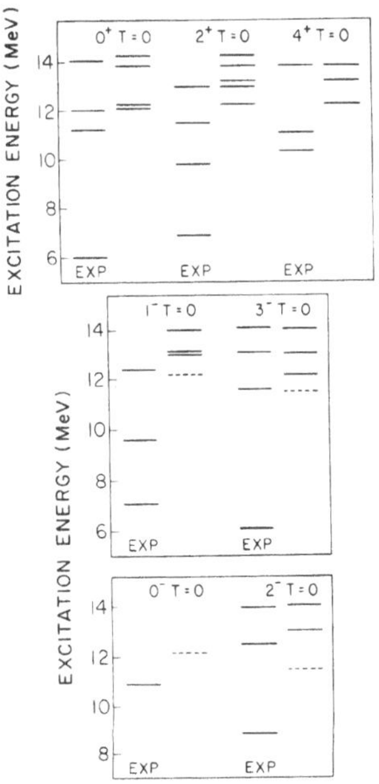

Figure 5. "Final" diagonalization for ^{16}O. For each J^{π} the left column gives the experimentally observed[18] levels. The right column contains the usual 1p1h unperturbed energies (dashed lines) as well as unperturbed two-phonon energies (full lines). In the ideal case of an exact calculation, the final diagonalization will produce the left columns from the "unperturbed" right columns.

A close correspondence between the number of states for each J^π can be observed both for positive parity as well as for negative parity states. The use of experimental energies here in the construction of "unperturbed" energies emphasizes the non-linear aspects of the formulation. One should keep in mind, however, that the skeleton diagram of Fig. 4 even with four bare interactions V, can be dressed by connecting both pairs of V (left and right) with full ph propagators containing the exact energies of the system. Therefore such a two-phonon contribution to the residual interaction always exists. Fig. 5 then illustrates how this contribution (plus all others) is responsible for obtaining the experimental energies (left columns) from these unperturbed ones (right columns).

Thus it is possible to understand the complicated structure of ^{16}O in terms of basic phonons (3^-, 1^-) much like in the work of Ref. 2. One also understands that the positive parity states are generated by a mixture of correlated 2p2h, 4p4h, etc. configuration as in Ref. 1. The advantage of the scheme proposed here is, of course, that it provides a unified description of the whole ^{16}O spectrum. Another consequence of this discussion is namely that the isovector spectrum should remain relatively simple. Since two $T = 0$ phonons can only couple to total isospin 0 and there is no attraction from a G matrix for isovector states,[6] one understands that the $T = 1$ states must remain relatively pure up to the lowest values of $E_{n_1}^{T=0} + E_{n_2}^{T=1} \approx 19$ MeV which is exactly what is experimentally observed.

Concluding, it is possible to understand the nuclear excitation mechanism for ^{16}O in a physically simple and appealing picture. At the same time a microscopic scheme has been established which contains all the necessary ingredients to understand the spectrum of ^{16}O or ^{40}Ca starting from a zero order Slater determinant picture. The formulation of the RPA here in terms of a^+a transition amplitudes is completely sufficient to describe e.g. inelastic electron scattering data. It seems, however, very unlikely that a wavefunction description of the physics described here, is possible.

Finally we would like to draw the attention to the connection of the present discussion for the ph interaction with a "conserving" treatment which constructs the ph interaction by breaking open lines in self-energy diagrams.[19] Inclusion of only RPA contributions to the self-energy already leads to the contribution of the contributions of Figs. 3 and 4 considered at the level of the first iteration step. Further work in clarifying this connection is needed in order to arrive at a more complete self-consistency formulation which includes the sp propagator and the ph propagator.

REFERENCES

1. G. E. Brown and A. M. Green, Nucl. Phys. **75**, 401 (1966).
2. H. Feshbach anf F. Iachello, Phys. Lett. **45B**, 7 (1973).
3. K. Amos, W. Bauhoff, I. Morrison, S. F. Collins, R. S. Henderson, B. M. Spicer, G. G. Shute, V. C. Officer, D. W. Devins, D. L. Friesel, and W. P. Jones, Nucl. Phys. **A413**, 255 (1984).
4. V. Gillet and N. Vinh Mau, Nucl. Phys. **54**, 321 (1964).
5. G. E. Brown and M. Bolsterli, Phys. Rev. Lett. **3**, 472 (1959).
6. P. Czerski, W. H. Dickhoff, A. Faessler, and H. Müther, Nucl. Phys. **A427**, 224 (1984).
7. K. Nakayama, S. Krewald, and J. Speth, Phys. Lett. **148B**, 399 (1984).
8. C. Yannouleas, M. Dworzecka, and J. J. Griffin, Nucl. Phys. **A397**, 239 (1983).

9. T. N. Buti, J. Kelly, W. Bertozzi, J. M. Finn, F. W. Hersman, C. Hyde-Wright, M. V. Hynes, M. A. Kovash, S. Kowalski, R. W. Lourie, B. Murdock, B. E. Norum, B. Pugh, C. P. Sargent, W. Turchinetz, and B. L. Berman Phys. Rev. C **33**, 755 (1986).

10. W. Hengeveld, W. H. Dickhoff, and K. Allaart, Nucl. Phys. **A451**, 269 (1986).

11. S. S. Wu, Scientia Sinica **16**, 347 (1973); **17**, 468 (1974).

12. W. H. Dickhoff, Nucl. Phys. **A399**, 287 (1983).

13. M. W. Kirson, Ann. Phys. **66**, 624 (1971).

14. S. Babu and G. E. Brown, Ann. Phys. **78**, 1 (1973).

15. G. E. Brown, *Lecture Notes in Physics* (Springer, Berlin, 1979), page 119, 1

16. W. H. Dickhoff, to be published.

17. S. P. Kamerdzhiev and V. N. Tkachëv, Yad. Fiz. **36**,73 (1982) [Sov. J. Nucl. Phys. **36**, 43 (1982)].

18. F. Ajzenberg-Selove, Nucl. Phys. **A375**, 1 (1982).

19. G. Baym and L. P. Kadanoff, Phys. Rev. **124**, 287 (1961).

ROLE OF VIRTUAL DOUBLE DELTA COMPONENTS IN NUCLEI*

Steven A. Moszkowski†

Department of Physics
University of California, Los Angeles
Los Angeles, CA 90024

ABSTRACT

A quark-model-based meson-exchange picture leads to a nonlocal short range nucleon-nucleon attraction, due to virtual double delta components, in addition to the well known π, σ, ρ, and ω exchange interactions. In contrast to traditional approaches to nuclear forces, there are no strong short range anti-correlations between baryons in spite of the ω induced repulsion, though there may well be anti-correlations between nucleons alone. This approach leads to significant repulsive three-body interactions in nuclei, and also to momentum-dependent short range effective NN interactions. It is expected that Hartree-Fock calculations for finite nuclei will be simpler than with traditional interactions.

1. TRADITIONAL PICTURE OF NUCLEON-NUCLEON INTERACTIONS

1.1. One Boson Exchange Model

The One Boson Exchange Model (OBEP) (Machleidt et al., 1987) generally involves exchange of four mesons: a. The One Pion Exchange (OPEP) describes long range part of the interactions. OPEP is attractive for even L, repulsive for odd L, and it dominates for high L at small energy (large impact parameter). b. Sigma exchange describes medium range attraction which is nearly spin-independent. Unlike the π, the σ is not a well defined meson with a sharp mass. Rather it is a correlated 0^+ 2π state. However, at least crudely, we can fit the NN phase shifts by assuming that the σ has m = 500 MeV. c. Omega exchange. The S-wave phase shift changes sign at a lab energy of about 250 MeV. This requires that the potential is mainly repulsive above this energy. A simple way to accomplish this is to have a short range repulsion. In early studies of nuclear forces, an infinite hard sphere repulsion was used. More recent studies take a Yukawa repulsion which can arise very naturally from omega exchange, analogous to photon exchange between like charges. The ω is known to have a rather narrow width of about 10 MeV, and it can also be regarded as a strongly correlated 3π system. d. Finally we must also

*Invited paper presented at the XIth International Workshop on Condensed Matter Theories, July 27-August 1, 1987, Oulu, Finland.
†Partly supported by a grant from the US Army Research Office.

include rho-meson exchange in order to obtain a realistic strength of the tensor force. e. The η meson is not important for the NN force, as it is a pseudoscalar with rather large mass.

1.2. Phase shifts and short range correlations

The short range repulsion reduces the NN wavefunction at short distances, even for low momenta where the overall effect of the interaction is attractive. This can be seen analytically if the Born approximation is valid, but it holds qualitatively for strong potentials as well. The Born scattering length a_B is proportional to the volume integral of the potential:

$$a_B = (m/4\pi\hbar^2) \int V(s)\, d^3s \quad .$$

On the other hand, the zero energy wavefunction, normalized such that

$$(r\psi)_{r\to\infty} \to r - a_B$$

has

$$(r\psi)_{r\to 0} \to \left(1 - (m/4\pi\hbar^2) \int V(s)\, s^{-1} d^3s\right) r$$

which weighs the short range region of the potential more strongly than does the scattering length.

At high energies, the Born S-wave phase shift also weighs short distances more:

$$\delta_S(k)_{k\to\infty} \to -(m/8\pi\hbar^2 k) \int V(s) s^{-2} d^3s.$$

For high partial waves, if both Born and semi-classical approximation are valid, then $\delta_L(k)_{k\to\infty}$ is proportional to $V(L/k)$.

If we assume that the potential is local, i.e., that it can be described as a function of r (and S and T), independent of energy, then a fit to the NN phase shifts definitely requires a short range repulsion, which gives rise to a correlation hole in the wavefunction.

1.3. Coupling to Delta Isobar

The intermediate range attraction can be roughly described as resulting from exchange of a scalar σ meson, and on a more fundamental level as resulting from exchange of two pions (TPEP). It has been known for more than 10 years that intermediate states containing one or more virtual deltas play an important role for this part of the interaction (Green, 1979). One of the main conclusions of the present work is that coupling to virtual $\Delta\Delta$ components essentially cancels weakens the short range NN anti-correlations due to ω exchange.

1.4. Why virtual double delta components might be important

In the two pion exchange model discussed above, the baryon intermediate states can be either NN, NΔ, or $\Delta\Delta$. (Of course there could be states involving other nucleon resonances, such as NN*, which we will not consider further here, though they could also play a role.) In first order, the excitation energy for $\Delta\Delta$ is twice as large as NΔ, if we neglect kinetic energies, and this is one of the reasons that $\Delta\Delta$ states are believed by some not to play an significant role. However, there are two effects, namely phase space and strong short range $\Delta\Delta$ attraction, both of which tend to enhance the role of the $\Delta\Delta$. For example, consider the deuteron ground state, which has $S=1$, $T=0$. According to traditional nuclear theory, it consists of a nucleon pair which is mainly in a $^3S_1$

state with some $^3D_1$ admixture. We can also have a ΔΔ system with these values of S and T, and indeed, according to the model discussed here, the deuteron contains about 1% ΔΔ admixture. However, it is not possible to couple a NΔ system to have T = 0, thus the deuteron ground state cannot have any NΔ component (provided crossed diagrams, where 2 pions are in the air at the same time, are neglected.) For an S = 0, T = 1 state, which forms the lowest (unbound) state of the nn or pp system, one can get an NΔ component due to the tensor force induced coupling between $^1S_0$ and $^5D_0$. Of course, the deuteron is a very atypical nucleus, due to its small binding energy. About 95% of the time, the neutron and proton are further apart than the average internucleon distance in normal nuclei. Thus, a 1% ΔΔ component in the deuteron translates into about 20% of ΔΔ in nuclear matter.

2. SHORT RANGE DELTA-DELTA ATTRACTION - THE DOUBLE DELTA FORCE

There are four ingredients which, when considered together, lead to this important effect. In the past, many authors have considered one or more of these, but not all four. They are:

1. The quark model for ratios of coupling constants,
2. the delta function term in the one pion exchange,
3. the larger Δ size compared to the nucleon size,
4. nonlocality.

Let us consider the effect of each of these.

2.1. Quark based NN and ΔΔ Coupled Channel Model

There is considerable evidence concerning the strength of the πNΔ coupling, but this is not the case for the πΔΔ coupling. However, since the quark model appears to give at least a ball park estimate for the former, it appears reasonable to use the quark model also to tell us something about the πΔΔ vertex. In fact, if we regard the N and Δ as three quark systems which differ only with respect to their spin and isospin coupling, then it is well known (Arenhovel, 1975) that:

$$f_{\pi qq} : f_{\pi NN} : f_{\pi N\Delta} : f_{\pi\Delta\Delta} = 1 : \frac{5}{3} : \sqrt{8} : \frac{1}{3}$$

Thus we find that:

$$f^2_{\pi NN} : f^2_{\pi N\Delta} : f^2_{\pi\Delta\Delta} = 25:72:1.$$

It might therefore be thought that the second coupling, namely $f_{\pi\Delta\Delta}$, is not important. As we will see, however, this is definitely **not** the case. Let us consider again the deuteron. Considering only relative S-states, we obtain the following coupled channel equations for the NN and ΔΔ components:

$$-(\hbar^2/M_n)u''_{NN} + V_{NN}\,u_{NN} + V_{Tr}\,u_{\Delta\Delta} = Eu_{NN}$$

$$-(\hbar^2/M_\Delta)u''_{\Delta\Delta} + 2\varepsilon_{N\Delta}\,u_{\Delta\Delta} + V_{Tr}\,u_{NN} + V_{\Delta\Delta}\,u_{\Delta\Delta} = Eu_{NN}$$

Here V_{Tr} denotes the transition potential $V_{NN\leftrightarrow\Delta\Delta}$, V_{NN} and $V_{\Delta\Delta}$ the diagonal potentials in NN and ΔΔ channels, and $E_{N\Delta}$ the nucleon-delta mass difference. According to the quark model, the three OPEP potentials are related by:

$$V_{NN} : V_{Tr} : V_{\Delta\Delta} = 1 : -\frac{16\sqrt{5}}{25} : -\frac{11}{5}$$

For the so called "B1 potential" (Dymarz and Khanna, 1986), $V_{NN}(r=0)$ and $|V_{\Delta\Delta}(r=0)| > 2\varepsilon_{N\Delta}$. Since V_{Tr} couples NN and ΔΔ channels, it leads to a lowering of the effective NN potential.

2.2. The one pion exchange potential

The central part of the OPEP potential in the deuteron channel can be written as (Machleidt et al., 1987)

$$\tilde{V}_{\sigma\tau}(q) = \frac{f^2_{\pi NN}}{4\pi} \times \frac{q^2}{m^2} \times \frac{1}{m^2 + q^2} \times \left(\frac{\Lambda^2 - m^2}{\Lambda^2 + q^2} \right)^2$$

in momentum space, where m refers to the pion mass, and monopole form factor $\frac{\Lambda^2 - m^2}{\Lambda^2 + q^2}$ is assumed at each meson nucleon vertex.

In the limit $\Lambda \to \infty$, which is equivalent to neglecting the finite size of the nucleon, we obtain the following:

$$\tilde{V}_{\sigma\tau}(q) = \frac{f^2_{\pi NN}}{4\pi} \times \left(\frac{1}{m^2} - \frac{1}{m^2 + q^2} \right) .$$

The corresponding coordinate space potential is:

$$V_{\sigma\tau}(r) = \frac{f^2_{\pi NN}}{4\pi} \times \left(\frac{4\pi}{m^2} \delta(\vec{r}) - \frac{e^{-mr}}{r} \right) .$$

Note that the contributions of the repulsive delta function and Yukawa attraction to the volume integral of the potential, $\tilde{V}_{\sigma\tau}(0) = \int V_{\sigma\tau}(r) d^3r$, just cancel.

The coordinate space potential, including a finite value of Λ, is:

$$V_{\sigma\tau}(r) = \frac{f^2_{\pi NN}}{4\pi} \times \left(\left(1 + \frac{\Lambda^2 - m^2}{2m^2} \Lambda r \right) \frac{e^{-\Lambda r}}{r} - \frac{e^{-mr}}{r} \right) .$$

For large r, this is just the usual Yukawa potential, while for small r:

$$V_{\sigma\tau}(r) = \frac{f^2_{\pi NN}}{4\pi} \times \left(\frac{(\Lambda - m)^2 (\Lambda + 2m)}{2m^3} - \frac{(\Lambda^2 - m^2)^2}{2m^3} r + \ldots \right) .$$

It is interesting to work out the distance r_c where $V_{\sigma\tau}$ changes sign. A rough analytic approximation is:

$$r_c \approx \frac{2}{\Lambda} \ln \frac{\Lambda}{m} \approx \frac{4}{\Lambda} .$$

Since $\Lambda \approx 1$ GeV $= 5$ fm^{-1}, we obtain $r_c \approx 1$ fm, which is much larger than might be thought on dimensional grounds alone. In fact, it appears that the short range repulsion due to the smoothed out δ-function in the π exchange potential is as important as the well known repulsion due to ω exchange.

Note that if there are strong short-range anti-correlations between nucleons, then higher order effects of the δ-function term tend to cancel the first order term, and the proper procedure is to drop it (Brown et al., 1985). However, if, as suggested by the work here, and also by results of quark model calculations (Oka and Yazaki, 1983), these anti-correlations are weak, then the δ-function should be kept.

The treatment of the δ-function term has a profound effect on the high momentum behavior of the pion exchange potential. Consider here the two 2-body diagrams illustrated in Fig. 1. In particular, let us make crude estimates for the average momentum carried by the mesons. For single meson exchange, the nucleon is in the Fermi sea in both initial and final state. The average momentum transfer is then of order $k_{AV} \approx \sqrt{(3/5)} k_F \approx 1$ fm^{-1}, where k_F denotes Fermi momentum. On the other hand, for two meson exchange, the intermediate baryon can be N or Δ (or N^*, etc.), and generally the transferred momentum can be large. The contribution of TPEP to the energy is proportional to the integral of $q^2 \tilde{V}^2_{\sigma\tau}(q)$ over q. This converges only by virtue of the cutoff Λ ($\gg m_\pi$) and has a maximum for q ≈

$\Lambda/\sqrt{3} \approx 3\ fm^{-1}$. The same argument applies for the two 3-body diagrams in the figure.

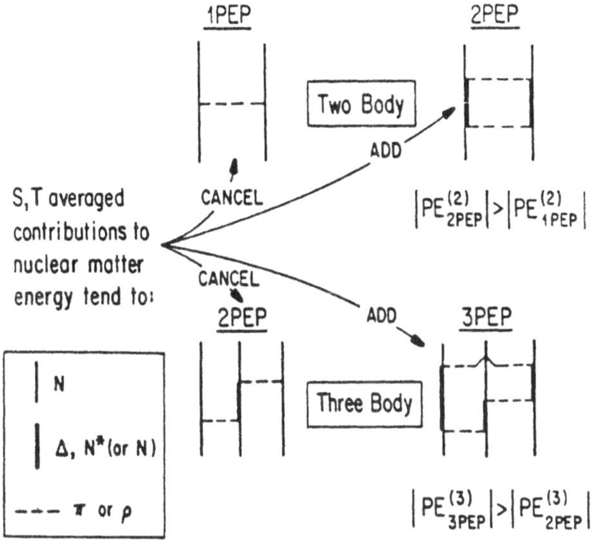

FIG. 1. Meson exchange model of two and three nucleon interactions.

The OPEP transition and ΔΔ potentials have the same form as the NN potential, but different coefficients, and opposite sign. Thus,

$$V_{\Delta\Delta}(\pi) = -\frac{11}{5} V_{NN}(\pi) ,$$

i.e., $V_{\Delta\Delta}(\pi)$ is repulsive at large distances, but attractive at short distances. The coefficient 11/5 obtained for π exchange attraction in the ΔΔ channel using the quark model is sufficiently large that it more than compensates for the ω exchange repulsion in this channel. Note that:

$$V_{\Delta\Delta}(\omega) = V_{NN}(\omega) .$$

In fact, at short distances (< 1/2 fm), the $V_{\Delta\Delta}$ attraction also dominates over the extra excitation energy $2\varepsilon_{N\Delta}$ of the ΔΔ system relative to NN. Since $V_{NN}(\pi)$ is repulsive in this region, we expect that for $r \approx 0$, the ΔΔ component of the two baryon wavefunction is actually **larger** than the NN component. At distances larger than 1 fm, we get the traditional picture of an attractive $V_{NN}(\pi)$ and a repulsive $V_{\Delta\Delta}(\pi)$, so that the two baryons are mainly NN in this region.

We note that there are two regions where V_{NN}^{eff} is attractive: at large distances, due to π and σ exchange, and at short distances, due to the ΔΔ admixture. In between, V_{NN}^{eff} is repulsive due to ω and π exchange. Classically, a baryon cannot travel from one region to the other, unless its energy is sufficient to surmount the barrier, but quantum mechanically, it can tunnel through the barrier. Clearly the probability of such tunneling depends sensitively on the details of the well. This point was discussed by Matveev and Sorba (1977), though they describe the short range region as a 6 quark state. These authors also discuss the close correspondence between the predominance of ΔΔ components at short distances and formation of a 6 quark bag. The diagonal potential V_{NN} is, however, repulsive at short distances. In the language of the quark model, this is due to the color magnetic interaction (Myhrer and Wroldsen, 1986). The change of sign for $V_{\Delta\Delta}$ can be interpreted as due to a change of sign in the color magnetic interaction.

If there were no transition potential, then the crossover between these two regions would be sudden, at the point where $2\varepsilon_{N\Delta} + V_{\Delta\Delta} = V_{NN}$. However, due to the presence of the transition potential, the crossover is gradual. The effective NN potential is obtained by diagonalization of the energy matrix:

$$\begin{vmatrix} V_{NN} & V_{Tr} \\ V_{Tr} & 2\varepsilon_{N\Delta} + V_{\Delta\Delta} \end{vmatrix}$$

which gives:

$$V_{NN}^{eff} = \frac{1}{2}\,(2\varepsilon_{N\Delta} + V_{\Delta\Delta} + V_{NN}) - \frac{1}{2}\left((2\varepsilon_{N\Delta} + V_{\Delta\Delta} - V_{NN})^2 + 4\,V_{Tr}^2\right)^{1/2} \quad .$$

Up to second order in V_{Tr}, we obtain:

$$V_{NN}^{eff} = V_{NN} - \frac{V_{Tr}^2}{2\varepsilon_{N\Delta} + V_{\Delta\Delta} - V_{NN}} + \ldots$$

This effective NN interaction due to π exchange is crudely sketched in Fig. 2. There are good reasons to believe that $\varepsilon_{N\Delta}$ increases with nucleon energy. Thus $\varepsilon_{N\Delta} = 300$ MeV corresponds to a low energy NN system "Low E", such as the deuteron. On the other hand, at high nucleon energies, "High E", $\varepsilon_{N\Delta}$ is probably much larger. The case of infinite $\varepsilon_{N\Delta}$ corresponds to a pure NN system without any Δ admixture.

The above expression for the effective potential are strictly valid only if V_{NN}, $V_{\Delta\Delta}$, and V_{Tr} all change slowly with distance. However, it qualitatively reproduces the results of integrating the full coupled channel equations for realistic potentials.

It should be emphasized that the short range $\Delta\Delta$ attraction occurs **only** in the S = 1, T = 0 or S = 0, T = 1 channels. For most other channels, the short range π exchange interaction between $\Delta\Delta$ is repulsive. For example, in the S = 3, T = 2 or S = 2, T = 3 channels, there is a strong repulsion (Oka and Yazaki, 1983), which can be associated with the Pauli principle, analogous to the well-known short range repulsion between neutral atoms.

2.3. Relative sizes of Δ and N and meson-baryon form factors

As has been pointed out, the region over which the $V_{\Delta\Delta}$ is attractive depends on the value of Λ. In most calculations done up to now, the **same** form factor was used in all channels. This means that we obtain the same radial dependence for $V_{\pi NN}$, $V_{\pi Tr}$, and $V_{\pi\Delta\Delta}$. Fits to NN phase shifts (Machleidt et al., 1987), stringent constraints on $\Lambda_{\pi NN}$, namely that $\Lambda_{\pi NN}$ = 1.3 to 1.5 GeV. However, the quark model also suggests that the Δ is slightly larger than the N, by about 15% (Dey and Dey, 1984). This is since the color magnetic one-gluon exchange potential is attractive for the N, but repulsive for the Δ. The same effect also accounts for the larger size of the ρ relative to the π. Very crudely, then we can write:

$$\Lambda_{\pi N\Delta} = (r_N/r_\Delta)\,\Lambda_{\pi NN};\; \Lambda_{\pi\Delta\Delta} = (r_N/r_\Delta)^2\Lambda_{\pi NN} \quad .$$

This implies a form factor $\Lambda_{\pi\Delta\Delta} \approx 1.0$ GeV which results in about 1% $\Delta\Delta$ component in the deuteron (Dymarz and Khanna, 1986). Clearly, the larger the value of $\Lambda_{\pi\Delta\Delta}$, the smaller is the region of the $\Delta\Delta$ attraction, and the admixture of $\Delta\Delta$ component. With $\Lambda_{\pi\Delta\Delta} \approx 1.3$ GeV, the $\Delta\Delta$ admixture is 1/2 %

or less. The ΔΔ admixture shows up as a high momentum component in the wavefunction. In order to tell which is correct, one must look at evidence for such high momentum components.

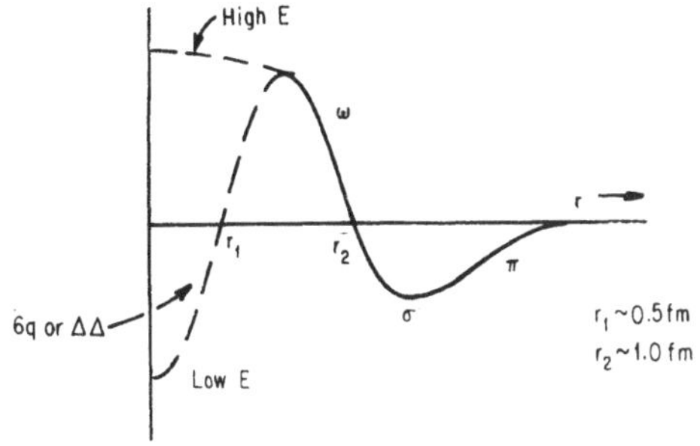

TUNNELLING PROBABILITY

$$P \sim e^{-2\int_{r_1}^{r_2} \sqrt{\frac{m}{\hbar^2}\left[V(r)-E\right]}\, dr}$$

FIG. 2. Effective NN potential with coupling to ΔΔ channel. The dashed lines indicate the expected energy dependence of the effective short range NN potential.

2.4. Nonlocality at Short Distances

Nucleon-nucleon potentials due to one-meson exchange are local, at least in the non relativistic approximation. Thus at distances larger than 1 fm, we do expect the NN potentials to be local. However, as has been pointed out many times in the literature, the NN potential is expected to be strongly nonlocal at distances less than 1 fm. The Paris potential (Lacombe et al., 1980) has this form, local at large r, but a short range nonlocality, expressed as a linear dependence of the short range interaction on energy. If we allow for short range nonlocality, then it is possible to fit phase shifts with a large variety of potentials, which differ greatly with respect to the short range NN correlations (Moravcsik et al., 1974). Indeed, it is not possible to get any reliable information regarding short range NN correlations from phase shifts alone.

The short range nonlocality is quite plausible in the Dirac description of the NN interaction (Cooper and Jennings, 1986). Thus, in the nonrelativistic limit, a Lorentz scalar interaction behaves like a σ exchange potential. However, the interaction energy involves not the particle density, but the scalar density, which contains an extra factor m/E. Thus the interaction is relativistically quenched. For small values of energy, the potential varies linearly with energy.

During the last few years, it has become clearly established that a fit of the spin dependent nucleon-nucleus scattering amplitudes (for incident nucleons of energies in the 100 to 500 MeV range) requires a strong nonlinear dependence of the single particle potential on densitiy (Clark et al., 1983). Such a density dependence can be easily generated if the nucleons are assumed to be Dirac particles and the single particle potential is assumed to be made up of an attractive Lorentz scalar plus a repulsive Lorentz vector, which happen to nearly cancel. The well-known transformation of the Dirac equation to an equivalent Schrödinger equation gives for the central part of the single particle potential (Anastasio et al., 1983; Amado et al, 1984).

$$U_{Eff} = \rho(-S + EV/Mc^2) + \rho^2(S^2-V^2)/2Mc^2 + \ldots$$

S and V are due to exchange of scalar and vector mesons. Using the relativistic theory, it now possible to understand the empirical saturation energy and density of nuclear matter (Brockmann and Machleidt, 1984). This had not been possible using the traditional non-relativistic theory, if one considered only two body interactions.

Perhaps a more fundamental mechanism for the short range nonlocality is provided by the quark model, as resulting from the composite structure of the nucleons. This has been worked out by several authors using the resonating group cluster model (See, for example, Hecht and Fujiwara 1986).

For the "Double delta force" discussed in this paper, the effective NN potential turns attractive at short distances. This picture may be quite consistent with the properties of the deuteron, which is essentially a zero energy system. However, it will not work, without modification, for a fit to NN phase shifts. This can be seen in a very simple way: For scattering at high energies, we are sampling the shortest range part of the potential. Thus the S-wave phase shifts should turn attractive again at high energies, contrary to what is found empirically. This, of course, presupposes a local N potential.

However, a coupled channel calculation with a linear energy dependence of $\varepsilon_{N\Delta}$ generates nonlocality in the effective NN potential, even without any explicit nonlocality in the potentials V_{NN}, V_{Tr}, and $V_{\Delta\Delta}$, themselves. The central short range part of the effective NN interaction can be written as:

$$V(r) + \frac{1}{2}\,[p^2W(r) + W(r)p^2] \quad .$$

The effect of energy dependence of $\varepsilon_{N\Delta}$ and of different form factors for N and Δ on NN phase shifts is currently being investigated by R. Dymarz. Preliminary results are encouraging.

3. APPLICATIONS TO NUCLEAR STRUCTURE

3.1. High Momentum Components

Results of electromagnetic form factors and also of deep inelastic results, say of electrons from D and from ^{3}He, indicate the presence of more high q components than expected for RSC or Paris potentials, but consistent with potentials which contain appreciable $\Delta\Delta$ or 6 quark components at short distances (Kisslinger, 1986). These conclusions are, however, subject to uncertainties regarding meson exchange currents.

3.2. Nucleon-Nucleon G-matrix

The model discussed here can be readily applied to nuclear structure calculations. Indeed, the standard Bethe-Brueckner G-matrix method can be used (Bethe, 1971), just as for traditional interactions which lead to short range anti-correlations. Even in the model discussed here, there must be correlations, but they involve mainly isobar excitations rather than a correlation hole. That there must be **some** kind of correlation can be seen by making a low density expansion of the G-matrix:

$$G_{NN}(\rho) = G_{NN}(0) + (U_{int} - U_i) \int (\psi-\phi)^2 \, d^3r \quad ,$$

where U_N and U_{int} denote single particle potentials in the fermi sea and for virtual intermediate states, respectively. $\phi(\psi)$ refers to unperturbed (perturbed) baryon-baryon wavefunction. $G_{NN}(0)$ is essentially proportional to free NN phase shifts. This term alone would overbind nuclear matter by about 20 MeV. Thus we need to consider the second term in G_{NN}, which is due to the nuclear medium effects. If we assume that the potential in intermediate states vanishes, then one obtains approximately:

$$G_{NN}(\rho) \approx G_{NN}(0) \frac{1}{1+2\kappa}$$

where

$$\kappa = \rho \int (\psi-\phi)^2 \, d^3r \approx \rho \int \Big([\psi(NN^3S_1)-1]^2 + \psi^2(NN^3D_1) + \psi^2(\Delta\Delta) \Big) d^3r \quad \text{for } T = 0,$$

$$\approx \rho \int \Big([\psi(NN^1S_0)-1]^2 + \psi^2(N\Delta^5D_0) + \psi^2(\Delta\Delta) \Big) d^3r \quad \text{for } T = 1,$$

and we have considered the most important sources of correlations for the deuteron (and set $\phi = 1$).

Detailed calculations (Brown et al., 1986) show that potential felt by a Δ in nuclear matter cannot be neglected, but is about half as large as that for a nucleon. This potential is mainly due to the interaction between Δ and N. For a Reid Soft Core potentials, the three contributions to κ are $0.05 + 0.1 + 0 = 0.15$. I.e., some short range repulsion, large tensor effect, but no isobar contribution.

For the kind of potential discussed here, for example, the B1 potential (Dymarz and Khanna, 1986), the contributions are estimated to be $0 + 0.05 + 0.1 = 0.15$. The total κ is about the same, though the density dependence is stronger for the B1 potential. (This is since the tensor correlations become weaker with increasing density, to a greater degree than do the repulsion or isobar correlations.) The quenching of the attraction at short distances, which gives rise to the factor $(1+2\kappa)^{-1}$ in the effective NN interaction, results from the fact that the $\Delta\Delta$ attraction becomes weaker with increasing density.

3.3. Nuclear three body interactions

The fact that nucleons are composite objects provides a natural explanation for the presence of three body interactions between nucleons. As is well known, two body interactions can be regarded from exchange of mesons or quarks between two nucleons. Similarly, when three nucleons are sufficiently close that their bags overlap, we can get non-additive interactions involving 3 meson or quark exchanges.

Figure 1 shows what I think are the important components of two and three body interaction between nucleons arising from meson exchange. First consider the well known case of the two body interaction. We have, of course, the one pi+rho exchange (One pion exchange two body force) denoted by 1PE2BF. As is well known, its spin-isospin average contribution to the nuclear matter energy is relatively small. Indeed, it cancels exactly in the Hartree approximation. The two meson exchange term 2PE2BF gives a much larger contribution to the nuclear matter energy, in spite of the fact that it is of higher order in the meson nucleon coupling than the 1PE2BF. In the 2PE2BF, the spin-isospin averaged contributions add coherently, and also most of the time the intermediate states are Δ's. The coupling $f^2_{\pi N\Delta}$ is about 4 times as large as $f^2_{\pi NN}$, which makes the 2PE2BF an intermediate range attraction, the dominant part of the nucleon-nucleon interaction.

Now let us look at the three body interaction. There is a two meson exchange contribution three body force (2PE3BF) (Coon et al., 1981; Friar et al., 1984), which was, in fact, used by Carlson et al. (1983) and by Wiringa (1983) in their calculations of nuclear matter and also of nuclei with A = 3 and 4. However, the same spin-isospin cancellations which decrease the 1PE2BF contribution act for the 2PE3BF as well, so that this part of the three body interaction contributes only 1 or 2 MeV/A to the nuclear matter binding energy. Finally consider the three meson exchange diagram 3PE3BF. The 3PE3BF interaction is the analogue of the Axilrod-Teller interaction for atoms (Axilrod and Teller, 1943). At first sight it seems reasonable that this diagram should be analogous to the 2PE2BF two body interaction, and should thus be quite large and repulsive. Indeed, I believe that this is the case. However, a more detailed calculation shows that a strong repulsive 3PEP contribution to nuclear matter results only if the short range anticorrelations between baryons are weak. With traditional NN potentials, the short range anticorrelations lead to a weak or even attractive third order energy (Friman and Nyman, 1978).

3.4. The Nuclear Shell Model Interaction. From the Double Delta Force to the Surface Delta Interaction

The density dependence of the G-matrix can be simulated by taking an effective $V_{NN} + V_{NNN}$ interaction, both of which have finite range. It should be pointed out here that the combination

$$V_{Bind.} = V_{NN} + \frac{1}{3} \rho V_{NNN}$$

enters in the total binding energy, while the different combination

$$V_{Shell} = V_{NN} + \rho V_{NNN}$$

determines the shell model interaction between valence nucleons. Thus, the three body interaction is three times as effective in reducing the correlations due to the attraction than in reducing the total binding (Sharp and Zamick, 1973).

The success of the nuclear independent particle model may be connected with the fact that at normal nuclear density, $V_{eff.}$ is quite small, i.e., the correlations due V_{NN} & V_{NNN} tend to cancel in the interior.

$$H_{shell} = E_0 + \sum h(i) + \sum\sum V_{Shell}(i,j) ,$$

$$h(i) = T(i) + U^{(2)}(i) + U^{(3)}(i) ,$$

i.e., both the 2 body and 3 body interactions contribute to the single particle hamiltonian. The fact that the contributions of these interactions nearly cancel in nuclear matter implies that the shell model interaction is most important at the nuclear surface, where the density is lower. This can be seen in another way by considering the matrix elements due to the nonlocal short range part of the effective NN interaction. Let us approximate the latter by a momentum dependent Gaussian:

$$V_{shell}(p,r) \approx -V_0 e^{-p^2/p_0^2} e^{-r^2/r_0^2} .$$

If $p_0 = \infty$, we have a conventional finite range Gaussian interaction. For $p_0 r_0 = 2\hbar$, we obtain a separable S-state interaction. It is interesting to note that a finite value of p_0 leads to matrix elements corresponding to a range smaller than r_0. This together with the momentum dependence leads to matrix elements similar to those due to the surface delta interaction (Plastino et al., 1966).

$$V_{SDI}{}^{ij} \approx \delta(r_{ij})\ \delta(r_i - R)\ \delta(r_j - R) \approx -4\pi G\ \delta(\Theta_{ij}) ,$$

which has proven quite successful for fitting general features of nuclear spectra. In fact, this correspondence is exact for certain values of p_0 and r_0 in the limit that the angular momenta are much less than the maximum allowed values.

4. SUMMARY AND FUTURE OUTLOOK

The main new result obtained here is the short range attraction between baryons (which must, however, become weaker with increasing momentum, so as to preserve the fit to NN phase shifts). While the NN wavefunction still has a wound (which can be associated with the effect of ω exchange or of color magnetic one gluon exchange), the coupling to other channels, especially $\Delta\Delta$ and probably also NN^*, fills in the hole in the baryon-baryon density at short distances. Truncating to the space of nucleons alone, the effective interaction between nucleons has a short range repulsion, but also a momentum dependent short range attraction. In addition, due to nuclear medium effects on the Δ self energy, there is a repulsive three body interaction of the Axilrod-Teller type. This NNN interaction tends to cancel the effect of the attractive NN interaction, especially in the nuclear interior.

About 15-20 years ago, considerable progress was made in calculating the properties of finite nuclei using the NN interactions available at the time, for example, the Reid soft core potential (Bethe, 1971). The use of the Thomas-Fermi method affords considerable simplification. (Bethe, 1968). Also, extensive Hartree-Fock calculations were made, in which the effect of density dependence was taken into account. (See, for example, Negele, 1970; Nemeth and Ripka, 1972.)

It appears likely that, in the view of the weaker correlations in the quark-based model discussed here, that the Hartree-Fock approximation will be easier to implement than it was in the traditional picture, though the calculations can be done along lines very similar to those in the past. Incidentally, the momentum dependence of the effective interactions plays a crucial role, and reproduces in a natural way some of the effects previously associated with exchange terms. (Bar-Touv and Levinson, 1967). For summaries of self consistent descriptions of finite nuclei with effective nuclear forces see for example, Quentin and Flocard (1978) and Brack, Guet and Hakansson, (1985). The quark-based model discussed here will, hopefully, provide a more microscopic theory for such effective nuclear forces, including significantly repulsive three body interactions between nucleons.

Ending on an optimistic note, the author hopes that perhaps the double delta can become something of a bridge between quark and nuclear physics.

REFERENCES

Amado, R. D., Piekarewicz, J., Sparrow, D. A., and McNeil, J. A., 1984, Phys. Rev. C, 29:936.

Anastasio, M., Celenza, L. S., Pong, W. S., and Shakin, C. M., 1983, Phys. Repts., 100:327.

Arenhovel, H., 1975, Z. Phys., A275:189.

Axilrod, B. M., and Teller, E., 1943, J. Chem. Phys., 11:299.

Bar-Touv, J., and Levinson, C. A., 1967, Phys. Rev., 153:1099.

Bethe, H. A., 1968, Phys. Rev., 167:879.

Bethe, H. A., 1971, Ann. Revs. Nucl. Sci., 21:93.

Brack, M., Guet, C., and Hakansson, H. B., 1985, Phys. Repts., 123:275.

Brockmann, R., and Machleidt, R., 1984, Phys. Lett., 149B:283.

Brown, G. E., Osnes, E., and Rho, M., 1985, Phys. Lett., 163B:41.

Carlson, J., Pandharipande, and Wiringa, R. B., 1983, Nucl. Phys., A401:59.

Clark, B. C., Hanna, S., Mercer, R. L., Ray, L., and Serot, B. D., 1983, Phys. Rev. Lett., 50:1643.

Coon, S. A., and Glockle, W., 1981, Phys. Rev. C, 23:1790.

Cooper, E. D., and Jennings, B. K., 1986, Nucl. Phys., A458:717.

Dey, J., and Dey, M., 1984, Phys. Lett., 138B:200.

Dymarz, R., and Khanna, F. K., 1986, Phys. Rev. Lett., 56:1448.

Friar, J. L., Gibson, B. F., and Payne, G. L., 1984, Ann. Rev. Nucl. Part. Sci., 34:403.

Friman, B. L., and Nyman, E. M., 1978, Nucl. Phys., A302:365.

Green, A. M., 1979, "Nuclear Physics with N's and Δ's: The Coupled Channel Method, in Mesons and Nuclei, Vol. 1," eds. M. Rho and D. H. Wilkinson, North Holland, Amsterdam, pp. 229-259.

Hecht, K. T., and Fujiwara, Y., 1987, Nucl. Phys., A463:255.

Kisslinger, L., 1986, The HQH Model for Two- and Three-Nucleon Systems, a Progress Report, in "Physics, Vol. 260," eds. B. L. Berman and B. F. Gibson, Springer, Berlin, pp. 432-442.

Lacombe, M., Loiseau, B., Richard, J. M., Vinh Mau, R., Côté, J., Pires, P., and de Tourreil, R., 1980, Phys. Rev. C, 21:861.

Machleidt, R., Holinde, K., and Elster, Ch., 1987, Phys. Repts., 149:1.

Matveev, V. A., and Sorba, P., 1977, Lett. al Nuovo Cim., 20:435.

Myhrer, F., and Wroldsen, J., 1986, Phys. Lett., B174:366.

Moravcsik, M. J., and Ghosh, P., 1974, Phys. Rev. Lett., 32:321.

Negele, J. W., 1970, Phys. Rev. C, 1:1260.

Nemeth, J., and Ripka, G., 1972, Nucl. Phys., A194:329.

Oka, M., and Yazaki, K., 1983, Nucl. Phys., A402:477.

Plastino, A., Arvieu, R., and Moszkowski, S. A., 1966, Phys. Rev., 145:837.

Quentin, P., and Flocard, H., 1978, Ann. Revs. Nucl. Part. Sci., 28:523.

Sharp, R., and Zamick, L., 1973, Nucl. Phys., A208:130.

Wiringa, R. B., 1983, Nucl. Phys., A401:86.

CHARACTERISTICS OF PION-CONDENSED NEUTRON STARS

PULSAR GLITCH AND SUPERNOVA 1987A

Tatsuyuki Takatsuka

College of Humanities
and Social Sciences
Iwate University
Morioka 020 Japan

1. INTRODUCTION

Pion condensation (PC), starting from the pioneering works by Migdal[1] and also by Sawyer and Scalapino[2], has been a subject of continuous interests in nuclear physics. At early stage attentions are paid mainly to the onset density and the theoretical framework. In recent years, interests are being directed to the possible consequences of PC on the phenomena related to neutron stars (NS) and energetic heavy-ion collisions.

Recent progress in observational side, in conjunction with theoretical developments, is providing us with important probes to the structure and thermal properties of NS and also to the equation of state (EOS) for dense nucleon matter. It is also suggesting the existence of superfluids and exotic phases such as PC and quark matter in NS. For example, observational data for NS masses or mass-radius ratios, extracted from binary systems[3] and X-ray[4] or γ-ray[5] bursters, impose an important constraint on the EOS. On the other hand, those for surface temperatures[6] and pulsar glitches[7] bring valuable informatins on NS interiors including the possible realization of exotic phases. In addition, the up-to-date observations of the neutrino (ν)-bursts from the supernova (SN) 1987A[8, 9] are expected to give fruitful informations, especially for the mechanism of NS formation.

In this paper, we make attempts to correlate PC with NS phenomena. Specifically we discuss the glitch phenomena of Vela pulsar by proposing the corequake model based on PC.[10] Also we present an idea to explain a particular time profile of the ν-bursts from SN 1987A.[11] Before going to these subjects, we briefly summarize the characteristics of NS with PC which give the basis for our discussions.

2. CHARACTERISTICS OF NS WITH PC

Typically, three types of PC can be considered in neutron matter; the charged pion condensation (CPC) of running wave mode, the neutral one (NPC) of standing wave mode and the combined condensation (CNPC) of CPC and NPC with making their condensed momenta orthogonal to each other. Much efforts have been paid to the realistic determination of the transition density ρ_t for PC, by taking acount of the various effects such as the short-range correlation, isobar Δ(1232) mixing, two nucleon interaction other than the one originating from the pion-nucleon P-wave interaction ($H_{\pi N}$) and the exchange-energy contributions. Although the ρ_t is not well determined yet, depending on the viewpoints of approach, we can expect that $\rho_t(CPC) \sim (1.5-2)\rho_0$ [12] and $\rho_t(NPC) \sim (2-4)\rho_0$,[13] with $\rho_0 \equiv 0.17 fm^{-3}$ being the standard nuclear density. Energetically CNPC is most favourable since energy gains from CPC and NPC are additive without serious conflict. In most cases $\rho_t(CPC) < \rho_t(NPC)$ and hence we have CNPC at densities $\rho \gtrsim \rho_t(NPC)$.

For later discussions, we pick up the particular aspects of PC. (i) Nucleons under NPC are characterized by a solid-like state sketched in Fig.1, with one-dimensional localization (in direction of spin quantization axis taken as $\hat{z}$ one) and specific spin-isospin ordering. This state is well described by the Alternating-Layer-Spin (ALS) model proposed by Tamiya, Tatsumi, Tamagaki and the present author.[14] (ii) Under CPC, nucleons are described by qusiparticles (quasineutrons) composed of neutrons and protons, but the Fermi gas (FG) nature is not changed from that in normal phase. Remarkable point is that the cooling of NS with CPC is dramatically accelerated by ν-emmision due to the pion-induced β-decays of qusineutrons (pion cooling). (iii) In CNPC, quasineutrons, in place of neutrons, form the ALS structure, with CPC occuring in direction ($\hat{r}_\perp$) perpendicular to the direction ($\hat{z}$) of NPC (see Fig. 1).[15] The CNPC has both properties of (i) and (ii). (iv) The nature of phase transition is considered to be of the first order for NPC and of the second order for CPC. (v) The EOS for nucleon matter is significantly softened accordig to the growth of PC.

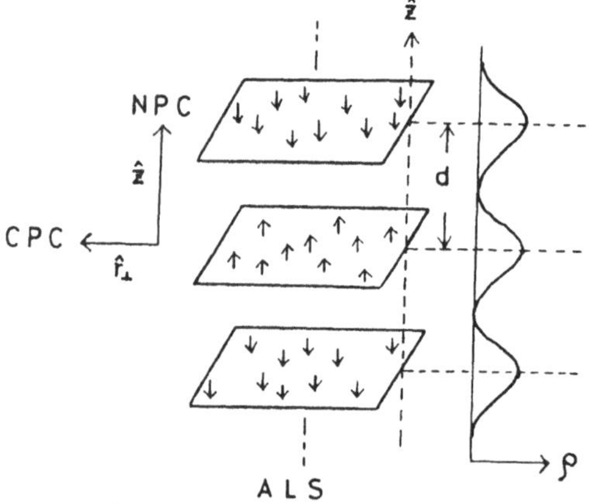

Fig.1. Alternating-Layer-Spin (ALS) structure of nucleons and pion condensations (PC). Charged PC (CPC) in $\hat{r}_\perp$ direction coexists with neutral one (NPC) in $\hat{z}$ direction in the case of combined PC (CNPC). ρ is the density and arrows denote spin-isospin expectation values $\langle \sigma_z \tau_3 \rangle$ of neutrons (quasineutrons) for NPC (CNPC).

Here we show some effects of PC on NS. As a reasonable approximation, the temperature T of NS is taken as T=0, except the moment of its birth. To avoid complexity, we make the following simplifications without loosing the essentials of PC. We use the EOS in which extra energy gains from PC, ΔE(CPC) and ΔE(NPC), are added to the standard EOS, the one from Bethe-Jhonson IH potential (BJ):

$$E = E(BJ) + \Delta E(CPC) + \Delta E(NPC). \quad (1)$$

Also the $\Delta E(CPC)$ and $\Delta E(NPC)$ are calculated within the framework of the mean field approximation (MFA) and the ALS model, for the simplified Hamiltonian $H=H_0+\tilde{H}_{\pi N}$ where H_0 represents the free parts of nucleons and pions and $\tilde{H}_{\pi N}$ is the effective $H_{\pi N}$ with the coupling constant $f^2_{eff}=f^2/2$ (0.5). They are expressed as*

$$\Delta E(CPC) \equiv \langle H\rangle_{CPC} - \langle H\rangle_{NORMAL} \cong -\frac{1}{4} m_\pi(\tilde{\rho}-1)^2/\tilde{\rho} \qquad \text{for } \hat{\rho}\geq 1, \quad (2)$$

$$\Delta E(NPC) \equiv \langle H\rangle_{NPC} - \langle H\rangle_{NORMAL}$$

$$\cong \frac{q^2_{\perp F}}{4m_N} + \frac{a}{4m_N} - \frac{1}{2} m_\pi \tilde{\rho} \frac{k_0{}^2}{k_0 + m_\pi{}^2} e^{-k_0{}^2/2a} - \frac{3}{5}\frac{q^2{}_F}{2m_N} , \quad (3)$$

where $q_F=(3\pi^2\rho)^{1/3}$, $q_{\perp F}=(4\pi\rho d)^{1/2}$, $\tilde{\rho}=2f^2_{eff}\rho/m^3{}_\pi$, $k_0=\pi/d$ and $m_N(m_\pi)$ is a nucleon (pion) mass, with d (layer spacing) and a (width parameter for Gaussian wave function) being the parameters in the ALS model. We have the CPC (NPC) when $\Delta E(CPC)<0$ ($\Delta E(NPC)<0$) and the CNPC when both $\Delta E(CPC)$ and $\Delta E(NPC)$ become negative.

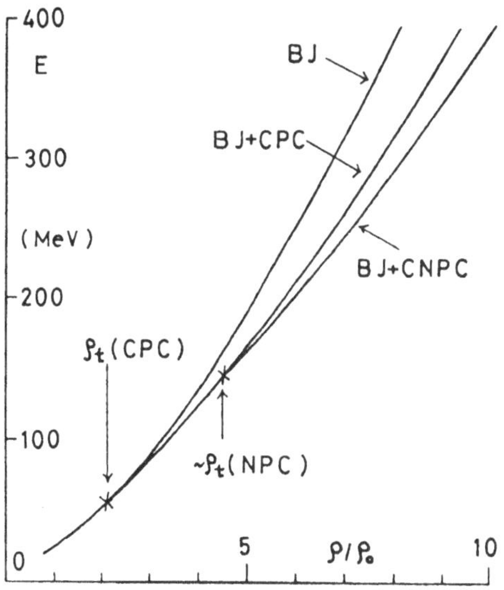

Fig.2. Energy per particle E versus density ρ for cases with (BJ+CPC, BJ+NPC) or without (BJ) PC. BJ denotes the case with use of Bethe-Jhonson two-nucleon potential. ρ_t indicates the transition density for the PC. ρ_0 is the standard nuclear density.

---

*) Throughout this paper, the units $\hbar=c=\kappa_\beta$(Boltzman constant)=1 are used.

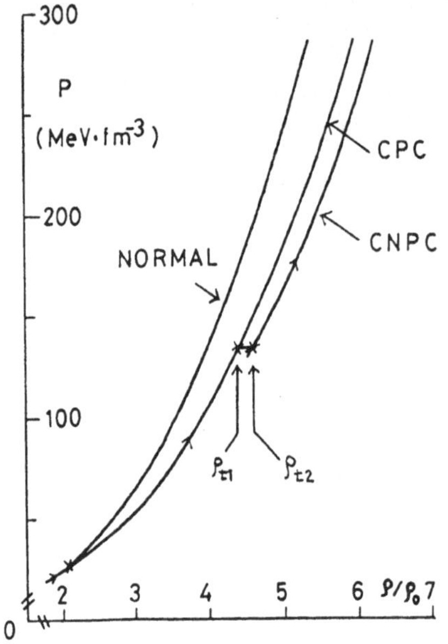

Fig.3. Pressure P versus ρ corresponding to E in Fig.2.

In Fig. 2, energy per perticle E is shown as functions of ρ. Also the pressure $P=\rho^2\partial E/\partial\rho$ is illustrated in Fig. 3. In our case, $\rho_t(\mathrm{CPC}) \cong 2.1\rho_0$, $\rho_t(\mathrm{NPC}) = \rho_{t1} \cong 4.4\rho_0$ and $\rho_{t2} \cong 4.6\rho_0$. The latter two are obtained by the double tangent construction. A remarkable softening of the EOS at $\rho>\rho_t(\mathrm{CPC})$ and first order phase transition at $\rho=\rho_t(\mathrm{NPC})$ are observed.

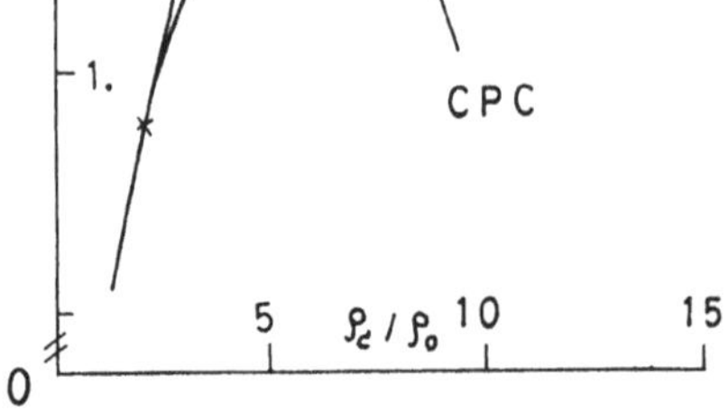

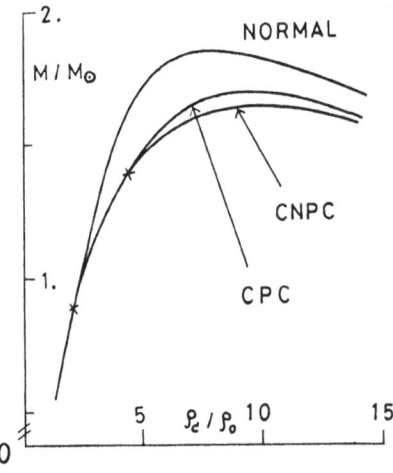

Fig.4. Mass M versus central density ρ_c of neutron stars for cases in Figs. 2 and 3. $M_\odot$ is the solar mass.

Based on the EOS, the NS models are obtained by solving the TOV equation. In Fig. 4, the NS mass M versus the central density ρ_c is illustrated. Due to the occurence of PC, the maximum mass M_{max} sustained by the corresponding EOS becomes smaller; $M_{max}/M_\odot$=1.85→1.69→1.65 as normal→CPC→CNPC, with $M_\odot$ being the solar mass, and ρ_c becomes higher, for example, ρ_c/ρ_0=4.2 →6.3→7.2 for $M/M_\odot$=1.6. The feature how the structure, together with the ρ_c and the radius R, depends on the EOS and the M is visualized by the cross section of NS sketched in Fig. 5. For the discussions below, we note the points; the portion of core superfluid depends crucially on M and the NS with $M \gtrsim 1.4M_\odot$ ($M < 1.4M_\odot$) contains (does not contain) NPC phase, namely, solid core.

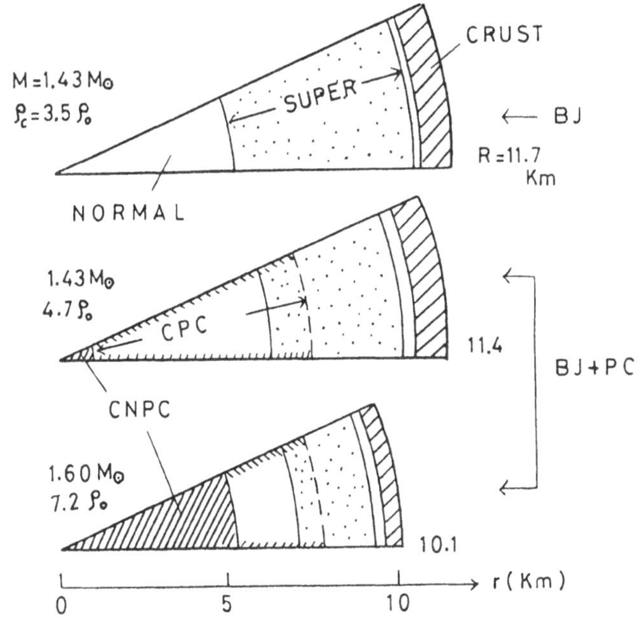

Fig.5. Cross sections of neutron stars to show how the structure and the radius R depend on M and equation of state.

3. COREQUAKE MODEL BASED ON PION-CONDENSED NS

Since the discovery of pulsar, the glich phenomena, sudden speed-up $\Delta\Omega_0$ of a rotating NS followed by its long relaxation time τ such as the angular frequency Ω versus time t relation sketched in Fig. 6, have gathered continuous attention as one of probes to the structure of NS. The giant glitches we concern here are observed five times for Vela pulsar ($\Delta\Omega_0/\Omega \sim 10^{-6}$) and two times for Crab pulsar ($\Delta\Omega_0/\Omega \sim 10^{-8}$). Let us go to the subject to explain the Vela pulsar glitches by the solid core provided by PC.[10]

Outline of Starquake Theory

In two-component starquake theory proposed by Baym, Pethich, Pines and Ruderman,[16] NS is divided into two components; the first one is the crust-

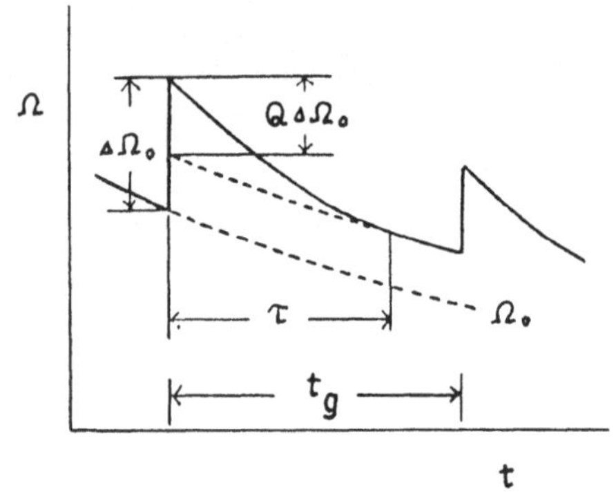

Fig.6. Time t dependence of angular frequency Ω for pulsar glitches. Ω_0 is the case without glitches.

-charged particles-normal neutron system (crustal component) and the second one is the superfluid neutrons (superfluid component). The former responses quite rapidly to the speed-up through the magnetic field, whereas the latter does very slowly to that, which is responsible for the long τ observed (loughly, ∿a week for Crab pulsar and ∿a year for Vela pulsar). The existence of superfluids in NS core and such a macroscopic time scale have been grounded theoretically by the coexistence of $^3P_2$-neutron superfluid[17] and $^1S_0$-proton one[18] at $\rho\cong(0.7\text{-}2.8)\rho_0$.

In starquake theory, the origin of glitch is attributed to the sudden release of elastic energy in the crust (crustquake) or the solid core (corequake). Historically, the crustquake model was quite successful for Crab but not for Vela. This mainly comes from that the glitch magnitude is larger by about two orders of magnitude in Vela than in Crab. To overcome the defects, Pines, Shaham and Ruderman[19] proposed the corequake model, assuming a baryonic solid core in NS. This model worked well for Vela because the big elastic energy of baryonic solid is consistent with the big $\Delta\Omega_0/\Omega$. Later on, however, following shortcomings showed up:
(A) Nonexistence of solid core; the solidification of neutron matter was denied from the theoretical side.[20]
(B) Heating up the NS; the strain energy released ΔE_s by a single glitch amounts to $\Delta E_s \sim 10^{44-45}$ erg which heats up the NS from inside. Since the glitch interval t_g is ∿(2-3) years and the age of Vela is $\sim 10^4$ years, the accumulation of heating up contradicts the observation of the surface temperature $T_e \sim 10^6$ K.

Revival of Corequake Model

Here we point out the possibility to overcome (A) and (B). The idea is to consider that ρ_c of Vela exceeds ρ_t (NPC) and therefore Vela contains CNPC. Then, firstly the (A) is solved by the existence of ALS-solid noted in (i) (as a macroscopic configuration of ALS, we take the spin quantization axis in radial direction of NS). Scondly the (B) is brought a solution by pion cooling described in (ii).

The energy per particle for the ALS solid is a function of d and is well approximated by $\alpha(d-d_0)^2$+const. with α, typically, ∿100 MeV·fm^{-2}. Then the strain energy E_s stored is estimated as

$$E_s \sim N_s\bar{\alpha}\bar{d}_0{}^2(\varepsilon-\varepsilon_0)^2 = \bar{B}_s(\varepsilon-\varepsilon_0)^2 \quad , \qquad (4)$$

where bar denotes the average over ρ, N_s the neutron number contained in the solid core and $\varepsilon(\varepsilon_0)$ the oblateness (reference oblateness). The shear modulus μ_s for the ALS solid is obtained as $\mu_s=\bar{B}_s/V_s \sim 2(3)\times10^{35}$ dyn·cm^{-2} by inserting $\bar{\alpha}$=100 MeV, $\bar{d}$ =1.25 fm, $N_s \sim 0.17(0.67)\times10^{57}$ and $V_s=1.9(6.2)\times10^{17}$ cm^3 for $M=1.5(1.6)M_\odot$. This μ_s is larger by $\sim10^5$ as compared to that in the crust and satisfies the one needed in corequake model.

How much energy is released? We estimate this for $M=1.5(1.6)M_\odot$ as

$$\Delta E_s = 2\bar{B}_s|(\varepsilon-\varepsilon_0)\Delta\varepsilon| \sim 0.5(1.1)\Theta_m\times10^{48} \text{ erg}, \tag{5}$$

depending on the strain angle Θ_m. Assuming a reasonable value,[21] $\Theta_m=5\times10^{-4}$, we have $\Delta E_s \sim 2(5)\times10^{44}$ erg. To what extent is NS heated up by ΔE_s? As a rough estimate, we have the internal-temperature rise ΔT from $T=T_i=10^8$ K ($\sim$0.01 MeV) to $T=T_f$ from the equation:

$$\Delta E_s = N\int_{T_i}^{T_f} \bar{C}dT \sim 10^{-2}\, N(T_f^2-T_i^2) \quad , \tag{6}$$

where N denotes the total nucleon number and the average specific heat per particle $\bar{C}$ has been taken as $\bar{C}$=0.02 T, by using low-temperature approximation. The resulting $\Delta T=T_f-T_i$ are $0.2(0.2)\times10^8$ K for $N=2.0(2.2)\times10^{57}$ corresponding to $M=1.5(1.6)M_\odot$. If we take $\Theta_m=5\times10^{-3}$, we have $\Delta T \sim 0.7(1.2)\times10^8$ K.

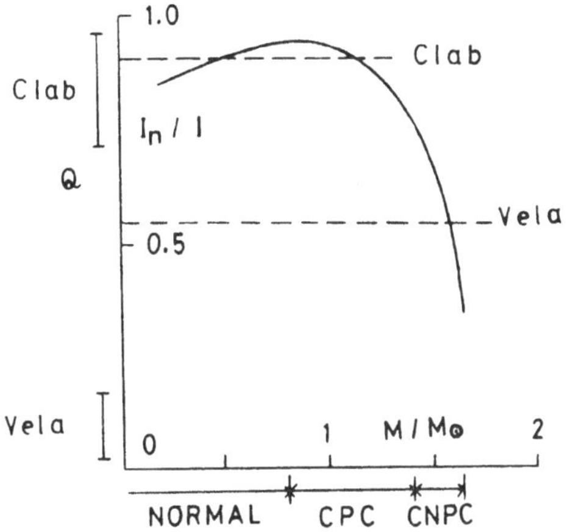

Fig.7. M dependence of superfluid portion of neutron stars with or without PC. $I(I_n)$ is the total (superfluid) moment of inertia. Observed structure factor Q are indicated by ⊢⊣ .

How long is the cooling time Δt for $T_f \to T_i$? Since the luminosity for pion cooling L_π predominates for $T>10^9$ K ($\sim$0.1 MeV) as compared with those for photons, URCA and ν-bremsstrahlung coolings, we concentrate on the cooling by L_π. It is given as

$$L_\pi \cong \eta(M/M_\odot)(\rho/\bar{\rho})T^6\times10^{57} \text{ MeV}\cdot\text{s}^{-1} \quad , \tag{7}$$

with T in MeV and $\bar{\rho}$ the average density of NS. η is in the range $1\text{-}10^{-2}$ depending on the type of PC and the calculational method: $\eta(CPC)=1$ for Maxwell et al.[22] whereas $\eta(CPC)=10^{-1}$ and $\eta(CNPC)\sim 10^{-2}$ for Tatsumi.[23] By using Eq. (7) we have

$$\Delta t = -N\int_{T_f}^{T_i} dT\bar{C}/L_\pi \cong (1\text{-}2)\times 10^{-2}\eta^{-1} \qquad \text{year} \tag{8}$$

corresponding to $\Theta_m=(5\times 10^{-4} - 5\times 10^{-3})$. This means that even for $\eta=10^{-2}$, the heating up of NS is cooled down within (1-2) years ($\lesssim t_g \sim (2\text{-}3)$ years), that is, before the next glitch. Thus we can revive the corequake model in terms of ALS solid provided by NPC.

So far, we have assumed the existence of CNPC, in other words, $M\gtrsim 1.4M_\odot$ for Vela. Is there another indication for this? According to starquake theory, the structure factor Q ($\cong(0.03\text{-}0.18)$ for Vela[24] and (0.71-0.94) for Crab[25]) deduced from observation is related to NS structure as

$$Q \cong \frac{I_n}{I}\left(1 - \frac{\Delta I_n/I_n}{\Delta I_c/I_c}\right) \tag{9}$$

with $I_c(I_n)$ the moment of inertia for crustal (superfluid) component ($I=I_c+I_n$). The I_n/I versus M for BJ+PC case is shown in Fig. 7 where I_n includes also the superfluids in the crust phase.[26] Eq. (9) indicates $Q\sim I_n/I$ for $I_n>>I_c$. It also suggest $Q\sim 2I_n/I-1$ for $I_n\sim I_c$ when we make a reasonable assumption that $\Delta I_n\cong\Delta I_c$ for $I_n\sim I_c$. If we take a mean value of Q as $Q\sim 0.9$ for Crab and ~ 0.1 for Vela, we have $I_n/I\sim 0.9$ for Crab and ~ 0.55 for Vela. This leads to $M\sim 1.1M_\odot$ for Crab ($M\sim 0.5M_\odot$ is also possible but is too small) and $M\sim 1.6M_\odot$ for Vela. Therefore the observation suggests that Crab does not have solid core but Vela does, which is nicely consistent with our preceding assumption.

4. ν-BURSTS FROM SUPERNOVA 1987A AND NS WITH PC

Very recent discovery of the supernova 1987A in the Large Magellanic Cloud is providing us with exciting problems. Observed data by Kamiokande II[8] indicates the characteristic time profile that the ν-burst has occured two times with the time interval $\Delta t\sim 10$s. In more detail, there seems to exist three bunches; five events in $t\cong(0\text{-}0.5)$s, three events in $t\cong(1.5\text{-}2)$s and three events in $t\cong(9\text{-}12)$s. But here we call the first two bunches ($t\lesssim 2$s) by the first ν-burst, since they are considered to be caused by the birth of NS. On the other hand, the last one which we call the second ν-burst, is very interesting because of the points, (a) the occurence about 10s after the first one and (b) the large energy release leading to ν-burst, both of them are hard to understand from the standard scenario of supernova explosion. It may suggest an exotic mechanism, such as the phase transition to PC or quark matter.[27] From only the energy consideration, such phase transition would be a promising candidate, but more important is how to explain (a) and (b) in an unified way. Here we give an idea to do this, by paing attention to particular properties of NS with PC.*

Explanation for Δt Between Two ν-Bursts

We begin with presenting the ρ_t-T relation (phase diagram) for neutron

*) Other ideas are also reported from different viewpoints[28].

matter at $T \gtrsim 0$, corresponding to hot NS at the birth. Qualitatively, the feature given in Fig. 8 can be expected. For NPC case, ρ_t increases as T.[29] This is mainly due to the characteristic: The entropy becomes smaller in the NPC phase (ALS phase) than in the normal phase (FG phase) since the former is an ordered state in contrast with the uniform one in the latter, which increases the free energy in the former as T goes high and push up ρ_t(NPC) to higher density side. On the other hand, such a difference in entropy does not arise for CPC case, because both CPC and normal phases are equally described by FG; the former is FG state of qusineutrons and the latter is the one of normal neutrons. Hence the ρ_t(CPC) is almost independent of T, as far as we use the MFA and neglect recoil effects of nucleons. At present we do not have the results calculated at T>0 with using the BJ-IH potential and so the ρ_t versus T in the Figure are only qualitative ones except the case at T=0. But this brings no problem since only the quantitative results at $T \cong 0$ are necessary in the following discussion.

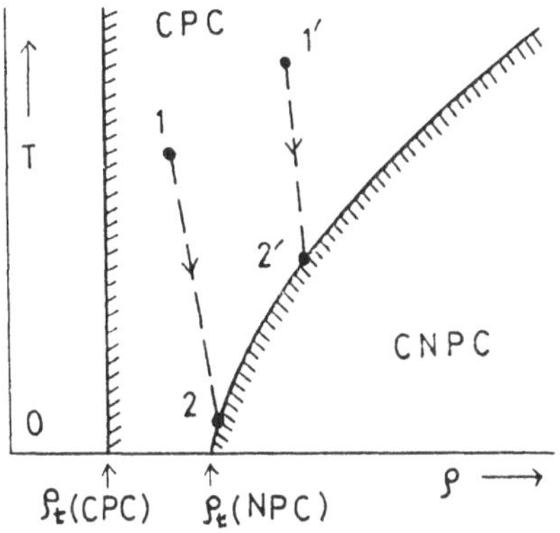

Fig.8. Conceptional phase diagram of pion condensates to explain the observed two ν-bursts from the supernova 1987A ; point 1=birth of a neutron star (the first ν-burst), 1→2 =pion cooling and 2=occurence of first order phase transition in the neutron star core, leading to the second ν-burst. The time needed for 1→2 is responsible for the time interval Δt between the two ν-bursts.

We consider the following scenario to explain Δt, in terms of Fig. 8. Firstly we assume the formation of a hot NS with the central density ρ_{c1} higher than ρ_t(CPC) and with the temperature T_1 , after the bounce of collapsing stellar core, although the relevant mechnism is not confirmed yet. We suppose that the first ν-burst originates from this process. Secondly this NS is cooled down very rapidly through the pion cooling (note that it contains CPC core) and arrive at the point 2 with (ρ_{c2} , T_2), as illustrated in Fig. 8 ($\rho_{c2} > \rho_{c1}$ since $T_2 < T_1$). Then the core of NS experiences the dramatic phase transition of first order into NPC phase (namely, CNPC one), with the sudden release of the latent heat together with the gravitational energy, which leads to the second ν-burst. Thus Δt between these two ν-bursts is the time demanded for $T_1 \to T_2$

From the cooling equation with use of Eq. (7), we have

$$\Delta t = - N\int_{T_1}^{T_2} d T\bar{C}/L_\pi \cong 0.03\eta^{-1}T_2^{-4}(1-(T_2/T_1)^4) \tag{10}$$

by inserting $M=1.4M_\odot$, $N=2\times10^{57}$ and $\bar{\rho}=3\rho_0$. To be consistent with the observed $\Delta t \sim 10s$, T_2 should be

$$T_2 \cong \eta^{-1/4} \times 0.23 \text{ MeV} \cong (0.2\text{-}0.7) \text{ MeV}. \tag{11}$$

We note the result is essentially independent of T_1 as far as $T_1>5$ MeV generally accepted. Therefore Δt imposes a strict constraint on the temperature at which the NS under consideration undergoes the phase transition from CPC phase to CNPC one. In the sense of nucleon matter at $\rho>\rho_0$, the T_2 is cold enough and hence we can rely on the results at T=0 in Figs. 3 and 4. The ρ_{c2} of this NS should be $\rho_{c2}\cong\rho_t(\text{NPC})\cong4.4\rho_0$, which leads to $M\cong1.4M_\odot$, consistent with the value adopted in Eq.(10). That is, Δt brings about an important information for the mass and structure of NS, in terms of the EOS used. Coversely, NS properties can be checked by the observation of Δt.

Released Energy

How much energy is released? Since NS is sustained by gravitation, the total energy release ΔE_T should be derived by taking account of the latent heat connected with the phase transition, gravitational energy, mechanical one and so on. Here we concern the first two. Then ΔE_T is obtained by noting that the conserved quantity is solely the total baryon number, namely, by comparing M before and after the phase transition under the conservation of N. For the present case, we have

$$\Delta E_T \sim 0.005 \, M_\odot \cong 9\times10^{51} \quad \text{erg} \tag{12}$$

which corresponds to $\sim$3% of the total binding energy of this NS. By considering that ΔE_T depends on the model, namely, the strength of the phase transition and the EOS, we can demonstrate from the above value that the energy release due to the phase transition is consistent with the estimated value for the second ν-burst (for instance, $\sim1.9\times10^{52}$ erg [27]). It should be noted that the latent heat (internal energy associated with nuclear system) takes part in ΔE_T only by few percents. This clearly shows that the release of gravitational energy is essential in ΔE_T and the phase transition plays an important role to drive the release by causing a contraction of NS. The time scale for such a minicollapse would be the one for free fall.

What happens when ΔE_T is suddenly released? One would be the shock wave propagation behind which a hot area is formed and from there energetic $\nu\bar{\nu}$-pair could be emitted, leading to the observed second ν-burst. On the other hand, continuous ν-emmision by pion cooling for the process 1→2 may well form the background ν. The other would be the heating up NS again. If we consider the case where ΔE_T is wholly converted into heat, the maximum temperature T_{max} after the thermalization is estimated as $T_{max} \sim 15$ MeV, quite a large value. This suggest the possibility that in some cases NS reheated makes a transition from CNPC to CPC, although quantitatively the situation depends crucially on the ρ_t(NPC)-T relation and the EOS at T>0. That is, we may conjecture that a certain NS just born oscilates repeating shrink and expansion, generating ν-burst more than two times.

In general , many paths are possible in Fig. 8. For example, a NS is formed at the point 1′ ($\rho_{c1}>\rho_t$(NPC) and $T_1 \gtrsim 10$ MeV) and arrive at the point 2′ whis $T_2 \sim 5$ MeV. In this case, $\Delta t \sim 0.002$s and only the first ν-burst is to be detected. Also if ρ_{c1} is considerably lower than ρ_t(NPC) (namely, M< $1.4M_\odot$), the situation is the same as just above. In this context, the supernova 1987A may be a rather particular case.

5. SUMMARY AND CONCLUDING REMARKS

We have presented the characteristics of pion-condensed NS and on the basis of these we have discussed the particular aspects of two phenomena, pulsar glitches and ν-bursts from the SN 1987A.

We have shown that the shortcomings inherent to the original corequake model is overcomed by the ALS solid due to NPC and by the rapid cooling due to CPC. Thus the giant glitches for Vela can be explained by the corequake model, while those for Crab by the crustquake model, assuming that the former contains the solid core due to NPC but the latter does not. This assumption is supported by the observed Q values and the difference of mass in these two pulsars is indicated; roughly speaking $M \sim (1.4-1.6)M_\odot$ for Vela and $\sim(1.0-1.2)M_\odot$ for Crab.

So far, we have treated the $^3P_2$-superfluid as responsible for the macroscopic τ. Indeed, this has been confirmed theoretically.[30] But recently the mechanism for its rapid response to crustal component has been pointed out[31] and another glitch model, based on the vortex creep theory, has been proposed.[32] In this connection, more detailed studies are necessary for our model. We want to stress, however, that even if the rapid response is true, our model is compatible with the vortex creep model,[32] in the point that the former provides the driving force for the sudden unpinning of vortex which is crucial for the latter.

We have proposed a scenario to understand the interesting aspect of the ν-bursts from the SN 1987A. In the scenario, the first ν-burst is attributed to the birth of NS and the second one is the phenomena related to the phase transition to NPC. The time interval between two ν-bursts, $\Delta t \sim 10$s, is the cooling time needed for the NS to make the phase transition. By this consideration, both of the total energy for second ν-burst and the Δt are consistently explained.

We have treated the phase transition to CPC as the second order one. It is so at T=0 but at T>0 it has a possibility to be the first order.[33] In that case, it might be possible that the first ν-burst (the birth of NS) could be caused by this phase transition which is itself or assists the supernova explosion.

In this article, we have simply assumed the rapid cooling of pion-condensed NS. This point, however, should be checked by taking account of the opacity, since hot NS just born ($T \gtrsim 10$MeV) is generally considered to be opaque. Quantitatively the approach such as the one from Burrows and Lattimer[34] has to be performed, with including the ν-emission, -absorption and -scattering under the circumstance of PC, which remains as a future problem. Here we note that even if the L_π is effectively reduced by, for instance, two orders of magnitude ($\eta \sim 10^{-4}$), the resulting T_2 is around 2

MeV and so our discussions are almost unchanged. Also it might be expected that within the time scale ∿10s, only the central region is cooled down ($T \sim T_2$) and the surface one remains hot ($T \gtrsim T_1$). We note that the situation is sufficient for the present scenario to hold, since the phase transition near the center of NS is essential.

Acknowledgements

The author wishes to thank Professor R.Tamagaki for his collaboration in the work and useful discussions. He is also grateful to Dr. T. Tatsumi, Professor. K. Sato and Dr. M. Takahara for their interests in the work and comments. Thanks are also due to Professor J. Hiura for his collaboration in the study of hot nucleon matter.

REFERENCES

1. A.B. Migdal, Zh. Eksp & Theor. Fiz. 61: 2210 (1971); ibid. 63: 1933 (1972).
2. R.F. Sawyer, Phys. Rev. Lett. 29: 382 (1972).
   D.J. Scalapino, Phys. Rev. Lett. 29: 386 (1972).
   R.F. Sawyer and D.J. Scalapino, Phys. Rev. D7: 953 (1972).
3. P.C. Joss and S.A. Rappaport, Ann. Rev. Astron. Astrophy. 22: 537 (1984).
4. T. Ebisuzaki and K. Nomoto, Astrophys. J. 305: L67 (1986).
   M.Y. Fujimoto and R.E. Taam, Astrophys. J. 305: 246 (1986).
5. E.P. Liang, Astrophys. J. 304: 682 (1986).
6. S. Tsuruta, Phys. Reports 56: 237 (1979).
   K. Nomoto and S. Tsuruta, Astrophys. J. 305: L19 (1986).
7. D. Pines, Cosmic Superfluidity: THe Evidence for Superfluidity in Neutron Stars, in: "From SU(3) to Gravity", E. Gotsman and G. Tauber ed., Cambridge Univ. Press, Cambridge (1985).
8. K. Hirata, T. Kajita, M. Koshiba, M. Nakahata, Y. Oyama, N. Sato, A Suzuki, M. Takita, Y. Totsuka, T. Kifune, T. Suda, K. Takahashi, T. Tanimori, K. Miyano, M. Yamada, E.W. Beier, L.R. Feldscher, S.B. Kim, A.M. Mann, F.M. Newcomer, R. Van Berg, W.Zhang and B.G. Cortez, Phys. Rev. Lett. 58: 1490 (1987).
9. R.M. Bionta, G. Blewitt, C.B. Bratton, D. Casper, A. Ciocio, R. Claus, B. Cortez, M. Crouch, S.T. Dye, S. Errede, G.W. Foster, W. Gajewski, K.S. Ganezer, M. Goldhaber, T.J. Haines, T.W. Jones, D. Kielczewska, W.R. Kropp, J.G. Learned, J.M. LoSecco, J. Matthews, R. Miller, M.S. Mudan, H.S. Park, L.R. Price, F. Reines, J. Schulz, S. Seidel, E. Shumard, D. Sinclair, H.W. Sobel, J.L. Stone, L.R. Sulak, R. Svoboda, G. Thornton, J.C. van der Velde and C. Wuest, Phys. Rev. Lett. 58: 1494 (1987).
10. T. Takatsuka and R. Tamagaki, Pion-Condensed Neutron Star and Pulsar Glitch: to be published in Proc. of XI International Conference, Kyoto, April 1987, on "Particles and Nuclei".
11. T. Takatsuka, Preprint HSIW-9 (April 7, 1987), submitted to Prog. Theor. Phys.
12. W. Weise and G.E. Brown, Phys. Lett. 48B: 297 (1974).
    T. Tatsumi, Prog. Theor. Phys. 68: 1231 (1982).
13. T. Kunihiro and T. Tatsumi, Prog. Theor. Phys. 65: 613 (1981).
    K. Tamiya and R. Tamagaki, Prog. Theor. Phys. 66: 948, 1361 (1981).
    T. Takatsuka, Y. Saito and J. Hiura, Prog. Theor. Phys. 67: 254 (1982)
    O. Benhar, Phys. Lett. 106B: 375 (1983); Nucl. Phys. A437: 590 (1985).
14. T. Takatsuka and R. Tamagaki, Prog. Theor. Phys. 58: 694 (1977).
    T. Takatsuka, K. Tamiya, T. Tatsumi and R. Tamagaki, Prog. Theor. Phys. 59: 1933 (1978).

R. Tamagaki, Nucl. Phys. A328: 352 (1979); Pion Condensation and Baryonic Structure in Nuclear Matter at High Density: in Proc. of International summer School, Changchun, July 1983, on "Nucleon-Nucleon Interaction and Nuclear Many-Body Problems", S.S. Wu and T.T.S. Kuo ed., World Scientific, Singapore (1984).

15. K. Tamiya and R. Tamagaki, Prog. Theor. Phys. 60: 1753 (1978).
16. G. Baym, C.J. Pethick, D. Pines and M. Ruderman, Nature 224: 872 (1969).
17. R. Tamagaki, Prog. Theor. Phys. 44: 905 (1970).
    M. Hoffeberg, A.E. Glassgold, R.W. Richardson and M. Ruderman, Phys. Rev. Lett. 24: 775 (1970).
    T. Takatsuka, Prog. Theor. Phys. 48: 1517 (1972).
18. N.C. Chao, J.W. Clark and C.H. Yang, Nucl. Phys. A179: 320 (1972).
    T. Takatsuka, Prog. Theor. Phys. 50: 1754 (1973).
19. D. Pines, J. Shaham and M. Ruderman, Nature Phys. Sci. 237: 83 (1972).
20. V.R. Pandharipande, Nucl. Phys. A217: 1 (1973).
    V. Canuto and J. Lodenquai, Phys. Rev. C12: 2033 (1975).
21. R. Smoluchowski, Phys. Rev. Lett. 24: 423 (1970).
22. O. Maxwell, G.E. Brown, D.K. Campbell, R.F. Dashen and J.T. Manassah, Astrophys. J. 216: 77 (1977).
23. T. Tatsumi, Pion Condensation and Neutron Star Phenomena: to be published in Proc. of the Yukawa International Seminar, Kyoto, April 1987, on "Mesosn and Quarks in Nuclei".
24. G.S. Downs, Astrophys. J. 249:687 (1981).
    P.M. McCulloch, P.A. Hamilton, G.W.R. Royle and R.N. Manchester, Nature 302: 319 (1983).
25. E.H.G. Lohsen, Astron. Astrophys. Suppl. Ser. 44: 1 (1981).
26. T. Takatsuka, Prog. Theor. Phys. 71: 1432 (1984).
27. K. Sato and H. Suzuki, Phys. Rev. Lett. 58: 2722 (1987).
28. T. Hatsuda, Preprint KEK-TH 157 (23 March 1987).
    T. Nakamura and M. Fukugita, Preprint RIFP-697 (March 1987).
29. T. Takatsuka and R. Tamagaki, Prog. Theor. Phys. 77: 362 (1987).
30. G. Baym, C. Pethick and D. Pines, Nature 224: 673 (1969).
    P.J. Feibelman, Phys. Rev. D4: 1589 (1971).
    J.A. Sauls, D.L. Stein and J.W. Serene, Phys. Rev. D25: 967 (1982).
31. M.A. Alpar, Stephan A. Langer and J.A. Sauls,Astrophy. J. 282: 533 (1984).
32. M.A.Alpar, P.W. Anderson, D. Pines and J. Shaham, Astrophys. J. 276: 325 (1984).
33. K. Kolehmainen and G.Baym, Nucl. Phys. A382: 528 (1982).
34. A. Burrows and J.M. Lattimer, Astrophys. J. 307: 178 (1986).

NN̄ ANNIHILATION IN TERMS OF QUARKS

A.M. Green

Research Institute for Theoretical Physics
University of Helsinki, Siltavuorenpenger 20 C
SF-00170 Helsinki, Finland
JNET% "GREEN at FINUHCB"

1. INTRODUCTION

The interest in NN̄ annihilation into two mesons is twofold. Firstly, on the experimental side there has been considerable progress from recent LEAR experiments at CERN which have determined several new $\bar{p}p \to M_1M_2$ branchings for stopped $\bar{p}$'s - see ref./14/. However, it should be emphasized that this progress is only partial and, hopefully, the Crystal Barrel and Obelisk proposals for the next series of LEAR experiments will clarify the situation to a much greater extent. On the theoretical side, for a quark description of NN̄ annihilation into two mesons it is necessary to introduce some mechanism for quark-antiquark ($q\bar{q}$) annihilation. This has resulted in many theoretical papers resembling the menu of a Chinese restaurant. From column A (see Fig. 1) a choice (or combination) is made between the two basic $q\bar{q}$ vertices:

a) The $^3P_0$-model, in which the $q\bar{q}$ annihilate into the vacuum (often called the Pair-Creation model) - Fig. 1a).

b) The $^3S_1$-model, in which the $q\bar{q}$ annihilate into a single (effective) gluon - Fig. 1b).

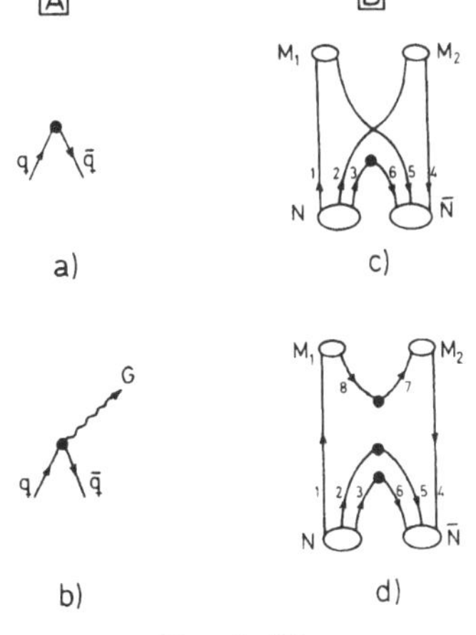

Fig. 1. The menu

Having made this choice, the number of orders from column B has to be decided - 'one' as in Fig. 1c) or 'three' as in Fig. 1d). Frequently it is then found that one combination can explain some experimental features and another can explain other features. It is the purpose of the present paper to convince the reader that the single combination Figs. 1a) plus 1c) is dominant.

2. The $^3P_0$-model versus the $^3S_1$-model

Before discussing $N\bar{N}$ annihilation it is useful to see how these two models for $q\bar{q}$ annihilation fare in their description of the meson decays $M_1 \rightarrow M_2M_3$, since this is presumably a simpler situation. For the $^3P_0$-model the most comprehensive study has been made by KOKOSKI and ISGUR /1/, who show that this approach is very successful for essentially all meson decays. Furthermore, they give a model based on the flux tube breaking of ISGUR and PATON /2/ that can explain this success. In section 5, this flux tube model will be discussed again in connection with $N\bar{N}$ annihilation. For the $^3S_1$-model there, apparently, has never been such a comprehensive study. However, as shown by GREEN and NISKANEN /3/ the $^3S_1$-vertex can be expressed in terms of the $^3P_0$-vertex as

$$\Theta(^3S_1) = F\Theta(^3P_0), \qquad F = [1+3/2(s_1(s_1+1)+2-s(s+1))] \tag{2.1}$$

where s_1 is the spin of the decaying meson (e.g. 1 and 0 for ρ and B) and s is the total spin of the decay products (e.g. 0 and 1 for $\pi\pi$ and $\omega\pi$). For the decays $\rho \rightarrow \pi\pi$ and $B \rightarrow \omega\pi$ the value of F becomes 7 and 1 respectively. Therefore, since the $\Theta(^3P_0)$ alone gives a good description of both of these decays the $^3S_1$-model will fail miserably.

Since the $^3S_1$-model is not the dominant mechanism for meson decays, why should it be so for $N\bar{N}$ annihilation? Of course, the effective gluon involved is different in the two cases. However, it seems unreasonable that the differences are so large that the $^3S_1$-model could be enhanced to such an extent that it now dominates over the $^3P_0$-vertex. In the view of the author, the main reason why the $^3S_1$-vertex is used in $N\bar{N}$ annihilation is because it is able to give, in lowest order, $N\bar{N}$ (S-wave)$\rightarrow$ Two s-wave mesons ($M_1^s M_2^s$) [e.g. $N\bar{N}(^{13}S_1) \rightarrow \pi^\pm\rho^\mp$] i.e. Figs. 1b) plus c). On the other hand, the usual treatment of the $^3P_0$-model in lowest order gives zero for Figs. 1a) plus c) - contrary to experiment. The $^3S_1$- advocates then say "Since the $^3S_1$-model gives a non-zero result, it must be better than the $^3P_0$-model" - an example of the "Anything" is better than "Nothing" philosophy.

3. The $^3P_0$-model in $N\bar{N}$ annihilation

Of course, advocates of the $^3P_0$-model have not been perturbed by the "conclusion" of the last subsection, since the $^3P_0$-vertex to third order [i.e. Figs. 1a) plus d)] readily gives $N\bar{N}$(S-wave)$\rightarrow M_1^s M_2^s$. Unfortunately, some of the branching ratios predicted do not agree with experiment. In particular, the model gives

$$R_s = \frac{N\bar{N}(^{31}S_0) \rightarrow \pi\rho}{N\bar{N}(^{13}S_1) \rightarrow \pi\rho} \approx 1 \tag{3.1}$$

whereas, experimentally $R_s << 1$. Also, the $^3P_0$-model in lowest order [i.e. Figs. 1a) plus c)] gives

$$R_p = \frac{N\bar{N}(^{33}P_{1,2}) \to \pi\rho}{N\bar{N}(^{11}P_1) \to \pi\rho} = 18 \tag{3.2}$$

whereas, experimentally $R_p << 1$. This observation of I=1 suppression in $N\bar{N}$ (S- and P-waves) is known as the $\pi\rho$ - puzzle. It has been suggested by DOVER et al. /4/ that $R_s << 1$ is not indicating a selection rule but is simply a reflection that in the $N\bar{N}(^{31}S_0)$ state $\pi^\pm\rho^\mp$ face a competition with annihilation into $\pi^\circ f$ resulting in a quenching of the $\pi^\pm\rho^\mp$ channel.

The above resolution of the S-wave $\pi\rho$ -puzzle leaves much to be desired. Firstly, the need to introduce three $^3P_0$-vertices is difficult to justify on any more basic approach such as the flux tube model /1,2/ - see section 5. Secondly, the need to employ a cancellation by competition with the $\pi^\circ f$ channel is not very "natural". However, as recently pointed out by GREEN /5/ there is an alternative where $R_s << 1$ can be achieved by simply Figs. 1a) plus c) i.e. the $^3P_0$-vertex to lowest order. As a first step in convincing the reader of this statement, it is necessary to show that Figs. 1c) and d) are in a certain sense "equivalent" to each other. This can be done by the chain of events

$$N\bar{N}(S) \xrightarrow{1} M^s_{14}\epsilon_{25}\epsilon_{36} \xrightarrow{2} M^s_{14}\epsilon_{25}(\ell = 0) \xrightarrow{3} M^s_{15}M^s_{24}(\ell = 1) \tag{3.3}$$

where the first step is to recouple the 3 quarks and 3 antiquarks of the $N\bar{N}(S)$ system into one s-wave meson and two ϵ-mesons (i.e. mesons with quantum numbers of the vacuum - $^{13}P_0$). These need not be considered as real mesons, but as objects with the quantum numbers of these mesons. In other words, this first step is simply a mathematical recoupling. The second step involves the annihilation of $\epsilon_{36}(=q_3\bar{q}_6)$ by the usual $^3P_0$-vertex to leave an s-wave meson M^s_{14} and the ϵ -meson (ϵ_{25}) in a relative s-wave. The final step can then be considered as yet another mathematical recoupling in which antiquarks $\bar{q}_4$ and $\bar{q}_5$ interchange to give two s-wave mesons M_{15} M_{24} in a relative p-wave - see ref. /5/ for details. The extent to which Figs. 1c) and d) are equivalent can be checked by comparing different branching ratios. The best compilation of ratios using Figs. 1a) plus d) is given by MARUYAMA et al. /6/. For the 11 possible branchings $N\bar{N}(S) \to M^s_1M^s_2$ [e.g. $N\bar{N}(^{13}S_1) \to \pi^\pm\rho^\mp$ etc.] the ratio

$$R = \frac{\text{Branching ratio ref./5/}}{\text{Branching ratio ref./6/}} = \frac{16}{81}$$

whereas for the 48 possible branchings $N\bar{N}(S) \to M^s_1M^p_2(\ell = 0,2)$ [e.g. $N\bar{N}(^{31}S_0) \to \eta\delta^\circ$, $N\bar{N}(^{11}S_0) \to \omega$ H etc] the value of R is 1/3. Possibly, the fact that R differs in the two cases (16/81 versus 1/3) is simply a matter of definition. Of course, it may be said that the above "equivalence" is obvious, since in Fig. 1d) the annihilation of $q_2\bar{q}_5$ and the creation of $q_7\bar{q}_8$ is represented in Fig. 1c) by a continuous $q\bar{q}$ state with the quantum numbers of the vacuum $(q_2\bar{q}_5)^{0+}$. However, between the two interpretations there is a fundamental difference which involves the question of strangeness. In Fig. 1d) the creation of $q_7\bar{q}_8$ from an SU(3) vacuum gives the combination $1/\sqrt{3}(u\bar{u} + d\bar{d} + s\bar{s})$. On the other hand Fig. 1c), since it is simply a recoupling of $N\bar{N}$ wavefunctions, never involves strange quarks. This offers an instant explanation of the observed K^+K^-

suppression, with the actual decay $N\bar{N}(^3S_1) \to K^+K^-$ being mediated by some higher order mechanism such as an effective $^3S_1$- vertex - see RUBINSTEIN and SNELLMAN /7/ and FURUI /8/. As shown in Ref. /9/ such a $^3S_1$-vertex need not have anything to do directly with gluons, but can emerge by simply forming an overlap of 3 quarks and 3 antiquarks each described by four component spinors. This is a natural extension of the usual $^3P_0$-vertex considered as the overlap of a single quark and a single antiquark - each described as four component spinors. There the leading term arises through the product of the "large" component in the q or $\bar{q}$ wavefunction with the "small" component of the $\bar{q}$ or q wavefunction.

The conclusion of the above is that $N\bar{N}$ (S-wave) can annihilate into two s-wave mesons using the $^3P_0$-vertex to only first order i.e. Figs. 1a)plus c).

At this stage it is convenient to point out a difference between the diagrams drawn in Figs. 1c) and d) and those drawn by GENZ /10/. For example, a possible configuration in Fig. 1c) after the $^3P_0$-vertex could be $[\rho_{14}\epsilon_{25}]^{J=1\ I=1}$, which is represented by four flavour combinations. Under the recoupling of step 3 in Eq. (3.3) some combinations of $M^s_{15}M^s_{24}$ could involve matrix elements such as $< d_1\bar{d}_4|\rho_{14} >< \rho_{14}|u_1\bar{u}_4 >$, which in the quark line rules of GENZ /10/ is called an annihilation diagram and is drawn in a way similar to Fig. 1d). The present interpretation of Fig. 1c) in terms of meson-like configurations of definite spin and isospin is, therefore, a specific combination of the rearrangement and annihilation diagrams of GENZ.

4. The $\pi\rho$-puzzle

Having convinced the reader that an alternative interpretation of Fig. 1d) is a recoupling of the configuration $M^s_{14}\epsilon_{25}(\ell = 0)$ as described in Eq. (3.3), the natural question to raise is the role of other configurations involving p-wave meson states. Such states $M^s_{14}\ M^p_{25}$ enter on the same footing as $M^s_{14}\epsilon_{25}$. For example the $N\bar{N}(^{11}S_0)$ state can be rewritten as

$$N\bar{N}(^{11}S_0) \to \sqrt{6}(\eta\epsilon)-\sqrt{2}(\pi\delta)+\sqrt{6}(\omega H)-\sqrt{2}(\rho B) \tag{4.1}$$

- others being given in ref. /5/. Equation (3.3) is now replaced by

$$N\bar{N}(S) \xrightarrow{1} \Sigma_p A(p) M^s_{14}M^p_{25}\epsilon_{36} \xrightarrow{2} \Sigma_p A(p) M^s_{14}M^p_{25}(\ell = 0)$$

$$\xrightarrow{3} \Sigma_p A(p) C\ M^s_{15}M^s_{24}(\ell = 1) \tag{4.2}$$

This now leads to the selection rule

$$N\bar{N}(S,\ \text{SPIN SINGLET}) \not\to M^s_1 M^s_2(\ell = 1) \tag{4.3}$$

At present this is only a numerical observation - but not a particularly suprising one when the amplitudes in Eq. (4.1) are seen to come in pairs that eventually cancel each other. This immediately explains the $\pi\rho$-puzzle in S-waves referred to in Eq. (3.1) without resorting to any competition with the $\pi^\circ f$ channel. Furthermore, as seen in table 1 some of the other singlet branchings are also reasonably small.

Table 1. Spin Singlet branchings

Branching	Expt. (see C. AMSLER /14/)
$N\bar{N}(^{11}S_0) \to \omega\omega$	$1\cdot4 \pm 0\cdot6$ %
$N\bar{N}(^{11}S_0) \to \rho^o\rho^o$	$0\cdot12 \pm 0\cdot12$
$N\bar{N}(^{31}S_0) \to \rho^o\omega$	$2\cdot1 \pm 0\cdot2$
	$0\cdot7 \pm 0\cdot3$
	$3\cdot9 \pm 0\cdot6$

For comparison the branching $N\bar{N}(^{13}S_1) \to \pi\rho$ is $4\cdot6 \pm 0\cdot4$ %. Of course, it must be remembered that a more meaningful comparison would require the inclusion of statistical factors, initial state interactions, phase space and form factor effects. However, as shown in ref. /11/, these tend to add overall support to the rule in eq. (4.3).

In table 2 a comparison is made between the rather meagre experimental results and the predictions of the present model for spin triplet states. In parentheses are given the predictions of Fig. 1d) i.e. simply using the $M^s_{14}\epsilon_{25}$ configuration. For $^{13}S_1$ and $^{33}S_1$ the theories are separately normalized to $\pi\rho$ and $\eta\rho$ respectively, since initial state interactions rule out at this stage any direct comparison between the two $N\bar{N}$ states.

Table 2. Spin Triplet branchings

Branching		Theory x 10^2		Expt. (see C. AMSLER /14/)
$N\bar{N}(^{13}S_1)$	$\to \pi\rho$	$4\cdot6$	$(4\cdot6)$	$4\cdot6 \pm 0\cdot4$ %
	$\to \eta\omega$	$13\cdot8$	$(1\cdot5)$	$1\cdot3 \pm 0\cdot2$
$N\bar{N}(^{33}S_1)$	$\to \eta\rho$	$0\cdot65$	$(0\cdot65)$	$0\cdot65 \pm 0\cdot16$
	$\to \pi\omega$	$0\cdot07$	$(0\cdot65)$	< 1 %
	$\to \pi^\pm\pi^\mp$	$0\cdot07$	$(0\cdot16)$	$0\cdot31 \pm 0\cdot03$
	$\to \rho^\pm\rho^\mp$	$0\cdot58$	$(4\cdot6)$	$< 9 \pm 3$

Again it should be emphasized that a meaningful comparison would require the inclusion of statistical factors, initial state interactions, phase space and form factor effects - see GREEN /11/. The only definite statement is that the addition of p-wave mesons other than the ϵ can result in massive changes in the predicted branching e.g. $N\bar{N}(^{33}S_1) \to \pi\omega$ decreases by an order of magnitude.

So far only the $\pi\rho$-puzzle in $N\bar{N}$ S-waves has been discussed. For the P-wave case mentioned after Eq. (3.2) it is necessary to consider the chain

$$N\bar{N}(P) \xrightarrow{1} \Sigma_{p_1p_2} A(p_1p_2)\, M^{p_1}_{14}M^{p_2}_{25}\epsilon_{36} \xrightarrow{2} \Sigma_p A(p_1p_2) M^{p_1}_{14}M^{p_2}_{25}(\ell=0) \qquad (4.4)$$

$$\xrightarrow{3} \Sigma_p A(p_1p_2)\, C\, M^s_{15}M^s_{24}(\ell=0)$$

As discussed more fully in ref. /11/ the ratio $R_p=18$ in eq. (3.2) from the usual rearrangement

$$N\bar{N}(P) \rightarrow \Sigma_{s_1 s_2} A(s_1 s_2)\, M^{s_1}_{14} M^{s_2}_{25} \epsilon_{36} \quad (4.5)$$

becomes $R_p=1/2$ - a significant step for understanding the experiment value of $R_p << 1$. It, therefore, remains to show why the rearrangement of eq. (4.5) is inhibited to such an extent that the chain in eq. (4.4) dominates P-wave annihilation.

It should be added that it is not unreasonable that p-wave mesons (or more precisely - p-wave meson-like configurations) are so important, when it is remembered that in the $p\bar{p}$ annihilation cross section P-waves already dominate over S-waves for p(Lab) > 200 MeV/c - see Fig. 15 in GREEN and NISKANEN /12/.

A potential problem, when introducing two p-waves into $N\bar{N}(S)$ is that nodes are unavoidably generated in the $N\bar{N}$ relative coordinate. These arise when the fourier transform of $k^2 Y_0(k)$ is made and come at $\approx 0 \cdot 5$ fm. However, it is not clear to what extent this is a real effect. Similar problems arose earlier in meson decays. There they were seen to be unrealistic and so various prescriptions were invoked to remove the nodes - see refs. /13/.

The inhibition $N\bar{N}(^{31}S_0) \not\rightarrow \pi\rho$ is not absolute since the branching is seen in $\bar{p}n \rightarrow \pi^- \rho^o$. However, this should presumably be considered as arising from a higher order effect of which there are several possibilities:

a) An effective $^3S_1$-operator could be induced as a higher order correction to the $^3P_0$-vertex /9/ as already mentioned at the end of section 3 in connection with K-meson production.

b) By one-pion-exchange the $N\bar{N}$(S-wave) state could become $N^*\bar{N}$(P-wave) where N^* is an odd parity baryon - as discussed in section 3.3.4 of GREEN and NISKANEN /12/.

c) In Eq. (4.1) the meson configurations could be considered more than simply a mathematical rearrangement of $N\bar{N}$. If these became true mesons - albeit in virtual configurations - then the mass differences (e.g. $M(\delta\pi) = 1.12$ GeV versus $M(B\rho) = 2$ GeV) could prevent an exact cancellation after the recoupling into two s-wave mesons.

5. In Search of a Theory for $N\bar{N}$ Annihilation

So far in $N\bar{N}$ annihilation no explicit mention has been made of gluons in the $^3P_0$-model - their presence being only implicit with the use of oscillator wavefunctions for the N, $\bar{N}$ and mesons. Of course, at the energies of interest it is the non-perturbative regime of QCD that enters, and so "complete solutions" cannot be expected. However, a step in the right direction could be the use of the flux tube model of ISGUR and PATON /2/ that has proved so successful for the description of meson decays /1/. A possible scenario would then be the one depicted in Fig. 2. The first step in Figs. 2a) to 2b) is similar to the meson decay problem, since it simply involves the breaking of the flux tube between junctions J_1 and J_2 by means of the $^3P_0$ operator.

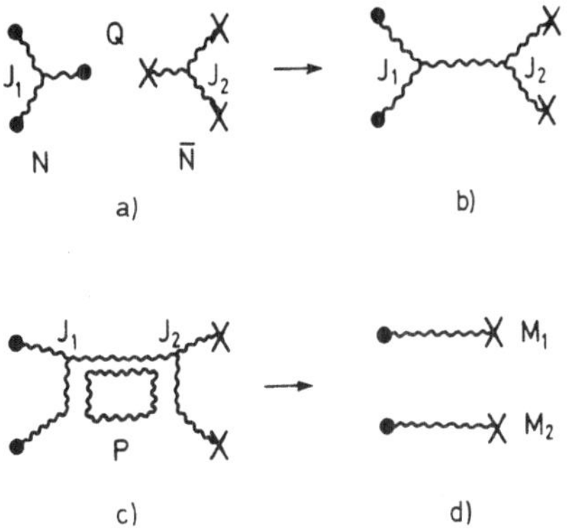

Fig. 2. Scenario for $N\bar{N}$ annihilation into two mesons M_1M_2 via Figs. 1a) and 1c)

a) Action of the $^3P_0$ operator (Q) to give:
b) Baryonium (B)
c) In baryonium the junctions J_1, J_2 can approach each other to be annihilated by the Plaquette operator (P) to give:
d) Two mesons M_1M_2.

The difficult step is that between Figs. 2c) and 2d). However, since the operation of the plaquette does not directly involve the two quarks and antiquarks, these remain in the same states of spin and flavour. Therefore, the effect of P is dominated by an effective interaction between the quarks and antiquarks that is spin and isospin independent. Schematically, the scenario of Fig. 2 can be written as

$$M = < M_1M_2|P|B > \frac{1}{\Delta E(B)} < B|Q|N\bar{N} > \tag{5.1}$$

$$\approx < M_1M_2|FQ|N\bar{N} > \tag{5.2}$$

where F is a spin-isospin independent factor simulating the role of the Baryonium and Plaquette. In other words, the model presented earlier in sections 3 and 4 is still valid from the point of view of spin and isospin. However, now because of F in Eq. (5.2) it should not be expected that the strength of the $^3P_0$-vertex in Figs. 1a) plus c) is the same as in meson decay. The above also shows that probably a better interpretation of Figs. 1a) and 1c) and Eq. (4.2) is that the recoupling of quarks (14) (25) into (15) (24) takes place at the step in Fig. 2 where the two junctions are annihilated.

The mechanism depicted in Figs. 1a) and 1c) is, therefore, seen to have a natural explanation in terms of flux tubes. However, the same cannot be said of Figs. 1a) and 1d). The repeated action of the $^3P_0$-vertex is unable to remove the junctions J_1 and J_2. As seen in Fig. 3, this results in a final two meson state where one of the mesons contains the initial two junctions in the form of a flux loop. Such a meson with an excited flux tube is called a hybrid and is not simply a π- or ρ-meson.

Needless to say, this approach has a very long way to go before it is on the same footing as the meson decay works of ref. /1/.

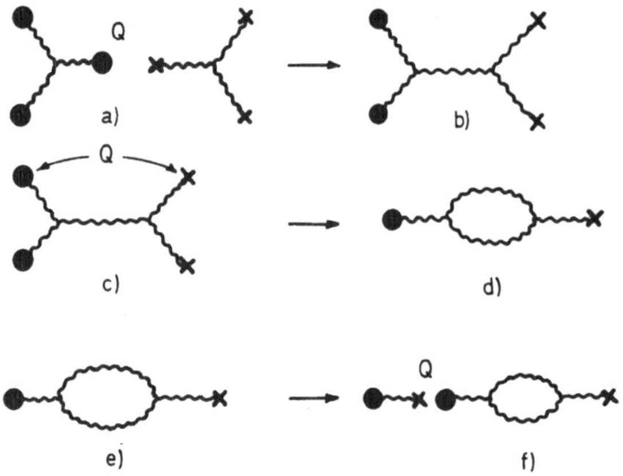

Fig. 3 Scenario for $N\bar{N}$ annihilation into two mesons M_1M_2 via Figs. 1a) and 1d)

a) Action of the $^3P_0$ operator (Q) to give:

b) Baryonium as in Fig. 2b)

c) In baryonium Q operates a second time to give:

d) A Hybrid meson.

e) In the hybrid meson Q operates a third time to give:

f) Two mesons one of which is still a hybrid.

6. Conclusions

a) For the main contributions to $N\bar{N}$ annihilation there is no justification or need for the use of the $^3S_1$-model, i.e. the vertex in Fig. 1a) dominates.

b) With the $^3P_0$-model there is no need to go to third order as in Fig. 1d), since this is essentially equivalent to the first order contribution of Fig. 1c). However, a recoupling of quarks and antiquarks is necessary.

c) As a consequence of b) it is necessary to include all p-wave meson-like configurations (M_2^p) in the rearrangement $N\bar{N}(S) \rightarrow M_1^s M_2^p \epsilon_3$.

d) At a more microscopic level involving explicit gluons, the removal of junctions is predominantly spin and isospin independent. Therefore, their effect can to some extent be simulated by modifying the strength of the $^3P_0$-vertex extracted from meson decays.

$N\bar{N}$ annihilation is dominated by the $^3P_0$-model in first order i.e. Figs. 1a) plus c).

References

1. R. Kokoski and N. Isgur, Phys. Rev. D35 (1987) 907

2. N. Isgur and J. Paton, Phys. Rev. D31 (1985) 2910

3. A.M. Green and J.A. Niskanen, Mod. Phys. Lett. A1 (1986) 441

4. C. Dover, Proc. 2nd Conf. on Intersections between Particle and Nuclear Physics, Lake Louise, Canada 1986 ed. D.F. Geesaman (AIP conference proceedings No. 150) p. 272

   C. Dover, P. Fishbane and S. Furui, Phys. Rev. Lett. 57 (1986) 1538

5. A.M. Green, to be published in Mod. Phys. Lett.

6. M. Maruyama, S. Furui and A. Faessler, Nucl. Phys. A472 (1987) 643

7. H. Rubinstein and H. Snellman, Phys. Lett. B165 (1985) 187

8. S. Furui, Z. Phys. A325 (1986) 375

9. A.M. Green, J.A. Niskanen and S. Wycech, Phys. Lett. B172 (1986) 171

10. H. Genz, Phys. Rev. D28 (1983) 1094 and Proc. of Workshop on the Elementary Structure of Matter, Les Houches, March 1987

11. A.M. Green, in preparation

12. A.M. Green and J.A. Niskanen, Prog. in Particle and Nuclear Physics Vol 18, ed. A. Faessler (Pergamon Press 1986), p. 93

13. R. Koniuk and N. Isgur, Phys. Rev. D21 (1980) 1868

    N. Törnqvist and P. Żenczykowski, Z. Phys. C30 (1986) 83

14. C. Amsler and G. Smith, Proc. of Workshop on the Elementary Structure of Matter, Les Houches, March 1987

QCD AS A MANY-BODY PROBLEM AND ASPECTS OF CONFINEMENT

D. Schütte

Institut für Theoretische Kernphysik der
Universität Bonn
Nussallee 14-16, D-5300 Bonn, West-Germany

1. INTRODUCTION

The numerical complexity of lattice Monte-Carlo calculations for the spectrum of QCD makes it desirable to investigate alternative formulations of gauge field theories. A natural alternative is the treatment of the field theoretical eigenvalue problem by suitable many-body approximation techniques. (Replace the Feynman path-integral by the Schrödinger equation).

It is the purpose of this talk to give a survey on the structure of such a field theoretical many-body problem, especially with respect to renormalization (regularization and check of scaling in the continuum limit) and with respect to the different choices of regularization for gauge theories (lattice regularization and complete gauge fixing).

A comparison of the structure of the many-particle problems within these two regularization schemes yields some new aspects of confinement.

2. GENERAL STRUCTURES OF QUANTUM FIELD THEORIES

A Poincaré invariant Lagrangian $L(\phi,\partial_\mu,\phi)$ (for simplicity, we discuss in this section only the scalar case) yields via canonical quantization operators

$$P^\mu(\phi,\pi),\ L^{\mu\nu}(\phi,\pi)\ ,\quad [\phi(\vec{x}),\pi(\vec{y})] = -i\delta(\vec{x}-\vec{y}) \qquad (1)$$

of energy-momentum and angular momentum which formally obey the commutation relation of the Poincaré algebra. In principle, the decomposition of this reducible, unitary representation of the Poincaré group should yield bound states (elementary particles) and scattering states as eigenstates of the Hamiltonian $P^0=H$. However, the pathological character of H for a quantum field theory with interaction does not allow to argue in such a direct way and renormalization has to be invoced to make the theory (hopefully) well-defined. Since non-perturbative methods must be used for the computation of bound states, a convenient strategy [1] for renormalization is a <u>regularization</u> of the (t=0) field operators:

$$\phi(\vec{x}) \rightarrow \phi_N(\vec{x}) = \sum_{\alpha=1}^{N} f_\alpha(\vec{x})q_\alpha \qquad (2)$$

and a <u>check of scaling</u> in the continuum limit. In (2), $\{f_\alpha(\vec{x}),\ .\alpha = 1,2...\}$ is a suitable single particle basis and $N = N(\Omega,M)$, Ω = volume, M = momentum cutoff. This yields the eigenvalue problem of the regularized Hamiltonian $H(g,\Omega,M) = H(p_1...p_n,\ q_1...q_n)$, $p_\alpha = -i\partial/\partial q_\alpha$ as that of a well-defined "standard" many-body system (The wave functions are $\Psi = \Psi(q_1...q_N)$ and $\langle\Psi,\Psi\rangle = \int d^Nq\Psi^*\Psi$).
In principle, the solution of this eigenvalue problem yields then observables $A(g,M,\Omega)$ whose "thermodynamical" limit $A(g,M) = \lim_{\Omega\to\infty} A(g,M,\Omega)$ will exist [2]. (Note, however, that the reliable determination of $A(g,M)$ is, in practise, a major computational problem). The function $A(g,M)$ behaves pathologically for $M\to\infty$. However, for a renormalizable field theory, a universal function $g(M)$, the "running coupling constant" [3] should exist such that

$$\lim_{M\to\infty} A(g(M),M) = \text{well-behaved for all observables } A \quad . \qquad (3)$$

For a Yang-Mills theory or massless QCD, the M-dependence of $A(g,M)$ is trivial

$$A(g,M) = M^D A(g,1) \qquad D = \text{dimension of } A \qquad (4)$$
$$= 1 \text{ for masses}$$
$$= 2 \text{ for string tension}$$

and (3) reduces to a predetermined "scaling" behaviour of $A(g,1)$

$$A(g,1) \simeq \exp(-D/\beta_0 g^2) \qquad \text{for} \quad g\to 0 \quad . \qquad (5)$$

This structure of $A(g,1)$ is observed (within errors) in lattice Monte Carlo calculations [4] which supports the confidence that the continuum limit (3) should be in order.

3. REGULARIZATION OF GAUGE FIELD THEORIES

In the case of gauge field theory, the construction of a regularized theory, given by (2) for the scalar case and yielding $A(g,\Omega,M)$, is more involved since now the physical variables are the "orbits" of gauge fields, i.e. the equivalence classes of fields $A_\mu(x)\in\text{Lie } SU(n)(\vec{x} = (t,x))$ where

$$A'_\mu \simeq A_\mu \iff \text{there exists } U(x) \in SU(n) \text{ with}$$

$$A'_\mu = U(A_\mu - i\partial_\mu)U^{-1} \ . \tag{6}$$

The convenient way to work with such equivalence classes is to choose a representative (fix the gauge) and to check that observables A(g,M) are independent from the choice of this representative. In the framework of canonical quantization, a convenient starting point is a partial gauge fixing by setting $A_0=0$. (This leaves open time-independent gauge transformations). Quantization of the gauge field Lagrangian yields then a formally covariant theory only on the "physical" states $\Psi(\vec{A})$ obeying the Gauss law [5,6]

$$\Psi(A) = \Psi(A') \quad \text{for any } \vec{A}' \text{ with}$$

$$\vec{A}'(\vec{x}) = U(\vec{x})(\vec{A}-i\vec{\nabla})U^{-1}(\vec{x}) \ . \tag{7}$$

This reduces the variables of the theory to the gauge orbits. There exist two regularizations of the gauge field theory which are consistent with the Gauss law, the lattice regularization and the regularization of equivalence classes using complete gauge fixing. (Hereby, the Coulomb gauge appears to be the most consistent choice).

3.1 Lattice Regularization

This regularization is described in detail in ref. [1]. The basic idea is to use gauge invariant wave functions $\Psi(A)$ related to the parallel transport along a closed loop κ

$$\chi_\kappa(A) = \mathrm{tr}\, P \exp i \int_\kappa \vec{A}d\vec{x} \tag{8}$$

$\vec{A}(\vec{x})$ is regularized with the characteristic functions related to a (3-dimensional, finite) lattice (with links $\kappa_1 \ldots \kappa_N, \kappa_\alpha$ goes $\vec{x}_\alpha$ to $\vec{y}_\alpha$)

$$\vec{A}(\vec{x}) = \sum_{\alpha=1}^{N} \sum_{a=1}^{R} \vec{f}_\alpha(\vec{x})\lambda_a q_{\alpha a} \tag{9}$$

$$\{\lambda_1 \ldots \lambda_R\} = \text{Basis Lie } SU(n).$$

$\chi_\kappa(A)$ is then computable for lattice loops and is just the trace of the product of link-parallel transports (we assume $\varepsilon = 1$, ε = lattice constant)

$$U_\alpha = P\exp i \int_{\kappa_\alpha} \vec{A}d\vec{x} = \exp i\lambda_a q_{\alpha a} \ . \tag{10}$$

Introducing U_α as new variables leads then to the Kogut-Susskind Hamiltonian form of standard lattice QCD [7,8] where (without quarks)

$$H_{ks} = \sum_\alpha \frac{g^2}{2} \operatorname{tr} E_\alpha^2 - \frac{1}{2g^2} \sum_\Box \operatorname{tr}(P_\Box + P_\Box^{-1} - 2) \qquad (11)$$

(E_α is canonically conjugate to U , $P_\Box$ is the parallel transport along a plaquette). The wave functions (describing a quantum many-top problem) must obey the Gauss law on the lattice ($g(\vec{x}) \in SU(n)$, arbitrary)

$$\Psi(U_1' \ldots U_N') = \Psi(U_1 \ldots U_n); \quad U_\alpha' = g(\vec{x}_\alpha) U_\alpha g^{-1}(\vec{y}_\alpha) \qquad (12)$$

The computation of observables related to the eigenfunctions of H_{ks}, but framed in the language of Feynman path integrals, leads to the standard lattice Monte-Carlo QCD [4]. Attempts to solve the eigenvalue problem of H_{ks} directly using many-body techniques are discussed in ref. [9].

3.2 Regularization in the Coulomb gauge

In the framework of complete gauge fixing, a specific description of the orbit space of gauge fields is used by choosing representatives.

Herefore, one relates any $\vec{A}(\vec{x})$ to a (unique) representative $\vec{F}_Q(\vec{x})$ ($Q \in I$ = suitable index set) such that a (modulo the center of SU(n) unique) gauge field $U(\vec{x})$ exists with [1]

$$\vec{A} = U(\vec{F}_Q - i\vec{\nabla})U^{-1} \qquad (13)$$

This allows to change coordinates in the functional space $\vec{A} \Leftrightarrow (U,Q)$ and the Gauss law may be trivially fulfilled for $\Psi = \Psi(Q)$ (independent of U). The set $\vec{F}_Q, Q \in I$ is conveniently determined via a gauge fixing condition. Translational and rotational invariance favours

$$\vec{\nabla}\vec{F}_Q = 0 \text{ , the Coulomb gauge .} \qquad (14)$$

The regularization is then introduced on these representatives:

$$\vec{F}_Q(\vec{x}) = \sum_{\alpha=1}^{N} \vec{f}_\alpha(x) Q_\alpha \qquad (Q_1 \ldots Q_N) \in R^N \qquad (15)$$

Since the set of $F_Q, Q \in I \subset R^N$, should be unique representatives for the fields $\vec{A}(\vec{x})$ one is lead to the condition that $Q = (Q_1 \ldots Q_N)$ should be restricted to a "Gribov-domain" I which is bounded by the Gribov horizon [10]. Quantitative estimates of Gribov [10] and Cutkosky

[11] yield that this horizon is well approximated by an ellipsoid in R^N, i.e. I is to a good approximation probably given by

$$I = \{Q \in R^N | \sum C_\alpha^2 Q_\alpha^2 < 1\} . \qquad (16)$$

The precise definition of I is given in refs. [1,10].

The regularized Coulomb gauge Hamiltonian $H_C(Q,\partial/\partial Q)$ of a Yang-Mills theory emerging from the non-linear coordinate transformation $\vec{A} \leftrightarrow (U,Q)$ is rather complicated [1,5,6]. The important feature of H_C is that it is singular at the Gribov horizon. Attempts to cope with the eigenvalue problem of H_C have been undertaken by several groups [11,12]. Like in the case of the lattice regularization, no many-body approximation has been found up to now which yields observables fulfilling the rather stringent scaling condition (5).

E.g., standard non-perturbative many-body techniques (exp S, cluster expansions) were applied to compute glue-ball spectra [12], but these techniques turned out to be suitable only for large momenta, the small momentum behaviour of the wave functions, where the Gribov-horizon is expected to play an important role, has still essentially to be improved.

4. ASPECTS OF CONFINEMENT

The singularity of the Coulomb gauge Hamiltonian H_C at the Gribov horizon suggests a special confinement scenario [1]. Herefore, one introduces static quarks and studies the structure for the strong coupling limit and its transition to the continuum limit. Within the lattice regularization, confinement is trivial in the strong coupling limit (due to the Gauss law), the weak coupling case is not accessible analytically. Lattice Monte-Carlo calculations, however, strongly suggest that confinement is still valid: the string tension is different from zero for $g \to 0$ and scales (at leasts up to 20%) [4].

Within the Coulomb gauge, confinement is non-analytic even in the strong coupling case. However, certain necessary conditions [1] (induction of non-trivial vacuum polarization effect as an indication of a string formation) hold because of the singularity of H_C at the Gribov horizon.

Assuming that confinement is true in the strong coupling limit, some possible conclusions for the nature of confinement for $g \to 0$ can be drawn. It is suggested that the eigenfunctions $\Psi(Q_1 \ldots Q_N)$ for $g \to 0$ behave with respect to the variable Q_α ($\alpha = (\vec{k},r,a)$ = (momentum, polarization, colour) [2]) for <u>small</u> $\vec{k}_\alpha$ ($|\vec{k}_\alpha| < M_0$) just like in the strong coupling case $g \to \infty$, whereas one should have perturbative structures for $|\vec{k}_\alpha| \gg M_0$. The reason, herefore, is that the Gribov horizon shrinks for small k_α (e.g. $C_\alpha^{-1} \approx |k_\alpha|^m, m>1$).

It is expected that $M_0 = M_0(g)$ scales (i.e. $M_0 \approx \exp(-1/\beta_0 g^2)$) but since it remains finite for any finite g, the strong coupling nature of $\Psi(Q)$ for Q_α with $|\vec{k}_\alpha| < M_0$ is sufficient to keep confinements in the continuum limit. (For details see ref. [1]). Numerical investigations aiming at supporting these suggestions are in progress.

Referenes

1. D. Schütte, The Structure of Canonically Quantized Gauge Field Theories within Different Regularization Schemes and Aspects of Confinement, Preprint, Bonn 1987, to be published
2. D. Schütte, Phys. Rev. D31 (1984) 810
3. J.B. Kogut, Rev. Mod. Phys. 55 (1983) 775
4. B.J. Pendleton, Non-Perturbative QCD, Contribution to the Proceedings of the Les Houches Workshop 1987
5. N.H. Christ and T.D. Lee, Phys. Rev. D22 (1980) 939
6. H. Cheng and E.C. Tsai, Canonical Quantization of Non-Abelian Gauge Field Theories and Feynman Rules, MIT preprint, to appear in Nucl. Phys. B
7. J. Kogut, L. Susskind, Phys. Rev. D11 (1975) 395
8. M. Creutz, Quarks, Gluons and Lattices (Cambridge University Press, 1983)
9. D. Horn et al, Phys. Rev. D31 (1985) 2589
   W. Furmanski, A. Kalowa, Yang-Mills Vacuum - an Attempt of Lattice Loop Calculus, (Preprint 1986) CALT-68-1330
   S.A. Chin, O.S. van Roosmalen, E.A. Umland and S.E. Koonin, Phys. Rev. D31 (1985) 3201
10. V.N. Gribov, Nucl. Phys. B139 (1978) 1
11. R.E. Cutkosky, Phys. Rev. D30 (1984)447
12. B. Faber, H. Nguyen-Quang and D. Schütte, Phys. Rev. D34 (1986) 1157
    H. Nguyen-Quang, Dissertation Bonn (1986)

NUCLEON MEAN-FREE PATH IN NUCLEAR MATTER*

Janusz Dabrowski

Institute for Nuclear Studies, Hoza 69, 00-681, Warsaw, Poland, and
Department of Physics, University of Arizona, Tucson, AZ 85721, U.S.A.

1. INTRODUCTION

The mean-free path (m.f.p.) λ of a nucleon in nuclear matter (NM) is one of the fundamental characteristics of NM. Directly related to λ is the imaginary part W of the nuclear optical potential ($\lambda \sim 1/W$). There exist numerous calculations of W in NM (see, e.g., refs. 1-3) and references therein), which start from the NN interaction and lead to results for both the real and imaginary parts, V and W, of the optical potential. Calculations of this type are complicated since they involve the whole machinery of the many-body theory. In the case of V, this appears unavoidable. If however, one is interested only in W (and λ), one may follow a much simpler procedure which goes back to the early paper by Lane and Wandel [4]), and relate W and λ directly to the NN cross section σ.

In sect. 2, I start from the Brueckner theory and then derive an approximate expression for W (and λ) in terms of σ. In sect. 3, I test this approximate expression against the "exact" Brueckner theory results of J.-P. Jeukenne et al. [5]). In sect. 4, I present the results for λ, discuss how the nucleon momentum distribution affects λ, and compare the results with experiment. Our procedure may also be applied to other particles, and sect. 5 outlines a calculation of the m.f.p. of a Σ hyperon, and of the width of a Ξ hyperon in NM.

The present talk is based on results presented in refs. 6-8.

2. APPROXIMATE EXPRESSIONS FOR W AND λ

We start from the Brueckner theory expression for the absorptive potential of a nucleon "0" moving with momentum $\vec{k}_0$ (in units of $\hbar$) through symmetric (N=Z) NM of density ρ:

$$W(k_0) = 4(2\pi)^{-3} \int d\vec{k}_1\, n_0(k_1)\, \mathrm{Im}\langle\vec{k}|\mathcal{K}|\vec{k}\rangle \qquad (2.1)$$

where $\vec{k}_1$ is the momentum of the nucleon "1" of NM, $\vec{k} = (\vec{k}_0 - \vec{k}_1)/2$ is the "0"-"1" relative momentum, and $n_0(k_1) = \theta(k_F - k_1)$ is the distribution of nucleon momenta k_1 in NM (k_F = Fermi momentum). The factor 4 in (2.1) takes care of the four spin-isospin states of a given momentum $\vec{k}_1$ -- otherwise spin and isospin is suppressed in our notation.

*Research supported by Polish-U.S. Maria Sklodowska Curie Fund under Grant No. P-F7FO37P.

The reaction matrix $\mathcal{K}$ is defined by

$$\mathcal{K} = v + v\,[Q/(\alpha + i\eta)]\mathcal{K} \quad , \tag{2.2}$$

where v is the NN potential,

$$Q = Q(k_1', k_0') = [1 - n_0(k_1')]\,[1 - n_0(k_0')] \tag{2.3}$$

is the exclusion principle operator,

$$\alpha = e(k_1) + e(k_0) - e(k_1') - e(k_0') \quad , \tag{2.4}$$

and k_0', k_1' are nucleon momenta in the intermediate states. The s.p. energy $e(k_N)$ is assumed to be of the form

$$e(k_N) = \begin{cases} \hbar^2 k_N^2/2M\mu + D & \text{for } k_N < k_F \quad , & (2.5a) \\ \hbar^2 k_N^2/2M\nu + C & \text{for } k_N > k_F \quad , & (2.5b) \end{cases}$$

where the ratio of the effective to the real nucleon mass, M^*/M, is denoted by μ (ν) for $k_N < k_F$ ($k_N > k_F$).

Equation (2.2) implies the optical theorem

$$\mathrm{Im}\langle\vec{k}|\mathcal{K}|\vec{k}\rangle = -\frac{1}{2}(2\pi)^{-2}\int d\vec{k}'\; Q|\langle\vec{k}'|\mathcal{K}|\vec{k}\rangle|^2\;\delta(\alpha)$$

$$\cong -2(\hbar^2/M)\,k\,\sigma^{NM}\,(4\pi)^{-1}\int d\hat{k}'\int dk'k'\;Q\;\delta(\alpha) \quad , \tag{2.6}$$

where $\vec{k}' = (\vec{k}_0' - \vec{k}_1')/2$, and where we introduce the NN cross section in NM

$$\sigma^{NM} = (k'/k)\,(M/4\pi\hbar^2)^2\int d\hat{k}'\;|\langle\vec{k}'|\mathcal{K}|\vec{k}\rangle|^2 \quad , \tag{2.7}$$

in which k' is determined by energy conservation, $\alpha = 0$. The approximate treatment of the $\hat{k}'$ integration in the last step in (2.6) means that we replace $Q\,\delta(\alpha)$ by its angle average.

With the sharp momentum distribution n_0, we have $k_0' > k_F$, $k_1' > k_F$, and for these states (see (2.5b)) $\alpha = -\hbar^2 k'^2/M\nu$ + terms independent of k'. Thus we have

$$(4\pi)^{-1}\int d\hat{k}'\int dk'k'\;Q\;\delta(\alpha) = (M\nu/\hbar^2)\bar{Q} \quad , \tag{2.8}$$

where $\bar{Q}$ is the angle average of Q.

Now, we make the crucial approximation

$$\sigma^{NM} \simeq \sigma = (\sigma_{np} + \sigma_{nn})/2 \quad , \tag{2.9}$$

which enables us to express W through the total cross sections σ_{np} and σ_{nn} for free nn and np scattering:

$$W(k_0) = -4\hbar^2\nu\int\frac{d\vec{k}_1}{(2\pi)^3}\,n_0(K_1)\,\bar{Q}\,\frac{k}{M}\,\sigma = -\frac{\hbar^2}{M}\,\nu\rho\,\langle\bar{Q}k\sigma\rangle \quad , \tag{2.10}$$

where $\langle\;\rangle$ denotes averaging over momenta of the nucleon "1" in NM.

To determine the m.f.p., we write the s.p. Schrödinger equation,

$$\{ - (\hbar^2/2M\nu)\,\Delta + C + i\,W\}\,\psi = e\psi \quad . \tag{2.11}$$

The appearance of the effective mass $M\nu$ is the result of the momentum dependence of the real s.p. potential V (= e - ϵ in eq. (2.6)), which is important in calculating λ [9,10]. By inserting the solution $\psi = \exp[i(k_0 + i\lambda/2)z]$, we get (for $k_0 >> 1/\lambda$) for k_0:

$$\hbar^2\, k_0^2/2M\nu + C = e \quad , \tag{2.12}$$

and for λ

$$\lambda = - \hbar^2 k_0/2M\nu W = k_0/2\nu^2\rho\, \langle Qk\sigma\rangle \quad . \tag{2.13}$$

This modified semiclassical expression takes into account the exclusion principle (through the Q operator) and the dispersive effects, i.e., the momentum dependence of the s.p. energies (through the factor ν, and the form of the energy conservation equation $\alpha = 0$). If we disregard the exclusion principle and binding effects (and neglect the motion of nucleons in NM), (2.13) goes over into the semiclassical expression: $\lambda = 1/\rho\sigma$.

To use expression (2.13), we have to know the s.p. energies [eq. (2.5)]. Let us note that eq. (2.5b) implies that $V - e - \epsilon = (1 - \nu)\, e + \nu C$. On the other hand, the optical model analysis [11]) suggests that at the center of nuclei (i.e., for $\rho = \rho_0 = 2k_{F0}^3/3\pi^2$, where ρ_0 is the equilibrium density of NM) $V \cong 0.3e{-}50$ MeV. Thus, we have $\nu(\rho_0) = \nu_0 = 0.7$ and $\nu_0 C(\rho_0) \cong 50$ MeV. For the dependence of ν on ρ, we use the form [12])

$$\nu = 1/[1 + (\nu_0^{-1} - 1)\, \rho/\rho_0] \quad . \tag{2.14}$$

The remaining parameters of the s.p. energies are determined by assuming continuity of $e(k_N)$ at $k_N = k_F$, and by fitting $e(k_N < k_F)$ to the "empirical" energy per nucleon in NM [for details, see ref. 6].

3. TEST OF THE APPROXIMATION

To test our approximation, we apply it to the case of the Reid hard-core NN interaction [13]), and compare our results for W [eq. (2.10)] with those of Jeukenne, Lejeune and Mahaux [5]) who calculated W using Brueckner theory with this interaction. I shall refer to the results of ref. 5 as the "exact" results.

Our test calculation was performed for $k_{F0} = 1.4$ fm^{-1}, the value used in ref. 5. The σ_{nn} and σ_{np} cross sections were obtained from the nuclear bar phase shifts of the Reid hard-core potential. Similarly as in ref. 5, all S and D partial waves with $J \leq 2$ and P waves were included. To stay within the range of $E_L < 350$ MeV, considered in ref. 13, we keep $k_0 \lesssim 2.7$ fm^{-1}, i.e., $k_0/k_{F0} \lesssim 1.9$.

The agreement between the results of our approximation and the "exact" results, shown in fig. 1, is very satisfactory.

A similar test of approximate expressions for W (and also for V) was presented by Köhler [14]).

4. RESULTS

For the NN cross sections, we use the effective range approximation for nucleon laboratory energies $E_L < 20$ MeV, and the parametrization of Metropolis et al.[15] for $E_L >$ 20 MeV. This parametrization is correct up to $E_L \sim 400$ MeV, and thus sufficient to calculate W and λ for $e \lesssim 200$ MeV, i.e., for $k_0 \lesssim 3$ fm.

The role played by different factors in our approximation are visualized in fig. 2, which shows $W(k_0)$ calculated in our approximation (2.16) and three other curves. By putting $Q = 1$ and $\nu = 1$ in (2.16), we get the upper curve. The curve $\mu = \nu = 1$ differs from the upper curve by the presence of the Q operator. Obviously the mere effect of the Pauli blocking is the most important factor in reducing $|W|$. The curve $\mu = \nu$ is obtained by applying the same form of the s.p. energy [eq. (2.5b)] for all nucleon momenta. Here $k' = k$, and W differs from W ($\mu = \nu = 1$) by the factor $\nu = 0.7$. A further reduction in $|W|$ is introduced in our approximation in which $k' < k$, and the final nucleons are slowed down, which enhances the effect of Q.

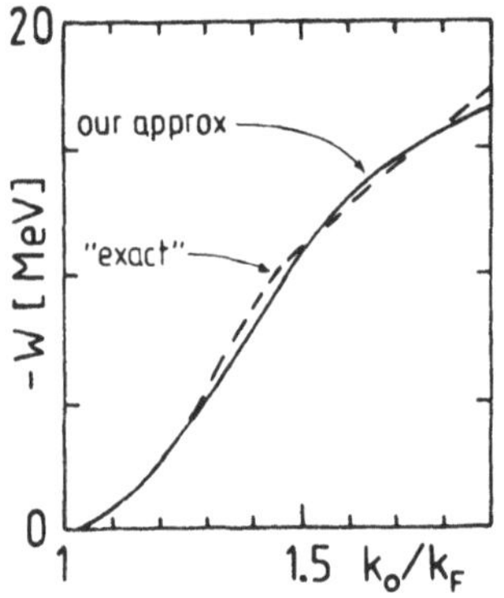

FIGURE 1
Our results for W for the Reid hard-core potential, compared with the "exact" results.

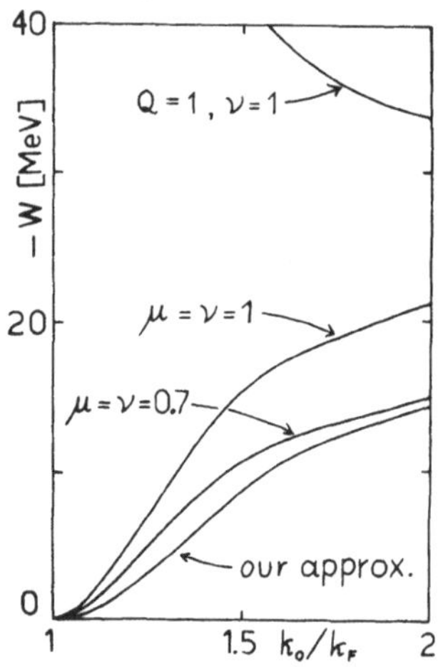

FIGURE 2
Results for $W(k_0)$ at $k_F = k_{F0} = 1.35$ fm^{-1}.

So far we have used sharp momentum distribution $n_0(k_N) = \theta(k_F - k_N)$. The whole calculation of W and λ may also be performed with a realistic diffused distribution $n(k_N)$, for which eq. (2.8) and also expression (2.10) are slightly modified (the first part of (2.13) remains unchanged). In our calculation, we use for $n(k_N)$ the distribution determined by Fantoni and Pandharipande [16]), shown in fig. 3. In presenting our results obtained with $n(n_0)$ we use the notation $\lambda(\lambda_0)$.

Our results for λ and λ_0 at $\rho = \rho_0 = 0.160$ fm^{-3} ($k_{F0} = 1.33$ fm^{-1}) and at $\rho = \rho_0/2$ are plotted in fig. 4 as functions of the nucleon energy e, connected with k_0 by relation (2.5b). At high energies, λ is only slightly smaller than λ_0. As e diminishes, λ does not change very much, whereas λ_0, especially at $\rho = \rho_0$, increases rapidly as $e \to 0$, which is the result of a strong Pauli blocking connected with the sharp distribution n_0. For the same reason, at low energies, $\lambda(\rho_0)$ is only slightly bigger than $\lambda(\rho_0/2)$ whereas $\lambda_0(\rho_0) >> \lambda_0(\rho_0/2)$. In short, λ does not depend on energy and density as drastically as λ_0.

In fig. 5, our results for λ and λ_0 are compared with empirical estimates. The results of ref. 11 are obtained from the compilation of the early results for W (via relation (2.13)). Similarly, the results of ref. 17 are obtained from the results for W from recent proton scattering on Pb and Ca. In the same way, the horizontally shaded band is obtained from the results for W of ref. 18 (based on a real potential of a wine-bottle shape). The vertically shaded band denotes the range of λ values determined in ref. 17 from p-Ca reaction cross section. The result of ref. 19 is an estimated upper limit of λ

inferred from inclusive proton-induced spectra. The estimate of ref. 20 was obtained by fitting λ to angle-integrated proton singles spectrum in (p,x) experiments. The estimate of ref. 21 is the result of adjusting the collision term in describing heavy ion collisions.

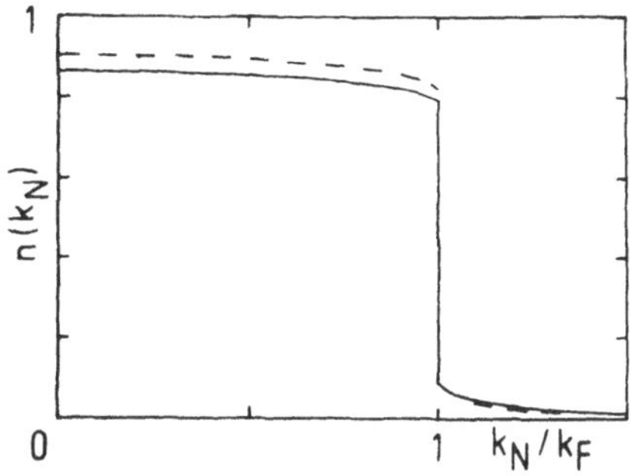

FIGURE 3
Distribution $n(k_N)$ of ref. 16 at $\rho=\rho_0=0.16$ fm^{-3} (solid curve) and $\rho=\rho_0/2$ (broken curve).

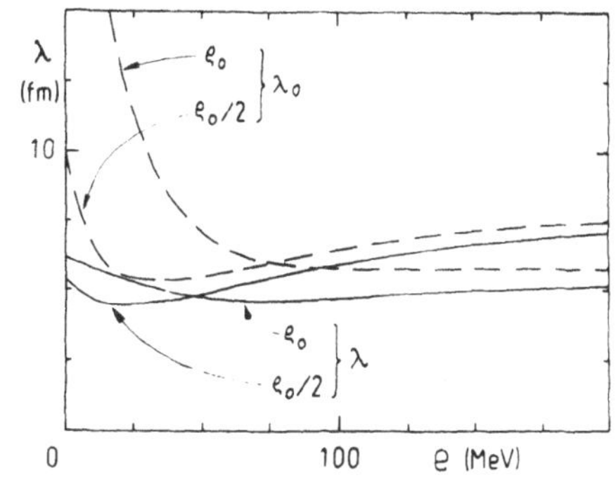

FIGURE 4
$\lambda(e)$ and $\lambda_0(e)$ calculated at $\rho=\rho_0=0.16$ fm^{-3} and $\rho=\rho_0/2$.

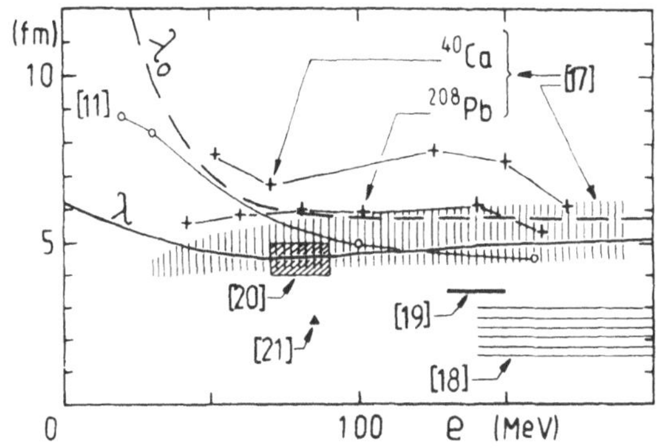

FIGURE 5
$\Lambda(e)$ and $\lambda_0(e)$ calculated at $\rho=\rho_0=0.16$ fm^{-3} compared with empirical estimates.

In general, our results agree better with the direct experimental estimates of λ than with those obtained from empirical values of W. The best agreement is achieved with the result of recent proton reaction cross-section measurements by Nadasen et. al. [17]) (and also with the less direct estimate of ref. 20). In spite of the considerable spread of the empirical points, none of them for $e \lesssim 50$ MeV is compatible with our results for λ_0 which increases rapidly with decreasing e. In this respect, all the experimental results favor the m.f.p. λ calculated with the realistic momentum distribution. Our results suggest that $\lambda = 5 \pm 1$ fm in the whole energy range considered, $0 \lesssim e \lesssim 200$ MeV.

5. HYPERONS IN NUCLEAR MATTER

Our approximate procedure may also be applied to other particles in NM, in particular to hyperons. Most interesting is the case of the Σ hyperon. Let us consider, e.g., Σ^-, which may undergo elastic scattering (to which we include the charge-exchange scattering $\Sigma^- p \rightarrow \Sigma^0 n$) and the $\Sigma\Lambda$ conversion $\Sigma^- p \rightarrow \Lambda n$. In applying the procedure of sect. 2, we assume for the s.p. energies of the Y hyperon ($Y = \Sigma, \Lambda$) the form

$$e_Y(k_Y) = \hbar^2 k_Y^2 / 2M_Y \nu + D_Y \quad , \tag{5.1}$$

with $D_Y = -30$ MeV. For the absorptive potential W_Σ and for the m.f.p. λ_Σ of Σ^- with momentum k_Σ, we get

$$W_\Sigma = W_e + W_c \ , \qquad 1/\lambda_\Sigma = 1/\lambda_e + 1/\lambda_c \quad , \tag{5.2}$$

where the subscript $x = e, c$ denotes the respective contribution of elastic and conversion processes. We have

$$W_x = -(\hbar^2/2\mu_{\Sigma N})\, \nu\rho\, \langle \overline{Q}_x\, k_{\Sigma N}\, \sigma_x \rangle \quad , \tag{5.3}$$

$$\lambda_x = -\hbar^2 k_\Sigma / 2M_\Sigma \nu W_x \quad , \tag{5.4}$$

where $\overline{Q}_x$ is the exclusion principle operator for final nucleons in the x processes, and σ_x is the total cross section for these processes.

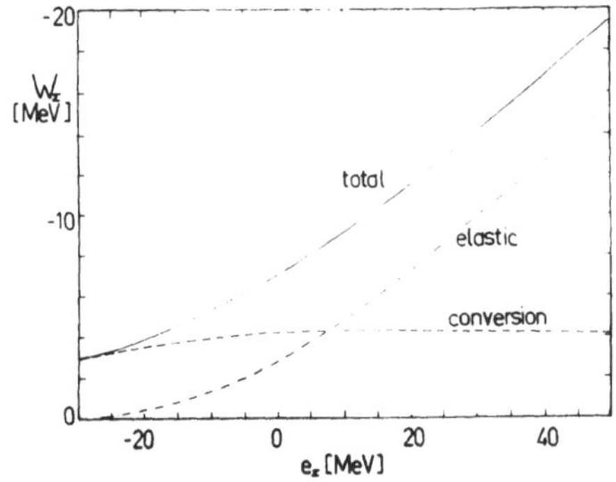

FIGURE 6
$W_\Sigma(e_\Sigma)$ calculated at $k_F = k_{F0} = 1.35\ \mathrm{fm}^{-1}$.

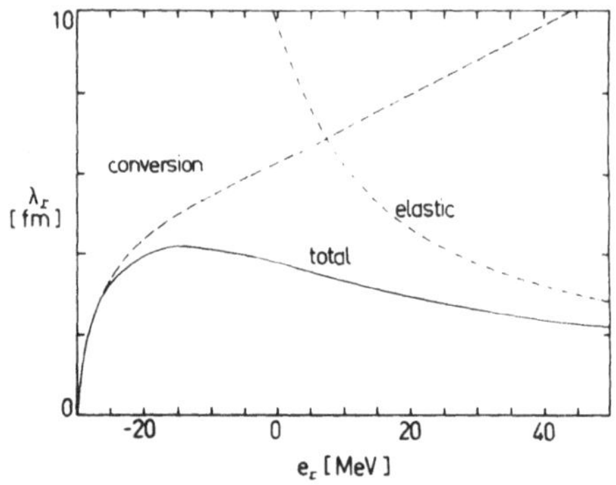

FIGURE 7
$\lambda_\Sigma(e_\Sigma)$ calculated at $k_F = k_{F0} = 1.35\ \mathrm{fm}^{-1}$.

Results for W_Σ and λ_Σ are shown in figs. 6 and 7, together with separate contributions of elastic and conversion processes. The $\Sigma\Lambda$ conversion process has a decisive effect on W_Σ and λ_Σ, especially for $k_\Sigma \rightarrow 0$, i.e., $e_\Sigma \rightarrow -30$ MeV [see eq. (5.1)]. Here, it plays a dominant role, and we have $W_\Sigma \cong W_c \cong -3$ MeV; thus, the corresponding width of the Σ ground state in NM is $\Gamma_\Sigma = -2W_\Sigma \cong 6$ MeV. This result agrees with the narrow widths observed in Σ hypernuclei. Notice that without the Pauli blocking and the binding effects, we would get 23.5 MeV for Γ_Σ.

An unsolved problem in hypernuclear physics is a possible existence of Ξ hypernuclei [22]. In the absence of any direct ΞN data, I shall rely on the theoretical Nijmegen baryon-baryon interaction [23,24], which suggest the binding energy of Ξ in NM of 23-30 MeV [25,26]. To estimate the width $\Gamma_\Xi = -2W_\Xi$ due to the strong process $\Xi^- + p \to \Lambda + \Lambda$, we may apply our approximation

$$W_\Xi = -(\hbar^2/2\mu_{\Xi N})\, \nu\rho\, \langle k_{\Xi N}\, \sigma_c \rangle \quad , \tag{5.5}$$

where σ_c is the total cross section for the process. Since no nucleons are produced in the process, no Q operator appears in (5.5). With the value of σ calculated with the Nijmegen interaction [24], we get $\Gamma_\Xi = 1.7$ MeV (with $\nu = M^*_\Lambda/M_\Lambda = 0.7$), a value sufficiently small for a possible detection of Ξ hypernuclei.

ACKNOWLEDGEMENTS

This paper was written while I was visiting the Physics Department of the University of Arizona, and I express my gratitude to Professor P. Carruthers, the Head of the Department, for his kind invitation. It is a pleasure to thank Professor Köhler for several discussions and most valuable suggestions concerning the topics of this paper, and for his extraordinary hospitality in Tucson.

REFERENCES

1) S. Fantoni, B. L. Friman and V. R. Pandharipande, Nucl. Phys. **A386** (1982) 1.
2) F. A. Brieva and J. R. Rook, Nucl. Phys. **A291** (1977) 299.
3) J.-P. Jeukenne, A. Lejeune and C. Mahaux, Phys. Rep. **25C** (1976) 83.
4) A. M. Lane and C. F. Wandel, Phys. Rev. **98** (1955) 1524.
5) J.-P. Jeukenne, A. Lejeune and C. Mahaux, Phys. Rev. **C10** (1974) 1391.
6) J. Dabrowski and W. Piechocki, Acta Physica Polonica **B16** (1985) 1095.
7) J. Dabrowski, Acta Physica Polonica **B17** (1986) 1001.
8) J. Dabrowski, Phys. Lett. **139B** (1984) 7.
9) S. Fantoni, B. L. Friman and V. R. Pandharipande, Phys. Lett. **104B** (1981) 89.
10) J. W. Negele and K. Yazaki, Phys. Rev. Lett. **47** (1981) 71.
11) A. Bohr and B. Mottelson, Nuclear Structure, Vol. I (W. A. Benjamin, Inc., 1969).
12) B. Friedman and V. R. Pandharipande, Phys. Lett. **100B** (1981) 205.
13) R. V. Reid, Ann. Phys. **50** (1968) 411.
14) S. Köhler, Nucl. Phys. **A415** (1984) 37.
15) N. R. Metropolis et al., Phys. Rev. **110** (1958) 204.
16) S. Fantoni and V. R. Panharipande, Nucl. Phys. **A427** (1984) 473.
17) A. Nadasen et al., Phys. Rev. **C23** (1981) 1023.
18) H. O. Meyer and P. Schwandt, Phys. Lett. **107B** (1981) 353.
19) R. E. Segel et al., Phys. Rev. **C32** (1985) 721.
20) E. F. Redish and D. M. Schneider, Phys. Lett. **100B** (1981) 101.
21) J. Aichelin and H. Stöcker, Phys. Lett. **163** (1985) 59.
22) C. B. Dover and A. Gal, Ann. Phys. **146** (1983) 309.
23) M. M. Nagels et al., Phys. Rev. **D12** (1975) 744; ibid. **15** (1977) 2547; ibid. **20** (1979) 1663.
24) W. M. Macek et al., preprint Nijmegen, 1982.
25) H. Bandō, Essence of Hypernuclear Physics, INS-PT-31, 1982; Y. Yamamoto, Prog. Theor. Phys. **75** (1986) 639.
26) C. B. Dover and A. Gal, Prog. Part. Nucl. Phys. **12** (1984) 171.

SATURATION IN NUCLEAR MATTER: A NEW PERSPECTIVE*

A. Ramos and A. Polls

Departament de Fisica Teorica, Universitat de Barcelona
E-08028 Barcelona, Spain

W. H. Dickhoff
Department of Physics, Washington University
St. Louis, Missouri 63130, U.S.A

Abstract: The nucleon self-energy in nuclear matter is calculated including particle-particle (pp) as well as hole-hole (hh) ladder contributions. It is shown that a proper treatment of the analytic structure of the ladder summed effective interaction Γ^B, requires the numerical use of dispersion relations in order to calculate the self-energy correctly. These results in principle allow a fully self-consistent treatment of the ladder equation and the Dyson equation for the single-particle propagator including the full energy dependence of the self-energy. As a first step towards such a complete solution the single-particle energy is calculated self-consistently from the real on-shell self-energy which contains both forward(pp)- and backward(hh)- going terms. The contribution of the hh contribution to the sp energy is repulsive for all momenta and larger than the increased attraction from the pp contribution leading to less binding energy. This effect increases strongly with density and therefore leads to a saturation mechanism which has not been identified previously. First results for the v_2 homework potential are discussed.

I. INTRODUCTION

The present status of nuclear many-body theory in a non-relativistic description is not completely satisfactory. The saturation properties of nuclear matter can not be described satisfactorily up to now when only two-body forces are considered.[1,2] This is true in the conventional hole-line expansion[3,4] as well as in the variational description of nuclear matter.[5,6] It is possible to obtain roughly the correct saturation properties together with a satisfactory description of the A=3,4 nuclei when a phenomenological three-body force is introduced which acts attractive at very low densities and repulsive at higher densities.[7] Obviously there is a need for a saturation mechanism which will provide attraction at low density and repulsion at high density whether this must be completely ascribed to many-body forces or not.

In recent years there have been attempts to implement certain features of relativistic dynamics into the description of nuclear matter.[8–11] The adaption of

*This research was supported in part by NATO under Grant No. RG.85/0684 and by the Condensed Matter Theory Program of the Division of Materials Research of the U.S. National Science Foundation under Grant No. DMR–8519077 (at Washington University).

conventional wisdom to the relativistic case has led to the relativistic Brueckner-Hartree-Fock prescription. It turns out that in this approach one has no problem to obtain the correct saturation properties of nuclear matter, the coupling to the antiparticles leads naturally to increasing repulsion with increasing density. Disturbing is the fact that one obtains these results using only two-body interactions, thereby contradicting the need for many-body forces. The missing attraction at lower densities is however not obtained in this approach.

Historically, the approach based on perturbation theory to the nuclear matter problem, has always been formulated from the point of view of Goldstone diagrams. Massive resummations have been performed leading to the so-called hole-line expansion.[12] The main concern in this nuclear matter theory is the calculation of the energy per particle as a function of the density. This is on the other hand not the only interesting quantity one would like to be informed about. Other important properties of the system are reflected in the dynamical response of the system, sp strength functions, the effective mass, optical potential, Landau parameters, momentum distribution and so on. An exact theory will give *all* these properties. In view of the present status of experimental nuclear physics, it might be fruitful to consider approximate solutions to the many-body problem which address all these quantities at the same time.

The Green function approach has this advantage. Practical applications have been restricted to a limited set of problems and no sophisticated numerical calculations based on this approach exist although certain attempts have been made in this direction.[13,14] In section II it will be shown how one can treat the coupling of the sp propagator to the two-particle propagator consistently, when the conventional ladder summation is performed. The inclusion of hole-hole terms in the ladder summed vertex function Γ^B introduces a more complex analytic structure. In this case not only forward (pp) propagating terms with corresponding poles in the lower half of the energy plane are present but also backward (hh) propagating contributions with poles in the upper half plane. The calculation of the self-energy requires a separation of these contributions which is achieved by the numerical use of dispersion relations. This technique allows the solution of the coupled ladder self-energy problem taking the full complexity of the energy dependence of the self-energy into account. In this paper the first step towards such a solution is made by solving the problem self-consistently at the level of the sp energy. The sp energy is then calculated from the real part of the on-shell self-energy which includes both pp and hh contributions. Results of such calculations for the v_2 homework potential are presented in section III. In this section the full self-energy is also discussed as well as the resulting sp strength functions and the momentum distribution.

II. GREEN FUNCTION METHOD AND THE LADDER SUMMATION

Consider the sp propagator g and the corresponding Dyson equation for this propagator which connects it to the two-particle Green function as displayed in Fig. 1. The first equation in Fig. 1 relates the sp Green function (double line propagator) to the proper self-energy Σ^* and the unperturbed propagator $g^{(0)}$. The second equation relates the proper self-energy to the two-particle Green function. This coupling is represented here by the four-point vertex function Γ. Fig. 1 represents a complete solution for the sp propagator when the bare interaction V is of two-body nature. Any approximation to the self-energy Σ^* leads to an approximation to g. It is clear from Fig. 1 that one has to solve a coupled, non-linear problem since the solution g is used in the calculation of Σ^*.

At present this problem has only been solved completely on the Hartree-Fock level. The Hartree-Fock approximation is obtained in Fig. 1 by only retaining the contribution from V and discarding the contribution with Γ, i.e. the two-particle propagator is approximated by two non-interacting propagators. For nuclear physics problems it is relevant to keep direct and exchange contribution together, therefore the V diagram in Fig. 1 represents both direct and exchange together. This approximation to the many-body problem leaves in tact the single-particle description, it only finds the optimum independent particle solution. The reason for this is that in this approximation Σ^* is static, i.e. energy independent and related to this it is real. Any many-body calculation which goes beyond the Hartree-Fock approximation necessarily has to include higher order contribution to the self-energy. In the nuclear case one has to go to infinite order since any realistic two-body interaction contains a strong repulsive core which necessitates at least the summation of the particle-particle ladder diagrams. This is the familiar Brueckner result. This approximation to Γ is shown in Fig. 2 and denoted by Γ^B. If we now consider this approximation in the calculation of g we have an obvious extension of the Hartree-Fock concept. Not only the sp propagator is calculated self-consistently however, also the vertex function Γ^B must be calculated self-consistently since in the calculation of Γ^B one has to use the full propagators g. The word *full* refers here to the full solution of the *approximation* to g displayed in Fig. 2.

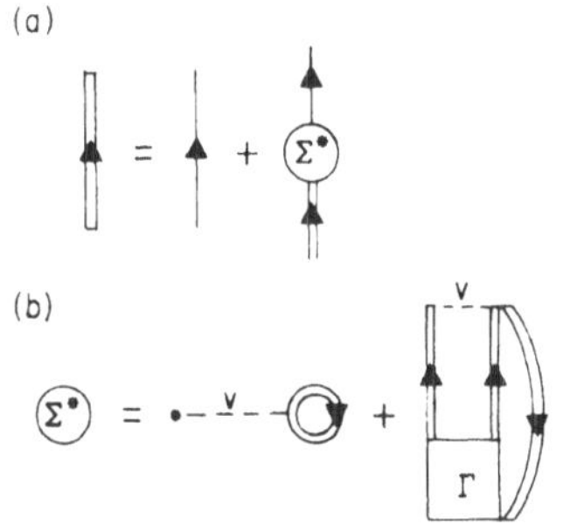

Figure 1. Diagrammatic representation of the Dyson equation (a) for the sp propagator. The structure of the proper self-energy Σ^* is given in (b) where the coupling to the vertex function Γ is shown.

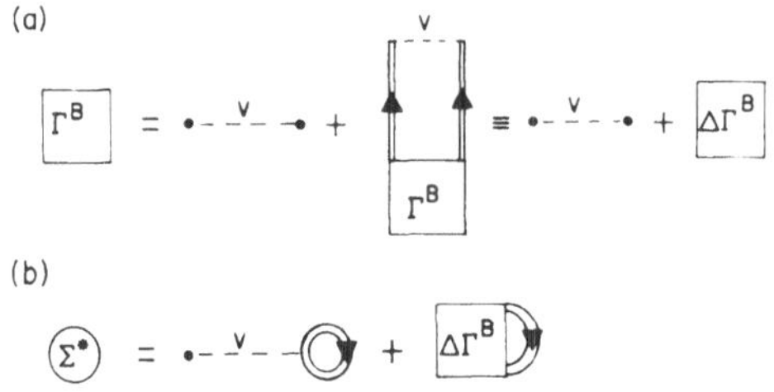

Figure 2. Ladder approximation to Γ (a) and the corresponding self-energy (b).

A few things can be observed immediately, the first is that Σ^* becomes complex and energy dependent. This represents the mixing of the single-particle (hole) states with the 2p–1h (1p–2h) and more complicated states which depends on the energy of propagation. A complex and energy dependent self-energy leads to a sp propagator which has the following structure for a normal system

$$g(\mathbf{k},\omega) = \int_{\varepsilon_F}^{\infty} d\omega' \frac{S_p(\mathbf{k},\omega')}{\omega - \omega' + i\eta} + \int_{-\infty}^{\varepsilon_F} d\omega' \frac{S_h(\mathbf{k},\omega')}{\omega - \omega' - i\eta} \quad . \tag{1}$$

With this propagator one then has to solve for Γ^B (see Fig. 2) until self-consistency is achieved in a similar way as in Hartree-Fock.

Supposing that one starts this procedure with the calculation of Γ^B using unperturbed propagators $g^{(0)}$, one obtains the following integral equation

$$\begin{aligned} &<\mathbf{k}\,|\,\Gamma^B_{(0)}(\mathbf{K},\Omega)\,|\,\mathbf{k}'> \\ &= <\mathbf{k}\,|\,V\,|\,\mathbf{k}'> + \frac{1}{2}\int \frac{d\mathbf{q}}{(2\pi)^3} <\mathbf{k}\,|\,V\,|\,\mathbf{q}> g^{II}_{(0)}(\mathbf{q};\mathbf{K},\Omega) <\mathbf{q}\,|\,\Gamma^B_{(0)}(\mathbf{K},\Omega)\,|\,\mathbf{k}'> \\ &\equiv <\mathbf{k}\,|\,V\,|\,\mathbf{k}'> + <\mathbf{k}\,|\,\Delta\Gamma^B_{(0)}(\mathbf{K},\Omega)\,|\,\mathbf{k}'> \quad . \end{aligned} \tag{2}$$

The usual decomposition in cm momentum ($\mathbf{K}$) and relative momenta ($\mathbf{k},\mathbf{k}'$ and $\mathbf{q}$) has been made, the spin and isospin indices have been suppressed and $g^{II}_{(0)}$ is given by

$$\begin{aligned} &g^{II}_{(0)}(\mathbf{q};\mathbf{K},\Omega) \\ &= \frac{\theta(|\tfrac{1}{2}\mathbf{K} + \mathbf{q}| - k_F)\theta(|\tfrac{1}{2}\mathbf{K} - \mathbf{q}| - k_F)}{\Omega - \varepsilon(\tfrac{1}{2}\mathbf{K} + \mathbf{q}) - \varepsilon(\tfrac{1}{2}\mathbf{K} - \mathbf{q}) + i\eta} - \frac{\theta(k_F - |\tfrac{1}{2}\mathbf{K} + \mathbf{q}|)\theta(k_F - |\tfrac{1}{2}\mathbf{K} - \mathbf{q}|)}{\Omega - \varepsilon(\tfrac{1}{2}\mathbf{K} + \mathbf{q}) - \varepsilon(\tfrac{1}{2}\mathbf{K} - \mathbf{q}) - i\eta} \end{aligned} \tag{3}$$

It is possible to solve the integral equation for $\Gamma^B_{(0)}$ numerically for nuclear matter. This is done in a partial-wave momentum representation which can be used after the conventional angle-averaging for the unperturbed pp propagator $g^{II}_{(0)}(\mathbf{q};\mathbf{K},\Omega)$. Further details of the numerical calculations will be published elsewhere.[15] The difficulty appears when one calculates the self-energy contribution from $\Delta\Gamma^B_{(0)}$ with these numerical results:

$$\Sigma^{\Delta^*}(\mathbf{k},\omega) = \int \frac{d\omega'}{2\pi i} \int \frac{d\mathbf{k}'}{(2\pi)^3} \, \Delta\Gamma^B_{(0)}(\tfrac{1}{2}(\mathbf{k} - \mathbf{k}');\mathbf{k} + \mathbf{k}',\omega + \omega')g^{(0)}(\mathbf{k}',\omega') \quad . \tag{4}$$

Since only diagonal matrix elements of $\Delta\Gamma^B_{(0)}$ are required, the relative momentum, $\mathbf{k}_{rel} = (\tfrac{1}{2}(\mathbf{k} - \mathbf{k}'))$ is only given once in this equation. The energy integration in this contribution which corresponds to the second term in Fig. 2b requires knowledge of the analytic structure of $\Delta\Gamma^B_{(0)}$, otherwise this integral cannot be performed correctly. The crucial observation is, that it is possible to obtain this information on $\Gamma^B_{(0)}$ by using dispersion relations. Using the Lehmann representation for the pp propagator one can show that $\Delta\Gamma^B$ can be separated into two parts each having poles in one half of the complex energy plane. These two parts are obtained from a dispersion integral over the imaginary part of $\Gamma^B_{(0)}$, the pp part with poles in the lower half plane is obtained by integrating from $2\varepsilon_F$ to ∞ whereas the hh part corresponds to an integration from $-\infty$ to $2\varepsilon_F$.

$$\begin{aligned} &\Delta\Gamma^B_{(0)}(\mathbf{k}_{rel};\mathbf{K},\Omega) \\ &= -\frac{1}{\pi}\int_{2\varepsilon_F}^{\infty} d\Omega' \frac{\mathrm{Im}\Gamma^B_{(0)}(\mathbf{k}_{rel};\mathbf{K},\Omega')}{\Omega - \Omega' + i\eta} + \frac{1}{\pi}\int_{-\infty}^{2\varepsilon_F} d\Omega' \frac{\mathrm{Im}\Gamma^B_{(0)}(\mathbf{k}_{rel};\mathbf{K},\Omega')}{\Omega - \Omega' - i\eta} \\ &\equiv \Delta^{\downarrow}\Gamma^B_{(0)}(\mathbf{k}_{rel};\mathbf{K},\Omega) + \Delta^{\uparrow}\Gamma^B_{(0)}(\mathbf{k}_{rel};\mathbf{K},\Omega) \quad . \end{aligned} \tag{5}$$

Note that the separation in the imaginary part is determined by $2\varepsilon_F$, the separation of the real part is obtained by performing the relevant principle value integrals in Eq. (5). With this separation of $\Delta\Gamma^B_{(0)}$ it becomes possible to calculate correctly the self-energy contribution from $\Delta\Gamma^B_{(0)}$

$$\Sigma^{\Delta^*}(\mathbf{k},\omega) = \int \frac{d\mathbf{k}'}{(2\pi)^3}\, \Delta^{\downarrow}\Gamma^B_{(0)}(\tfrac{1}{2}(\mathbf{k}-\mathbf{k}');\mathbf{k}+\mathbf{k}',\omega+\varepsilon(\mathbf{k}'))\theta(k_F - |\mathbf{k}'|)$$

$$- \int \frac{d\mathbf{k}'}{(2\pi)^3}\, \Delta^{\uparrow}\Gamma^B_{(0)}(\tfrac{1}{2}(\mathbf{k}-\mathbf{k}');\mathbf{k}+\mathbf{k}',\omega+\varepsilon(\mathbf{k}'))\theta(|\mathbf{k}'| - k_F) \quad . \tag{6}$$

Another way of visualizing this procedure is to note that the original integral equation will generate two types of diagrams, those which have to be linked to a particle line or those which have to be linked to a hole line in order to obtain a valid self-energy contribution. This is illustrated for the second order contribution in Fig. 3. In this figure the upward going lines refer to particle states and the downward going lines to holes, diagram a) can only be closed with a hole line and diagram b) only with a particle line. This illustrates in lowest order the necessity of separating these two contributions which is mathematically achieved by calculating the corresponding dispersion integrals.

Figure 3. Schematic illustration of the analytic structure of $\Delta\Gamma^B$. In (a) only a hole line can be attached to this contribution and similarly in (b) only a particle line to obtain a valid self-energy contribution.

In order to achieve this separation one must know $\Gamma^B_{(0)}$ for all energies. Present investigations show that it is indeed possible to obtain these separate pieces numerically which added together give the original solution of the integral equation for $\Gamma^B_{(0)}$. Obviously this presents a substantial increase in the numerical effort and one has to consider the relevance of this investment. Although the actual solution of the full problem represented by Fig. 2, is the ultimate goal of the present work, one gets already a flavor of its usefulness by a simple extension of the continuous prescription advocated by Mahaux *et al.*[16] To solve the problem approximately in a first step, one can start with an extension of the on-shell prescription to include the hole-hole terms for the single-particle energy and calculate this self-consistently

$$\varepsilon(\mathbf{k}) = \frac{\mathbf{k}^2}{2m} + \mathrm{Re}\ \Sigma^*(\mathbf{k},\varepsilon(\mathbf{k})) \quad . \tag{7}$$

Having determined the sp spectrum self-consistently one can then calculate other relevant quantities like the effective mass, sp strength functions and the momentum distribution from the full self-energy which increases the scope of the calculation beyond calculating the binding energy. One should observe at this point that Eq. (7) is only an intermediate step towards solving the problem using full propagators (see Eq. (1)) in the ladder equation and in that sense Eq. (7) is the closest prescription in terms of a sp energy which approaches this full problem. It also means that with this interpretation of the coupling of the sp and the two-particle propagator there is no ambiguity in the choice of the sp spectrum, it will be continuous across k_F and eventually include the full dynamic content of the self-energy.

Note that one can write Σ^* in the following way (in obvious notation)

$$\Sigma^*(\mathbf{k},\omega) = \Sigma^{V^*}(\mathbf{k}) + \Sigma_{\updownarrow}^{\Delta^*}(\mathbf{k},\omega) + \Sigma_{\uparrow}^{\Delta^*}(\mathbf{k},\omega) \quad . \tag{8}$$

The on-shell contribution from $\Sigma_{\updownarrow}^{\Delta^*}$ including pp propagation as given in Fig. 3a and its generalizations to higher order are known to give a strongly attractive contribution to the sp energy. In the conventional hole-line expansion only those terms are included and they lead to the binding of the system. The contribution to Eq. (7) from hole-hole propagation can be analyzed in terms of the following dispersion integral (see e.g. Ref. 17)

$$\Sigma_{\uparrow}^{\Delta^*}(\mathbf{k},\omega) = \frac{1}{\pi} \int_{-\infty}^{\varepsilon_F} d\omega' \frac{\mathrm{Im}\, \Sigma^{\Delta^*}(\mathbf{k},\omega')}{\omega - \omega' - i\eta} \quad . \tag{9}$$

The imaginary part of the self-energy is positive in this domain of integration, for $|\mathbf{k}| = k_F$ the on-shell prescription (Eq. (7)) leads to $\omega = \varepsilon_F$ which implies that also the denominator is always positive and therefore this contribution to the sp energy is *repulsive*. Results for the self-energy and the sp energy will be discussed in section III.

It should be noted that the results discussed in section III are obtained by determining the sp energy according to Eq. (7) self-consistently, i.e. after solving Eq. (2) the separation of $\Delta\Gamma^B_{(0)}$ was performed according to Eq. (5) which was used to calculate the on-shell self-energy contributions according to Eq. (6) until the sp spectrum was self-consistent. At this point the sp propagator still has the analytic structure of the free propagator $g^{(0)}$ and the binding energy per particle is obtained from

$$B/A = \frac{2}{\rho} \int \frac{d\mathbf{k}}{(2\pi)^3} \int \frac{d\omega}{2\pi i} e^{i\omega\eta} \left(\frac{k^2}{2m} + \omega \right) g(\mathbf{k},\omega)$$

$$= \frac{4}{\rho} \int \frac{d\mathbf{k}}{(2\pi)^3} \left(\frac{k^2}{2m} + \tfrac{1}{2}\varepsilon(\mathbf{k}) \right) \theta(k_F - |\mathbf{k}|) \quad . \tag{10}$$

After establishing self-consistency the full self-energy was calculated from which one can calculate the strength functions S_p and S_h (see Eq. (1)) according to

$$S_p(\mathbf{k},\omega) = -\frac{1}{\pi} \,\mathrm{Im}\, g^f(\mathbf{k},\omega) \tag{11a}$$

for $\omega > \varepsilon_F$ and

$$S_h(\mathbf{k},\omega) = \frac{1}{\pi} \,\mathrm{Im}\, g^f(\mathbf{k},\omega) \tag{11b}$$

for $\omega < \varepsilon_F$. The propagator g^f is then calculated from

$$g^f(\mathbf{k},\omega) = \left(\omega - \frac{k^2}{2m} - \Sigma^*(\mathbf{k},\omega) \right)^{-1} \quad . \tag{12}$$

Note that this g^f is the starting point of the iteration scheme to solve the coupled ladder self-energy problem exactly.

From $S_h(\mathbf{k},\omega)$ one can then calculate the momentum distribution according to

$$n(\mathbf{k}) = \int_{-\infty}^{\varepsilon_F} d\omega S_h(\mathbf{k},\omega) \quad . \tag{13}$$

It should be noted that it is now in principle possible to attack the full solution of the coupled problem shown in Fig. 2 including the full energy dependence of the self-energy. One observes that the propagator $g^{\Pi}_{(0)}$ becomes a complex matrix when propagators are used which have the structure of Eq. (1). The solution for Γ^B is then no more difficult than for $\Gamma^B_{(0)}$. A similar separation of the full $\Delta\Gamma^B$ is then again possible as in Eq. (5) after which the full self-energy can be calculated correctly analogous to Eq. (6). This extension is presently being pursued. It will allow for the first time the complete solution of the coupled ladder self-energy problem which in turn means that one e.g. determines the momentum distribution self-consistently within the present ladder approximation.

III. RESULTS and DISCUSSION

In order to compare the present method with other available calculations, we have chosen the v_2 homework potential for which many results are available.[4,18–22] The results discussed here are obtained for three densities corresponding to $k_F = 1.6$, 1.8 and 2.0 fm^{-1}. Results for the pp contribution to the self-energy, $\Sigma^{\Delta^*}_{\downarrow}(k_F,\omega)$ are given in Fig. 4. These results are obtained for $k_F = 1.8\ fm^{-1}$ after the sp spectrum was calculated self-consistently by iterating the sp spectrum according to Eq. (7) and the ladder equation (2). The imaginary part of $\Sigma^{\Delta^*}_{\downarrow}$ starts at ε_F and has been given an extra sign as in Ref. 17. The corresponding real part is shown in the same figure with the correct sign. This figure can be directly compared with Fig. 3.9 of Ref. 17 where similar results are shown for a model interaction which has been used to calculate the second order contribution to $\Sigma^{\Delta^*}_{\downarrow}$. The *only* essential difference with this result is the scale which is needed for the present calculation. We observe e.g. that the on-shell contribution roughly corresponds to 1100 MeV attraction. The imaginary part is also large and goes to zero only for very high energies, this implies that the real part which can be viewed as the principle value integral over the imaginary part (similar to Eq. (9) and see Ref. 17) will become positive for even larger values of ω than shown in the figure, before it finally goes to zero. The reason for the importance of these high energy contributions is fairly obvious since the v_2 interaction contains the strong repulsive core of the central $^3S_1$ nucleon-nucleon interaction as constructed by Reid.[23] This strong core implies that high momentum components are mixed into the relative wavefunction of two particles which in turn implies that important contributions are received from very high energies as is evident from the resulting self-energy contribution shown in Fig. 4.

In Fig. 5 we show the results for $\Sigma^{\Delta^*}_{\uparrow}(k_F,\omega)$ at the same density. In this case the imaginary part is shown with the correct sign. Again one can compare this result directly with the second order model calculation shown in Fig. 3.15 of Ref. 17. In this case also the scale is comparable which confirms the smallness of the hh contribution to the self-energy as compared to the pp one. An argument which has been used to neglect the hh contributions in the standard hole-line expansion. Before addressing this question in more detail, we note that the on-shell contribution to the sp energy is indeed positive (see Eq. (9)) and furthermore we again observe a very similar functional dependence of this part of the self-energy for both the real part and the imaginary part as in the second order calculation[17] which confirms the conclusions reached in that work also when these contributions are considered to infinite order.

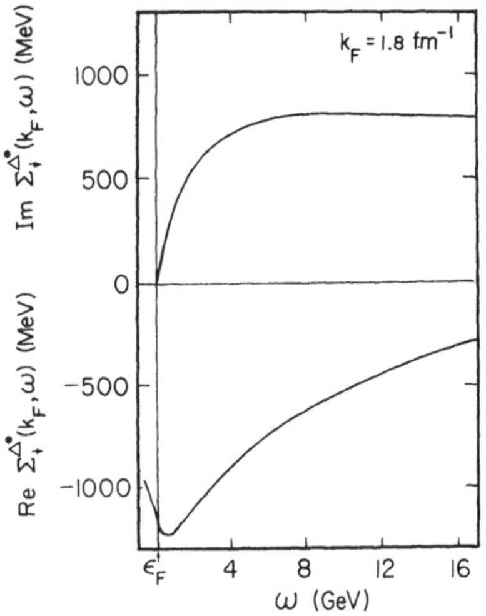

Figure 4. Real and imaginary (with opposite sign) part of the pp contribution to the self-energy, $\Sigma_{\downarrow}^{\Delta^*}(\mathbf{k},\omega)$ for $|\mathbf{k}| = k_F = 1.8\ \mathrm{fm}^{-1}$.

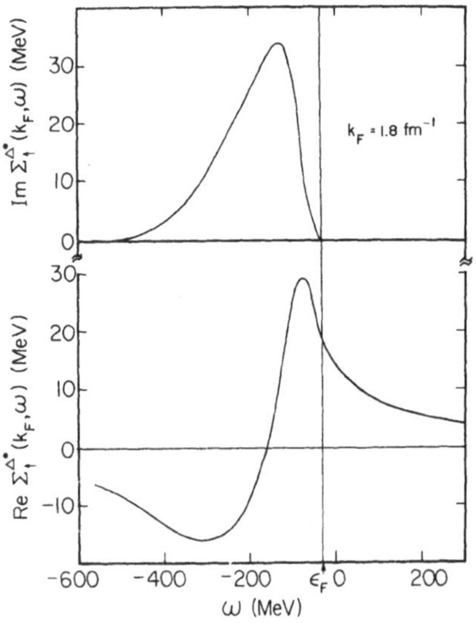

Figure 5. Same as Fig. 4 for $\Sigma_{\uparrow}^{\Delta^*}(\mathbf{k},\omega)$. In this figure the imaginary part has the correct sign.

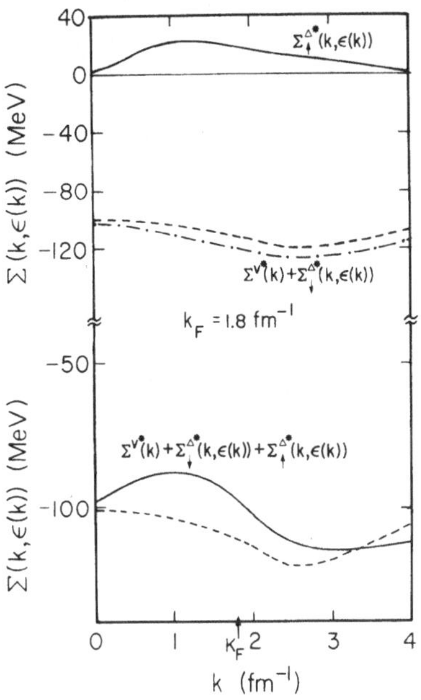

Figure 6. Different contributions to the sp energy as discussed in the text. In the upper part separate contributions to the on-shell self-energy are shown. The full curve corresponds to $\Sigma_{\uparrow}^{\Delta^*}$, the dash-dot curve to the sum $\Sigma^{V^*}+\Sigma_{\updownarrow}^{\Delta^*}$. The dashed curve represents this last sum for the case without hh propagation. The lower part of the figure contains the same dashed curve on a larger scale and the combined result $\Sigma^{V^*} + \Sigma_{\updownarrow}^{\Delta^*} + \Sigma_{\uparrow}^{\Delta^*}$ (solid curve).

In Fig. 6 the different on-shell self-energy contributions to the sp energy are compared for $k_F = 1.8\ fm^{-1}$. The dashed line is the result of a calculation in which the hh terms were not included and is therefore equivalent to the continuous choice prescription for the sp energy as advocated by Mahaux et al..[16] This contribution includes also the repulsive Hartree-Fock contribution Σ^{V^*} which is comparable in magnitude to $\Sigma_{\updownarrow}^{\Delta^*}$ but of opposite sign. The inclusion of hh terms makes the contribution from $\Sigma^{V^*}+\Sigma_{\updownarrow}^{\Delta^*}$ somewhat more attractive due to the additional phase space which becomes available, it is shown by the dash-dot curve in Fig. 6. The contribution of $\Sigma_{\uparrow}^{\Delta^*}$ to the sp energy turns out to be repulsive for all momenta, not only for k_F. It is given by the full curve in Fig. 6. The magnitude of this contribution is now certainly not negligible compared to the *combined* Σ_V^* and $\Sigma_{\updownarrow}^{\Delta^*}$ terms. The total results are compared in the lower half of the figure where the dashed line again refers to the calculation without hh propagation and the full line includes the hh effect. This result implies that the resulting binding energy calculated from Eq. (10) is substantially less with inclusion of holes at this density.

This result is shown in Fig. 7 where the binding energy for the v_2 potential is shown for various calculations as a function of the Fermi momentum. The results labeled BB(2) and BB(3) are taken from Ref. 4 and represent the two and three hole-line estimate for the binding energy calculated with the conventional gap in the sp spectrum. We have recalculated the BB(2) results and show them for densities up to $k_F = 2.4\ fm^{-1}$. The curve labeled "Mahaux" represents the result when only pp propagation is considered but the sp spectrum is chosen continuous across k_F as advocated by the Liège group. We observe that the saturation point for this potential with this choice lies at much higher density than the full three hole-line calculation which gives similar results as the FHNC calculation of Ref. 18 and the Monte Carlo calculation of Ref. 19. The results of the inclusion of the hh propagation are labeled "present" and demonstrate that the correct inclusion of hh effects leads to a substantial repulsion with increasing density. The calculated binding energies are -3.88 MeV for $k_F = 1.6\ fm^{-1}$, -5.54 MeV for $k_F = 1.8\ fm^{-1}$ and -5.40 MeV for $k_F = 2.0\ fm^{-1}$. Physically one can argue that the admixing of pp and hh amplitudes in the ladder equation which is enhanced by the continuous sp spectrum, leads to a depletion of normally occupied states and therefore results in a sp spectrum which reflects this increased kinetic energy (and possibly less potential energy). The density dependence of the effect can simply be understood from Fig. 3b where the integration over the hole states will become increasingly important as the density increases.

This saturating effect is observed when compared to the continuous choice results including only pp propagation. Since these results are very different compared to the standard hole-line results (BB(2)) it is premature to draw definite conclusions for realistic NN interactions. Nevertheless we observe that for a realistic interaction the difference between BB(2) results and the continuous choice are much less dramatic and lead to some additional attraction for the continuous choice.[24] The effect we have discussed here will not change very much when the interaction changes and we therefore expect a real saturating effect for a realistic NN interaction when the hh effects are properly treated. This will also help to correct the result of Ref. 14 where the real part of the self-energy was not separated correctly and thus an unpleasant saturation curve including substantial overbinding was obtained for normal ^{3}He. A comparison with the calculations of Ref. 22 is not made since these results are very similar to the continuous choice results with only pp propagation. The comparison with the BB(3) results shows that our results show somewhat more attraction as a function of density. It will be very interesting to obtain the binding energy from the complete solution of the coupled ladder self-energy problem which determines the momentum distribution self-consistently, and compare it with the other results in Fig. 7. As a final remark on the saturation mechanism we observe that the coupling to hh terms can be compared to the coupling to antiparticles in the relativistic Brueckner-Hartree-Fock calculations[8–11] and therefore might provide a non-relativistic alternative explanation to saturation in nuclear matter.

The sp strength function is one of the important results one can derive from knowledge of the full self-energy (see Eq. (11)). In Fig. 8 the full sp strength function for $|\mathbf{k}| = 1.51\ fm^{-1}$ is shown. The function gives $S_h(\mathbf{k},\omega)$ for $\omega < \varepsilon_F$ and similarly for $\omega > \varepsilon_F$ it shows $S_p(\mathbf{k},\omega)$ multiplied by a factor 10. Since this momentum is already

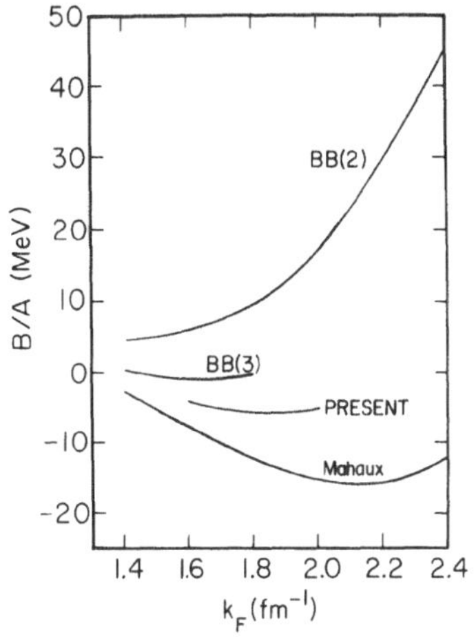

Figure 7. Binding energy of nuclear matter for the v_2 potential as a function of the Fermi momentum. The curves labeled BB(2) and BB(3) refer to the two and three hole-line contribution,[4] the curve labeled "Mahaux" also allows only propagation of particles in the ladder approximation but with a continuous sp spectrum, the "present" result refers to the calculations which are discussed in the text and take into account the hh scattering effects self-consistently in the sp energy.

quite close to k_F one observes a narrow peak in the hole strength function. For further applications towards a full solution of the coupled ladder self-energy problem it is essential that the sum rule for the strength function

$$\int_{-\infty}^{\varepsilon_F} d\omega S_h(\mathbf{k},\omega) + \int_{\varepsilon_F}^{\infty} d\omega S_p(\mathbf{k},\omega) = 1 \tag{14}$$

is fulfilled for each $|\mathbf{k}|$. This integral is difficult to perform directly for values of $|\mathbf{k}|$ in the neighborhood of k_F for obvious reasons. We have overcome this difficulty by performing the following separation of the sp propagator

$$g(\mathbf{k},\omega) = g^{QP}(\mathbf{k},\omega) + g^{BG}(\mathbf{k},\omega) \quad . \tag{15}$$

This quasiparticle-background separation is performed in the standard fashion

$$g^{QP}(\mathbf{k},\omega) = \frac{z(\mathbf{k})}{\omega - E(\mathbf{k}) - iz(\mathbf{k})W(\mathbf{k})} \tag{16}$$

where the quasiparticle strength is given by

$$z(\mathbf{k}) = \left(1 - \frac{\partial \mathrm{Re}\, \Sigma^*(\mathbf{k},\omega)}{\partial\omega}\right)^{-1} \quad , \tag{17}$$

the quasiparticle energy by

$$E(\mathbf{k}) = \frac{\mathbf{k}^2}{2m} + \mathrm{Re}\,\Sigma^*(\mathbf{k},E(\mathbf{k})) \tag{18}$$

and the width finally by

$$W(\mathbf{k}) = \mathrm{Im}\,\Sigma^*(\mathbf{k},E(\mathbf{k})) \quad . \tag{19}$$

This allows an analytical integration of the quasiparticle contribution to Eq. (14), the remaining background contribution which is obtained by subtracting the quasiparticle result from the full propagator can then be integrated numerically without further problem. With this procedure we obtain an accuracy for Eq. (14) on the level of 0.5% for all momenta $|\mathbf{k}|$.

The quasiparticle strength is related to the contribution to the effective mass due to the energy dependence of the self-energy. Using the decomposition of the self-energy (Eq. (8)) we have

$$\frac{m^*_\omega(\mathbf{k})}{m} = z^{-1}(\mathbf{k}) = \frac{{}^{\downarrow}m^*_\omega}{m} + \frac{{}^{\uparrow}m^*_\omega}{m} - 1 \quad . \tag{20}$$

The different parts of this effective mass contribution are shown in Fig. 9. One observes again a striking similarity with Fig. 3.24 of Ref. 17. This confirms first of all the numerical procedures used in the present calculations but on the other hand shows that the conclusions drawn in Ref. 17 have a wide range of validity.

As a final result we show in Fig.10 the momentum distribution which is obtained from the present calculation at a density corresponding to $k_F = 1.8\ \mathrm{fm}^{-1}$. We compare our results with results obtained from the BB(2) approximation[21] to $n(\mathbf{k})$ and the FHNC result.[20] Rather similar results are obtained except for the infinite slope of $n(\mathbf{k})$ at $|\mathbf{k}| = k_F$. We stress again at this point that in the present calculation this momentum distribution is not calculated self-consistently but only after the sp spectrum is determined self-consistently according to Eq. (7). In a complete solution of the coupled ladder and self-energy problem this would be the case, however and it should be interesting to see whether the momentum distribution changes significantly.

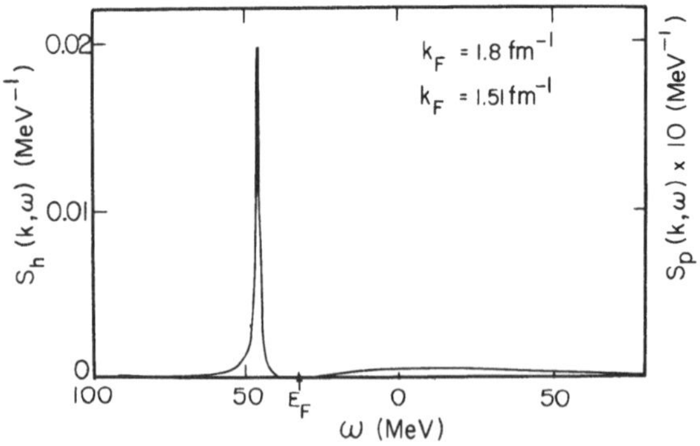

Figure 8. The single particle strength function for $|\mathbf{k}| = 1.51\ \mathrm{fm}^{-1}$ for a density corresponding to $k_F = 1.8\ \mathrm{fm}^{-1}$. Note the different scales for S_h (to the left) and S_p (to the right).

The present work approaches the solution to the many-body problem from the point of view of Green functions. As such it represents the first step which also would be taken if a parquet type summation of diagrams would be performed.[25] It should be clear from this work that many interesting as yet undiscovered results and insights can be obtained if this Green function approach is implemented with a proper treatment of the energy dependence in the problem which describes the dynamic properties of the system. As an example of this we have shown that a saturation mechanism exists due to the repulsive contribution of hh scatterings which increases with density. It therefore becomes urgent to perform these calculations also for realistic interactions. Applications of this work to neutron matter and the Helium liquids are also being considered.

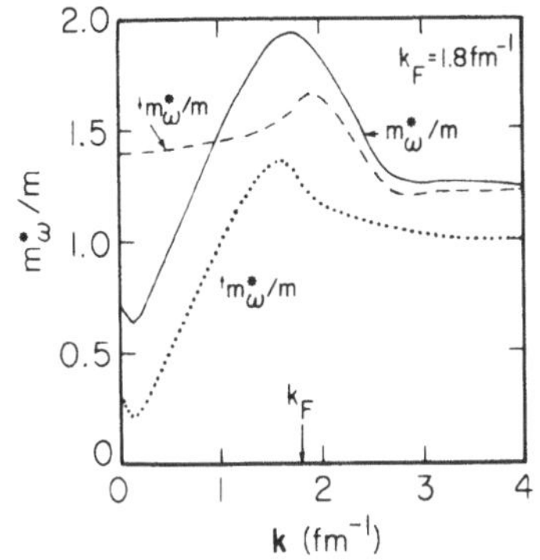

Figure 9. Different parts of the "energy" effective mass as discussed in the text. The solid curve corresponds to Eq. (20). The separate contributions ${}^{\downarrow}m^*_\omega/m$ corresponding to $\Sigma^{\Lambda *}_{\downarrow}$ and ${}^{\uparrow}m^*_\omega/m$ corresponding to $\Sigma^{\Lambda *}_{\uparrow}$ are shown by the dashed and the dotted curve respectively.

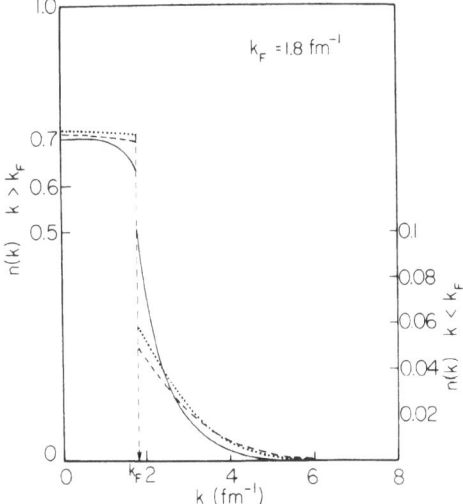

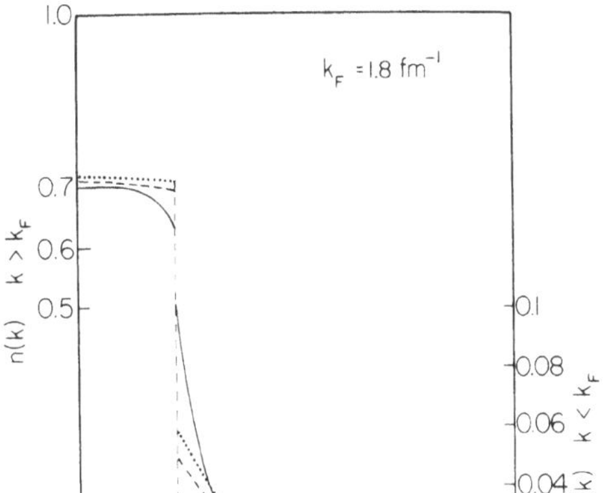

Figure 10. The momentum distribution for $k_F = 1.8$ fm^{-1} for the present work (solid curve) compared with results from Refs. 20 (dashed curve) and 21 (dotted curve). In the present work one obtains the infinite slope at k_F.

REFERENCES

1. B. D. Day, Comm. Nucl. and Part. Phys. **11**, 115 (1983).
2. A. D. Jackson, Ann. Rev. Nucl. Sci. **33**, 105 (1983).
3. B. D. Day, Phys. Rev. Lett. **47**, 226 (1981).
4. B. D. Day, Phys. Rev. C **24**, 1203 (1981).
5. V. R. Pandharipande and R. B. Wiringa, Rev. Mod. Phys. **51**, 821 (1979).
6. B. D. Day and R. B. Wiringa, Phys. Rev. C **32**, 1057 (1985).
7. J. Carlson, V. R. Pandharipande, and R. B. Wiringa, Nucl. Phys. **A401**, 59 (1983).
8. M. R. Anastasio, L. S. Celenza, W. S. Pong, and C. M. Shakin, Phys. Reports **100C**, 327 (1983).
9. C. J. Horowitz and B. D. Serot, Phys. Lett. **137B**, 287 (1984).
10. R. Brockmann and R. Machleidt, Phys. Lett. **149B**, 283 (1984).
11. B. ter Haar and R. Malfliet, Phys. Reports **149C**, 207 (1987).
12. H. A. Bethe, Ann. Rev. Nucl. Sci. **21**, 93 (1971).
13. F. Weber and M. K. Weigel, Phys. Rev. C**32**, 2141 (1985).
14. H. R. Glyde and S. I. Hernadi, Phys. Rev. B**28**, 141 (1983).
15. A. Ramos, A. Polls, and W. H. Dickhoff, in preparation.
16. J. P. Jeukenne, A. Lejeune, and C. Mahaux, Phys. Reports **25C**, 83 (1976).
17. C. Mahaux, P. F. Bortignon, R. A. Broglia, and C. H. Dasso, Phys. Reports **120C**, 1 (1985).
18. J. G. Zabolitzky, Phys. Rev. A**16**, 1258 (1977).
19. D. Ceperley, G. V. Chester, and M. H. Kalos, Phys. Rev. B**16**, 3081 (1977).
20. M. F. Flynn, J. W. Clark, R. M. Panoff, O. Bohigas, and S. Stringari. Nucl. Phys. **A427**, 253 (1984).
21. B. D. Day, private communication to J. W. Clark.
22. T. T. S. Kuo, Z. Y. Ma, and R. Vinh Mau, Phys. Rev. C**33**, 717 (1986).
23. R. V. Reid, Ann. Phys. **50**, 411 (1968).
24. P. Grangé and A. Lejeune, Nucl. Phys. **A327**, 335 (1979).
25. A. Landé and R. A. Smith, Phys. Lett. **131B**, 253 (1983).

TRANSITION TO CHAOS IN ASYMMETRIC NEURAL NETWORKS

K.E. Kürten

Institut für Theoretische Physik

Zülpicher Str. 77, 5000 Köln

West Germany

Abstract

The dynamical evolution of a version of the Mc Culloch and Pitts model[1], describing sparsely connected networks composed of N binary threshold neurons with randomly chosen asymmetrical couplings, is solved exactly in the thermodynamic limit. Analytical results based on the time evolution of the Hamming distance of two different initial configurations predict the possibility of the existence of two phases upon varying any of the network parameters: a frozen phase and a chaotic phase.

Introduction

It has long been known that there is an intimate connection between the dynamics of neural networks and the dynamical properties of spin glasses, posessing a highly structured phase space, which enables them to act as content addressable memories, thus exhibiting features of learning, memory and pattern recognition[2]. One crucial assumption for these theories is the symmetry of the coupling matrix. In biological networks the synaptic efficiencies have a high degree of asymmetry and on the average, the interconnectivity of the networks is rather sparse. Thus, the influence of the effects of asymmetry and dilution is of considerable interest. Moreover, it has been shown that in many cases models with asymmetrical and diluted interactions are even easier to study in the thermodynamic limit [3,4,5]. These powerful results can easily be applied to similar models with symmetric or non-symmetric distributions of bonds or multispin interactions.

I The model

The network consists of N interacting formal neurons, which can take two possible states ($\sigma_i = +1$ or $\sigma_i = -1$ for $1 \le i \le N$). Each neuron is supposed to have the same number K of inputs ($1 \le K \le N$), chosen differently at random among all neurons of the network. The time evolution of the system can be given in terms of the firing function of the i-th neuron

$$(1) \qquad F_i(t+\tau) = \sum_j c_{ij}\, \sigma_j(t) + h \quad , \; i = 1,\ldots,N \; ,$$

where the first term collects the total synaptic input from all cells of the net; the second term h represents a threshold, which is assumed to be the same for all neurons. The quantity τ defining the delay time for signal transmission from one neuron to another is usually supposed to be of the order of one msec. The strengths of the interactions among the neurons are specified by the quantities c_{ij} chosen randomly according to a distribution $\varrho(c_{ij})$. A positive (negative) coupling coefficient c_{ij} means that the postsynaptic effect of neuron j on neuron i is excitatory (inhibitory). At each time step, the state of each neuron is updated according to

$$(2) \qquad \sigma_i(t+\tau) = \begin{cases} +1 & \text{if } F_i(t+\tau) \ge 0 \\ -1 & \text{if } F_i(t+\tau) < 0 \end{cases} \quad , \; i=1,..,N$$

Thus, the dynamics of the network is synchronous and fully deterministic.

II Cycling activity

Starting from any initial condition, since phase space consists only of 2^N different states, the same state will eventually be repeated, establishing a limit cycle. Cyclic modes are of biological interest in terms of their possible correspondence to the active phase of the memory process. They have often been proposed as carriers of short-term memory traces, or "trains of thoughts".[6] Thus, the study of the length of the periods of limit cycles as a function of the network parameters is of fundamental interest. Since in the thermodynamic limit the average period of the system is always infinite, the question of the dependence of the period with the system size arises. Rather crude numerical studies already indicate that there exist two phases, a chaotic phase and a frozen phase[7]; in the latter the average period increases much slower with N than in the former one. A quantity which sheds more light into this phenomenon is the time evolution of the normalized Hamming distance H(t)[8], representing the fraction of spins being different in two initial configurations $\underline{\sigma}^{(1)}(t)$ and $\underline{\sigma}^{(2)}(t)$, defined by

$$(3)\quad H(t) = 1/(4N) \sum_i^N \{\sigma_i^{(1)}(t)-\sigma_i^{(2)}(t)\}^2 \quad .$$

In the frozen phase an infinitesimally small initial distance either remains confined or eventually disappears, such that H(t) approaches the value zero for sufficiently large t and N. On the other hand, in the chaotic phase even an infinitesimally small initial distance will evolve into a finite distance after infinite time such that H(t) approaches a finite value. For a symmetric distribution of the coupling coefficients $\varrho(c_{ij}) = \varrho(-c_{ij})$ an analytic expression for the time evolution of the Hamming distance $H_K(t)$ (3) can be derived in the thermodynamic limit for finite K. For K << N the quantity $H_K(t)$ can then be written as[9]

$$(4)\quad H_K(t+1) = \sum_\nu^K \binom{K}{\nu} H_K^\nu(t)[1-H_K^{K-\nu}(t)] * J_{K\nu}(h) \qquad \text{with}$$

$$(5)\quad J_{K\nu}(h)=\int.\int dx_1 .dx_K \varrho(x_1).\varrho(x_K)\Theta\{(x_{\nu+1}+.+x_K+h)^2-(x_1+.+x_\nu)^2\},$$

where $\Theta(x)$ represents the Heaviside function. The existence of a phase transition then depends uniquely on the nature of the fixed point $H_K\equiv 0$. If it is attractive, two configurations differing by an infinitesimal fraction of spins will become almost identical, whereas if it is repulsive, they will produce diverging trajectories. Thus, the relevant phase can be determined from the sign of the following quantity

$$(6)\quad S_K(h) = \left.\frac{dH_K(t+1)}{dH_K(t)}\right|_{H_K\equiv 0} - 1 = K\{1 - I_K(h)\} - 1 \quad \text{with}$$

$$(7)\quad I_K(h) = \int..\int dx_1..dx_K\varrho(x_1)..\varrho(x_K)\Theta[(x_2+..+x_K+h)^2- x_1^2] \;,$$

and the critical threshold h_c is that threshold h, where the fixed point $H_K\equiv 0$ changes its stability from attraction to repulsion.

III *Applications*

In the following section we will show that whenever one of the network parameters is varied, two possible phases can be observed.

One can easily show that the quantity $I_K(h)$ is symmetric with respect to h, increasing with increasing positive threshold parameter h, and satisfies the inequality $0 \leq I_K(h) \leq 1$. Since $I_1(0) = 0$ and $I_2(0) = 0.5$, $S_1(h) \leq 0$ and $S_2(h) \leq 0$ hold for any threshold h. Consequently no phase transition for K = 1 and K = 2 exists. Note that this result is independent of the choice of the distribution $\varrho(c_{ij})$, as long as it is symmetric with respect to the coupling coefficients. For a Gaussian distribution $\varrho(c_{ij}) = (\alpha\pi)^{-1/2}\exp(-\alpha c_{ij}^2)$, $I_3(0)$ takes the value $2-\sqrt{2}$, the corresponding positive quantity $S_3(0)$ signalling that the system is in its chaotic phase. Thus, the system shows quite different behavior for $K \leq 2$ and $K > 2$. The same behavior has been found in Kauffman's original random Boolean networks [10,11].

In a second step we keep K=3 constant and vary the threshold parameter h. The nonvanishing coupling coefficients c_{ij} are chosen randomly according to the following distribution

$$(8)\quad \varrho(c_{ij}) = 1/2\ [\ \delta(c_{ij}-1) + \delta(c_{ij}+1)\ ] \quad .$$

In this case the multidimensional integral $I_K(h)$ can be written as

$$(9)\quad I_K(h) = 1 - 1/(2^{K-1})\sum_{\nu=0}^{K-1}\binom{K-1}{\nu}\ \Theta\{(\ -(K-1) + 2\nu + h)^2 - 1\}$$

representing a step function in the interval [-K,+K] with discontinuities at $h_\nu = K - 2\nu$, $\nu = 0,..,K$. Thus, the quantity $S_K(h)$ can only take K discrete values, besides the two trivial values of unit magnitude at the boundaries.

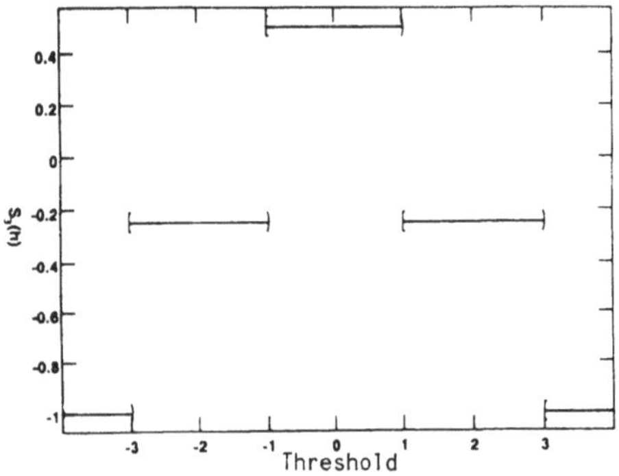

Fig. 1 Theoretical $S_3(h)$ from eq. (6) ac according to distribution (8).

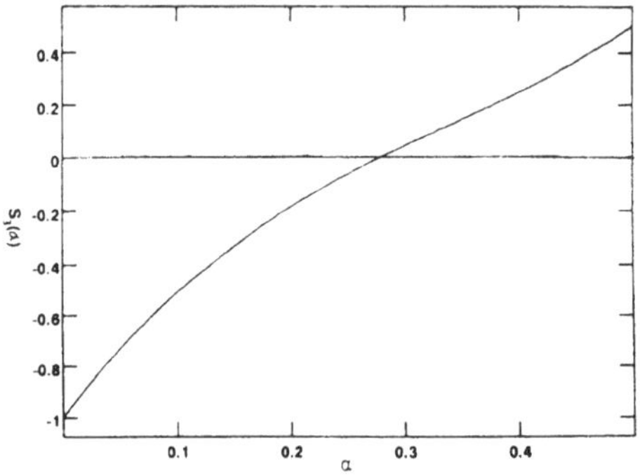

Fig. 2 Theoretical $S_3(\alpha)$ from eq. (6) for zero threshold according to distribution (10).

One can see from Fig.1, representing the quantity $S_K(h)$ for K=3 as a function of the threshold h, that only in the threshold regime $-1 \leq h < 1$ chaotic behavior prevails.

Finally we vary the following parameterized distribution of the coupling coefficients, choosing K=3 and zero threshold,

$$(10)\quad \varrho(c_{ij}) = \alpha\delta(c_{ij}-1) + \alpha\delta(c_{ij}+1) + [1-2\alpha]\delta(c_{ij})$$

such that the quantity α serves as a convenient control parameter, defining the degree of dilution of the interconnectivity. In this case the 3-dimensional integral $I_3(h)$ takes the form of a step function with discontinuities at h = -3,-1,+1,+3 . After a few algebraic manipulations $I_3(\alpha)$ takes the following form

$$(11)\quad I_3(\alpha) = -\,4\alpha^3 - 4\alpha^2(1-2\alpha) - 2\alpha(1-2\alpha)^2 + 1 \quad .$$

Fig. 2 shows the corresponding quantity $S_3(h)$ as a function of the control parameter α. The critical value is $\alpha_c \approx 0.275$. It is remarkable that for $\alpha = 1/3$, which corresponds to an average connectivity of $< K > = 2$, the system is in its chaotic phase, whereas for $K = 2$ with distribution (8), which corresponds to exactly two inputs per cell, the system is in its frozen phase. Thus, fluctuations in the connectivity of the network favor disorder. Fig. 3a and 3b show the mean cycle length as a function of N for two values of α from computer simulations. The chaotic phase is characterized by an exponential increase of the mean cycle length, whereas in the frozen phase the mean cycle length increases with a power of the total number of cells(3). It is interesting to note that similar results have been found in computer simulations within a Kauffman model, based on nearest neighbor interactions(12), where to date no theoretical predictions exist.

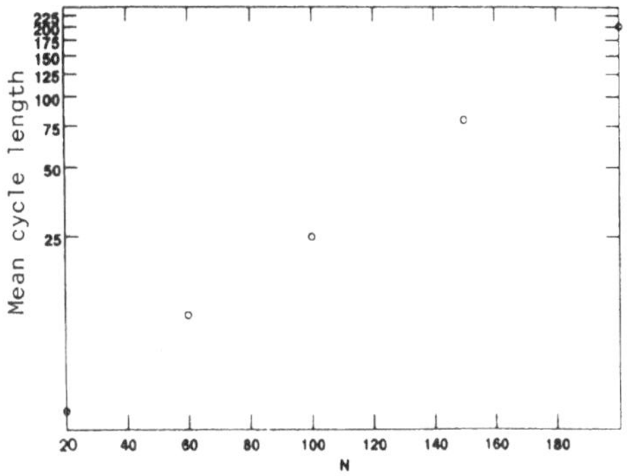

Fig. 3a Mean cycle length as a function of N for K=3, zero threshold and $\alpha = 1/3$.

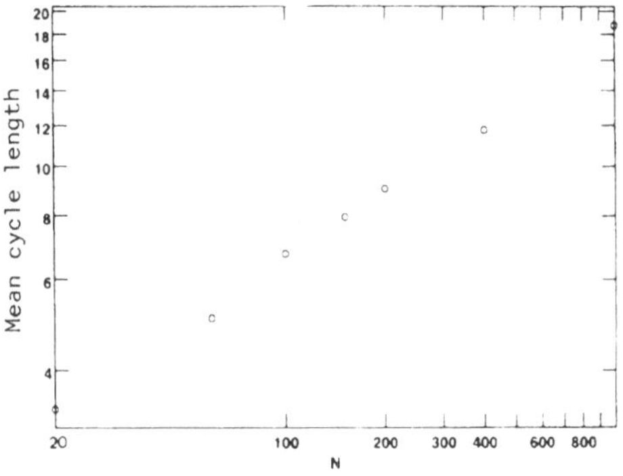

Fig. 3b Mean cycle length as a function of N for K=3, zero threshold and $\alpha = 1/4$.

Conclusion

In this work we have shown that the above deterministic model exhibits phase transitions from a frozen to a chaotic regime with respect to all net parameters: the connectivity K, the threshold h, and the distribution of the coupling coefficients $\varrho(c_{ij})$. Though one should not attach too much biological significance to these chaotic modes in terms of temporally structured memories, there might be important biological implications(13,14).

Note that the model obviously shows common characteristic features with the Kauffman model, which has recently found renewed interest as a model of biological evolution in terms of natural selection as well as in the field of disordered cellular automata.(11,12,15)

Acknowledgements

The author would like to acknowledge useful and stimulating discussions with B. Derrida during several visits. Thanks are also due to G. Senger for much fruitful advice. It is a pleasure to thank U. An der Heiden, J.W. Clark, H.Flyvbjerg, P.M. Lam, P. Peretto, M. Schreckenberg and D. Stauffer for informative discussions.

References

[1] W.S. Culloch and W.H. Pitts (1943), A logical calculus of ideas immanent in nervous activity. *Bull. Math. Biophys.*,5,115-133.

[2] J.J. Hopfield (1982), Neural networks and physical systems with emergent collective computational abilities. *Proc. U.S. Nat. Acad.Sci.*,79,2554-2558.

[3] K.E. Kürten (1987), Phase transitions in quasirandom neural networks, *Proceedings of "IEEE First Annual International Neural Network Conference" (San Diego,1987)*

[4] G. Toulouse (1987), Understanding physicists' brains, *Nature*, 327,662.

[5] B. Derrida (1987), Dynamical phase transition in non-symmetric spin glasses, *preprint.*

[6] J.W. Clark, J. Rafelski and J.W. Winston (1985), Brain without mind: Computer simulation of neural networks with modifiable neuronal interactions,*Physics reports*,123,215-273.

[7] J.W. Clark, K.E. Kürten and J. Rafelski (1987), Access and stability of cyclic modes in quasirandom networks of threshold neurons obeying a deterministic synchronous dynamics. *Cambridge University Press, Proceedings of "Computer Simulation in Brain Science" (Copenhagen,1986).*

[8] B. Derrida and G. Weisbuch (1986), Evolution of overlaps between configurations in random Boolean networks. *J.Phys*.47,1297-1303.

[9] B. Derrida (1987), *private communication.*

[10] B. Derrida and Pomeau (1986), Random networks of automata: A simple annealed approximation, *Europhys.Lett.*,1,(2),44-49

[11] S.A. Kauffmann,(1984). Emergent properties in random complex automata. *Physica*,10D,145-156.

[12] L. de Arcangelis and D.Stauffer (1987), Period distribution for Kauffman cellular automata, *preprint.*

[13] K.E. Kürten and J.W. Clark (1986), Chaos in neural systems, *Phys. Lett.*,114A,413-418.

[14] K.E. Kürten and J.W. Clark (1987), Exemplification of chaotic activity in nonlinear neural networks obeying a deterministic dynamics in continuous time. *Cambridge University Press,Proceedings of "Computer Simulation in Brain Science" (Copenhagen,1986).*

[15] G. Weisbuch and D. Stauffer (1987), Phase transition in cellular random Boolean nets. *J.Phys.*,48,11-18.

LOCAL DYNAMICS, CORRELATION, AND PHASE TRANSITIONS N-BODY VERSUS NONLINEAR QUANTUM OPTICS†

R. G. Brown and M. Ciftan

Duke University Physics Department
Durham, NC 27707 Bitnet: DJULIA@TUCC

ABSTRACT

A microscopic theory of the quantum optics of N two level atoms interacting with a resonant radiation field is applied to the problems of superradiance, photon echoes and absorptive bistability. This linear theory correctly describes these phenomena which were previously explained in terms of disparate, nonlinear theories and also predicts new ones that mean field theories are inherently incapable of treating. A new, stable pseudospin locked phase that could lead to ultrafast optical switches follows from a model hypothesis. This microscopic approach is possible because of a fundamental ansatz we make which has its origins in the quantum theory of measurement.

I. INTRODUCTION

The field of quantum optics provides us with an ideal opportunity to compare results obtained from single body, nonlinear theories to those obtained from N body, linear theories. The second quantized, nonrelativistic Hamiltonian that accurately describes the interaction of N two level atoms with each other and with a resonant field is well known. The resulting equations of motion have been solved in various approximations and the solutions accurately describe a wide variety of optical phenomena.

There are certain features of these approximate solutions that are unsatisfactory. The equations of motion for the atomic operators, when correlations are accounted for and broadening included phenomenologically, seem to describe a continuous, non-unitary evolution although we know from correlation experiments that the actual time evolution of the system has both continuous and discrete elements, but is strictly unitary. In addition to phenomenological corrections, the usual theories contain one or more Markovian, mean field, semiclassical, or statistical approximations, and introduce collective atomic

---

† -- This work was supported by the Army Research Office.

operators. These approximations reduce the N atom equations of motion to a single set of equations. When the response of the field to these collective operators is factored into these equations, the resulting equations for the collective atomic operators and the related fields are nonlinear.

These approximations are undesirable; each one washes out some element of the dynamics that may be important, but they do result in analytically tractable theories. The remarkable success these theories of quantum optics have enjoyed results from applying them to problems where their approximations are valid. For example, one could not expect them to describe an optical system with a large local atom-atom interaction.

Theories that do not treat the different atoms explicitly are intrinsically incapable of yielding certain information we are interested in even where they are otherwise successful. Deviations from collectively describable states cannot be consistently considered. Non-Markovian elements of the time evolution (such as phase transitions) can only be inferred from stability analysis rather than directly observed.

To date, no one has (to our knowledge) attempted to solve the equations of motion that result from the N atom Hamiltonian *without* making these approximations, even for a relatively small number of atoms. The reason for this is that the (Heisenberg) equations of motion mix field and atomic operators in a very tortuous manner. One must untangle this mixing for small time steps in such a way that the resulting equations of motion make sense. Attempts to do this so far have resulted in a description of broadening that neglects certain quantum correlations and is no longle unitary.

In a recent work [1] we explicitly account for the mixing of operators to first order, apply a pair of ansätze to restore the unitarity of the fully broadened time evolution, and derive a set of operator equations that consistently describe N atoms interacting with a quantized field. The result is a complete theory (with its own assumptions and approximations) that is quite different from the usual collective, mean field theories. The price we pay for treating the N atoms explicitly is the need to solve 3N+1 coupled equations of motion instead of 3; we must give up the hope of obtaining analytic solutions. The payoff we expect is the ability to explicitly treat local interactions or inhomogeneities, which is impossible in the mean field.

In all cases where there is no significant inhomogeneity or local interaction we expect the two approaches to give essentially identical results. This will serve as a test of the correctness of our N atom equations, and will additionally give us the first microscopic "look" at the organization of the states that lead to macroscopic nonlinear behavior. For the first time we will have a theoretical look at a non-equilibrium system in the middle of a first order phase transition.

We begin by briefly reviewing the correct Maxwell-Bloch equations for N two level atoms interacting with a coherent field derived in [1]. The equations include the effects of spontaneous emission (SpE), inhomogeneous and collision broadening, and allow for the presence of a cavity that traps radiation. They also include the resonant dipole-dipole interaction between atoms that is neglected in most treatments.

We solve these equations numerically for systems of 2-300 atoms to study superradiance, photon echoes and absorptive optical bistability. In all cases the N atom theory gives the correct macroscopic results and also shows us the time-dependent microscopic state in detail. It is the first theory, to date, that can consistently describe all three phenomena without introducing ad hoc mechanisms.

Finally, we discuss the dipole-dipole interaction and a model intended to account for relaxation processes. Using this model, we study an optically bistable system that takes the dipole-dipole interaction into account. A radically different organization of the bistable states results that could be very important technologically. If suitable materials are found that exhibit the appropriate relaxation behavior, optical "phase" switches in the sub-picosecond regime that do not require an etalon or cavity may be constructed out of a few thousand atoms.

For readers who are unfamiliar with quantum optics we provide the following list of texts, paper collections and conference proceedings. This list by no means exhausts the available literature, but it does contain many tutorial references, including several review papers. An excellent review of the fundamental theory of quantum optics (in the radiation reaction formulation we use here and in [1]) can be found in Allen and Eberly (AE)[2]. Other useful texts are [3]-[7]. Collected works include Mandel and Wolf [8], Barut[9], Feld and Letokhov[10], Bowden, Ciftan and Robl[11], and Gibbs et. al. [12]. Gibbs[13] also has an excellent tutorial text on optical bistability that we recommend. We will cite a number of particular papers from these collections.

II. THE THEORY OF N TWO LEVEL ATOMS

The derivation of our N atom Maxwell-Bloch equations is given in [1]. Here we briefly review the operator definitions (commutation rules) that hold at t = 0 and their action on the stationary states, give the N atom Hamiltonian, and proceed to describe the actual equations of motion we will solve.

i,j are atomic labels. We designate the ground state of the ith atom as $|-,i\rangle$, the excited state as $|+,i\rangle$, and define the action of a set of "pseudospin" operators by

$$R_3(i)|\pm,i\rangle = \pm\frac{1}{2}|\pm,i\rangle \qquad 2.1$$

and

$$[R_1(i),R_2(j)] = iR_3(i)\delta_{ij}. \qquad 2.2$$

This implies that there is no significant overlap of atomic wavefunctions; the atoms must be at least a few radii apart. Similarly, we designate the field number state $|n_k\rangle$ where $k = (\vec{k},\hat{\epsilon})$ represents both wavenumber and polarization. The field operators satisfy the commutation rules:

$$[a_k,a^{\dagger}_{k'}] = \delta_{kk'} \qquad 2.3$$

so that

$$a^{\dagger}_k a_k|n_k\rangle = n_k|n_k\rangle. \qquad 2.4$$

The vector potential operator of the free field is thus:

$$\vec{A}(\vec{r},) = \sum_k \left\{\frac{2\pi\hbar c^2}{\omega_k V}\right\}^{\frac{1}{2}} \hat{\epsilon} \left\{e^{i\vec{k}\cdot\vec{r}}a_k + e^{-i\vec{k}\cdot\vec{r}}a^{\dagger}_k\right\}. \qquad 2.5$$

Finally, the field operators and atomic operators commute:

$$[a^{(\dagger)}_k,R_m(i)] = 0. \qquad 2.6$$

The N atom Hamiltonian, including a "static" dipole-dipole interaction between atoms is given by (see, for example, Agarwal[3], eqn. 2.15):

$$H = \sum_i \hbar\omega_0 R_3(i) + \sum_k \hbar\omega_k a_k^\dagger a_k + \sum_i 2\left\{\frac{\omega_0 d}{c}\right\} A(\vec{r}_i)R_2(i)$$

$$+ \sum_{i \neq j} V_{ij}\left\{R_1(i)R_1(j) + R_2(i)R_2(j)\right\} \qquad 2.7$$

where
$$V_{ij} = \frac{d^2}{r_{ij}^3}\left\{1 - 3\cos^2\theta\right\} \qquad 2.8$$

with $r_{ij} = |\vec{r}_j - \vec{r}_i|$ and θ the angle between $(\vec{r}_j - \vec{r}_i)$ and the z axis. In this expression we have assumed that both the field and the atoms are polarized only in the z-direction for the relevant transition.

A general, normalized single atom state is given by

$$|\theta,\phi,i\rangle = \cos\frac{\theta}{2}\, e^{-i\phi/2}|-,i\rangle + \sin\frac{\theta}{2}\, e^{i\phi/2}|+,i\rangle \qquad 2.9$$

where θ, ϕ are spherical polar coordinates of the R_1, R_2, R_3 "pseudospinor". A normalized state of the kth field mode is,

$$|k\rangle = \sum_n a_n |n\rangle_k. \qquad 2.10$$

We *choose* the initial state of the combined system of atoms and field modes to be a product state:

$$|\Psi\rangle = \prod_i |\theta_i,\phi_i,i\rangle \times \prod_k |k\rangle . \qquad 2.11$$

Thus at t = 0 the atoms and the field modes are all independent.

This state is correlated only via the independent atomic and field coordinates; there are no *quantum* correlations between atoms, field modes, or the atoms and the field. However, quantum correlated atomic states such as Dicke states [14] or atomic coherent states [15],[16] are frequently used in quantum optics. A measurement performed on one part of a quantum correlated system uniquely determines the state of the other by nonlocally "collapsing" a collective wavefunction.

The success of our N atom theory suggests that quantum correlated collective atomic states are not essential to an accurate description of known phenomena. Their use greatly complicates a local description and so we avoid them. This point is discussed in greater detail in [1] and later in this paper.

From the Hamiltonian 2.7 and the Heisenberg equation:

$$\frac{d\hat{O}}{dt} = -\frac{i}{\hbar}\left[O,H\right] \qquad 2.12$$

we obtain the equations of motion for each operator. The result (which is we do not write down here) is a set of 3N+2 coupled ordinary differential equations for the 3N atomic *operators* and the 2 mutually conjugate field *operators*. The time evolution of these operators is manifestly unitary.

However, these equations are not in a form that is suitable for conversion to c-# differential equations (using the initial state $|\Psi\rangle$). If we do the first integral of the field operator equations and substitute the result into the atomic operator equations, products of single atom operators that do not commute appear. These products do not semiclassically factor and must be treated before taking expectation values ([17]).

Whitney [18], was the first (to our knowledge) to evaluate these products and remove them correctly from the equations of motion. They become the single atom SpE terms (with rate γ_0) in the atomic operator equations that are usually introduced phenomenologically. These terms produce quantum correlations *between each atom and the field, not between different atoms.* This is by far the most important kind of correlation to correctly account for.

If one ignores the photon correlations between atom and field states that results from this analysis, one obtains a description of radiative damping in terms of the atomic variables and a "semiclassical" field that is non-unitary. We instead introduce an ansatz to account explicitly for the quantum *photon* correlations of the process of single atom SpE *as measured in an experimental apparatus.* This ansatz is based on the well known fact that SpE photons emitted from a weakly driven single atom are antibunched. This means that the measurement of a SpE photon in the detecting apparatus is correlated with the atom going into the ground state [4].

We therefore interpret the atomic operator decay term resulting from the first integral of the atomic field as the following unitary *rule.* This rule follows from restoring discrete photon statistics to the field in which SpE photons appear instead of leaving it in the form of a continuous differential equation. The field is coupled to the detector and is, therefore, exactly known (or knowable) while the detector introduces a stochastic element into the field and hence the atomic operators that is not known or knowable.

In each time step Δt (small relative to γ_0) we evaluate the probability $\gamma_0\Delta t$ that each atom underwent a SpE process. This probability is independent of the state of the atom (it depends on the unknown state of the detector). We generate a random number between 0 and 1 and use it to determine whether or not the atom decayed. If it decays, it goes directly to the ground state (expressed in either picture) from whatever excitation state it was in. This process may or may not be accompanied by the emission of a photon that is detected in our measurement apparatus, with a probability given by the relative excitation of the atom and a phase determined by the time of emission.

This ansatz is unitary: the atomic state and the field are correctly normalized before and after the transition. The SpE process is now discrete and stochastically initiated and hence is incoherent. Our differential equations (with the continuous SpE part removed and this rule substituted) are no longer strictly deterministic -- they contain a part that correctly describes the *indeterminacy* that arises via the interaction of the system with the measuring apparatus. Note that energy is conserved by this rule on the average, but for any particular event there is dispersion because when a partially excited atom decays it must emit either a whole photon or no photon. This is consistent with the theory of quantum mechanics.

SpE is not the only stochastic, incoherent process occurring. Collisions (between atoms or between atoms and phonons) also cause a degradation of the coherent phase. We extend our ansatz to a suitable N atom rule for elastic collision broadening: Collisions occur at the rate γ_1 and randomize the azimuthal (polarization) phase of the atoms in a single timestep. This rule is also unitary, and does not change the excitation of the atom. The rule could clearly be generalized to allow for collisional excitation.

Finally, the atoms may be inhomogeneously broadened (by either its velocity or a stress field); microscopically this means that each atom has its own resonant frequency selected from the appropriate distribution of frequencies centered on the rest frame resonant frequency of the two level atom. This results in the continuous, *reversible* degradation of polarization phase known as free induction decay.

One can now convert the Heisenberg equations of motion for the atom-field system into c-# differential equations (the N atom Maxwell-Bloch equations) that include our two discrete, stochastic ansätze and the effect of inhomogeneous broadening. This is done in detail in [1] so here we just give the result. The variables are described below. Please note that this is *not a semiclassical or neoclassical theory* (at least to first order); full account of atom-field correlations is made by the SpE ansatz. The 3N + 1 coupled equations are:

$$\frac{du(i)}{dt} = \Delta(i)\ v(i) + \sum_{j\neq i} V_{ij} v(j) w(i) \tag{2.13}$$

$$\frac{dv(i)}{dt} = -\Delta(i)\ u(i) - \kappa E(i)\ w(i) - \sum_{j\neq i} V_{ij} u(j) w(i) \tag{2.14}$$

$$\frac{dw(i)}{dt} = -\sum_{j\neq i} V_{ij}\Big\{u(i)v(j) - v(i)u(j)\Big\} + \kappa E(i)\ v(i) \tag{2.15}$$

and

$$\frac{dE}{dt} = -\gamma_c(E - E_0) - \kappa\eta\sum_{j\neq i} v(j) + \chi(\tau_{SpE})). \tag{2.16}$$

These N atom Maxwell-Bloch equations are expressed in the (u,v,w,E) c-# variables that are common in the literature and defined in AE [2] (they result from canonical and gauge transformations applied to the operators and potential). (θ,Φ) are now the spherical polar angles of the (u,v,w) pseudospinor:

$$u = \sin\theta\cos\Phi, \quad v = \sin\theta\sin\Phi, \quad w = -\cos\theta$$

where $u = \langle\theta,\Phi|\hat{u}|\theta,\Phi\rangle$, etc. and (0,0,-1) is the ground state. The normalization condition for these states is:

$$u^2 + v^2 + w^2 = 1$$

and unitary evolution processes must preserve this relationship.

The (θ_i,Φ_i) thus uniquely define a point on a unit "Bloch sphere" for the ith atom. If one solves the c-# Heisenberg equations below for $u_i(t), v_i(t), w_i(t)$ the solution can be referred back to the Schrödinger states $|\theta_i(t),\Phi_i(t)\rangle$ *without ambiguity* as long as the unitarity condition is satisfied. We may thus *switch pictures freely* and preserve *both* our t = 0 commutation rules and product states. Quantum correlations between atoms *cannot grow to first order!*

We have also made the rotating wave approximation (RWA) assuming a resonant external field of the form

$$E_0(t) = E_0\cos\omega_0 t$$

(where E is polarized in the z-direction) and extracted the slowly varying parts (with respect to ω_0) of all quantities. We are neglecting the dispersive Maxwell equation and the spatial variation of the field in this work; we will treat them in a later paper. The RWA frequency difference of the ith atom (including its Lamb shift) is $\Delta(i)$.

These are just the usual RWA Maxwell-Bloch equations for each atom (see [2]), with an additional part that describes the resonant dipole-dipole interaction. E(i) is the total field at the ith site (from 2.16) *minus* the near self-field, which is counted separately in the SpE algorithm. In practice we approximate E(i) by E, which introduces negligible errors for more than a few atoms.

We have also introduced a field damping time γ_c that is the average time required for resonant mode light to propagate out of the sample (Bonifacio and Lugiato [19]). γ_c can be adjusted by, for example, placing the sample in a tuned cavity to trap resonant radiation.

E_0 is the slowly varying amplitude of the external free field and η is a constant proportional to the inverse volume of the sample. $\chi(\tau_{SpE})$ is the field fluctuation term arising from the SpE process. We evaluate it by counting the number of SpE events in a given timestep and discretely increasing the field by the correponding number of $\langle R_3 + \frac{1}{2}\rangle$ weighted "photons" emitted times a small fraction that describes the solid angle and frequency overlap of the SpE pattern and the resonant field mode. Note that SpE (at rate γ_0) and collisions (at rate γ_1) are to be discretely included in the numerical solution of the atomic equations as previously described.

A typical trajectory of a single atom being driven by a constant resonant field is shown below on a (rotating frame) Bloch sphere. The effect of the different driving forces and decay processes is clearly marked. The decay processes are randomly initiated. If we recalculate the trajectory of the atom a second time from identical initial conditions the deterministic parts will obey the same equations of motion, but the discrete jumps will occur at different times. This picture must therefore represent a *possible* quantum trajectory of a given atom in a classical statistical ensemble and hence may be viewed as a single, appropriately weighted "fiber" in the fiber bundle of all trajectories. Our use of product states and our ansatz precludes interference between different atoms in different fibers.

Any indeterminacy in our knowledge of the single atom states arises from the stochastic broadening processes. These, in turn represent irreversible interactions with degrees of freedom over which we have no control, such as the the rest of the universe or internal collisions between atoms. We lack the information necessary to include these degrees of freedom in our "system" and hence can only consider their effect statistically.

If we repeat a caculation many times to obtain a large number of distinct fibers, we should be able to obtain arbitrarily accurate averages for macroscopic quantities even in non-equilibrium situations. If only a few atoms are considered, then the predicted values for most "macroscopic" quantities (like total field) will be subject to large fluctuations that are unpredictable. With that understood, the results of applying these equations numerically are not to be thought of as a "simulation" or a "model", but rather as a *stochastic representative* of the actually measured results -- a different fiber from the same fiber bundle.

One of the most interesting features of this N atom theory is that it allows us to calculate non-equilibrium dynamics. As we shall see when we discuss absorptive bistability, we can actually follow an optical system through a first order phase transition and obtain a meaningful description of the system at all stages of the transition. Generalizations of this idea to spin systems, glasses, and other non-equilibrium systems of interest should result in a new understanding of systems in transition.

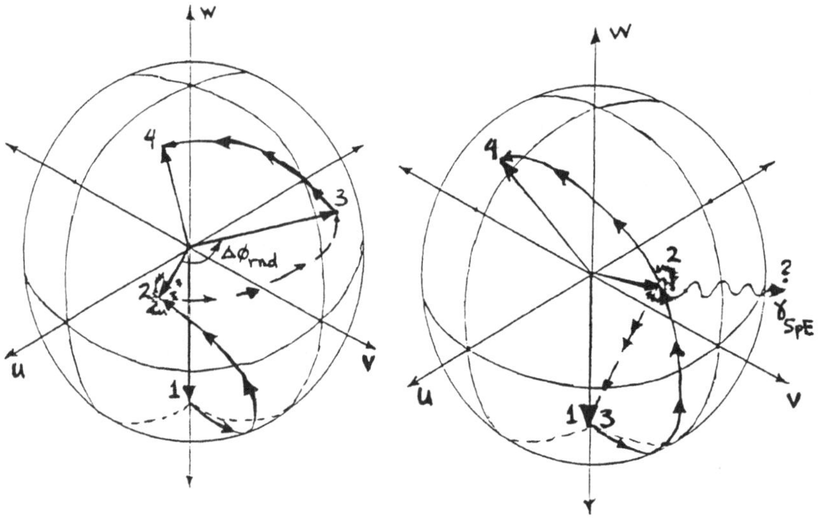

Figure 1. Typical trajectories arising from the solution to our independent atom Bloch equations above. In a) an atom with doppler shifted frequency $\omega < \omega_0$ is driven by a resonant field through part of a Rabi cycle (1-2). When it arrives at 2 it collides with another atom, instantly changing its phase but not its excitation (2-3). It then continues to evolve freely (3-4). In b) the sequence is the same except that at a random time a SpE event occurs that drives the atom back into the ground state (2-3). The atom then continues to freely evolve, which causes it to retrace its path through 2 and beyond (3-4). A photon (γ_{SpE}) may or may not be emitted (depending on w) and may or may not be in the resonant mode.

III. NUMERICAL RESULTS

In this section we show the results obtained from applying the equations 2.13-2.16 to small systems of atoms and compare them with the results obtained from nonlinear theories and from experiment. To correctly compare the results we must ignore the dipole-dipole interaction since it is neglected in the usual nonlinear theories.

Superradiance

We study the initiation and subsequent development in time of the superradiant process ([14],[20]-[27]) when the system is completely inverted at t = 0. We take a sample of N = 150 atoms in a small volume, and assume that only one resonant mode of the field is relevant to the superradiant process; the solid angle overlap of the sample with the resonant field mode and the single atom SpE fluctuations are therefore small.

We present the results of three calculations that represent

three different experimental situations. In all three figures, the time development of the emitted field (not intensity) is shown, and the superradiance process is initiated by SpE photons. τ_r is the inverse width of the superradiant pulse and γ_2 is the width of the gaussian doppler frequency distribution of the $\Delta(i)$'s.

Figure 2 represents a "classical" superradiance pulse. We set $\gamma_0 = \gamma_1 << 1/\tau_r < \gamma_2$, so that inhomogeneous decay of the collective polarization occurs *faster* than the development of the superradiant pulse. The long delay between inversion and the ultimate emission of the pulse is characteristic of "self-starting" superradiance and has been experimentally observed. The sample is in a "bad cavity" ($\gamma_c = 1.0$).

Figure 3 displays the same system but now we have put the sample in a "good" cavity with $\gamma_c = 0.05$. The cavity traps radiation; this causes "ringing" in the superadiant pulse. The collective pseudospin moment of the atoms behaves like a "rigid pendulum". The initially inverted atoms superradiate into the field as their pseudospin moments swing coherently down in the vw plane. They are then are driven *through* the ground state by the emitted field before the field is damped out of the cavity and reabsorb their own radiation. This polarizes the system with opposite phase, and the atoms again superradiate into the field; oscillation ensues that is damped by cavity losses and free induction decay. This is very close to what is observed in superradiance experiments [28]-[31].

In figure 4 we are very close to the limit of where superradiance can occur. In this system $\gamma_0 = \gamma_1 = \gamma_2/25$, and the single atom decay rate is comparable to the development time of the superradiant pulse. Consequently, a large number of atoms reach the ground state before the pulse is significantly large and are thus in a position to reabsorb the emitted superradiant pulse. The atoms are again in a bad cavity.

There is very little delay between inversion and the emission of the pulse because of the large number of SpE events. As the pulse grows, noise is clearly visible. About halfway up the leading edge of the pulse radiation trapping is evident as deviation from a smooth sech shape. The sample becomes very sensitive to fluctuations as the SpE noise is "amplified" in its effect on the time evolution. There are chaotic features in the trailing edge, and ringing is clearly observed as re-excited atoms superradiate when they come into phase.

The results in figures 3 and 4 are consistent with the current explanation of the experimentally observed ringing: radiation trapping [25]. We observe ringing due to radiation trapping whether the trapping is due to a cavity or atomic re-excitation. These figures qualitatively (and quantitatively to the extent that one can expect at this stage in our work) reproduce both the experimental results on superradiance to date and the best theoretical studies done in the mean field. We therefore conclude that the N atom + field equations accurately describe the microscopic picture of the superradiance process, including the quantum process of pulse initiation that is generally modelled by a "noise term" in other approaches.

Photon Echoes

We begin with a sample of 150 atoms again, this time initially in the ground state. We turn on a strong resonant

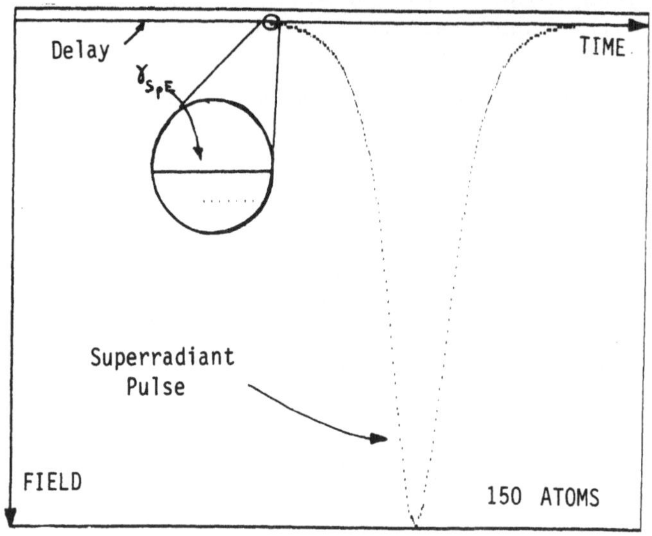

Figure 2. A single SpE photon (see detail) initiates the superradiance process after a long delay. The sample is in a "bad cavity" and the (negative) pulse is emitted before a significant number of atoms can reach the ground state, so radiation trapping is negligible.

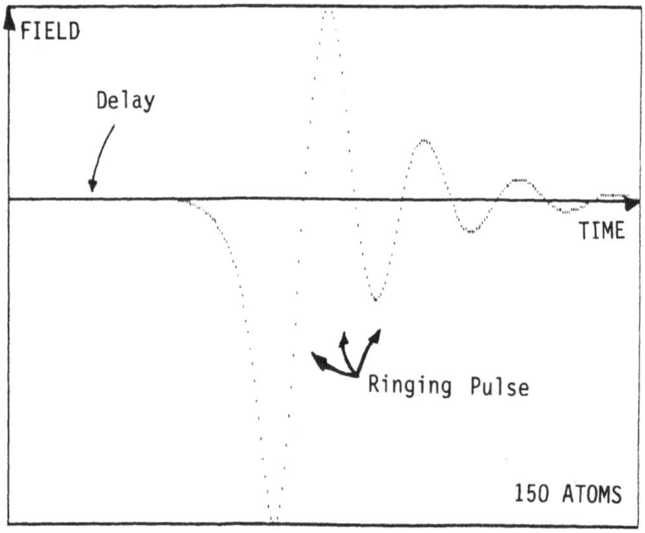

Figure 3. A superradiant pulse from a system of atoms similar to that in figure 5, but in a cavity that traps radiation (γ_c = 0.05). Note ringing as energy oscillates between the atoms and the cavity field, and the long delay between sample inversion and the emission of the superradiant pulses. This figure is very similar to what has been experimentally observed.

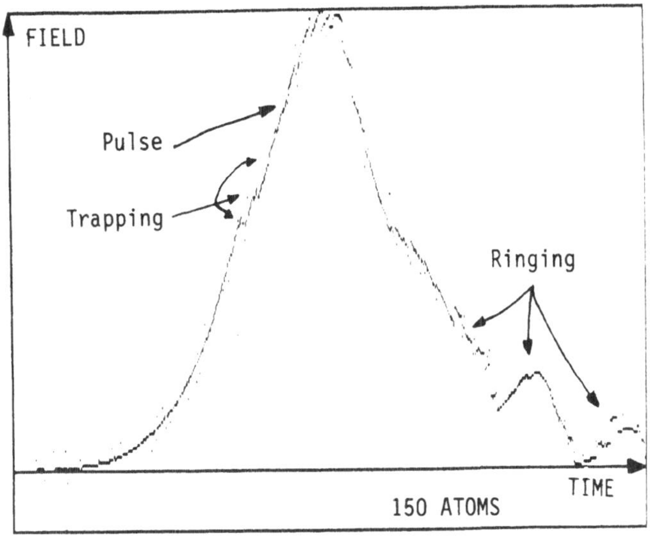

Figure 4. Superradiance when the SpE rate is close to $1/\tau_r$. Note sensitivity to SpE noise, the radiation trapping evident as the pulse grows, and the "ringing" that results as radiation reabsorbed by ground state atoms is superradiated in turn.

field in a bad cavity ($\gamma_c = 1.0$) at time $t = 0$ and rapidly drive the atoms into a polarized state with $\theta = \pi/2$ (a $\pi/2$ pulse). $\gamma_0 = \gamma_1 = \gamma_2/200$, so the time dependence of the system is dominated by free induction and the polarization moment decays before too much energy is superradiated away.

After a suitable time T ($\gamma_2^{-1} \ll T \ll \gamma_0^{-1}$) a resonant π pulse is applied to the system (all the moments rotate around the u-axis by π. Free induction then proceeds in the same sense and the mirror image polarization moments rephase after an additional time T. The reconstructed collective polarization moment radiates the "echo" pulse seen on the right.

This exactly reproduces the classical photon/spin echo results ([32],[33]), and *requires that the atoms be in an independent atom product state!* Quantum correlated atomic states cannot be sensibly defined if the atoms are distinguished by independent, randomly distributed frequencies that cause the phases of the different product states to rapidly change in time. This is even more true when the phases are degrading due to irreversible processes like SpE and collisions.

We feel that photon echoes provide a powerful argument for the use of product states in quantum optics. The use of quantum correlated states in contexts where photon echoes could conceivably be measured is inconsistent. The motivation for constructing collective atomic states disappears when the atoms

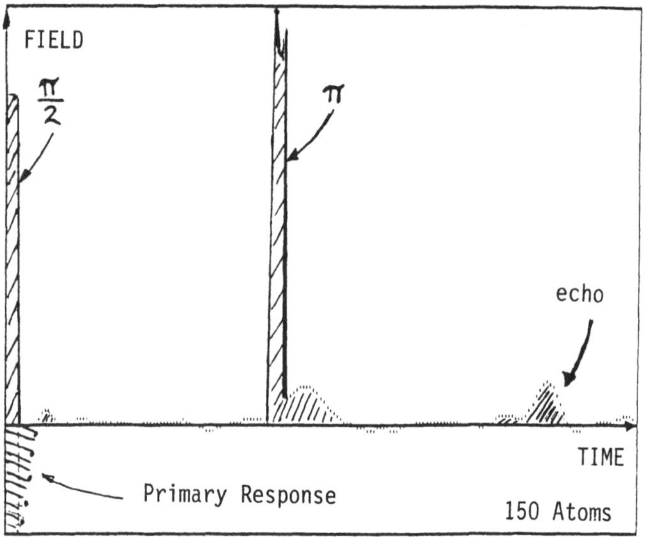

Figure 5. The field emission from a collection of atoms polarized via a resonant $\pi/2$ pulse that decays via free induction decayand is irradiated with a π pulse at time T. An echo pulse occurs an additional time T later. This phenomenon is easily explained only if the atoms are in an independent atom product state.

have distinct resonant frequencies or different environments.

Absorptive Bistability (AB)

Optically bistable systems have a hysteresis cycle in the intensity of transmitted light as a function of the incident intensity ([19],[34]-[37]). The two branches of the hysteresis cycle are characteristically a low transmittivity (o-opaque) branch and a high transmittivity (t-transparent) branch.

The accepted theoretical model for AB [19] is based on mean-field Maxwell-Bloch equations for the *collective* atomic system with phenomenological decay and radiation trapping due to the cavity (via γ_C). The system of N atoms and the field is thus modelled by a set of three coupled differential equations for the atomic polarization and inversion interacting with the mean field. The radiation trapping enhances the non-linear reaction field generated by the collective polarization. No information concerning the actual microscopic state of the system is available in this model; spectral information and the properties of the system undergoing the phase transition are inferred from stability analysis and quantum regression.

We applied our system of equations to the AB of 300 atoms in a cavity with $\gamma_C = 0.1$, treating spontaneous emission and broadening discretely and stochastically in each timestep as described above. In figure 6 below we show the hysteresis cyle in transmittivity and polarization that results when we neglect inhomogeneous broadening: we set $\gamma_2 = 0$, $\gamma_0 = \gamma_1 = 0.02$, $\gamma_C = .1$ in units of inverse timesteps. $\kappa = 0.25$, $\kappa\eta/N = 0.33$, and we

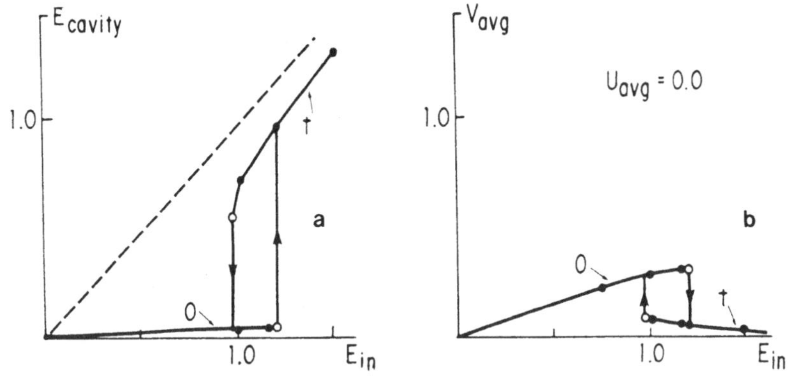

Figure 6. Hysteresis cycle in field transmittivity (a) and the u and v components of the polarization (b) in AB.

neglect x_{SpE}. Figure 6 generally agrees with what is obtained by the mean field theory in [19].

However, our theory also allows us to examine the distributed independent atom states at all times. Below (figures 7a and 7b) we represent typical opaque and transparent states on a Bloch sphere of radius N. The dark vector represents the collective pseudospin moment; the small vectors represent a few of the individual atomic components of the collective moment, drawn to a larger scale.

Our theory gives us even more information than these "equilibrium" states. Since we can follow the system timestep by timestep through the hysteresis cycle, we can for the first time describe the non-equilibrium dynamics of the phase transition itself in detail. We begin in the o-branch.

A resonant field in equations 2.13-2.16 rotates the pseudospin vectors of each two-level atom around the u-axis at the Rabi frequency κE (where E is the cavity field). Radiation from the atoms constructively interferes and generates a cavity enhanced equilibrium reaction field equal to $\frac{\kappa\eta}{\gamma_c}\sum v(j)$ that is *out of phase* with the applied field and hence *reduces* it. The total cavity field (and hence transmitted field) will thus be small as long as this quantity is comparable to E_0.

Collisions randomize the azimuthal phase angles Φ_i of the individual atomic pseudospin vectors at rate γ_1, causing the collective polarization to decay. However, the process of spontaneous emission carries atoms back into the ground state at a random time and starts atoms over in the Rabi cycle *with the absorptive Φ phase!* The cavity field rotates these atoms into an absorbing state and rebuilds the collective v-moment. If the rates balance, the o-branch is dynamically *stable*.

As the intensity of the applied field is increased, the individual moments rotate further before decaying. The collective vector shrinks (broadening is more effective at higher excitations) at the same time its v-moment grows. The enhanced reaction field continues to cancel most of the cavity field. A few of the atoms may progress far enough in the Rabi cycle to begin coherent emission (from a negative v-component) before decaying spontaneously.

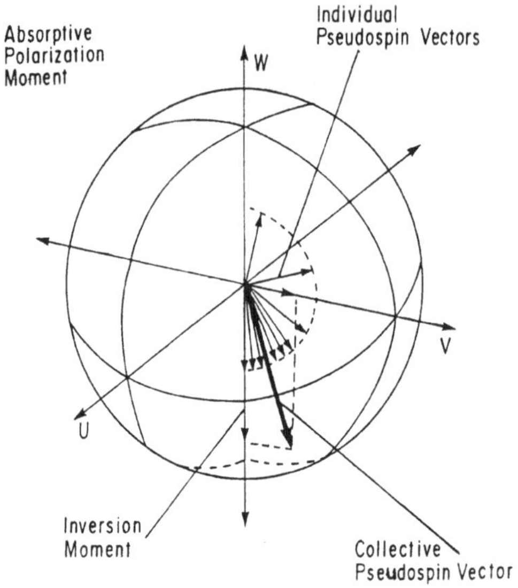

AB

Figure 7a. In the opaque state, the individual moments are driven around the u axis by the resonant cavity field. The sum of their v-components generates an absorptive reaction field that keeps the cavity field and thus the Rabi frequency low. SpE causes the individual atoms to begin the Rabi cyle over before they make it into the emitting phase, and the picture is stable.

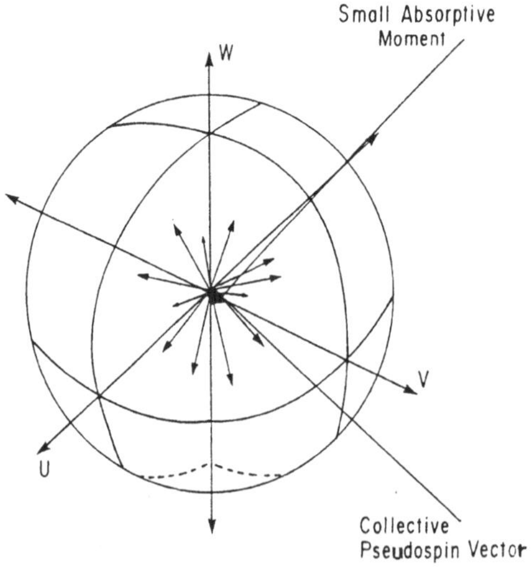

Figure 7b. In the transparent state, the individual moments are in a strong field and Rabi oscillate around the u axis rapidly. SpE and collisions randomize both θ and Φ so that the net v-moment (and reaction field) remains nearly zero. This situation persists until differential growth in the reaction field becomes possible.

If the incident field goes beyond a critical point, the system exhibits differential gain in the cavity field from the atoms that make it into the emitting part of the Rabi cycle. This in turn increases the rate that atoms are carried into the emitting phase, which further increases the cavity field, and the collective polarization vector begins to rotate around the u-axis (Rabi oscillate).

The system undergoes a first order phase transition. While the individual atomic pseudospin vectors rotate around the u-axis the various decay processes randomize their phases; the collective dipole moment and the reaction field rapidly decay to almost zero. The system becomes transparent. Since the atoms Rabi oscillate more than once (on the average) before decaying the collective v-moment cannot rebuild and the t-branch is *stable*.

As one lowers the driving field, at some t-to-o critical intensity the system exhibits differential loss in the reaction field. Spontaneous emission creates atoms with the absorbing phase faster than dynamical processes destroy their phase, reconstructing a collective polarization vector. Its reaction field further reduces the Rabi frequency in the cavity and the system undergoes a first order phase transition to the o-branch.

There are two new features of interest in these results. The first is that we now understand the mechanism that maintains the "cooperativity" (in quotes because the atoms here are *independent*) of the o-branch in the AB cycle: The *process of spontaneous emission* prepares atoms in the ground state at a fixed rate; these atoms are then driven into collectively absorbing state that rebuilds faster than decay degrades it. The radiation reaction field from the absorptively phased independent atoms is effectively amplified by the cavity relative to the incident field so that the system is stable.

If the overlap of the SpE modes and the resonant cavity mode is small, virtually all the SpE photons escape the active volume incoherently and carry energy from the system. This mechanism is obvious in the N independent atom equations with our ansatz concerning the correct treatment of the SpE process, but to our knowledge has not been previously discussed. The second feature of interest is that we now have a believeable microscopic picture of the state of the system *at all times*. We can follow the system *through* the phase transition in both directions, study size extensivity and fluctuations, and investigate switching speeds. This cannot help but be of use to those attempting to design an AB system, or to better exploit the mechanism itself. Note that no unstable, unphysical middle branch ever appears in this theory. One can directly observe such phenomena as critical slowing down, critical opalescence, and the predicted spectrum without recourse to linearization, regression or stability analysis.

Even if this picture is not entirely correct in all the particulars (for example, we neglected γ_2 in this study, although we will include it in future studies) it still yields a far better understanding of the basic processes and critical phenomena occurring in AB than does a mean field theory.

IV. THE MODEL DIPOLE-DIPOLE INTERACTION

The equations of motion 2.13-2.16 contain an analytically derivable dipole-dipole term that is generally neglected in N atom theories. This term was discussed in some detail in [1] and a *model* to describe a dipole-dipole relaxation was introduced. This model is intended to correct for retardation, radiation damping of the pair interaction (which causes a mutual precession of pseudovectors), and other effects such as higher order interactions, wavefunction overlap, and electron exchange that might contribute to relaxation but are beyond our ability to calculate. In the rotating frame the Maxwell-Bloch equations become:

$$\frac{du(i)}{dt} = (\Delta_0(i) + \xi(i))\, v(i) + \sum_{j\neq i} V_{ij} v(j) w(i) \qquad 4.1$$

$$\frac{dv(i)}{dt} = -(\Delta_0(i) - \xi(i))\, u(i) - \kappa E\, w(i) - \sum_{j\neq i} V_{ij} u(j) w(i) \qquad 4.2$$

$$\frac{dw(i)}{dt} = -\sum_{j\neq i} V_{ij}\left\{u(i)v(j) - v(i)u(j)\right\} + \kappa E\, v(i) \qquad 4.3$$

and

$$\frac{dE}{dt} = -\gamma_c(E - E_0) - \kappa\eta\sum_{j\neq i} v(j) + \chi(\tau_{SpE})). \qquad 4.4$$

We treat SpE and collisions discretely as previously discussed.

In these equations, $\xi(i)$ is an effective frequency shift that models the relaxation process:

$$\xi(i) = \sum_{j\neq i} \frac{\beta_{ij}}{4}\left\{u(i)v(j) - v(i)u(j)\right\}. \qquad 4.5$$

β_{ij} is the appropriate pseudospin-pseudospin relaxation rate. The frequency shifts in 4.5 exert an azimuthal "torque" on the pseudospin vector of the ith atom that aligns (or antialigns) its (u,v) moment with that of its neighbors. This may be considered a dynamical ferroelectric (or antiferroelectric) effect.

If we assume β_{ij} is proportional to V_{ij}, then it should be extremely large for nearest neighbors and should drop off rapidly with separation. We feel that if one restricts the shift interaction to nearest neighbors or next-nearest neighbors then one will include the greater part of the relevant effect, at least for qualitative studies like this one.

Recall that the sign of V_{ij} depends on the geometry of the neighbors. It is entirely possible that substances exhibiting a *net* ferroelectric effect have unusual geometries or rely on a mechanism other than the direct dipole-dipole interaction to drive relaxation. This point is discussed again below.

Pseudospin Ordered Bistability: The Phase Locked State

We applied equations 4.1 to 4.5 to the same system of atoms described earlier in the study of absorptive bistability. Our purpose was to see if the dipole-dipole relaxation process would introduce any *qualitative* change into the bistability cycle; we view this (so far) as a *model* calculation. As before we set $\gamma_0 = \gamma_1 = .02$, $\kappa = 0.25$, $\gamma_2 = 0$, $\gamma_c = .1$ and $\kappa\eta/N = 0.333$ in inverse timesteps with $N = 300$.

To simplify the problem, we neglected the mutually induced precession that results from the direct dipole-dipole terms in the equations of motion (this is one way of "averaging" over the precession cycles, which occur much more rapidly than secular changes for nearest neighbor separations of around 10 atomic radii) and studied only the relaxation process. The pseudospin locking that plays an important role below was therefore due strictly to the resonant cavity field.

We restricted the sums in $\xi(i)$ to the nearest neighbors of the ith atom. We chose $\beta_{nn} = 0.25$ so that relaxation proceeded much faster than homogeneous decay but at a rate comparable to Rabi oscillation with a unit field. It should be easy to improve the model (include more neighbors and the mutual precession, etc.) in a future calculation, but we feel that the essential features of the relaxation are preserved by these assumptions.

The model system was bistable, as before, but the internal organization of the transparent branch was radically different from that observed in AB. The pseudospin moments of the individual moments underwent a saturated "second order" phase transition after they went through a first order phase transition from opaque to transparent. The hysteresis cycle is pictured below in figure 8; note especially the bifurcated u-moment in the upper branch.

The dynamical evolution of this "pseudospin-ordered bistability" (POB) system can be described as follows: A resonant field in the cavity rotates the pseudospin vectors of each two level atom out of the ground state around the u axis at a rate κE, where E is the total cavity field. This creates a collective v-polarization whose reaction field reduces the cavity field.

As before, collisions randomize the azimuthal phase angles ϕ_i of the individual atoms, causing the collective v-moment to decay. However, the process of SpE prepares atoms in the ground state where they can be coherently re-excited and hence is responsible for the continuous reconstruction of the absorptive v-moment and the o-branch is stable.

The o-branch is additionally stabilized by the model relaxation process, but while this alters quantities like the o-to-t critical point and fluctuation properties it does not affect the qualitative equivalence of the lower transmittivity microstate in the AB and POB. This can be seen by comparing figure 9a below with figure 7a.

As the intensity of the applied field is increased, the collective pseudospin vector is driven up in the v-w plane. At the critical point, the system exhibits differential gain from the atoms that make it into the emitting phase before spontaneously decaying and begins to Rabi oscillate. The collective moment rapidly decays due to both SpE and collisions, and the system becomes transparent.

However, as this *first order* transition is proceeding, atoms that happen to have their moments nearly aligned with either the +u or the -u axis (because of a collision, for example) *are pseudospin locked* ([38],[39])*!* These randomly selected atoms function as *phase attractors* to neighboring atoms, which tend to align their ϕ-phases via the β relaxation and hence stabilize themselves. They in turn are phase locked (we use this phrase to mean pseudospin locked in a transparent, stable state) by the large cavity field and act as attractors. A process of nucleation and growth ensues in which atoms form spatial clusters, with opposite azimuthal pseudospin locked

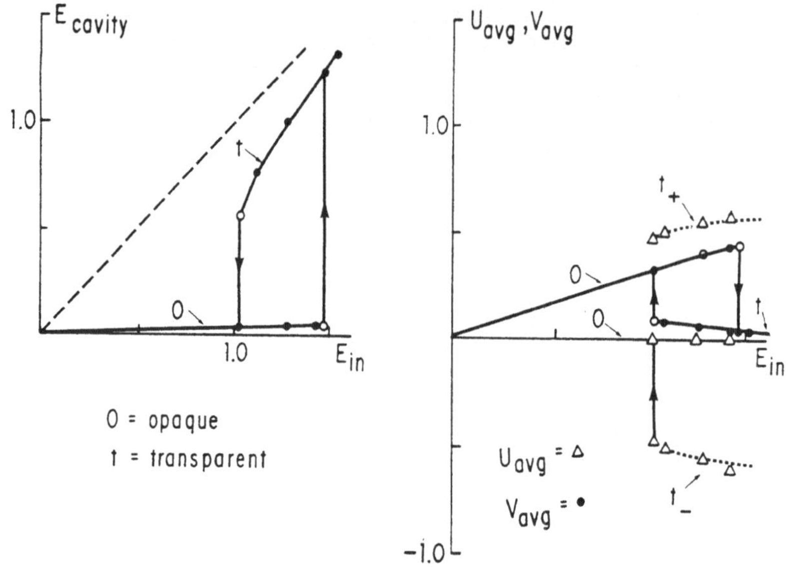

Figure 8. The hysteresis cycle in transmittivity and polarization for POB. Note the bifurcated u-moment in the t-branch corresponding to the spontaneously broken symmetry of the ferroelectric state.

phases, that compete for atoms along their boundaries until one phase or the other wins out. In figure 9b we present a typical equilibrium t-branch state on the Bloch sphere. In figure 10 we present a series of computer generated figures that follow the process of nucleation and growth.

In figure 10, + sites are mostly aligned with the +u axis, - sites are mostly aligned with the -u axis, and 0 atoms are nearly in the vw plane. We do *not* force atoms into ±u eigenstates in each timestep; this is merely a convenient pictorialization of appropriately binned Φ phases. Note that granulated clusters are immediately formed, and that atoms along the boundaries (which are somewhat "fluid" because of SpE and collisions) are eventually pulled into the locally dominant cluster. Surface-to-volume effects and fluctuations determine which phase eventually wins out (although a "pseudospin-glass" situation where the time required for one phase to win out is large relative to all other times is quite conceivable).

This resembles exactly a saturated "second order" phase transition to a dynamically ferroelectric phase locked state that follows the first order phase transition from the opaque to the transparent phase. Atoms that are dephased from the large, collective u-moment by SpE or collisions are quickly "phase healed" by the relaxation process.

As before, when one lowers the driving field, at some t-to-o critical intensity the system exhibits differential gain in the reaction field (SpE creates atoms with an absorbing phase faster than dynamical processes degrade it). The large collective phase locked vector dynamically rotates into an absorptive phase and the o-branch dynamics resume.

If the reaction field is sufficiently enhanced by means of the cavity gain or a high density then the two critical points

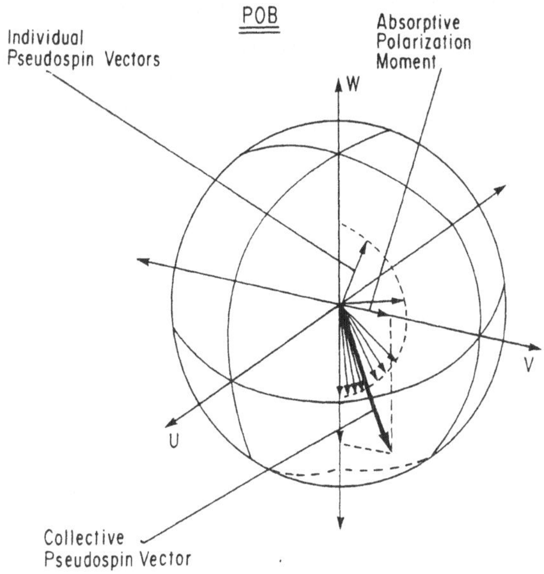

Figure 9a. The opaque state in pseudospin ordered bistability (POB) is qualitatively similar to the opaque state in AB. A large collective v moment is dynamically maintained by a competition between SpE and the evolution of atoms through the Rabi cycle. The pseudospin relaxation mechanism changes the critical points (compare figures 6 and 8) and broadens the bistable region.

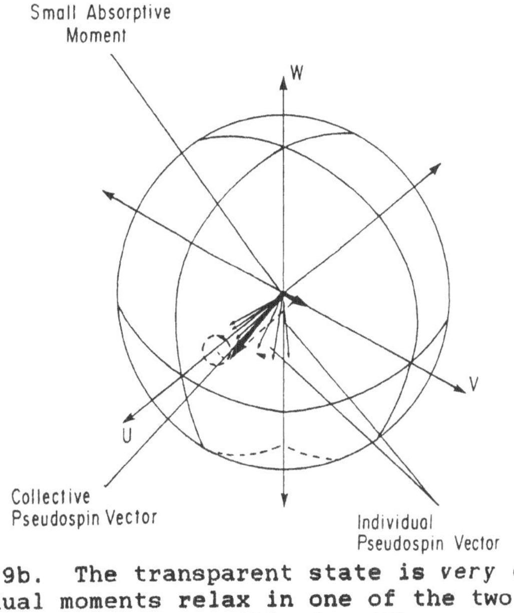

Figure 9b. The transparent state is *very* different. The individual moments relax in one of the two points that are stable against Rabi oscillation (the ± u axis). These moments are "pseudospin locked" by the field and stabilized against SpE and collisions by the pseudospin relaxation mechanism. This kind of transparent state has some very important technical advantages described later.

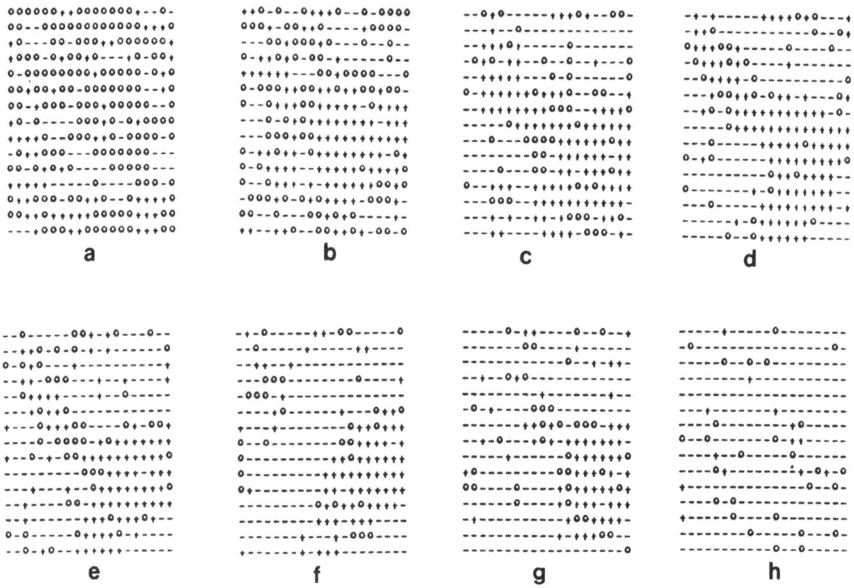

Figure 10. A time-ordered sequence of pictures that represent the evolution of the phase locked state shown in figure 9b. After a the first order transition (first panel) pseudospin locked atoms act as phase attractors to the others via the model relaxation force. These in turn are pseudospin locked and a process of nucleation and growth rapidly pulls the atoms into one phase or the other. Boundary fluctuations (which proceed much more slowly) cause one phase or the other to win out.

are separated in such a way that "pseudospin ordered bistability" (POB) results.

The self-ordering of the electric pseudospins in the o-to-t phase transition is analogous to the spontaneous self-ordering of magnetic spins in a paramagnet cooled below the Curie temperature. However, in POB this is a truly *dynamical phase transition*, that is, it is *driven* by dynamical ordering processes, and the stable transparent states have ordered *dynamically induced* dipole moments. Furthermore, the o-to-t phase transition combines features of both first and second order phase transitions; it has a "latent energy" like AB (and hence is a first order transition exhibiting bistability, critical slowing down, critical opalescence, etc.), but on the other hand it is the induced electric dipole equivalent of a saturated second order ferromagnetic phase transition and exhibits the dynamics of spontaneously broken symmetry! It is thus a very curious object and one of considerable abstract interest.

To conclude our discussion of POB we note that the existence of a macroscopic, phase locked polarization in the t-branch can be exploited to reduce the time required to go through a complete optical hysteresis cycle, which is the minimum switching time for an optical device. Also, the existence of a dynamically switchable ferroelectric state means that two distinct information storage mechanisms are possible for such a system.

If a suitable "detuned $\pi/2$ pulse" is applied to the POB system in the bistable region of the o-branch, the collective moment will move directly into the u-v plane and swing azimuthally around to line up with the +u or -u axis. In this manner the system is moved *directly into the stable transparent branch!* (See figure 11). To make the reverse transition one can apply a similar pulse.

A short time, typically on the order of β^{-1} (which is, by hypothesis, much less than γ_1^{-1} for POB to work at all), is required before the system is stable. This should be *less* than the shortest relaxation times for AB systems (which are on the order of γ_1^{-1} or γ_2^{-1}). A dynamically switched POB system may be able to cycle an order of magnitude or more faster than an AB system. Note that this mechanism is quite *different* from the switching of a dispersive bistable etalon by momentarily detuning the field. It requires that a carefully shaped field be used to rotate the collective Bloch vector in a highly specific way to achieve transparency.

Finally, we can exploit directly the bifurcation of the transparent state (and *tri*stability of the PO system)! To do this we do *not require that the system be in a cavity or bistable*! After driving a system of ferroelectric atoms to a tranparent, pseudospin locked state one can non-destructively test whether its collective polarization is aligned with the +u or the -u axis (using, for example, the scattering of a weak polarized light pulse or through the use of appropriately sequenced interrogatory pulses). One can dynamically switch the u-alignment by applying a detuned pulse for a time sufficient to azimuthally rotate the collective vector by π around the w axis.

In this way one can store and recover binary information in a very small collection of atoms maintained in a pseudospin-locked transparent state. Since *no relaxation time at all* is required by the switching process and all switching is done with fast pulses the switching time is limited only by the spectral broadening of ultra-short pulses and the slowly varying approximation (i.e. ω_0^{-1}). This may be the fastest optical information storing mechanism possible with atoms.

Since the output (intensity) curves arising from AB and POB look qualitatively similar, it is possible that some of the experimental observations of bistability attributed to the absorptive mechanism may in fact have been due to the ferroelectric phase locking mechanism. The major difference between the two is in the organization of the t-branch, with its macroscopic dipole moment. We feel that it is important to repeat experiments on materials that have exhibited bistability to determine if they have a macroscopic polarization in the t-branch.

In addition, a search for materials that might exhibit POB because of their known or calculated dynamical dielectric properties is in order. In order for the atoms to relax in a ferroelectric configuration it is necessary for β_{ij} to be positive. If β_{ij} and V_{ij} have the same sign or the opposite sign the optically active atoms would have to be distributed either in two dimensional layers perpendicular to the field polarization (with the distance between layers much greater than the nearest neighbor distance) or in filaments parallel to the field polarization (with a large distance between filaments). Since we are assuming no wavefunction overlap, in a solid the optically active atoms must be embedded in lattice of neutral atoms that simply shift the background index of refraction.

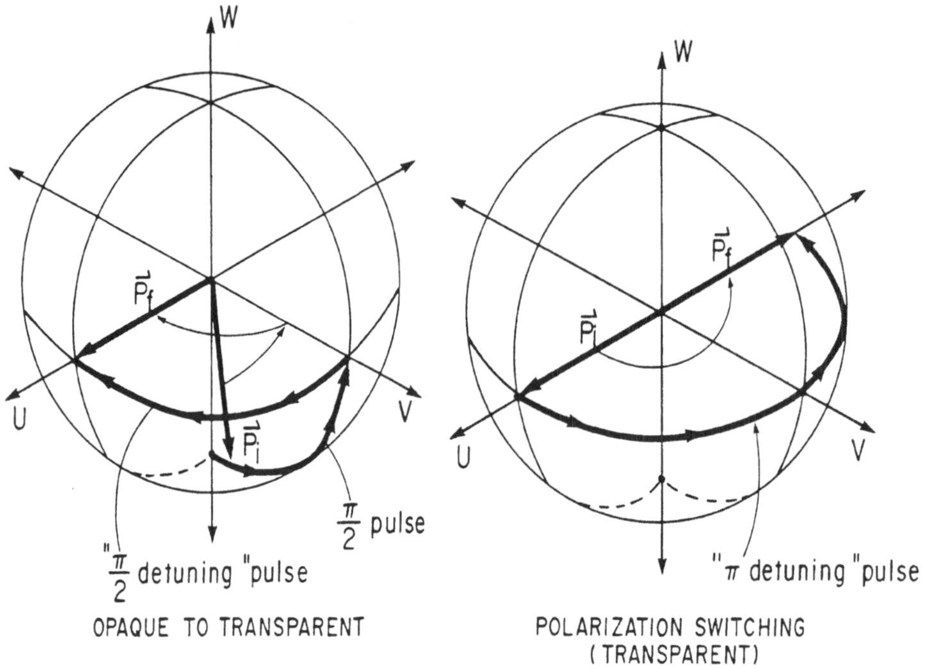

Figure 11. Two mechanisms for enhancing the speed of switching for both AB and POB. a) A detuned $\pi/2$ pulse can carry an AB or POB system directly from the opaque to the transparent branch (although the AB system must then dephase to become completely stable). This takes advantage of the fact that a phase locked state is transparent. b) A phase locking system that exhibits a strong phase relaxation can also be used to make an extremely fast *phase switch*. The continuously driven system is phase locked by the field and stabilized by the relaxation process. Binary information can be stored in the relative phase of the collective moment. The state can be switched by a "π detuning pulse" from one orientation to another, and the information can be queried by an appropriate sequence of detuned pulses or from the attenuation or amplification of a phase shifted resonant pulse.

If β_{ij} carries a sign that is independent of the sign of V_{ij} (perhaps because the underlying mechanism lies involves interaction with the lattice or electron exchange instead of the direct dipole-dipole force) then the material may be an intrinsic dynamic "ferroelectret" or "antiferroelectret". We will present a study of all of these possibilities and make more remarks on spontaneously phase locking systems in a later publication.

To conclude our discussion of POB, our model calculations have demonstrated the possibility of a novel mechanism for optical bistability in materials that shows great promise in the design of ultra-fast optical switching and information storage devices. We hope this stimulates an experimental search for what may be a very exciting phenomenon.

V. CONCLUSIONS

We have developed a very old theory and a very old set of equations into a new theory that appears to be capable of at least qualitatively reproducing the prominent features of the experiments and theories of modern quantum optics. These equations differ from the older, mean field or master equations theories in three ways. First, we solve the complete N atom + field equations numerically. Second, we account for unitary self-interaction (SpE) and broadening with our ansätze and obtain equations of motion in which no quantum correlations between atoms evolve. The resulting picture of radiation reaction is consistent with a quantum mechanical version of the contact interaction theory of Wheeler and Feynman [40]. Third, we treat the dipole-dipole force between atoms explicitly and use a model interaction to study locally correlated states.

The new results we have shown are: a microscopic picture of the initiation process of superradiance; a microscopic picture of classical AB where we show how SpE is responsible for maintaining the opacity of the lower branch and we follow the non-equilibrium dynamics of the first order phase transition; a microscopic picture of pseudospin-ordered AB where the model dipole-dipole interaction spontaneously creates a pseudospin-locked, transparent, bifurcated upper branch.

We have explored this last possibility in some detail as it is a completely new result. If pseudospin ordered, phase locked states of this sort are experimentally shown to exist, then we will be able to exploit them to build optical phase switches that store information in the relative phase between the pseudospin locked state and the reference field. Switching can be done by detuning a constant field for a short time or by using an auxiliary pulse. The switching times are limited only by the slowly varying approximation and the bandwidth of the switching pulses: switch times of less than a picosecond may be possible. *No cavity is necessary* to enhance the reaction field and *only a few atoms* are needed to produce a statistically stable system.

We feel that these considerations make it very important to search for, and to try to construct, suitable "filamented" or "layered" materials that might reasonably exhibit this property. Further theoretical work is needed to put the model interaction on a sounder footing and to study the dynamics. This work is in progress and we hope to report additional results soon.

REFERENCES

[1] R. G. Brown and M. Ciftan, submitted to Phys. Rev. (1987).

[2] L. A. Allen and J. H. Eberly, Optical Resonance and Two-Level Atoms, John Wiley and Sons, New York, 1975.

[3] G. S. Agarwal, Quantum Statistical Theories of Spontaneous Emission and their Relation to Other Approaches, Springer-Verlag, New York, 1974.

[4] R. Loudon, The Quantum Theory of Light, Clarendon Press, Oxford, 1983.

[5] A. Corney, Atomic and Laser Spectroscopy, Clarendon Press, Oxford, 1977.

[6] P. L. Knight and L. A. Allen, Concepts of Quantum Optics, Pergamon Press, New York, 1983.

[7] W. H. Louisell, Radiation and Noise in Quantum Electronics, McGraw-Hill Book Co., New York, 1964.

[8] Coherence and Quantum Optics, ed. L. Mandel and E. Wolf, Plenum Press, New York, 1973.; Coherence and Quantum Optics IV, ed. L. Mandel and E. Wolf, Plenum Press, New York, 1977; Coherence and Quantum Optics V, L. Mandel and E. Wolf, Plenum Press, New York, 1983.
[9] Quantum Electrodynamics and Quantum Optics, ed. A. O. Barut, Plenum Press, New York, 1983.
[10] Coherent Nonlinear Optics, ed. M. S. Feld and V. S. Letokhov, Springer Verlag, New York, 1980.
[11] Optical Bistability, ed. C. Bowden, M. Ciftan, and H. Robl, Plenum Press, New York, 1980.
[12] Optical Bistability III, ed. H. M. Gibbs, P. Mandel, N. Peyghambarian and S. D. Smith, Springer Verlag, New York, 1986.
[13] H. M. Gibbs, Optical Bistability: Controlling Light with Light, Academic Press, New York, 1985.
[14] R. H. Dicke, Phys. Rev. 93, 99 (1954).
[15] J. M. Radcliffe, J. Phys. A 4, 313 (1971).
[16] F. T. Arecchi, E. Courtens, R. Gilmore, and H. Thomas, Phys. Rev. A 6, 2211 (1972). Atomic coherent states are also called Bloch states.
[17] J. H. Eberly, in Coherence and Quantum Optics, ([8a]). p. 635.
[18] K. G. Whitney, ibid (CQO,[8a]), p.767.
[19] R. Bonifacio and L. A. Lugiato, a) Opt. Comm. 19, 172 (1976); b) Phys. Rev. A 18, 1129 (1978). These authors, in particular, have written many papers on bistability in a mean field optical theory.
[20] R. H. Lehmberg, a) Phys. Rev. A 2, 883 (1970); b) Phys. Rev. A 2, 889 (1970).
[21] N. E. Rehler and J. H. Eberly, Phys. Rev. A 3, 1735 (1971).
[22] R. Bonifacio, P. Schwendimann and F. Haake, a) Phys. Rev. A 4, 302 (1971); b) Phys. Rev. A 4, 854 (1971).
[23] C. R. Stroud, J. H. Eberly, W. L. Lama, and L. Mandel, Phys. Rev. A 5, 1094 (1972).
[24] R. Bonifacio and L. A. Lugiato, a) Phys. Rev. A 11, 1507 (1975); b) Phys. Rev. A 12, 587 (1975).
[25] J. C. MacGillivray and M. S. Feld, a) Phys. Rev. A 14, 1169 (1976); b) M. S. Feld and J. C. MacGillivray, in Coherent Nonlinear Optics, ed. M. S. Feld and V. S. Letokhov, Springer-Verlag, New York, pp. 7-57 (1980).
[26] R. Glauber and F. Haake, Phys. Lett. 68A, 29 (1978).
[27] F. Haake, H. King, G. Schröder, J. Haus, and R. Glauber, Phys. Rev. A 20, 2047 (1979).
[28] N. Skribanowitz, I. P. Herman, J. C. MacGillivray, and M. S. Feld, Phys. Rev. Lett. 30, 309 (1973).
[29] M. Gross, C. Fabre, P. Pillet and S. Haroche, Phys. Rev. Lett. 36, 1035 (1976).
[30] A. Flusberg, T. Mossberg and S. R. Hartmann, Phys. Lett. 58A, 373 (1976).
[31] H. M. Gibbs, Q. H. F. Vehren and H. M. J. Hikspoors, Phys. Rev. Lett. 39, 547 (1977).
[32] E. L. Hahn, Phys. Rev. 80, 580 (1950).
[33] N. A. Kurnit, I. D. Abella, and S. R. Hartmann, Phys. Rev. Lett. 13, 567 (1964).
[34] H. Seidel, "Bistable optical circuit using bistable absorber within a resonating cavity", U. S. Patent 3,610,731 filed May 19, 1969 granted October 5, 1971.
[35] A. Szöke, V. Daneu, J. Goldhar, and N. A. Kurnit, Appl. Phys. Lett. 15, 376 (1969).

[36] S. L. McCall, Phys. Rev. A 9, 1515 (1974).
[37] H. M. Gibbs, S. L. McCall, and T. N. C. Venkatesan, Phys. Rev. Lett. 36, 1135 (1976).
[38] A. G. Redfield, Phys. Rev. 98, 1787 (1955).
[39] E. T. Sleva, I. M. Xavier, and A. H. Zewail, J. Opt. Soc. Am. B 3, 483 (1986).
[40] J. A. Wheeler and R. P. Feynman, Rev. of Mod. Phys. 17, 157 (1945).

SOLITONS IN NUCLEAR PHYSICS - A REVIEW

Ernst F. Hefter

Springer-Verlag
Tiergartenstrasse 17
D-6900 Heidelberg 1
FRG

INTRODUCTION

Most of you will associate a certain picture with nuclear physics which should be precise enough for our purposes. But we should be more careful with the word soliton. The working definition which we are going to use is:
Solitons are analytical solutions of (nonlinear partial) differential equations that behave like particles. In particular they emerge unchanged from collisions with each other - except for a possible phase shift in their relative positions - regaining for large times their asymptotic initial amplitudes, speeds and shapes.

This "mathematical" definition is more restrictive than the "physical" definition often adopted in the discussion of physical phenomena occurring in plasma physics, field theory and also in the context of the presently rather fashionable Skyrmions. Physical solitons, solitary waves or solitonoids (as we would prefer to name them) are also well localized (analytical) solutions, however, they do experience permanent changes (dissipation) when colliding with each other.

Under the appropriate simplifying approximations most physical problems may be reduced to simple special cases that sometimes do even allow for analytical solutions. A typical example appearing almost everywhere is the harmonic oscillator which reflects nicely some prominent features of the underlying (linear) systems. Within the realm of nonlinear physics its counterpart seems to correspond to the mentioned analytical soliton solutions. A distinct feature of the latter is that they account in contrast to the harmonic oscillator still for the nonlinear characteristics of the original nonlinear system. When these soliton solutions were first discovered and eventually understood it was a real surprise to find out that rather complicated nonlinear partial differential equations do support such simple analytical solutions.

The first qualitative mention of the phenomenon of solitons

was made some 150 years ago but it was only in 1967 that it was possible to solve the respective equations analytically [1]. Only around 1980 there appeared the first papers mentioning this new concept in the context of nuclear physics. In the majority of these earlier contributions solitons or soliton-like formfactors appeared more as a by-product without really making use of their particularities. Such a negligence seems to be largely due to misinterpretations, personal prejudices, a lack of experience with this concept and also to the restriction to rather particular or academic cases with little practical relevance.

We shall not be able to do justice to all these earlier contributions. Hence, we shall concentrate on those that had a larger impact on the subsequent development or that are for other reasons of more general interest. But we shall at least make reference to the other papers so that the reader will have no problems in locating the (hopefully) complete biography. To our knowledge the so-called inverse mean field method (Imefim), see the fourth section, is the only one of these approaches that facilitated a direct comparison of computed with heuristic and experimental data. In addition to that it seems to be more comprehensive and thus more prospective than the other ones because it incorporates their prominent features. Thus it is natural that a good portion of space will be reserved to this approach.

To keep mathematical structures and general features of solitons separate from the particulars of the different models we give in the next section first a discussion of the famous nonlinear Korteweg - de Vries equation (KdV) and of its soliton solutions. This includes extensions of the KdV that account for the effects of external forces and dissipation. In the section to follow we discuss in some detail how traditional methods lead to solitons. In particular that refers to simplifying assumptions on the interaction and to hydrodynamical approaches. The inverse mean field method is dealt with in the fourth section containing also direct applications of this approach. The final part puts the different approaches in relation to each other and gives an outlook in regard to possible future developments.

A CLASSICAL EXAMPLE: THE KORTEWEG - DE VRIES EQUATION

The real world in which we are living and which we intend to describe is three-dimensional (x,y,z plus time t). However, for spherically symmetric systems and for cases with a preferred direction of motion the respective three-dimensional equations may (with the appropriate care and boundary conditions) be reduced to one-dimensional equations. So far the concept of solitons as such is well established for one-dimensional systems (x+t). More-dimensional solitons and related concepts are still the subject of intense research. To remain on safe grounds, it has therefore to be stated that the following considerations hold for spherically symmetric systems in three dimensions only with some reservations (this is even more pronounced for other symmetries). But without any limitations they are valid for (x+t)-dimensional systems.

The Galilean invariant nonlinear Korteweg - de Vries equation (KdV)

$$\partial_t U(x,t) = 6U\partial_x - M\partial_{xxx}U; \qquad M=\text{const}; \tag{1}$$

has originally been derived for water waves. [2] (With $U(x,t)$ standing for the amplitude of a wave and M for the dispersion constant of the liquid.) Its particular solution for a single lump (or soliton) is

$$U_1(x,t) = -U_{01}\text{sech}^2(\sqrt{U_{01}/2M}\cdot(x-v_1t)) \tag{2}$$

where v_1 denotes the speed of this soliton and U_{01} its amplitude.

Before proceeding further we should pause to make a few remarks in respect to some peculiarities of (1):
Sometimes it is given with an additional term of the form $\pm v_0\partial_x U$ where v_0 is a constant speed. This term does not contain any physics, a simple Galilean transformation may be used to eliminate it. A change of sign in front of the coefficient M in (1) would not change its character but only yield upwards facing solutions for U in contrast to the downwards facing one given by (2). Another point is that fluid dynamical boundary conditions imply (upon neglect of surface tension) a direct relation between speed and amplitude of such a soliton, a feature which does not hold for the most general form of (1).

For more than fifty years the KdV remained an obscure almost forgotten curiosity. Towards the end of the fifties it reemerged in the context of numerical simulations attempting to explore the time evolution of heat pulses propagating in nonlinear lattices. Computer experiments leading to particle-like (soliton) solutions of (1) spurred further research and efforts trying to solve (1) analytically. But traditional methods did not give rise to any progress. The crucial step towards the solution of such equations was done in 1967 with the discovery [3] of a formal relation between (1) and the linear eigenvalue problem

$$-M\partial_{xx}\Psi(x,t) + U(x,t)\Psi = E\Psi. \tag{3.1}$$

Wavefunction Ψ and (energy) eigenvalue E should be interpreted as one-solumn matrices

$$\Psi(x,t) = \begin{pmatrix} \psi_1(x,t) \\ \vdots \\ \psi_N(x,t) \end{pmatrix} \quad \text{and} \quad E = \begin{pmatrix} E_1 \\ \vdots \\ E_N \end{pmatrix} \quad \text{with } n=1,\ldots,N \tag{3.2}$$

with the (single-particle) wavefunctions ψ_n and the related (single-particle) eigenvalues E_n. Alternatively (3) could be cast into the form

$$-M\partial_{xx}\psi_n(x,t) + U(x,t)\psi_n = E_n\psi_n; \qquad n=1,\ldots,N. \tag{4}$$

Knowing the interrelation between (1) and (3), (4) the recently developed inverse methods could be employed to solve (1) by the aid of the well-known solutions of (3,4). The analytical soliton solutions of (1) turned out to be in a one-to-one correspondence with the bound states of the related eigenvalue problem. The general solution of (1) can only be obtained formally or numerically - not analytically. It is proportional to the squares of the real wavefunctions $\rho_n(x,t) = \psi_n^2(x,t)$. Retaining only the self-energies of the N bound states we restrict ourselves to the analytical N-soliton solution

$$U_N(x,t) = \sum_{n=1}^{N} U_{Nn}(x,t) = \sum_{n=1}^{N} -4\sqrt{-E_n M}\psi_n^2(x,t) = \sum_{n=1}^{N} -4\sqrt{-E_n M}\rho_n(x,t) \qquad (5)$$

of (1) where we used the above shorthand notation for the square of the wavefunction. Insertion of (5) into (3), (4) yields immediately a real cubic nonlinear Schrödinger-type equation.

The most peculiar feature of the above system of equations is that the eigenvalues E_n remain unchanged as long as the "potential" $U(x,t)$ evolves according to the KdV (1). This corresponds to purely elastic soliton collisions. Thus these objects serve readily as model for idealized particles. Their dynamics are rather well explored so that we refer the reader for details to the literature [4-7] and the references found therein.

It is of interest that the phase shift due to N colliding solitons may be given analytically [7]. For the simple case of two colliding solitons it reads

$$\sigma_{ij} = \sqrt{2M/U_{0i}} \cdot \ln\left[\left(\sqrt{U_{0j}} - \sqrt{U_{0i}}\right) / \left(\sqrt{U_{0j}} + \sqrt{U_{0i}}\right)\right] \qquad (6)$$

for the phase shift experienced by soliton i=1 in a collision with soliton j=2 (or vice versa if i and j are interchanged). These analytical results are slightly affected by the presence of dispersive waves [8]. In the cases of intermediate interest to us these (quantitative) modifications are, however, of minor interest.

A particularly useful feature of the system of equations (1) plus (4) is that it allows through the use of inverse methods for the solution of the Cauchy or initial value problem. In other words, it is possible to follow the time evolution of arbitrary initial disturbances at $t=0$ towards large times $t=\infty$. This is widely exploited in the analysis of fluid dynamical experiments and has been suggested to be of relevance to quantum mechanical equilibration processes [9].

The equations discussed so far correspond to conservative systems. The effect of external forces (or of a changing water depth if one takes a more pictorial application) may be accounted for by taking M to depend on this quantity, say E_p, or by adding to (1) a term proportional to $U(x,t)$. Then the soliton amplitude $U_{0i}=U_0$ is to depend on the strength of the external force (or on the depth of the sea bed h) [10,11]. To a good approximation that dependence is depicted by

$$U_0(E_p) = U_0(E_0) \cdot (1+eE_p)^{-3}; \qquad e=\text{const}; \qquad (7)$$

where E_0 (or h_0) denotes a constant reference energy (depth) and e a material constant characteristic for the medium under consideration.

Dissipation is introduced by adding to (1) a term proportional to $\partial_{xx}U(x,t)$. If this term acts for all times then it does gradually destroy the solitons (lead to a decay of the N bound states).

The full equation that emerges in such a way from (1) is the forced Korteweg - de Vries Burgers equation (FKdVB) which reduces for C=0=M to the nonlinear Burgers equation (BE), the linearized version of which is just the famous Fokker-Planck equation. Taking C=0, the (partially analytical) solutions of the KdVB range from the particle-like solutions like (2) for D=0 to shock-wave type solutions for M=0. The transitional region between the two is characterized by the ratio of M to D and may also exhibit particular shock-waves with additional oscillatory structures.

The above equations and relations may be derived and discussed without making mention of quantum mechanics, yet, they will turn out to be of utmost importance for the subsequent discussion of nuclear physics which is to follow below.

Interpreting (3,4) indeed as the quantum mechanical Schrödinger equation, the ψ and ψ_n would be the wavefunctions of the associated singel-particle states and their squares as used in (5) would be the associated densities. At least for the solution (5) we could most conveniently rewrite the forced Korteweg - de Vries Burgers equation

$$\partial_t U(x,t)=6U\partial_x U-M\partial_{xxx}U+D(x,t)\partial_{xx}U+C(t)U \tag{8}$$

in terms of these densities to obtain

$$\begin{aligned}\partial_t \Sigma K_j\rho_j(x,t)=&-M[24(\Sigma_j K_j\rho_j)(\partial_x \Sigma K_j\rho_j)+\partial_{xxx}\Sigma K_j\rho_j] \\ &+D(x,t)\partial_{xx}\Sigma K_j\rho_j+C(t)\Sigma K_j\rho_j\end{aligned} \tag{9}$$

where the notation $K_j^2=-E_j/M$ has been introduced. For a single bound state (9) simplifies to

$$\partial_t \rho(x,t)=-M[24\sqrt{-E/M}\cdot\partial_x\rho+\partial_{xxx}\rho]+D(x,t)\partial_{xx}\rho+C(t)\rho \tag{10}$$

if we suppress the index "1" everywhere. Expressing the last term in a slightly different way this is readily recognized as the Fokker-Planck equation - except for the nonlinear and third derivative terms in square brackets.

Thus we have completed our excursion into soliton physics and are now ready to discuss applications to nuclear physics.

TRADITIONAL METHODS AND SOLITONS

"Traditional methods" means in this context methods the usage of which corresponds to the long standing practice in nuclear physics. To our understanding the most prominent ones of them are the "potential approach" (i.e., one makes simplifying assumptions in respect to the macroscopic potential or to the microscopic interparticle forces) and fluid dynamical methods. Hence, we concentrate below very much on these two without going into much detail as far as the others are concerned. However, we shall attempt to mention their main points and to give the relevant references. As usual their negligence is to a large extent simply a matter of personal priorities and prejudices.

Potential Approach

Modern textbooks like [12] provide a good picture of contemporary nuclear physics and give also details in respect to potentials and interparticle forces that are useful and/or convenient. At least within the framework of the famous (time dependent) Hartree Fock approach (TD)HF the Skyrme forces are most frequently used. Essentially they are nothing more but a suitable combination of appropriately weighed δ-functions. The success of such simplified potentials and even more the fact that contact potentials yield often analytical solutions of the associated Schrödinger and (TD)HF problems lead to their use in model calculations.

In here we shall focus our attention on the (TD)HF, however, as discussed in [13], the Thomas-Fermi approach and the equations of nuclear fluid dynamics do lead to almost identical results when used in liaison with Skyrme forces.

The true nucleon-nucleon interaction V_{nn} may be written in the form

$$V_{nn}(x_i,x_j)=-V_0\delta(x_i-x_j)+\Delta V_{nn};\quad V_0>0;\ i,j=1,\ldots,N \tag{11}$$

which is correct as long as we are not forced to specify ΔV_{nn}. For a one-dimensional many-particle system with such an attractive contact potential with $\Delta V_{nn}=0$ there are analytical solutions in reach. Thus the many-boson as well as the many-fermion problems are both exactly soluble in the Schrödinger [14] as well as Hartree-Fock [15] theories. For the case of N bosons (or 2N fermions with spin 1/2 giving rise to bound dimers) the respective static HF equations are

$$-M\partial_{xx}\phi_n(x)-2V_0\Big[\sum_{m=1}^{N}\phi_m^2(x)\Big]\phi_n=E_n\phi_n \tag{12}$$

where the second term stands for the self-consistent mean field acting on the particle in state "n" and position "x", due to all other particles. The simplicity of this term, however, comes from the simplicity of the pair interaction (11) in which we ignored possible contributions from ΔV_{nn}. It is certainly of interest to note that with

$$\phi_n(x)=\sqrt{2K_nM/V_0}\cdot\psi_n(x)\quad\text{and } K_n^2=-E_n/M;\ M=\hbar^2/2m; \tag{13}$$

we may convert (12) into exactly the same real cubic nonlinear Schrödinger-type equation as arrived at by inserting (5) into (4) and demanding t=0. The significance of this coincidence will later on become more obvious.

Similar equations as (11,12) occur at different places in the literature [14-18], however, even in places in which mention of the similarity of the solitons of the KdV with the solutions of the HF equations has been made, this was just left as an additional remark without exploiting possible consequences. In cases in which further discussions of (11,12) were provided cases were considered that were not representative for real physical situations. To some extent that does apparently also apply to recent attempts trying to formulate a comprehensive model based on such simplified interactions [19]. To come really close to solitons a time dependent description has to be considered, e.g., TDHF as done in [20].

sidered, e.g., TDHF. In this spirit Nogami and Warke [20] use the δ-interaction of (11) with $\Delta V_{nn}=0$ to obtain the TDHF equations

$$-\frac{\hbar^2}{2m}\partial_{xx}\psi_n(x,t)-2V_0\sum_{m=1}^{N}|\psi_m|^2\psi_n=i\hbar\partial_t\psi_n \ . \tag{14}$$

(Actually they consider a spin and isospin saturated system with a slightly different strength of the contact potential, however, as illustrated by the discussion of (12,13) this does not affect the principle.) The ansatz

$$\psi_n(x,t)=\phi_n(x)\ \exp(-iE_n t/\hbar) \tag{15}$$

reduces (14) to the stationary HF equations (12). For n=1 the explicit form of $\phi_n=\phi$ is readily verified to be given by

$$\phi_n(x)=\sqrt{K/2}\ \text{sech}(Kx) \tag{16}$$

(corresponding also to the result for $\psi_n(x,0)$ implied by (2,5) with $t=0$, $M=\hbar^2/2m$ and $U_{01}=\hbar^2K^2/m$). Subsequently it is indicated [20] that (14) has also travelling wave solutions of the type

$$\psi(x,t)=\phi(x-vt)\ \exp[i(mv/\hbar)x-i\omega t] \tag{17}$$

where ϕ is defined by (16) and $\hbar\omega=E+mv^2/2$. This solution is taken to represent an isolated model nucleus moving with the constant speed "v". As mentioned in [13] and discussed in a by far more profound and detailed manner by Burt [21], (17) is essentially a one soliton solution of the TDHF equations.

Within TDHF one may then continue to study the time evolution of the colliding densities of model nuclei. In [20] that is done in a very nice investigation (a minor interpretational error is corrected in [6]). However, no attempt is made to develop the model to a stage facilitating a direct comparison with experiment.

Another approach, the <u>time dependent mean field S-matrix theory</u> [22] leads under the same conditions as mentioned above also to soliton solutions of the same type. They were only exploited to bring about a correspondence of this approach with TDHF, and to discuss the phases within these two models, i.e., again without attempting to exploit the soliton properties within a more comprehensive discussion or without coming close to evaluating data suitable for a comparison to experiment.

The Pöschl-Teller [23] potential

$$V(x)=-V_o\text{sech}^2(ax)\ ; \quad a=\text{const}; \tag{18}$$

leads to an analytical soluble Schrödinger equation. This holds even more for the particular case

$$V(x)=j(j+1)\cdot\text{sech}^2(x)\ . \tag{19}$$

Both of them have been discussed in the context of group theoretical approaches showing that bound and scattering states of certain one-dimensional potentials can be related to unitary representations of a certain group and its analytic continuation to a noncompact group, respectively. Such considerations are certainly of interest to other systems [25] and not just to atomic nuclei. At the moment mention of these approaches and of the potentials (18,19) appears to be unrelated to the preceding - except for their formfactors which we know already from (2). However, further below-within the context of the inverse mean field method-all of them will be seen to be interrelated.

Coming to the end of this section we find two remarks appropriate:
1) It is known [26] that the nonrelativistic limit of the ϕ^4 model coincides essentially with the above soliton supporting HF equations.
2) The interaction (11) - in particular with $\Delta V_{nn}=0$ corresponds to two-body forces. If three-body forces are considered then the respective HF equations exhibit a nonlinearity proportional to ϕ^5. According to the present understanding one should either use state independent n-body forces (leading to higher nonlinearities in (12)) or state dependent two-body forces. Usually the former is done as indicated above.

The Hydrodynamical Approach

The usage of classical hydrodynamical concepts within quantum mechanical physics has a long-standing practice. Within nuclear physics it dates back to the old days of the simple, yet, suprisingly successful liquid drop model. As should be expected, phenomenological approaches based on fluid dynamical considerations lead in general to nonlinear Schrödinger-type equations [27]. In the present context we are only interested in cases where the soliton concept has been involved. Thus the work of Khodel at al. [28], of Kartavenko [29] and of Weiner et al. [30] will be of particular interest to us.

In [28] it is attempted to explore the evolution of finite-amplitude waves in cold nuclear matter. This approach gives rise to the KdVB, i.e., (8) with C=0 and does not so much rely on classical hydrodynamics but is closer to theory of Fermi liquids. Hence, we shall not follow it in detail.

Nuclear fluid dynamics is at the roots of [29] and nonlinear soliton supporting Schrödinger equations emerge only after the insertion of Skyrme forces. Alternatively another route may be chosen to obtain the nonlinear KdV (1) as an equation for density disturbances. By its nature this approach combines some features of hydrodynamics with the potential picture of the preceding section.

Both of the above approaches are rather close to the hydrodynamical picture, but mixed with other considerations and methods. To-date none of them allowed for a direct qualitative and quantitative comparison with experiments. They give only rise to tentative qualitative results. Unfortunately the same holds for [30], however, it relies entirely on the hydrodynamical picture.

Weiner et al. [30] set out to describe the time evolution of a density elevation (hot spot) created within a nucleus involved in a high energy nucleus-nucleus collision. It is argued that it is reasonable to apply fluid dynamical concepts. Rather lengthy derivations lead then eventually to the KdV (1) as the appropriate evolution equation for such a density disturbance ρ'. The spatial changes in the nuclear density $\rho(r)$ that occur in the surface region necessitate the inclusion of a term proportional to ρ' and dissipation requires an additional $\partial_{xx}\rho'$-term. Hence, one ends up with the forced KdVB of (8) for ρ'.

We do not have the space to repeat the derivations of [30], suffice it to mention that the starting point is classical hydrodynamics with

$$\partial_t \rho(\vec{x},t) + \vec{\nabla}(\rho\vec{v}) = 0 \quad \text{and} \quad d_t \vec{v}(\vec{x},t) = \partial_t \vec{v} + (\vec{v}\cdot\vec{\nabla})\vec{v} = -\vec{\nabla}\Phi \tag{20}$$

where ρ is the local density and $\vec{v}$ the local velocity field. The potential Φ is taken to be

$$\Phi = \left[c_1\rho' + c_2\nabla^2\rho' + \dots\right]/\rho_0 \tag{21}$$

where ρ_0 is the (central) equilibrium density and $\rho' = \rho - \rho_0$ the devation of the local density from ρ_0; c_1, c_2 are abitrary constants. If only c_1 is retained in the expansion (21) then it is to be identified with the speed of sound in nuclear matter. Retaining also c_2 makes this interpretation a doubtful one without giving a specific hint for the physics contained in c_2.

The first two terms in (21) are retained, a coordinate transformation is performed and the preferred direction of motion is taken to coincide with the x-axis. In the subsequent transformation

$$\bar{x} = (x' - t')\sqrt{\varepsilon}; \qquad y = y; \qquad z = z; \qquad \bar{t} = t'\varepsilon^{3/2} \tag{22}$$

the arbitrary but small constant ε is introduced. Now the higher derivatives into all other directions but the preferred one are neglected. Expanding then the quantity

$$\rho' = \rho/\rho_0 = 1 + \varepsilon\rho_1' + \varepsilon^2\rho_2' + \dots \tag{23.1}$$

and the components of the velocity

$$v^i = \varepsilon v_1^i + \varepsilon^2 v_2^i + \dots \tag{23.2}$$

yields upon the collection of terms of the same power in ε eventually the nonlinear KdV in the density disturbance

$$\partial_{\bar{t}}\rho_1'(\bar{x},\bar{t}) + \frac{3}{2}\rho_1'\partial_{\bar{x}}\rho_1' + \sqrt{c_2/4c_1^3}\;\partial_{\bar{x}\bar{x}\bar{x}}\rho_1' = 0 \tag{24}$$

This procedure follows certainly the traditional train of thought but it involves quite a bit of algebra and leaves the interpretation of c_2 to be determined.

Further extensions are appropriate to take dissipation, changing density in the nuclear surface and flux going into other directions into account. In line with the other soliton models, this one does yield particular qualitative features (in respect to hot spots) but it does not yet facilitate a direct experimental verification via specific predictions or quantitative results. This unsatisfactory feature is shared with all other soliton models we are aware of - except for the one to be discussed below.

THE INVERSE MEAN FIELD METHOD (IMEFIM)

All the approaches mentioned so far evolved along rather well established lines (using simplified contact interactions, hydrodynamics, etc.). Now we would like to follow a different train of thought characterized by the application of inverse methods. This approach has been termed the inverse mean field method (Imefim) and claimed [31] to pave the way towards a comprehensive view at nuclear physics. It has been shown to yield qualitative and quantitative results which may not just be compared to other theoretical results but as well to experiment.

First we shall try to persuade you that it is not unreasonable to apply inverse methods to the single-particle equations

$$-\frac{\hbar^2}{2m}\,\partial_{xx}\psi_n(x)+U(x)\psi_n=E_n\psi_n;\qquad n=1,\ldots,N; \tag{25}$$

(which could be viewed as the radial equations for spherically symmetric three-dimensional systems).

Of course, it has to be stated that the traditional direct method of solution is extremely successful yielding a surprising wealth of results and predictions. It starts with the microscopic nucleon-nucleon interaction V_{nn} (or the macroscopic mean field U) as the imput. But, though V_{nn} is rather well known, it is neither derived from first principles nor from experiment. Its detailed form originates from phenomenology. - In here we would like to start with well established and precise experimental data (the total nuclear masses or binding energies $B_t(A)$) to derive from them the fundamental nucleon-nucleon interaction. We are not so much interested in producing precise results but rather in showing that it is possible to follow this philosophy. Consequent and consistent extension of the results will then be shown to lead us to a novel consistent time dependent description.

The programme we intend to follow for stationary problems is given by

$$B_t(A)\rightarrow E_n\rightarrow \psi_n,\ U(x)\rightarrow V_{nn}. \tag{26}$$

Time dependent potentials are also to be considered.

Since Imefim yields essentially the same nonlinear Schrödinger-type equation as Hartree-Fock (HF), the latter has been invoked to do the first step. This leads to [32]

$$B_t(A)=0.35\sum_i n_i E_i \tag{27}$$

where n_i is the degeneracy of the orbit "i". The quality of the resulting data is of the same order as the one of Hartree Fock calculations [32].

The second step in (26) is the solution of the Schrödinger problem (25) via inverse methods. This requires as input data the bound state energy eigenvalues E_n and information from the continuous part of the spectrum. The equation to be solved is the Gelfand-Levitan-Marchenko integral equation [33]. This can only be done formally and numerically. But in the particular case in which we retain only the self-interactions of the occupied ground states one obtains even the analytical solution

($U(x)$ and $\psi_n(x)$ are both real). They are given by (5) with $M=\hbar^2/2m$ and $t=0$.

The characteristic feature of the different contributions U_{Nn} to the total potential is that their amplitudes appear again in the arguments of their formfactors. This is most obvious in the case N=1, see (2). The same holds obviously also for the respective densities as demonstrated for N=1:

$$\rho(x)=\rho_0 \operatorname{sech}^2(2\rho_0 x) \quad \text{with} \quad \rho_0=\sqrt{-Em/2}\hbar \tag{28}$$

This feature has been used [34] to obtain approximate formulas for absolute $\langle r^2\rangle^{1/2}$ and relative nuclear radii

$$\delta\langle r^2\rangle=\langle r^2(A)\rangle-\langle r^2(A_0)\rangle=6.56\left[A^{2/3}/|B(A)|-A_0^{2/3}/|B(A_0)|\right] \tag{29}$$

which agree no worse with the experimental data than the findings of sophisticated approaches like HF. The complete analytical solution of the inverse problem compares also rather well with model potentials that have been used to test it [31, 35, 36].

So far we have treated the stationary problem (25) and solved it via inverse methods. Before going over to the last step of (26) let us now <u>look for a time dependent description</u>.

In the spirit of field theoretical approaches we consider (25) to be our basic equation, allow the potential to depend on the time t

$$-\frac{\hbar^2}{2m}\partial_{xx}\psi_n(x,t)+U(x,t)\psi_n=E_n\psi_n; \quad n=1,\dots,N \tag{30}$$

and look for an evolution in the mean field $U(x,t)$. First we look for its conservative version. The mathematics developed in the last ten years [37] tell us now that if $U_N(x,0)$ corresponds to (5) and if we consider a conservative system with $\partial_t E_n=0$ then one can <u>derive</u> the KdV (1) as the unique field equation for $U(x,t)$. An arbitrariness in the time scaling left open by the derivation is removed by the physical boundary conditions characterizing bound systems.

Further considerations attempting to cater for the physical effects of dissipation and external forces lead [31] to the extension of (1) given by (8). The pair of equations (30,1) is expected to contain the complete information on the system.

Applications related to the optical model are based on

$$V(r,A,E_p)=U(r,A)\cdot f(E_p) \tag{31}$$

where the function $f(E_p)$ contains the entire projectile energy dependence of V. Imefim allows for an exact evaluation of the respective volume integral over U_n divided by the number of nucleons involved which is to a good approximation represented by

$$J_U(A)=\frac{1}{A}\int d\vec{x}\; U(r,A)=-187\sqrt{|B(A)|}. \tag{32}$$

The dependence of (7) has been argued [11] to be associated with the energy dependence $f(E_p)$ of the optical potential for nucleon-nucleus scattering [11]. The volume change induced by varying E_p has been associated with the imaginary potential. Thus one arrives at the predictions (with $f(E_p)=(1+0.0027E_p)^{-3}$)

$$J_V(E_p)=-187\sqrt{|B(A)|}f(E_p);\ J_{W_V}(E_p)=J_V(0)[1-f(E_p)]/3 \tag{33}$$

for real and imaginary nuclear volume potentials.

In the literature uniqueness [40] of (5) and applicability and convergence of approximating potentials in terms of U_N have been discussed numerically [35,36] and formally [42]. As recalled in [37] and - as yet unpublished studies of ours - the results are quite encouraging. The situation for the nuclear physicist's favourite toy, a Saxon-Woods well, is as follows: The first few deepest lying energy eigenvalues determine to a surprisingly large extent the global characteristics of the potential well. Adding the further eigenvalues up to the Fermi level E_F results in a rather close approximation of U_N to the exact Saxon-Woods potential. The U_N-approximation displays the typical wiggles also observed in other numerical studies [35]. Adding still further information on this model system in terms of bound state energy eigenvalues that are located between E_F and $E=0$ gives rise to further improvements, however, they are rather small. The effect of other contributions than the self-interaction of the system is very small so that the restriction to the self-interactions is quite reasonable.

In the outline of our programme in (26) we promised to give also indications about the way in which one may eventually extract some information related to the nucleon-nucleon interaction. The gist of the method put forward in [42] is straightforward:
The quantities given in (33) describe for nucleon-nucleus scattering the average contribution of a single effective target nucleon to the total scattering potential. (Already because of the implicit assumption of a point particle for the projectile, (33) can only be an approximation.) Taking now (38) to correspond to the integration over such a single average nucleon with a soliton-formfactor like (2) - which appears in view of [38] to be quite justified - we arrive rather quickly at the explicit expressions for real V_{nn} and imaginary W_{nn} contributions to this effective nucleon-nucleon interaction

$$V^t_{nn}(A,E_p)=V_{nn}+iW_{nn}=-V_0\operatorname{sech}^2(r\sqrt{V_0 m}/\hbar)\cdot(f(E_p)+\tfrac{i}{3}[1-f(E_p)])$$

$$\text{with}\quad V_0=5.27\cdot|B(A)|\quad\text{and}\quad f(E_p)=(1+0.0027E_p)^{-3} \tag{34}$$

As appropriate for structure calculations this is a real quantity for $E_p=0$. Applied to collisions with $E_p\neq 0$, $V^t_{nn}(E_p)$ is complex with the imaginary part accounting as usual for the flux transferred from the elastic channel into the inelastic ones. From the functional form of this effective interaction and from its derivation it is obvious that it is no worse than most of the currently available central interactions. But it is also clear that more detailed considerations are to lead to quantitative modifications of (34) and that they are to introduce an additional state dependence into V^t_{nn}. Nevertheless, let us bear in mind that our main objective was simply to show that it is possible to complete the programme outlined in (26) - precise results certainly do demand a lot more efforts.

The combination of (1) with (25) allows to follow in detail the time evolution of arbitrary initial disturbances as well as the one of colliding solitons representing nucleons and/or nuclei. This is well in line with TDHF, compare the results of [5,6] and [20], and two-center shell model. Rather simple approximations to Imefim lead in a few lines [9] to the equations derived by Weiner et al. [30] in a rather laborious way. Apparently Imefim is quite useful for discussing the equilibration of quantum mechanical systems [9] (e.g., an individual distribution of the form (19) is to evolve into j solitons). The solitons of (1) have also been used to model nucleons and alpha particles with dynamics rather similar to TDHF but providing also via (6) the elastic scattering phase shifts for such collision [5,6].

In this account of Imefim it has - due to limitations in space - only been attempted to draw attention to some of its peculiarities. Hopefully that provides some indications that this is indeed a viable model which does deserve further attention. A really comprehensive review on it is to be given elsewhere.

OUTLOOK

Some details on a soliton supporting equation were given to go then over to sketch some features of soliton approaches to nuclear physics. The reasons for some of the mentioned connections between different models and studies will unfortunately remain unclear to the uninitiated reader and it will require some further efforts to make them transparent to him. This it not really satisfactory but within such a short presentation it was only possible to give the most important key words and references facilitating independent studies.

To our understanding the soliton concept provides us with a viable and efficient tool. Most of the existing models are very interesting and instructive but they do unfortunately not pursue matters up to a level were direct confrontation with experiment is possible. Imefim seems to be the only exception and it does also provide a comprehensive view of the subject so that its further development is expected to yield a nice and consistent description of rather different phenomena in nuclear physics and possibly also in other branches of nonrelativistic quantum mechanics.

ACKNOWLEDGEMENTS

For the time spent on discussions and collaborations related to nonlinear dynamics and solitons I am very grateful to K.A. Gridnev, V.G. Kartavenko, M. de Llano, I.A. Mitropolsky, B. Fuchsteiner and R.M. Weiner. For their encouragement and helpful comments I am very much obliged to H. Kümmel and G. Rowlands.

REFERENCES

1. R.K. Bullough and P.J. Caudrey (eds.), "Solitons" Springer, Berlin Heidelberg New York (1980);
G. Eilenberger, "Solitons. Mathematical Methods for Physicists" Springer, Berlin Heidelberg New York (1983).
2. D.J. Korteweg and G. de Vries, Phil. Mag. Ser. 5 39:422 (1895).
3. C.S. Gardener, J.M. Greene, M.D. Kruskal, and R.M. Miura, Phys. Rev. Lett. 19:1095 (1967).
4. P. Lax, Commun. Pure and Appl. Math. 27:97 (1968).
5. E.F. Hefter, Z. Physik C14:87 (1982);
E.F. Hefter, Nuovo Cim. 59A:275 (1980).
6. E.F. Hefter and K.A. Gridnev, Prog. Theor. Phys. 72:549 (1984).
7. M. Wadati and M. Toda, J. Phys. Soc. Japan 32:1403 (1972);
E.F. Hefter, Prog. Theor. Phys. 69:329 (1983).
8. P. Schuur, "Asymptotic Analysis of Soliton Problems" Springer, Berlin Heidelberg New York (1986).
9. E.F. Hefter, Lett. Nuovo Cim. 32:9 (1981);
E.F. Hefter, in "Local Equilibrium in Strong Interaction Physics" D.K. Scott and R.M. Weiner, eds., World Scientific, Singapore (1985);
E.F. Hefter and V.G. Kartavenko, Rapid Commun. JINR 3(23):29 (1987).
10. J.W. Miles, J. Fluid Mech. 91:181 (1979);
D.J. Kaup and A.C. Newell, Proc. R. Soc. London A361:413 (1978).
11. E.F. Hefter and K.A. Gridnev, Z. Naturforsch. 38a:813 (1983).
12. P. Ring and P. Schuck, "The Nuclear Many-Body Problem" Springer, Berlin Heidelberg New York (1980).
13. E.F. Hefter, Nuovo Cim. 89A:217 (1985).
14. J.B. McGuire, J. Math. Phys. 5:622 (1964);
M. Gaudin, Phys. Lett. 24A:55 (1967).
15. F. Calogero and A. Degasperis, Phys. Rev. A11:265 (1975);
E.H. Lieb and M. de Llano, J. Math. Phys. 19:860 (1978).
16. L. Dolan, Phys. Rev. D13:528 (1976).
17. B. Yoon and J.W. Negele, Phys. Rev. A16:145 (1977).
18. M. de Llano, Nucl. Phys. A317:183 (1979).
19. H. Kuratsuji, Prog. Theor. Phys. 74:433 (1985);
H. Horiuchi, Prog. Theor. Phys. 74:66 (1985).
20. Y. Nogami and C.S. Warke, Phys. Lett. 59A:251 (1976);
Y. Nogami and C.S. Warke, Phys. Rev. C17:1905 (1978).
21. P.B. Burt, Phys. Lett. 71A:19 (1979);
P.B. Burt, "Quantum Mechanics of Nonlinear Waves" Harwood, Chur London New York (1981).
22. H. Reinhardt, Phys. Lett. 121B:9 (1983);
H. Reinhardt, in: "Hartree-Fock and Beyond" K. Goeke and P.G. Reinhard, eds., Springer, Berlin Heidelberg New York (1982).
23. G. Pöschl and E. Teller, Z. Physik 83:143 (1933).
24. A. Frank and K.B. Wolf, Phys. Rev. Lett. 52:1737 (1984);
Y. Alhassid, F. Gürsey, and I. Iachello, Ann. Phys. (N.Y.) 148:346 (1983).
25. R.D. Levine and C.E. Wulfman, Chem. Phys. Lett. 60:372 (1979); C.E. Wulfman, J. Phys. Chem. 90:2264 (1986).
26. D.K. Campbell, in: "Nuclear Physics with Heavy Ions and Mesons" R. Balian et al., eds., North Holland, Amsterdam (1978).
27. J.J. Griffin and K.-K. Kan, Rev. Mod. Phys. 48:467 (1976);
K. Mikulas, K.A. Gridnev, E.F. Hefter, V.M. Semjonov, and

V.B. Subbotin, Nuovo Cim. 93A:135 (1986);
E.A. Spiegel, Physica D1:236 (1980); and references.

28. V.A. Khodel, N.N. Kurilkin, and I.N. Mishustin, Phys. Lett. 90B:37 (1980);
V.N. Kurilkin, I.N. Mishustin, and V.A. Khodel, Pis'ma Zh. Eksp. Teor. Fiz. 30:463 (1979);
Yad. Fiz. 32:1249 (1980) and 36:95 (1982).
29. V.G. Kartavenko, Yad. Fiz. 40:377 (1984);
V.G. Kartavenko, Dubna-Report P4-83-461 (1983).
30. G.N. Fowler, S. Raha, N. Stelte, and R.M. Weiner, Phys. Lett. 115B:286 (1982);
S. Raha and R.M. Weiner, Phys. Rev. Lett. 50:407 (1983);
S. Raha, K. Wehrberger, and R.M. Weiner, Nucl. Phys. A433:427 (1984);
E.F. Hefter, S. Raha, and R.M. Weiner, Phys. Rev. C32:2201 (1985).
31. E.F. Hefter, J. de Phys. 45,C6:67 (1984);
E.F. Hefter, Acta Phys. Pol. A65:377 (1984);
E.F. Hefter, M. de Llano, and I.A. Mitropolsky, Anales de Fisica (Madrid) 81A:185 (1985).
32. E.F. Hefter and I.A. Mitropolsky, LNPI-Report-860, Leningrad Nuclear Physics Institute, Gatchina (1983);
Nuovo Cim. 95A:63 (1986).
33. K. Chadan and P.C. Sabatier, "Inverse Problems in Quantum Scattering Theory", Springer, Berlin Heidelberg New York (1977);
B.N. Zakhariev and A.A. Suzko, "Potentials in Quantum Scattering - Direct and Inverse Problems" Springer, Berlin Heidelberg New York (in preparation).
34. E.F. Hefter, M. de Llano, and I.A. Mitropolsky, Phys. Rev. C30:2042 (1984); E.F. Hefter, Phys. Rev. A32:1205 (1985).
35. C. Quigg and J.L. Rosner, Phys. Rev. 23D:2625 (1981);
A. Asthana and A.N. Kamal, Z. Physik C19:37 (1983).
36. V.G. Kartavenko and P. Mädler, Dubna-Report-P4-87-156 (1987)
37. B. Fuchssteiner, (Univ. Paderborn, FRG) priv. commun. and to be published.
38. E.F. Hefter, Phys. Lett. 141B:5 (1984).
39. E.F. Hefter, Phys. Rev. 34C:1588 (1986).
40. I. Kay and H.E. Moses, J. Appl. Phys. 27:1503 (1956).
41. I. Sabba-Stefanescu, J. Math. Phys. 23:2190 (1982).
42. E.F. Hefter, Verhandl. DPG (VI)21:490 (1986).

PROBABILISTIC NEURAL NETWORKS: IN OR OUT OF EQUILIBRIUM?

John W. Clark

Physics Division, Argonne National Laboratory

Argonne, Illinois 60439, USA

INTRODUCTION

The long-term behavior of neural networks following exposure to external stimuli is central to attempts at modeling brain activity and to the design of physical systems imitating biological mechanisms of memory storage and recall. A set of stimuli all eliciting the same final operating condition of the network defines an equivalence class of experiences which are said to be associated with the same stored memory. A given network may exhibit one, several, or many such attractors, while a given attractor may correspond to a fixed point, a terminal cycle, multiperiodic motion, or chaos. If the dynamical law by which the system updates its state is probabilistic rather than deterministic, one considers an ensemble rather than a particular system. The initial preparation of the system attendant to the imposition of a temporary stimulus is reflected in the specification of an initial probability distribution $\{p_i(0)\}$ over microscopic system states i. One is then concerned with the behavior of the state occupation probabilities $p_i(t)$ at asymptotically large times t.

Here we shall examine an especially simple class of probabilistic neural network models, simple in the senses that (a) the large-t occupation probabilities are steady, implying a fixed-point attractor, and (b) this final condition is unique, i.e. independent of initial preparation, for given single-neuron properties and interneuronic couplings. This class of models includes that of Little,[1] which may be taken as the prototype. Thus, to be definite, we consider a microscopic dynamics in which time is quantized in units of a universal delay τ and, at each time-step, each neuron makes a stochastic choice of whether or not to fire, biased by the signal it receives from neurons which fired at time $t-\tau$. More broadly, the models in question have a simple mathematical characterization in terms of aperiodic, irreducible, homogeneous, finite Markov chains, and hence their behavior is in principle well known. Perhaps their most essential (or least trivial) feature is that any microscopic state can be reached from any other microscopic state with finite probability in a finite number of time steps. Another important property of these systems is that if it is possible to go from state j to state i in one time-step, the reverse transition is also allowed with finite probability.

To make the discussion concrete, we shall work in terms of Little's model. However, it must be emphasized that many of our results are much more general.

SPECIFICATION OF THE MODEL

1. The network consists of N (formal) neurons labeled $\nu = 1, \ldots N$. Each neuron ν is assigned a state variable σ_ν which takes on the value +1 if ν is firing and −1 if it is not (representing the "all-or-none" character of the action potential).

2. The neurons are permitted to fire only at instants belonging to a discrete set $\{0, \tau, 2\tau, \ldots n\tau, \ldots\}$, where τ is some elementary time interval. (The time quantum τ might be taken as the absolute refractory period of a neuron or the delay time for signal transmission from presynaptic to postsynaptic neuron.)

3. The stimulus felt by neuron ν at time $t = n\tau$, n integral, due to synaptic input(s) from neuron μ, is written $V_{\nu\mu}\pi_\mu(t-\tau)$, where $\pi_\mu = (\sigma_\mu + 1)/2$. The coupling matrix $(V_{\nu\mu})$, with $V_{\nu\mu}$ a real number, embodies the pattern of synaptic connections between neurons and the efficacies of these junctions in information transfer. A positive [negative] value of $V_{\nu\mu}$ means that the synapse (or generally the set of synapses) from μ onto ν is excitatory [inhibitory] in effect, and $V_{\nu\mu} = 0$ if there is no synapse from μ onto ν. In general we must consider $V_{\mu\nu} \neq V_{\nu\mu}$ if we wish to simulate biological nerve nets; i.e., the neuronic interactions are ordinarily not symmetrical, since (for example) a synapse of μ onto ν does not imply the presence of a reciprocal synapse of ν onto μ. This clashes with our experience with ordinary physical systems, where Newton's third law holds.

4. After an initial excitation which turns on a selected subset of neurons, or initiates the probability distribution over states of the system, the network is isolated from any external stimuli. Thus we study the *autonomous* operation of the neural net. Otherwise an additional *control term* U_ν must be added to the right side of Eq. (1) below.

5. The decision of a neuron ν whether or not to fire at time $t = n\tau$ is *spontaneous* (or probabilistic). The firing function

$$F_\nu(t) = \sum_\mu V_{\nu\mu}\pi_\mu(t-\tau) - \theta_\nu , \tag{1}$$

of the neuron at t measures the amount by which its algebraic stimulus at that time exceeds its threshold θ_ν. Various stochastic effects inherent in the electrochemical synaptic transduction mechanism[2–5] may be simulated by assuming that even if $F_\nu < 0$, there is a finite probability that neuron ν will fire, and even if $F_\nu \geq 0$, there is a finite probability that it will not. For large positive [negative] values of F_ν, the firing probability should approach unity [zero], whereas for F_ν near zero the response should be less predictable. Following Little, we make the specific ansatz

$$\rho_\nu(\sigma_\nu(t)) = \left\{1 + \exp[-\beta_\nu\sigma_\nu(t)F_\nu(t)]\right\}^{-1} \tag{2}$$

for the conditional firing probability. Here $\rho_\nu(+1)$ [resp. $\rho_\nu(-1)$] denotes the probability that ν will fire [fail to fire] at the specified time t, given the states $\sigma_\mu(t-\tau)$ or $\pi_\mu(t-\tau)$ of the neurons one time-step earlier and hence the firing function $F_\nu(t)$. The nonnegative parameter β_ν^{-1} is a measure of the noisy character of signal transmission to neuron ν. We note that $\rho_\nu(-1) = 1 - \rho_\nu(+1)$ coincides, *formally*, with a Fermi distribution at temperature $k_B T = \beta_\nu^{-1}$. In the limiting case $\beta_\nu^{-1} = 0$ for all ν (zero temperature), the dynamics becomes deterministic, if we exclude the measure-zero case that the firing function is exactly zero. An effort to justify the choice (2) on the basis of the "quantal" nature of information transfer at synapses (whereby excitation is carried by packets of neurotransmitter molecules[2]) has been made in Ref. 3.

6. The *state of the network* as a whole is represented at time t by the set $\{\sigma_\nu(t), \nu=1,\ldots N\}$ of individual neuronal state values, i.e. by the *firing pattern* at that time. There are exactly 2^N distinct states of the model, which will be labeled $i=1,2,\ldots 2^N$. The analogy to a spin system is apparent. We note that the conditional firing probabilities ρ_ν at time t, given by (2) with (1), are not correlated with one another, as they do not depend on the σ_η realized by the other neurons at that time, but instead depend only on the state j occupied by the system one time-step earlier. Accordingly, the probability per unit time that the net will undergo a transition from state j at time $t-\tau$ to state i at time t may be written as the product

$$T_{ij} = \tau^{-1}Q_{ij} = \tau^{-1}\prod_{\nu=1}^{N}\left[1+\exp\left\{-\beta_\nu\sigma_\nu^{(i)}\left[\sum_\mu V_{\nu\mu}\pi_\mu^{(j)}-\theta_\nu\right]\right\}\right]^{-1} . \tag{3}$$

For extensive discussion of the properties of this model, see Refs. 1,3,5-8.

STATISTICAL TIME DEVELOPMENT OF THE NETWORK

With the model specified as above, the time development of the probability distribution over system states, i.e. the time development of the state occupation probabilities $p_i(t)$, is given rigorously by

$$p_i(t) = \sum_j Q_{ij}\, p_j(t-\tau) \quad , \tag{4}$$

in terms of the matrix (Q_{ij}) defined by Eq. (3). This dynamics defines a Markov chain with stochastic matrix (Q_{ij}), indeed, a finite, homogeneous, irreducible, aperiodic Markov chain.[9] Alternatively, but approximately, we can describe the statistical development of the system by a master equation

$$\frac{dp_i(t)}{dt} = \sum_j [T_{ij}\, p_j(t) - T_{ji}\, p_i(t)] = \sum_j W_{ij}\, p_j(t) \quad , \tag{5}$$

where $W_{ij} = T_{ij} - \delta_{ij}\sum_k T_{ki}$, the transition rate per unit time, T_{ij}, being specified by Eq. (3). This approximate equation of motion becomes accurate as the time delay τ becomes much smaller than the time scale of appreciable change in the $p_i(t)$. Hence the master equation will provide an acceptable description in the asymptotic regime where the solution $\mathbf{p}(t) = \{p_i(t)\}$ of either (4) or (5) approaches a constant. The steady-state solutions $\hat{\mathbf{p}}$ of (4) and (5) are obviously identical. We may further remark that $\Sigma_i W_{ij} = 0$ by the definition of the matrix (W_{ij}) and hence $\det(W_{ij}) = 0$. The latter property implies $\mathrm{rank}(W_{ij}) \le 2^N - 1$, so that at least one steady-state solution of (5) *does* exist. It also implies that $\Sigma_i p_i(t)$ is a constant of motion, as may be verified by summing the master equation on i. By virtue of the linearity of (5), this constant of motion may be normalized to unity provided the p_i are bounded and nonnegative.

The analysis of Eq. (5) is facilitated by a graphical representation in which each state of the system is represented by a point or vertex (thus, 2^N points for an N-neuron system) and each nonzero transition rate T_{ij}, from state j to state i, is represented by a line or edge bearing an arrow which points from j to i. Drawing all state vertices and all allowed edges, we arrive at the *basic graph* G of the system. In his formal theory of linear master equations describing the effects of transitions among a finite set of states, Schnakenberg[10] was able to derive some powerful results which are directly applicable to the class of neural networks under consideration. It is safe to assume that if there is a finite transition probability in one direction, then there is also a finite transition probability in the reverse sense, i.e., $T_{ij} > 0$ implies $T_{ji} > 0$. We are concerned with the existence of a steady-state solution $\hat{\mathbf{p}}$ of Eq. (5) *within the physical region* circumscribed by $\Sigma_i p_i = 1$ and $0 \le p_i \le 1$, all i, conditions which must be met if the p_i are to be probabilities. One of Schnakenberg's theorems establishes that if the basic graph G is *connected* (meaning that it is possible to get from any point in the graph to any other point by an appropriate sequence of allowed transitions), then such a solution is guaranteed to exist, with $0 < p_i < 1$, all i. This steady solution is known as the *Kirchhoff* solution. An elegant graph-theoretic expression for it is provided in Refs. 10,5. A second important result is that this solution is *unique* in the physical region: there is only one steady solution of (5) such that the p_i may be interpreted as state-occupation probabilities.

Turning to the nature of time-dependent solutions of the master equation, two further theorems of Schnakenberg are relevant. The first asserts that if the initial set of p_i is in the physical region, the solution of (5) *remains* in the physical domain, provided that the basic graph G is connected. The second result concerns approach to the Kirchhoff steady state: Again assuming a connected G, the matrix (W_{ij}) has a *nondegenerate* maximum eigenvalue 0, and all other eigenvalues have *positive* real parts. The nondegeneracy of the 0 eigenvalue implies that in fact $\mathrm{rank}(W_{ij}) = 2^N - 1$. (N.B. the inequality sign can be removed from the above bound on the rank of the W matrix.) The nature of the other eigenvalues implies that any time-dependent solution initiated at any point in the physical domain *eventually relaxes exponentially to the Kirchhoff steady state*.

Thus we know that the Kirchhoff steady state is unique in the physical region, and that, as an equilibrium point, it is absolutely stable with respect to all physical solutions of the master equation. When we think in terms of ordinary thermodynamics, the question naturally arises: does the Kirchhoff solution correspond to *thermodynamic* equilibrium? In order to address this issue, we must decide what thermodynamic equilibrium is to mean in this rather abstract context, where the brain (as represented by our neural network model) is not just regarded as "a piece of meat," but rather as an information-processing system with its own special brand of dynamics, operating on a plane distinct from that of mundane physical systems. Evidently, we want to identify thermodynamic equilibrium with a condition of *detailed balance* in the master equation, where each term in the sum over j on the right in (5) vanishes independently of the others,

$$T_{ij}\hat{p}_j - T_{ji}\hat{p}_i = 0 \quad , \qquad \text{for all } ij \quad . \tag{6}$$

In principle, there are of course many ways in which the right-hand side of (5) could be zero, due to cancellations between terms with different j values. One therefore suspects that rather special choices of neuronic couplings and/or neuronal parameters will be required, if the Kirchhoff solution is to represent thermodynamic equilibrium. It will be shown below that this suspicion is correct.

ENTROPY PRODUCTION IN THE NEURAL NET

Pursuing an analogy with an open but homogeneous system of 2^N reacting chemical species, the rate of entropy production in the neural network takes the bilinear form

$$\mathbb{P} = \frac{1}{2}\sum_{ij} J_{ij}A_{ij} \tag{7}$$

in terms of *generalized thermodynamic fluxes* $J_{ij} = T_{ij}p_j - T_{ji}p_i$ and *generalized thermodynamic forces* (or affinities) $A_{ij} = \ln[T_{ij}p_j/T_{ji}p_i]$. (For detailed argumentation, see Refs. 10,5.) The entropy of the system itself is given by

$$S = -\sum_i p_i \ln p_i \quad , \tag{8}$$

i.e., by the information entropy. The entropy production rate (7) splits into two components, $\mathbb{P} = \mathbb{P}_1 + \mathbb{P}_2$, where $\mathbb{P}_1$ is just the rate of change dS/dt of the entropy assigned to the system, and the remainder,

$$\mathbb{P}_2 = \frac{1}{2}\sum_{ij} J_{ij} \ln [T_{ij}/T_{ji}] \quad , \tag{9}$$

is ascribed to the action on the system of external thermodynamic forces which (potentially) keep it from reaching thermodynamic equilibrium. In the case of the neural network, which we suppose to be in autonomous operation, these external forces are a convenient fiction of the formalism; in fact they are inherent in the internal microscopic dynamics of the model and are ultimately expressible in terms of the transition probabilities T_{ij}. We note that in general neither $\mathbb{P}_1$ nor $\mathbb{P}_2$ is required to be nonnegative; however, $\mathbb{P} = \mathbb{P}_1 + \mathbb{P}_2 \geq 0$ applies rigorously. Obviously, $\mathbb{P}$ vanishes in a steady state corresponding to thermodynamic equilibrium.

It should be pointed out explicitly that three levels of description are involved in our analysis. At the bottom level we have the *microscopic dynamics*, where attention is paid to the changing firing states σ_ν of the individual neurons. The next level up is that of *statistical mechanics*, where we follow the time development of the state occupation probabilities p_i. The third or highest level is the analog of thermodynamics, generally *nonequilibrium thermodynamics*; at this level the description is framed in terms of certain *macroscopic forces and fluxes* (analogous, say, to temperature gradients and concomitant heat flows). A bridge from the first to the second level has been provided by the master equation (5) (or by the dynamics (4) of the Markov chain defined by the network model). To link the second and third levels, we shall equate the relevant expressions for the rate of entropy production of the network in the asymptotic steady state, denoted $\hat{\mathbb{P}}$. We already have an expression for $\hat{\mathbb{P}}$ at the level of statistical mechanics: just substitute the Kirchhoff solution $\hat{\mathbf{p}}$ into (7).

MACROSCOPIC FORCES AND FLUXES

At the macroscopic level of nonequilibrium thermodynamics, the steady-state entropy production must assume the bilinear form[11]

$$\hat{\mathbb{P}} = \sum_{f=1}^{F} J(\mathbf{C}_f) A(\mathbf{C}_f) \quad , \tag{10}$$

in a set of $2F$ *macroscopic forces and fluxes*, a force $A(\mathbf{C}_f)$ and a flux $J(\mathbf{C}_f)$ being present for each member of a *fundamental set of cycles* of the basic graph G. A fundamental set of cycles is characterized by the property that an arbitrary cycle (i.e., closed loop of oriented edges) occurring in G may be expanded in such a set, according to rules similar to those for expanding an arbitrary vector in a basis for the linear vector space in which it resides. The required brevity precludes a development of the graph-theoretic aspects of expression (10), which is carried through at length in Refs. 10,5. For our purposes it is sufficient to realize that upon setting this macroscopic version of $\hat{\mathbb{P}}$ equal to the statistical version constructed above, one may determine the macroscopic variables $A(\mathbf{C}_f)$, $J(\mathbf{C}_f)$ in terms of the microscopic transition rates T_{ij} and the steady-state values $\hat{p}_i$ of the statistical variables p_i. It is sobering to note that if τ is interpreted as a uniform absolute refractory period of the neurons of the system, then the number of thermodynamic variables is $2\times[2^N(2^N-1)/2 - 2^N + 1]$, a huge number for any sizeable N. (For the brain, N would be at least of order 10^{10}.)

We quote three beautiful results from the graph-theoretic analysis:

(A) The condition $A(\mathbf{C}_f) = 0$ for all f holds if and only if $J(\mathbf{C}_f) = 0$ for all f; either is a necessary and sufficient condition for detailed balance (6) and and accordingly for the Kirchhoff solution to correspond to thermodynamic equilibrium.

(B) The forces $A(\mathbf{C}_f)$ (and hence the forces $A(\mathbf{C})$ associated with *arbitrary* cycles C) are independent of the state-occupation probabilities $\hat{p}_i$.

(C) The force $A(\mathbf{C})$ around an arbitrary cycle $\mathbf{C}$ may be computed as the log of the ratio of the product of transition rates going in the forward direction around $\mathbf{C}$ to the product of transition rates going in the reversed direction around $\mathbf{C}$. (Thus, in thermodynamic equilibrium, "forward" and "backward" pressures balance and $A(\mathbf{C}) = 0$.) For a cycle $j \to k \to i \to j$, one has in particular

$$A(\mathbf{C}) = \ln \frac{T_{ji} T_{ik} T_{kj}}{T_{jk} T_{ki} T_{ij}} \tag{11}$$

The findings (A)-(C) permit an efficient explication of the conditions on neuronic couplings which permit the attainment of thermodynamic equilibrium.

CONDITIONS FOR THERMODYNAMIC EQUILIBRIUM

In this section we state and prove our principal results.

Theorem I. The symmetry properties

$$\beta_\nu V_{\nu\mu} = \beta_\mu V_{\mu\nu} \ , \qquad \mu,\nu = 1, \ldots N \ , \tag{12}$$

to be obeyed by the spontaneity parameters and off-diagonal couplings, provide necessary and sufficient conditions for the steady state of a stochastic N-neuron network specified by 1.-6. to correspond to thermodynamic equilibrium, i.e. satisfy Eq. (6), independently of neuronal thresholds and diagonal couplings.

Subject to the assumption that all the β_ν coincide, this criterion has already been asserted by Peretto.[8] The proof outlined below rests on the fact [Result (A) above] that thermodynamic equilibrium is equivalent to the vanishing of $A(\mathbf{C}_f)$ for all members $f = 1, \ldots F$ of *some* fundamental set of cycles (any will do). In exploiting this result, Property (B) is obviously crucial. Rule (C) is invoked to express the balance condition as an equality of the product of transition rates going forward around an arbitrary fundamental cycle (left-hand side) and the product of transition rates going backward around the cycle (right-hand side). An important feature of this expression is that for every s factor appearing on the left, there is an s factor on the right with exactly the same argument (apart from the sign of the argument), and vice versa; this property is assured by the fact that the same *initial* states appear, each once and only once, on the left and on the right. In the detailed manipulations, extensive use is made of the identity

$$s[\pm x] = (1 + e^{\pm x})^{-1} = \frac{1}{2} e^{\mp x/2} \operatorname{sech}(x/2) \tag{13}$$

and the additivity of exponents.

(i) That the relevant conditions on the network parameters must be independent of the thresholds θ_ν may be traced to the fact that the same system states appear as *final* states on the left and right in the generic balance equation. For then a given θ_ν enters with the same number of minus signs on the left and right, and the same number of plus signs. Hence, in applying (13) and combining exponents, the θ_ν's all cancel out.

(ii) That the conditions we seek must be independent of the diagonal couplings may be seen as follows. Suppose neuron ν remains excited for n successive state points around an arbitrary cycle, just before which and just after which it is not excited. Then in using (13) and adding exponents, $\beta_\nu V_{\nu\nu}$ will enter $n-1$ times with a minus sign and once with a plus sign when going *either* forward *or* backward around the cycle. Transitions from states in which ν is not excited obviously contribute no terms in $\beta_\nu V_{\nu\nu}$. Extension of the argument to cycles in which ν is alternately excited for n_1 successive vertices, de-excited for the next n_2, then excited for the next n_3, etc., is trivial.

(iii) To affirm the necessity and sufficiency of the symmetry conditions (12), it is convenient (as well as permissible) to choose a very simple set of fundamental cycles, each member of which contains the dead state 0 (i.e., the state with all neurons "off") and only three edges. (That such a set can always be formed is easily seen in terms of elementary graph theory[12,10,5]: just take a maximal tree with the 2^N-1 edges $0i$, $i \neq 0$.) Thus we may restrict our considerations to the generic cycle $0ji$, a triangle with vertices 0, j, and i, where i and j are any states that (directly) communicate with one another, 0, j, and i being distinct. Accordingly, thermodynamic equilibrium hinges on the balance condition

$$T_{0i}T_{ij}T_{j0} = T_{0j}T_{ji}T_{i0} \quad . \tag{14}$$

Appealing to results (i) and (ii) above, we may make the further simplification of setting all thresholds θ_ν and diagonal couplings $V_{\nu\nu}$ equal to zero, without sacrifice of generality. With zero thresholds, the transition probabilities T_{j0} and T_{i0} with 0 as initial state both reduce to 1. The next step is to pair off the remaining T elements on the left and right sides of (14) having the same initial state (i.e., T_{0i} with T_{ji} and T_{ij} with T_{0j}) and cancel the common s factors they contain. Factors corresponding to neurons "off" in state j [in state i] are identical in T_{0i} and T_{ji} [in T_{ij} and T_{0j}]. But we may go further: Let $\nu_1 \ldots \nu_n$ label the neurons which are *on* in state i but *off* in state j; and let $\mu_1 \ldots \mu_m$ denote the neurons which are *on* in j but *off* in i. Then the exponential factors involving *only* the "common" neurons, $\eta \neq \nu_1 \ldots \nu_n, \mu_1 \ldots \mu_m$, are clearly the same on the left and right of (14) and hence may be removed. The ensuing balance condition is best reduced--having again appealed to (13) and the feature that the s-function arguments occurring on the left of (14) are repeated on the right (apart from signs)--to a condition on exponents of the form $\pm\beta_\xi V_{\xi\omega}$. The result is

$$\sum_{p=1}^{n} \beta_{\nu_p} \sum_{t=1}^{m} V_{\nu_p\mu_t} = \sum_{t=1}^{m} \beta_{\mu_t} \sum_{p=1}^{n} V_{\mu_t\nu_p} \quad . \tag{15}$$

Consider now the special case $n=1$, $m=1$. The neurons ν_1 and μ_1 are distinct but otherwise completely arbitrary; thus they may be relabeled ν and μ, respectively. For this case (15) just reproduces (12), and the symmetry condition stated in Theorem I is indeed *necessary*. To show that it is likewise *sufficient*, we return to the general case and verify straightaway with a mere renaming of summation indices that (12) implies (15).

In arriving at (15) we are assuming that neither of the sets $\{\nu_1 \ldots \nu_n\}$, $\{\mu_1 \ldots \mu_m\}$, is null. Suppose instead that the ν set is empty; in that case (and similarly when the μ set is empty), the necessary condition (15) is to be replaced by

$$\sum_{\eta} \beta_\eta \sum_{t=1}^{m} V_{\eta\mu_t} = \sum_{t=1}^{m} \beta_{\mu_t} \sum_{\eta} V_{\mu_t\eta} \quad . \tag{15'}$$

But again the condition (12) derived from (15) guarantees that this equality will hold.

Remark. It should be emphasized that in establishing Theorem I we have relied on certain special properties of the conditional firing probability defined by (2). This probability may be conveniently expressed as

$$p_\nu(\sigma_\nu) = s(-\beta_\nu \sigma_\nu F_\nu), \tag{16}$$

wherein the special choice of (13) has been made for the function s. Other choices for s of sigmoid character could have been made in setting up the network model, without affecting the formal development which preceded Theorem I. (For example, we could have adopted $s(x) = 1 + (1/\pi)\tan^{-1}(\pi x/2)$.) However, our proof of Theorem I will only go through if s has the form

$$s[\pm x] = e^{\mp cx}\, y(x) \quad , \tag{17}$$

where c is an arbitrary constant and $y(x)$ some even function of x. This *does not* mean that our strongly negative conclusions regarding the occurrence of thermodynamic equilibrium are peculiar to (13). Quite the contrary: with Result (A) imposing F relations (not necessarily independent) among the $N^2 + 2N$ structural parameters $V_{\nu\mu}$, θ_ν, β_ν of the network, *highly restrictive* conditions for detailed balance will still prevail. But now there is the added complication that, in general, these conditions involve not only the $\beta_\nu V_{\nu\mu}$ with $\nu \neq \mu$, but also the quantities $\beta_\nu V_{\nu\nu}$ and $\beta_\nu \theta_\nu$. Moreover, the form of these conditions will depend on N, in contrast to the universality of relations (12) within the context of Theorem I. The trivial assumption of noninteracting neurons always works, irrespective of the choice of s.

Higher-order synapses are known to exist in the nerve networks of invertebrates[13] and in the peripheral nervous systems of vertebrates. Presumably they also play a significant role in the central nervous systems of vertebrates, considering the observed density and complexity of synaptic junctions.[14,15] A kth-order synapse is one at which neuroanatomical processes (axon terminals, dendrites) from k presynaptic neurons μ_1, $\mu_2 \ldots, \mu_k$ come together to influence the action of postsynaptic cell ν; that is, the change in membrane potential of neuron ν attributable to such a synapse depends on the constellation of neurons of the set $\{\mu_1, \mu_2, \ldots \mu_k\}$ which were simultaneously active during a relevant interval of the recent past (cf. Ref. 16). In the framework of synchronous modeling, there will be a nontrivial dependence of the input from this synaptic complex at time t on the correlated activity of the presynaptic neurons at time $t-\tau$. Using the language of many-particle theory, the neurons experience interactions of up to $(k+1)$-body [($(k+1)$-neuron] character. This increased realism and increased richness may be accommodated in our model by a straightforward generalization of the firing function F_ν of Eq. (1):

$$F_\nu(t) = \sum_{l=1}^{k} \sum_{\mu_1 \cdots \mu_l} V_{\nu\mu_1 \cdots \mu_l} \pi_{\mu_1}(t-\tau) \cdots \pi_{\mu_l}(t-\tau) - \theta_\nu \quad , \tag{18}$$

where the $l+1$-body interactions are described by the quantities $V_{\nu\mu_1 \cdots \mu_l}$, which define tensors of order $l+1$. We are interested in determining the constraints on these tensors (in particular, the symmetries) which are imposed by thermodynamic equilibrium. For the simple case of binary interactions, $k = 1$, the answer is given by Theorem I. We may elaborate the same reasoning to extend this theorem to cases where synapses of arbitrary order occur.

Theorem II. In the presence of higher-order synapses as described by the firing function (18), and assuming a conditional firing probability (16) obeying (17), the following relations among the neuronic interactions are necessary and sufficient for the ultimate attainment of thermodynamic equilibrium:

$$\beta_\nu \bar{V}_{\nu\mu} = \beta_\mu \bar{V}_{\mu\nu} \quad , \tag{19}$$

$$\sum_{(\mu_1 \cdots \mu_l)} \bar{V}_{\nu\mu_1 \cdots \mu_l} = 0 \qquad (2 \leq l \leq k \ , \ k \geq 2) \quad . \tag{20}$$

Here ν, μ, $\mu_1, \ldots \mu_k$ are distinct but otherwise generic neuron labels, and the sum in (20) runs over *all permutations* of the indices $\mu_1 \ldots \mu_l$. The barred coupling $\bar{V}_{\nu\mu_1 \cdots \mu_l}$ is the original coupling $V_{\nu\mu_1 \cdots \mu_l}$ plus a suitably normalized sum of all partially diagonal couplings bearing the same set of indices (see below).

We shall sketch the proof for $k = 3$, i.e., for a firing function containing two-, three-, and four-body interactions. Argument (i) of the proof of Theorem I applies without alteration and argument (ii) is easily generalized; thus the salient conditions are again independent of the thresholds θ_ν and the diagonal couplings $V_{\nu\nu}$, $V_{\nu\nu\nu}$, etc. Argument (iii) proceeds as before, up to the point of explicating the relations imposed by Result (A) on the exponents of the s functions appearing on the left and right of the balance condition. In place of (15) we obtain

$$\sum_{p=1}^{n} \beta_{\nu_p} \left[\sum_{t=1}^{m} V_{\nu_p\mu_t} + \sum_{t,t'=1}^{m} V_{\nu_p\mu_t\mu_{t'}} + \sum_{t,t',t''=1}^{m} V_{\nu_p\mu_t\mu_{t'}\mu_{t''}} \right]$$

$$= \sum_{t=1}^{m} \beta_{\mu_t} \left[\sum_{p=1}^{n} V_{\mu_t\nu_p} + \sum_{p,p'=1}^{n} V_{\mu_t\nu_p\nu_{p'}} + \sum_{p,p',p''=1}^{n} V_{\mu_t\nu_p\nu_{p'}\nu_{p''}} \right] \quad . \tag{21}$$

Choices of states i, j such that $n = 1$ and $m = 1$ yield a necessary condition of the form (19), with

$$\bar{V}_{\nu\mu} = V_{\nu\mu} + V_{\nu\mu\mu} + V_{\nu\mu\mu\mu} \quad . \tag{22}$$

In the case $n = 1$, $m = 2$ (or the converse, upon relabeling) we are led, under (19), to the further necessary condition

$$\bar{V}_{\nu\mu_1\mu_2} = -\bar{V}_{\nu\mu_2\mu_1} \quad , \tag{23}$$

wherein

$$\bar{V}_{\nu\mu_1\mu_2} = V_{\nu\mu_1\mu_2} + \frac{1}{2}(V_{\nu\mu_1\mu_1\mu_2} + V_{\nu\mu_1\mu_2\mu_1} + V_{\nu\mu_2\mu_1\mu_1} + V_{\nu\mu_1\mu_2\mu_2} + V_{\nu\mu_2\mu_1\mu_2} + V_{\nu\mu_2\mu_2\mu_1}) \quad . \tag{24}$$

This relation matches (20). We observe that while the barred two-body interaction must be symmetric, the barred three-body interaction tensor must be antisymmetric in its second and third indices; these properties must evidently be maintained by the original two- and three-body V's in the absence of couplings with two or more coincident indices ("redundant" or "self" interactions). Finally, we may consider $n = 1$, $m = 3$, and, after invoking (19) and (23), arrive at a necessary condition on the four-body interaction, again matching (20) in the present context:

$$V_{\nu\mu_1\mu_2\mu_3} + V_{\nu\mu_2\mu_1\mu_3} + V_{\nu\mu_1\mu_3\mu_2} + V_{\nu\mu_3\mu_2\mu_1} + V_{\nu\mu_2\mu_3\mu_1} + V_{\nu\mu_3\mu_1\mu_2} = 0 \quad . \tag{25}$$

It is readily seen that the three necessary conditions so derived, with generic neuron labels, are also sufficient. (In particular, they validate the counterpart of (15') as well as (21).) The pattern of the general proof, and the general definition of the barred V's, should now be apparent.

CONCLUSIONS

In answer to the title question, we stress that the dissipative, nonlinear networks of the class studied here *do* necessarily approach an equilibrium condition (which is, in fact, absolutely stable in the 2^N-dimensional space of the distributions $\{p_i\}$ over firing patterns i). On the other hand, this condition is generally not one of *thermodynamic* equilibrium, wherein the probability flow $T_{ij}p_j$ from state j to state i exactly balances the flow from state i to state j, for all pairs ij. Instead, the final stationarity is generally achieved only by cyclic processes involving triples, quadruples, ... of states. In this sense we may speak of the neural system as operating away from (thermodynamic) equilibrium.

The conditions on synaptic interactions under which thermodynamic equilibrium does prevail, established in detail in the preceding section, are highly restrictive and unlikely to be met by biological systems, except approximately in sharply defined contexts.[17] Thus, to the extent that the class of models studied here is relevant, it must be concluded that biological nerve nets typically do not obey the principle of detailed balance.

In closing, we should briefly address the issue of the information storage capacity of stochastic nets of the Little type, which would appear to be severly limited. As we have seen, there is only one final condition of a given network in $\{p_i\}$ space (the Kirchhoff steady state), *independent of initial conditions*. In this sense the network can only store one memory, since all stimuli elicit the same ultimate response. However, information can also be carried by the transient response of the system, which may persist for a very long time if there happen to be eigenvalues of the matrix (W_{ij}) which are nearly degenerate with its zero eigenvalue. With careful choice of the coupling matrix $(V_{\nu\mu})$, the system will be characterized by a number of different "persistent states" (Refs. 1,3,6), which may be excited by appropriate stimuli. Thus the *effective* memory capacity of networks of the sort considered here is potentially of reasonable size.

In the same connection it is worth mentioning that there exists another class of models based on a closer view of the stochastic nature of the "quantal" mechanism of information transfer at synapses. The discrete model of Taylor[4] may be regarded as the prototype of this class. In terms of our approach to the statistical evolution of the neural system (the basic dynamical variables being the state-occupation probabilities $p_i(t)$), this model gives rise to a *nonlinear* master equation. Accordingly, one anticipates a more elaborate menu of final conditions, which might include periodic modes and chaos along with a multitude of fixed points. The intriguing cognitive properties of such rich memory maps await exploration.

ACKNOWLEDGMENTS

The author acknowledges the hospitality of the Theoretical Division and the Center for Nonlinear Studies of Los Alamos National Laboratory and the Physics Division of Argonne National Laboratory, while on leave from the Department of Physics, Washington University, St. Louis, MO 63130. This research was supported by the U. S. Department of Energy, Nuclear Physics Division, under Contract W-31-109-ENG-38 and by the Condensed Matter Theory Program of the Division of Materials Research of the U. S. National Science Foundation under Grant No. DMR-8519077. A travel grant from the U. S. Army Research Office is gratefully acknowledged.

REFERENCES

1. W. A. Little, The existence of persistent states in the brain, Mathematical Biosciences **19**, 101 (1974).
2. B. Katz, *Nerve, Muscle, and Synapse* (McGraw-Hill, New York, 1966); B. Katz, *The Release of Neural Transmitter Substances* (Thomas, Springfield, 1969).
3. G. L. Shaw and R. Vasudevan, Persistent states of neural networks and the random nature of synaptic transmission, Mathematical Biosciences **21**, 207 (1974).
4. J. G. Taylor, Spontaneous behavior in neural networks, Journal of Theoretical Biology **36**, 513 (1972).
5. J. W. Clark, Statistical mechanics of neural networks, Physics Reports, in press.
6. W. A. Little and G. L. Shaw, A statistical theory of short and long term memory, Behavioral Biology **14**, 115 (1975); W. A. Little and G. L. Shaw, Analytic study of the memory storage capacity of a neural network, Mathematical Biosciences **39**, 281 (1978); G. L. Shaw, Space-time correlations of neuronal firing related to memory storage capacity, Brain Research Bulletin **3**, 107 (1978); G. L. Shaw and K. J. Roney, Analytic solution of a neural network theory based on an Ising spin system analogy, Physics Letters **74A**, 146 (1979).
7. J. W. Clark, J. Rafelski, and J. V. Winston, Brain without mind: Computer simulation of neural networks with modifiable neuronal interactions, Physics Reports **123**, 215 (1985).
8. P. Peretto, Collective properties of neural networks: A statistical physics approach, Biological Cybernetics **50**, 51 (1984).
9. N. T. J. Bailey, *Elements of Stochastic Processes with Applications to the Natural Sciences* (Wiley, New York, 1964), Chapters 3 and 5.
10. J. Schnakenberg, Network theory of microscopic and macroscopic behavior of master equation systems, Reviews of Modern Physics **48**, 571 (1976).
11. I. Prigogine, *Introduction to the Thermodynamics of Irreversible Processes,* Third Edition (Wiley, New York, 1967); P. Glansdorff and I. Prigogine, *Thermodynamic Theory of Structure, Stability, and Fluctuations* (Wiley, New York, 1971).
12. W.-K. Chen, *Applied Graph Theory* (North-Holland, Amsterdam, 1971).
13. E. R. Kandel, *Cellular Basis of Behavior* (Freeman, San Francisco, 1976); E. R. Kandel and L. Tauc, Heterosynaptic facilitation in neuron of the abdominal ganglion of Aplysia depilans, J. Physiol. **181**, 1 (1965).
14. G. H. Shepherd, *The Synaptic Organization of the Brain* (Oxford University Press, Oxford, 1979).
15. K. J. Roney, A. B. Scheibel, and G. L. Shaw, Dendritic bundles: survey of anatomical experiments and physiological theories, Brain Res. Rev. **1**, 225 (1979); G. L. Shaw, E. Harth, and A. B. Scheibel, Cooperativity in brain function: assemblies of approximately 30 neurons, Exp. Neurol. **77**, 324 (1982).
16. P. Peretto and J. J. Niez, Long term memory storage capacity of multiconnected neural networks, Biol. Cybern. **54**, 43 (1986).
17. J. J. Hopfield and D. W. Tank, Computing with neural circuits: A model, Science **233**, 625 (1986).

CONTRIBUTORS AND PARTICIPANTS

† invited speaker, ‡ co-author, § not present

Index